The Origins of the
Scottish Railway System
1722-1844

For Marjorie

The Origins of the Scottish Railway System 1722-1844

C. J. A. ROBERTSON
Lecturer in Economic and Social History,
University of St Andrews

JOHN DONALD PUBLISHERS LTD
EDINBURGH

© C. J. A. Robertson 1983

All rights reserved. No part of this publication
may be reproduced in any form or by any means
without the prior permission of the publishers,
John Donald Publishers Ltd., 138 St Stephen Street,
Edinburgh.

ISBN 0 85976 088

Exclusive distribution in the United States of
America and Canada by Humanities Press Inc.,
Atlantic Highlands, NJ 07716, USA.

Typeset by R D Composition Ltd., Glasgow
and printed in Great Britain by Bell & Bain Ltd., Glasgow

Preface

By comparison with their English counterparts, Scottish nineteenth-century railways have suffered from a degree of neglect by economic historians. There is certainly a substantial literature, mostly written for the railway enthusiast, concentrating mainly on topography, dates of opening, mechanical developments and episodes of an entertaining or eccentric kind, and typified by the works of John Thomas. Few of these books cover the whole of Scotland; Nock's *Scottish Railways* is essentially descriptive rather than historical, Thomas's *Regional History* is limited to the area south of the Forth, and other works are mainly treatment of single companies or of particular dramatic events. Most of the more academic studies are in journal articles, and there are as yet no Scottish parallels to Hawke's largely econometric analysis of railways and economic growth in mid-nineteenth century England and Wales, to Gourvish's work on railway management, or above all to the comprehensive study by Simmons of which the first volume has so far appeared. These books are consciously restricted, with the exception of Gourvish's references to Huish's work at Greenock, to events south of the border; other writers, claiming to cover the whole of Britain, too often either include very little Scottish material or ignore the country altogether. This study endeavours to fill some of the gaps for the earliest period of Scottish railway history. It covers the years from the first waggonway developments in the eighteenth century to the advent of the railway mania of the 1840s, and concentrates on the planning and formation of the various railways, the problems and achievements associated with their construction, and the financial record of the companies up to 1844.

I owe many debts of gratitude. I am particularly grateful to Professor Christopher Smout for reading and commenting on the manuscript; sections of it have also been read by Professor Norman Gash and Mr Bruce Lenman. Professor Roy Campbell gave me valuable advice on iron prices. For access to primary sources or permission to quote from them, my thanks go to the following: the Rt. Hon. the Earl of Airlie; His Grace the Duke of Buccleuch and Queensberry; Professor and Mrs S. G. Checkland; Sir John Clerk of Penicuik, Bt; the Rt. Hon. the Earl of Dalhousie; Sir William Gladstone, Bt; the Rt. Hon. the Earl of Leven and Melville; Mr A. M. M. Matheson; G. A. More Nisbett, Esq; the legal firm of Messrs Shiell and Small; Professor Anthony Slaven; Col. the Rt. Hon. the Earl of Stair; the Keeper of the Records of Scotland; the Clerk of the Records of Parliament; and the Trustees of the National Library of Scotland. Attempts to contact the owners of some other papers have been unsuccessful, and I apologise if any offence has been caused.

Further thanks are due to Mr Robin Gibb for drawing the maps; to my aunt, Mrs Doris Hunter, for not only typing the manuscript superbly but also keeping an eye on my inconsistencies of style and practice; to Mrs Isobel Fraser for ferrying material between author and typist; and to Mr John Tuckwell, a most helpful and relaxed publisher. The editors of the *Scottish Historical Review* have allowed me to include a version of part of my article on the sabbatarian problem in their October 1978 issue. The collection of material has been much eased by grants from the Travel and Research Funds of the University of St Andrews.

It has become an authors' cliché to attribute their greatest debt to their wives; it remains true that the dedication of this book reflects not only the fact that my wife has lived with this project for longer than either of us would care to think about, but also the recognition that without her encouragement and patience it might never have reached a conclusion.

Contents

	Page
Preface	v
Chapter 1. The Century of the Waggonways	1
I. The eighteenth-century background	1
II. Early waggonways: Tranent, Alloa and Lord Elgin's	6
III. The first railway?: Kilmarnock and Troon	21
IV. Coal, construction and costs	25
V. Growing ambition: long-distance projects	30
Chapter 2. The Coal Railways	44
I. The idea of the railway	44
II. The Monklands lines	51
III. Other early railways: Dalkeith, Newtyle and the Clyde	63
IV. The impact of the locomotive	70
V. Financing the coal railways	75
VI. The lines in operation	83
Chapter 3. From Town to Town: Planning and Authorisation	97
I. Expansion in the 1830s	97
II. The essential link: Edinburgh to Glasgow	99
III. Landowners' initiatives: north of the Tay	121
IV. Problems and profligacy: Granton, Leith and the route to Fife	127
V. From Glasgow to the Firth of Clyde	137
VI. A selection of failures	144
VII. Sources of capital: an examination of subscription contracts	147
Chapter 4. From Town to Town: Construction and Operation	163
I. Expectation and reality: the accounts of the inter-urban railways	163
II. Ordeal by Parliament	174
III. The acquisition of land	181
IV. Works, engineers and contractors	188
V. Other costs: rolling stock and road trustees	206
VI. The search for further capital	211
VII. The costs of operation: expenditure on current account	219
VIII. Current income and the setting of rates	230
IX. The passenger boom	236
X. Rendering unto Caesar: the passenger duty	246
XI. Rendering unto God: the sabbatarian question	252
XII. Relations with other transport	257

	Page
Chapter 5. The Battle for the Border	266
I. English trade and border topography	266
II. Joseph Locke and the Smith-Barlow Commission	274
III. Supporters and committeemen	282
IV. Western delays and the authorisation of the North British	288
V. The triumph of the Caledonian	294
Chapter 6. Summary and Conclusions	305
I. Scottish railways in 1884	305
II. The conveyance of coal	306
III. Diversification: general freight and passengers	309
IV. Connections and extensions	315
V. Costs and estimates	317
VI. Investors and promoters	323
VII. Managerial developments	327
VIII. English influences	329
IX. Forward to the mania	335
Appendix. Analysis by Location and Occupation of Subscriptions to Scottish Railway Companies authorised in 1837-38	339
Notes	352
Bibliography	396
Index	409

MAPS

	Page
1. Scottish Waggonways to 1824	16
2. Alloa Area Waggonways	20
3. Dunfermline Area Waggonways	20
4. Midlothian and East Lothian Waggonways	27
5. Waggonway Projects not Constructed	33
6. Railways Authorised to 1835	55
7. North Lanarkshire Railways	58
8. Edinburgh and Glasgow Projects	103
9. Edinburgh Railways to 1844	109
10. Glasgow Railways to 1844	110
11. Angus and Perth Railways	122
12. Fife Area Projects, 1840-41	134
13. Railways West of Glasgow	139
14. Engineers of Scottish Railways to 1844	193
15. Border Projects	273
16. Border Routes examined by the Smith-Barlow Commission	281
17. The Caledonian and its Rivals, 1844-45	301
18. Scottish Railways in 1844	312

FIGURES

1. Wishaw & Coltness Railway: Effects of Possible Rating Systems, 1843   94
2. Wishaw & Coltness Railway: Goods Traffic by Type and Distance, 1843   95
3. Price of Bar Iron, 1836-44   204
4. Movement of Share Prices, April 1841 – July 1845   337

ABBREVIATIONS USED IN REFERENCES AND IN NOTES TO TABLES

br	branch
DNB	Dictionary of National Biography
GUEH	Glasgow University, Department of Economic History
HC	House of Commons
HL	House of Lords
HLRO	House of Lords Record Office
HRJ	Herapath's Railway Journal
NLS	National Library of Scotland
NSA	New Statistical Account
OSA	Old Statistical Account
PICE	Proceedings of the Institution of Civil Engineers
PP	Parliamentary Papers
RT	Railway Times
S.C.	Select Committee
SRO	Scottish Record Office

A NOTE ON PROPER NAMES

Place names have have been spelt in accordance with modern usage except where they form part of a company name (e.g. Newtyle & Glammis Railway). Names of individuals have where possible been given in the spelling used by the individual concerned (e.g. Macneill, Airlie). For ease of reading, the intermediate commas in the names of railway companies have been omitted (e.g. Glasgow Paisley Kilmarnock & Ayr). Where details of a route are given in the tables, the insertion in parentheses of the name of another railway or waterway indicates that a connection was made with it at the place mentioned.

1
The Century of the Waggonways

I. The eighteenth-century background

THE tentative origins of developments which later assume substantial significance have often been of particular interest to historians. In Scotland at least, the eighteenth-century prehistory of the railways has been examined perhaps more closely than any subsequent period. Among other authors, Dott has catalogued most of the Scottish waggonways, Dendy Marshall has placed them in a British, and Lewis in a European, context, Baxter has examined them with the eye of an industrial archaeologist, and Duckham has paid them considerable attention in his study of the Scottish coal industry.[1] Although much material has thus been sifted, none of these studies has been concerned with the waggonways as part of the continuing development of the Scottish railway system; conclusions remain to be drawn both about the waggonways themselves and about their ancestry of the later railways.

The rationale for construction of the waggonways was quite simply coal. The eighteenth-century economy, particularly in the latter part of the century, was for various familiar reasons an expanding one, and this is not the place to re-examine the causes of the Industrial Revolution. In Scotland the increasing trading opportunities offered by the Act of Union led in particular to the tobacco fortunes made by Glasgow merchants, and later to the establishment of a cotton industry in Strathclyde. The coal and iron industries showed little expansion during the first half of the century, but from 1760 — the year of the foundation of the Carron Company — the growth was rapid, encouraged by the spread of coke smelting, the availability of merchant money, and the desire of landowners to develop their mineral resources.[2] For expanding industries, adequate supplies of fuel at reasonable cost were essential, and, although textile mills in suitable locations might use water power, for most firms this meant coal. A key concern of eighteenth century industrialists was satisfactory access to coal supplies; equally, on the supply side, coalowners were compelled to attend to the transport facilities which would allow them to extend their markets.

Early in industrialisation most countries have been faced by problems caused by inadequate transport, and of these Scotland had its share. For the carriage of goods, transport by water was generally preferable to carting by road. Adam Smith calculated that the conveyance of two hundred tons of goods from Edinburgh or Leith to London would require either one ship with six or eight men, or fifty waggons each with two men and eight horses; in either case a round trip would take about six weeks.[3] The country was therefore fortunate in having a long, if stormy, coastline, with few important towns at any distance from the sea. A substantial coasting trade distributed fuel, consumer goods and food around the country, while regular services linked Leith to London, Newcastle and the Scottish east-coast ports, and Glasgow to Lancashire and Ireland. The long and dangerous voyage round the north of Scotland, however, discouraged sea trade between the east and west coasts. Nor was Scotland well endowed with navigable rivers; ships plied up the Tay and Forth only as far as Perth and Stirling, the fast-flowing Tweed, Don and Spey could not be used at all, and until improved in the 1770s the Clyde was too shallow for any but the smallest vessels to reach Glasgow. Golborne's dredging work on the Clyde was the only substantial improvement to any Scottish river navigation, and no unnavigable river was ever made navigable.[4] Compared to developments in England, relatively little effort was made to extend the scope of water transport by canals: only two significant ones — the Forth & Clyde and the Monkland — were opened in the eighteenth century, both in its last decade.[5]

Canals and navigable rivers were in any case mainly of use to those in their immediate vicinity, and for the most part inland transport depended on the roads. The traditional view, which saw the roads of the eighteenth century as badly organised, badly constructed, and of limited economic significance, particularly for the movement of heavy goods, has in the last decade been considerably revised. Almost all the evidence for important and increasing flows of road traffic, based on an expanding and reasonably well co-ordinated turnpike system, has come from England and Wales.[6] Similarly detailed work on Scottish roads would be much welcomed — meanwhile it still seems probable that most Scottish roads, particularly in the earlier part of the century, were quite unfit for wheeled vehicles, with many north of the Tay being frequently impassable.[7] Even in the Glasgow area in the 1740s, carriage was conducted by packhorses or horse-drawn sledge: 'for all practical purposes the wheel might as well not have existed'. In the city itself personal transport for visitors in 1744 was reported to be limited to a few sedan-chairs.[8]

Much of the evidence for the state of the roads comes from the ministers who contributed to the *Old Statistical Account*. Allowance must be made for the enthusiasm with which many of them embraced the progress made in the twenty years before they wrote in the 1790s, which may have led them to exaggerate both the inadequacy of early eighteenth-century roads and the degree of improvement achieved thereafter. There is, however, general agreement that the six days per year of statute labour imposed by an act of 1669 had merely resulted in roads which were 'neither half made, nor half kept in repair'.[9] Travellers on the road from Strathaven to Muirkirk were reported to require the services of a guide, though the road was 'even then very dangerous and altogether impassable for any

carriage'. In Dumfriesshire:
> most of the roads were unmade, or had been repaired in a very superficial manner; and in that district of the county called Annandale, almost the whole of the roads were impassable during the winter season.[10]

Further north, 'previous to the year 1790, a great part of the interior of the Carse of Gowrie was perfectly inaccessible to carts for almost half the year'.[11]

Some improvement came when the requirement for statute labour was gradually commuted into a monetary payment after 1750, although in some parishes the labour requirement survived into the 1790s. In Whittingham, East Lothian, the parish roads were kept in good repair for £56 per year:
> a heavy tax upon the farmers, but it is generally paid with the greatest cheerfulness, from a thorough conviction of the great conveniency and advantage of good roads.[12]

The most important change, however, was the advent of the turnpike system. Here, Scotland lagged a long way behind England. The first Scottish turnpike trust was established for Edinburgh in 1714; by the time of the second, in Haddington in 1750, England and Wales already had 3386 miles of turnpiked road. Pawson's extensive list of trusts (which does, however, omit some which were authorised towards the end of the century) shows 24 authorisations in eighteenth-century Scotland, all in the Borders or the Forth-Clyde Valley.[13] Initial prejudice against the payment of turnpike tolls appears to have been soon overcome:
> When they were at first proposed, they met with keen opposition; but they have since been universally acknowledged to be of signal benefit to the country.[14]

John Naismith, reporting on the parish of Hamilton, gave perhaps the most judicious summary:
> They are generally kept in pretty good order; though, from the softness of the soil, and the scarcity of materials, hard enough to stand the fatigue of the many heavy carriages which pass, it is attended with considerable difficulty and expence. Nobody here entertains any doubt of the advantage of turnpike roads, since, at least, three times as much weight can be drawn in a carriage, as was sufficient to load it before they were made. If any objection can be made to the turnpike roads of this country, it is to the manner in which they have been laid out, being generally conducted over the summit of every eminence in their course; when with a little judgment and attention, a direction might have been found equally near, and incomparably more easy and convenient.[15]

Severe gradients would be especially inconvenient in an area like Hamilton, where there would be considerable coal traffic.

Undeniably, the turnpikes were an improvement. They were not, however, a financial success. The road from Perth to Crieff, for instance, in its early days attracted so little traffic that its receipts were not even sufficient to pay the interest on its construction costs.[16] By 1859 the 1400 miles of turnpike in the west of Scotland had accumulated a combined debt of £731,000, much of which they had managed to offload on to railway companies as the price for withdrawing their opposition to the railways' acts of authorisation.[17] Nor were they providing the complete and cheap transport system which was required, even by the end of the eighteenth century. The problem was not so much personal travel. This was in any case normally a matter of necessity rather than pleasure. The turnpikes had made coach journeys practicable at least in the south of the country, although the loss of time involved, the discomfort of travelling over a surface often still made up of the ruts of previous traffic, the fear of highwaymen and the exactions of innkeepers ensured that such travelling was not undertaken lightly. On other roads the rigours of horseback were unavoidable. For long journeys coastal passage-boats provided regular, if slow and uncomfortable, services to the ports of Britain and beyond. One way or another, the man with enough money to pay for his travel could usually achieve his journey. The poor man moving for a considerable distance — say from the Highlands to seek work in Glasgow — might find a cheap passage or work his way by boat; for shorter distances, and sometimes for long ones, he would walk.[18]

The continuing problem was the transportation of heavy goods. Dependence on road transport was not an absolute bar to economic expansion: enterprises such as the ironworks at Shotts and Wilsontown, or the cotton mills at New Lanark, were after all successfully established away from navigable water, and presumably found that the advantages of immediate access to coal, iron ore or water power compensated for any difficulties with transport. But the cost of moving heavy items by cart, even on a turnpike where loads could be two or three times as heavy as those on the old roads, meant that an individual producer's market was at best restricted, and at worst it could make extraction of minerals or industrial production non-viable. Turnpike management, and the improved methods of road construction introduced in the late eighteenth and early nineteenth centuries and associated with the names of Telford and McAdam, could offer only limited help. In Aberdour in 1791, for instance, even although the road was turnpiked, it was not in sufficiently good condition to permit profitable mining of the parish's coal. In the parish of Barr, in Ayrshire, inadequate roads prevented the inhabitants from obtaining coal, though there were coal mines only $4\frac{1}{2}$ miles away.[19] The minister of Hoddom, in Dumfriesshire, spoke for many coal-less parishes:

> No county, either in Scotland or in England, can boast of having better roads than the county of Dumfries... The great, almost the only drawback, which this parish sustains, is the want of coals. Our distance from these is about 16 miles, which renders their carriage by land very expensive...[20]

It may be true that, in overall national terms, transport inadequacies were not yet a major factor impeding economic growth, and that, as Campbell suggests, the concentration of Scottish industry and wealth in the central belt, giving producers a local market sheltered by distance from English rivals, meant that:

> transport improvements were less important for the industrial growth of Scotland in the eighteenth century than they were for the development of the heavy industries and the full exploitation of the country's natural resources in the nineteenth century.[21]

But even in the later eighteenth century the constraints imposed by transport costs were irking an increasing number of individual producers and potential producers, particularly in a coal industry which was already undergoing growth. Nef's well-known studies drew attention to the expansion of the British coal industry in the sixteenth and seventeenth centuries, and suggested that Scotland's proportion of the total remained reasonably steady through this period and beyond. Duckham has charted the continuing growth of the industry in the later eighteenth century, to reach an output of almost two million **tons per year at the end of the century**.[22]

The best answer to the transport problems of the coal industry was water, and most of the first Scottish collieries to be developed were close to the sea, in particular to the Firth of Forth. But even those areas of Fife, Clackmannanshire and the Lothians which had access to the sea found it difficult to compete with the highly organised sea trade of Tyneside, while the major coal reserves of Lanarkshire and Ayrshire remained undeveloped for lack of reasonably priced transport. The solution might have seemed to be to follow the example of the Duke of Bridgwater and bring water transport to the coal. But although the later part of the century saw canals spread throughout the northern and midland coal areas of England, in Scotland a great deal of discussion led to relatively little action. Of the few major canals opened before the end of the Napoleonic wars, the Aberdeenshire was far from the coalfields, the Glasgow, Paisley & Ardrossan might have been much concerned with coal if it had ever been built beyond Johnstone, and even the Forth & Clyde opened up no new coalfields and carried comparatively little coal. Only the Monkland Canal was primarily concerned with coal, as it successfully opened up a previously unworked area of North Lanarkshire and helped to solve the problem of supply to Glasgow. For reasons of both geography and economics, the canal boom of the 1790s did little to help most Scottish coalmasters.[23]

The other possibility offered by the English example was the use of rails to link the pits with the nearest navigable water. Such waggonways had been in use on Tyneside and in Shropshire from the seventeenth century,[24] and they had certain clear advantages for the coal owner. They were cheaper to construct than canals, and could be more flexible in use. They could be realigned or extended with reasonable ease as old shafts were closed and new ones opened. They would normally use horse power, but, since the navigable water would be downhill from the pitheads, the loaded waggons would be helped by gravity, and might even, as on the Tranent & Cockenzie waggonway, be run entirely by it: there was seldom much uphill traffic on coal waggonways, although some carriage of salt from Cockenzie was reported.[25] Since their value was in moving large quantities of heavy goods over a fixed route, their use was effectively limited to the coal and, to a lesser extent, the iron industries. From these aids to heavy industry the Scottish railway system eventually developed.

Plans for long-distance waggonways never came to fruition, although projects to deliver coal to areas like the central Borders, and even to link Glasgow and Berwick, were seriously canvassed in the early nineteenth century. A proprietor like the Earl of Dumfries, who had hoped to export coal from his Cumnock mines through Ayr to Ireland but had been defeated by the costs of sixteen miles of land haulage, had to await the arrival of the steam locomotive and the fully fledged railway.[26] Short waggonways, however, were invaluable as links from collieries to harbours or to nearby ironworks. Excluding lines entirely within the confines of a mine or works, some thirty such waggonways have been traced in the years up to 1824. On some the information is extremely shadowy, depending on isolated references or single indications on maps. Not the least obscure is the candidate uncovered by Lewis for the first possible Scottish waggonway: in 1606 Thomas Tulloch of Inveresk was granted a patent for transporting coal by 'ane work and ingyne nocht knowin in this Kingdome at na tyme before'.[27] Those for which some positive information is known are listed in Table 1. All but two had coal mines at one end, and the two exceptions — from Carron works to the Forth & Clyde Canal, and from Newbigging limeworks to the sea — may well also have carried some coal. Five lines took coal directly into ironworks: almost all the rest ran to navigable water.

II. Early waggonways: Tranent, Alloa and Lord Elgin's

Among the list of early Scottish waggonways, some deserve individual consideration. If we ignore the claims for Thomas Tulloch, the waggonway was brought from England by a group of 'adventurers with more hope than acumen or perseverance'.[28] The York Buildings Company, started in the previous century to supply water to parts of London, had changed ownership in 1719. In this and the following year the new partners had moved into property speculation, and contributed to the excitement preceding the South Sea Bubble by spending £308,913 on buying the greater part of the Scottish estates forfeited after the Jacobite rising of 1715. Among those purchases, which made the company the largest landowner in Scotland, was the estate of the Earl of Winton, including the coal mines of Tranent and the salt of Prestonpans. The new owners at least showed themselves willing to innovate. Murray credits them with the introduction of the first Scottish 'fire engine' at Elphinston colliery, and in 1722 they built a wooden waggonway from Tranent pits to the sea at Cockenzie.[29] The line was built to give a steady downhill incline to the sea, even though this required the construction of a substantial embankment, so that loaded trains of waggons could be sent down by gravity under the control of a brakesman, and horses would only be required for returning the empties.[30]

The Tranent & Cockenzie waggonway gained an accidental footnote in military history when Sir John Cope's cannon rested on it at the battle of Prestonpans, but it apparently failed in its primary purpose. The York Buildings Company invested £3,500 in their improvements, but were unable to make £500 per year on the coal and salt combined; they also had difficulty finding a tenant, and in 1729 they petitioned the Barons of the Exchequer for a reduction in the purchase price of the Winton estate.[31] Duckham considers that the dominance of north-east England in the sea-coal trade meant that investment capital in Scotland was unlikely to be spent, at least in the early eighteenth century, on linking collieries to tidal water: he implies that the company's decision to build their waggonway may have been due either to unjustified speculative optimism, or to ignorance of the coal industry.[32] It is however possible that a limited operation aimed perhaps at cutting transport costs to Edinburgh rather than competing for a wider market could have been a justifiable economic decision: the later Fife waggonways, after all, were reasonably successful working from this premise. And it is likely that the financial failings of the mid-eighteenth century were due to the incompetence of the York Buildings Company rather than to any inherent defects in the physical arrange-

Table 1. Waggonways constructed, 1722–1824

Waggonway and route	Date of opening	Length (miles)	Gauge	Rails	Engineer	Owner	Traffic	Later History	Other comments	Sources (see list after table)
1. Tranent & Cockenzie: Tranent collieries—Port Seton harbour	1722	2–2½	3'3"	Wood 1815 iron edge		York Buildings Company 1779 John Cadell	coal, salt	1886 Tranent–Meadowmill section taken into North British: rest abandoned		2,11,22,23,24
2. Knightswood: Woodside colliery—Yoker (R. Clyde)	pos. c.1753 def. by 1775	2½		Wood		John Dixon of Dumbarton Glassworks	coal		Dixon bought glassworks 1750	19,20,21,22,34
3. Leven: colliery–Firth of Forth	by 1760						coal			19
4. Carron works a: to Kinnaird and Carronhall collieries	?1760 prob. 1766	1½–2		Wood & iron strip 1767 poss. iron rails from Coalbrookdale By 1795 iron edge		Carron Iron Company: manager William Cadell Jr.	coal			2,19,21,22,25,26
b: to Quarrole colliery	by 1769	3		Wood & iron strip By 1795 iron edge		do.	coal		Carron Co. Leased Quarrole coal 1761	19,25
c: to Bainsford basin, Forth & Clyde canal	1810	2	4'8½"	Iron edge		do.	iron, coal		Replaced an earlier navigable cut	23,25
d: Lines inside works				Iron plate		do.		c.1860 abandoned, after branch line built to Edinburgh & Glasgow Railway 1858 1924 abandoned		1,2,10,21,22
5. Alloa: Sauchie collieries–Alloa harbour	poss. 1766 prob. 1768	1½	3'3"	Wood 1785 wood & iron strip		J. F. Erskine of Mar 1835 Alloa Coal Company	coal, iron			
extn. to Collyland	1771	1½	3'3"	do.						
6. Elgin (i.e. Lord Elgin's: from 1821 Dunfermline & Charlestown): Berrylaw–Limekilns harbour	c.1768	3	4'3"	Wood ?1794, def. by 1812, iron edge		5th Earl of Elgin	coal, limestone later also general goods, passengers	From 1812 main line Wellwood–Charlestown (6 miles) 1821 further realignment c.1856 link line to Halbeath railway 1861 to West of Fife 1862 to North British 1866 part abandoned		1,2,17,19,20,21,22
branch to Baldridge colliery etc.	1794	4								
branch to Charlestown kilns	b–1792	c.1½								

The Century of the Waggonways 7

8 The Origins of the Scottish Railway System

Waggonway and route	Date of opening	Length (miles)	Gauge	Rails	Engineer	Owner	Traffic	Later History	Other comments	Sources (see list after table)
: branch to Charlestown harbour	1801									
: branch to Pitfirrane colliery	c.1801	c.1			George Johnson					
: branches to Wellwood and Rosebank collieries	1812	c.3			Charles Landale					
7. Brora										
: colliery–harbour and salt pans	Early 1770s	1	1'8"	Wood ?after 1800 iron edge		John Williams 1810 Marquis of Stafford	coal	c.1810 closed: soon reopened on realigned route 1920 part abandoned 1947 abandoned		4,21,27
	1810	1¼		Iron edge			coal, bricks, tiles			
8. Fordell										
: Colton collieries– St. David's harbour, Inverkeithing	poss. c.1752 prob. c.1770	3–4½	4'	Wood 1835–38 iron edge		Sir Robert Henderson of Fordell 1817 Admiral Sir Philip Durham, Bt. (son-in-law of Sir John Henderson)	coal	1835–38 major reconstruction 1850 link to Edinburgh, Perth & Dundee 1868 steam locomotives introduced	St. David's harbour built 1752 but waggonway probably later	1,2,19,20 21,22,28
: extn to Menniwick	1798									
: extn to Fordell village	1808	c.1								
: extn to Crossgates road	1832									
: branch to Peathouse colliery	?	1								
: branch to Millstone Meadows quarries	?	¼								
: poss. two other branches	?									
9. Ayr										
a: Whitletts colliery– Ayr harbour	by 1775	c.1		by 1824 iron edge		George Taylor & Co.	coal	c.1832 abandoned		10,19,20
b: Newton–Ayr harbour	prob. by 1775 poss. 1786–90					do.	coal	c.1832 abandoned	may be branch of 9a; O.S.A. suggests built after 1786, but probably a rebuilding. Newton pits were closed 1781–86	1,6,10,20
10. Govan										
: Govan colliery– Springfield (R. Clyde)	1775–78	1		Wood		William Dixon	coal	1840 part to Polloc & Govan		15,19,20,21,22

The Century of the Waggonways

Waggonway and route	Date of opening	Length (miles)	Gauge	Rails	Engineer	Owner	Traffic	Later History	Other comments	Sources (see list after table)
11. Halbeath : Halbeath collieries–Inverkeithing harbour	1780–83	5		Wood by 1798 iron, perhaps strip on wood, 1811 iron edge	Alex. Thompson 1785 Campbell Morrison	Lloyd brothers & Co. Ltd.	coal, limestone	Regular changes of route: by 1847 Halbeath colliery not on line c.1856 link line to Dunfermline & Charlestown 1867 abandoned		1,17,19,20,22
: branch to Townmill and Appin collieries		c.3		Iron edge						
12. Wemyss : Wemyss colliery–Methill harbour	early 1790s	2				7th Earl of Wemyss	coal			1,21
13. Auchencruive : Achencruive colliery–Ayr	well before 1811	c.2½				Richard Oswald of Auchencruive	coal		No connection to Ayr harbour	6
14. Wilsontown : Climpy collieries–Wilsontown ironworks	1804–05	c.1				John Wilson	coal	1812–13 abandoned		20,21
15. Sauchie : Devon pits and ironworks, Sauchie-Clackmannan pier : several branches	by 1806	5¼				Earl of Mansfield 1806 Earl of Mar	coal, iron	c.1854 abandoned		20,21,22
16. Muirkirk : Wellwood collieries–Muirkirk ironworks and tar kilns	by 181.	1–1½				Muirkirk Iron Company	coal, limestone		Works also had canal for moving iron	6
17. Kilmarnock & Troon : Kilmarnock–Troon harbour	1812	9½	4'	Iron plate	William Jessop	4th Duke of Portland et al.	coal, general goods, passengers	1846 to Glasgow Paisley Kilmarnock & Ayr Abandoned by 1849, possibly 1846	1808 Act of Parliament, 48 Geo.III, c.46	20,21,22
: branch, Drybridge–Peatland colliery	1818	2½	3'4"	Iron edge	William Jessop	Sir William Cunningham	coal			
18. Venturefair : Venturefair colliery–Knabble Street, Dunfermline	1812	1				Symes	coal	Abandoned by 1854		20,21

Waggonway and route	Date of opening	Length (miles)	Gauge	Rails	Engineer	Owner	Traffic	Later History	Other comments	Sources (see list after table)
19. Legbrannoch : Legbrannoch collieries–Monkland canal	by 1813	c.1					coal			22,30
20. Shotts : Shotts collieries–Shotts iron works	by 1813			Iron plate, possibly flat		J. Baird	coal			7,21
21. Omoa : Newarthill collieries–Omoa ironworks	by 1813	c.1				Col. Dalrymple	coal			20,21
22. Pinkie : Pinkiehill collieries–Fisherrow harbour	1814	1¼–2		Iron edge Poss. first to have completely wrought iron rails	John Grieve	Sir John Hope	coal	Abandoned after 1832		10,13,20,21,22
23. Port Dundas Road, Glasgow	1816	c.½		Flangless iron plate	John Baird of Shotts ironworks	Forth & Clyde Canal Co.	coal, general goods		Succeeded 'waggon-road' authorised 1770, almost completely 1793, which may have possibly involved rails	10,19
24. Burntisland : Newbigging limeworks–Burntisland harbour	c.1817	¾				Newbigging limeworks	limestone			21,22
25. Edmonston or Newton : Newton colliery–Little France (Edinburgh–Dalkeith road)	1818	3¾		Iron edge	Robert Stevenson	John Wauchope of Edmonston Alex Laing of Shawfair Alex Stenhouse of Whitehill	coal	1831 part closed: rest linked to Edinburgh & Dalkeith		10,22,23
26. Hurlet : Hurlet collieries and chemical works–Glasgow, Paisley & Johnstone canal	c.1820	c.2¼		Flat iron bars	Neilson	Earl of Glasgow	coal, lime	1848 superseded by Glasgow Barrhead & Neilston Direct		22,31

Waggonway and route	Date of opening	Length (miles)	Gauge	Rails	Engineer	Owner	Traffic	Later History	Other comments	Sources (see list after table)
27. Kilwinning	c.1820					Earl of Eglinton	coal	c.1827 taken into Ardrossan Railway		22
28. Ardrossan	by 1824	½		Concave iron plate		Earl of Eglinton	coal	do.	may be same as no. 27 though Kilwinning is 5 miles from Ardrossan	10
29. New Cumnock	poss. by 1824 def. by 1838					Menteath of Closeburn	coal	1850 abandoned–superseded by Glasgow & South Western		22
30. Irvine : colliery–harbour	?c.1824	c.3				?Taylor	coal			20,21
31. Calder : Calder ironworks–Monkland canal	by 1824	c.½					coal, iron			18,30

Table 2. *Waggonways projected but not constructed*

Waggonway and route	Date of plan	Length (miles)	Engineer	Projectors	Remarks	Sources (see list after Table)
1. Bo'ness						
: colliery–harbour and salt pans	1754		William Brown of Throckley	Bo'ness colliery		19
2. Monkland Canal						
a: Woodhall basin, Old Monkland	c.1769		James Watt	Monkland Canal Co.		21
b: Glasgow–Port Dundas	1769–70	c.1	do.		Authorised by act, but not made; 'waggon–road' substituted for 1816 development. See Table 1: no.23	16,19
3. Leith–Edinburgh–Midlothian collieries	c.1801	c.12		Henry Seton Steuart		20
4. Glasgow–Greenock	1801–2			Greenock Committee		30,35,42
5. Glasgow–Peebles–Melrose–Berwick	1807–10	c.125	Thomas Telford	Committee under Sir James Stewart Denham of Coltness		4,7,8
6. Berwick & Kelso	1811	22	John Rennie	Committee of Berwickshire landowners etc.	Act 51 Geo.III c.133, 1811. Revival plan 1824 Abandoned 1838	3,32,33
7. Dumfries–Sanquhar	1811	30	Robertson Buchanan	Sir Charles Menteath of Closeburn et al.		5,38
8. Union Canal	1813		Hugh Baird	Union Canal Co.		11
: branches from canal to						
Broxburn		2				
Camps		2				
East Calder		2				
Bathgate		3½				

Waggonway and route	Date of plan	Length (miles)	Engineer	Projectors	Remarks	Sources (see list after Table)
9. Edinburgh–Glasgow	1817	c.45	Robert Stevenson		Level line	18
10. Midlothian : Edinburgh–Dalkeith–Leith etc.	1817–21	$8\frac{1}{2} + 2\frac{1}{4}$ branches	Robert Stevenson		Possible level line	7,31
11. Roxburgh & Selkirk : Edinburgh–border towns	1817, 1820–1	44	Robert Stevenson	Committee of border landowners		8
12. East Lothian : Cairnie (Midlothian Ry)–Haddington–Dunbar	1818, 1825	$27\frac{1}{2}$	Robert Stevenson	Committee under Marquis of Tweedale		18,40,41
13. Brechin : Brechin–Montrose	1818	9	Robert Stevenson	Committee under Provost Guthrie of Brechin		21,32,36
14. Fife : Dunfermline–Kirkcaldy–Falkland–Perth br: Newport	1819		Robert Stevenson			18
: Dunfermline–Kirkcaldy–Newburgh	1820		Robert Stevenson			37
15. Perth–Crieff–Stirling–Lock 20, Forth & Clyde Canal	1819–20		Robert Stevenson			18,37
16. Strathmore Crieff–Perth–Aberdeen br: Forfar–Arbroath br: Brechin–Montrose	1820–7	104	Robert Stevenson		Level line	9,37
17. Glasgow–Peebles–Kelso–Berwick	1821		Robert Stevenson			18

Waggonway and route	Date of plan	Length (miles)	Engineer	Projectors	Remarks	Sources (see list after Table)
18. West Lothian Ryal (Union Canal)– Whitburn – Shotts br: Bathgate br: Bangour and Silvermines	1824–5	$14\frac{1}{2}$ 4 $4\frac{1}{8}$	Hugh Baird	Union Canal Co.	Act 6 Geo IV c.169 1825	12
19. Port Annan–Brampton (Newcastle & Carlisle)	1825			Committee of Dumfriesshire landowners		32

Sources for Tables 1 and 2:

1. *Old Statistical Account* 2. *New Statistical Account* 3. J. Rennie, *Report respecting the Proposed Rail-way from Kelso to Berwick* 4. T. Telford, *Report relative to the Proposed Railway from Glasgow to Berwick-upon-Tweed* 5. R. Buchanan, *Report relative to the Proposed Rail-way from Dumfries to Sanquhar* 6. W. Aiton, *General View of the Agriculture of the County of Ayr* 7. R. Stevenson, *Report relative to various Lines of Railway from the Coal-field of Mid-Lothian to the City of Edinburgh and Port of Leith* 8. R. Stevenson, *Report on the Roxburgh and Selkirk Railway* 9. R. Stevenson, *Report relative to the Lines of Railway surveyed from the Ports of Perth, Arbroath, and Montrose, into the Valley of Strathmore* 10. R. Stevenson (ed.), 'Essays on Rail-Roads', *Trans. Highland Soc.*, VI (1824) 11. H. Baird, *Report on the Proposed Edinburgh and Glasgow Union Canal* 12. H. Baird, *Report on the Proposed Railway from the Union Canal at Ryal, to Whitburn, Polkemmet, and Benhar: or, the West Lothian Railway* 13. C. MacLaren, *Railways compared with Canals and Common Roads* 14. HLRO Deposited Plan, HC 1824, Monkland & Kirkintilloch Railway 15. J. Priestley, *Historical Account of the Navigable Rivers, Canals, and Railways, of Great Britain* 16. T. Grainger & J. Miller, *Report to the Proprietors of, and Traders on the Canals and Railways terminating on the North Quarter of Glasgow* 17. P. Chalmers, *History and Statistical Account of Dunfermline* 18. D. Stevenson, *Life of Robert Stevenson* 19. M. J. T. Lewis, *Early Wooden Railways* 20. B. F. Duckham, *A History of the Scottish Coal Industry* 21. B. Baxter, *Stone Blocks and Iron Rails* 22. G. Dott, *Early Scottish Colliery Waggonways* 23. C. F. Dendy Marshall, *A History of British Railways down to the Year 1830* 24. D. Murray, *The York Buildings Company* 25. R. H. Campbell, *Carron Company* 26. W. Nimmo, *The History of Stirlingshire* 27. I. D. O. Frew, 'The Brora Colliery Tramway', *Railway Mag.*, 106 (1960) 28. J. C. F. Inglis, *The Fordell Railway* 29. C. Highet, *The Glasgow and South-Western Railway* 30. J. Thomas, *A Regional History of the Railways of Great Britain, VI: Scotland* 31. C. E. Lee, *The Evolution of Railways* 32. W. W. Tomlinson, *The North Eastern Railway* 33. H. G. Lewin, *Early British Railways* 34. J. A. Fleming, *Scottish and Jacobite Glass* 35. Glasgow Town Council Minutes, 6 April 1802 36. SRO, BR/PROS(S)/1/1: *Projected Railway between the Harbour of Montrose and the City of Brechin* 37. R. Stevenson, *A Memorial regarding the Propriety of Opening the Great Valleys of Strathmore and Strathearn by means of a Railway or Canal* 38. *Railway Times*, 22 July 1838 39. Glasgow, Paisley, Kilmarnock & Ayr Railway, Report of Directors to Shareholders' Meeting, 21 Aug. 1845 40. SRO, BR/LIB(S)/6/224 p.123: anon. cutting, c.1820 41. *Scotsman*, 8 Jan 1825 42. Anon., 'Early Scottish Railways', *Three Banks Rev.* 74 (1967)

The Origins of the Scottish Railway System

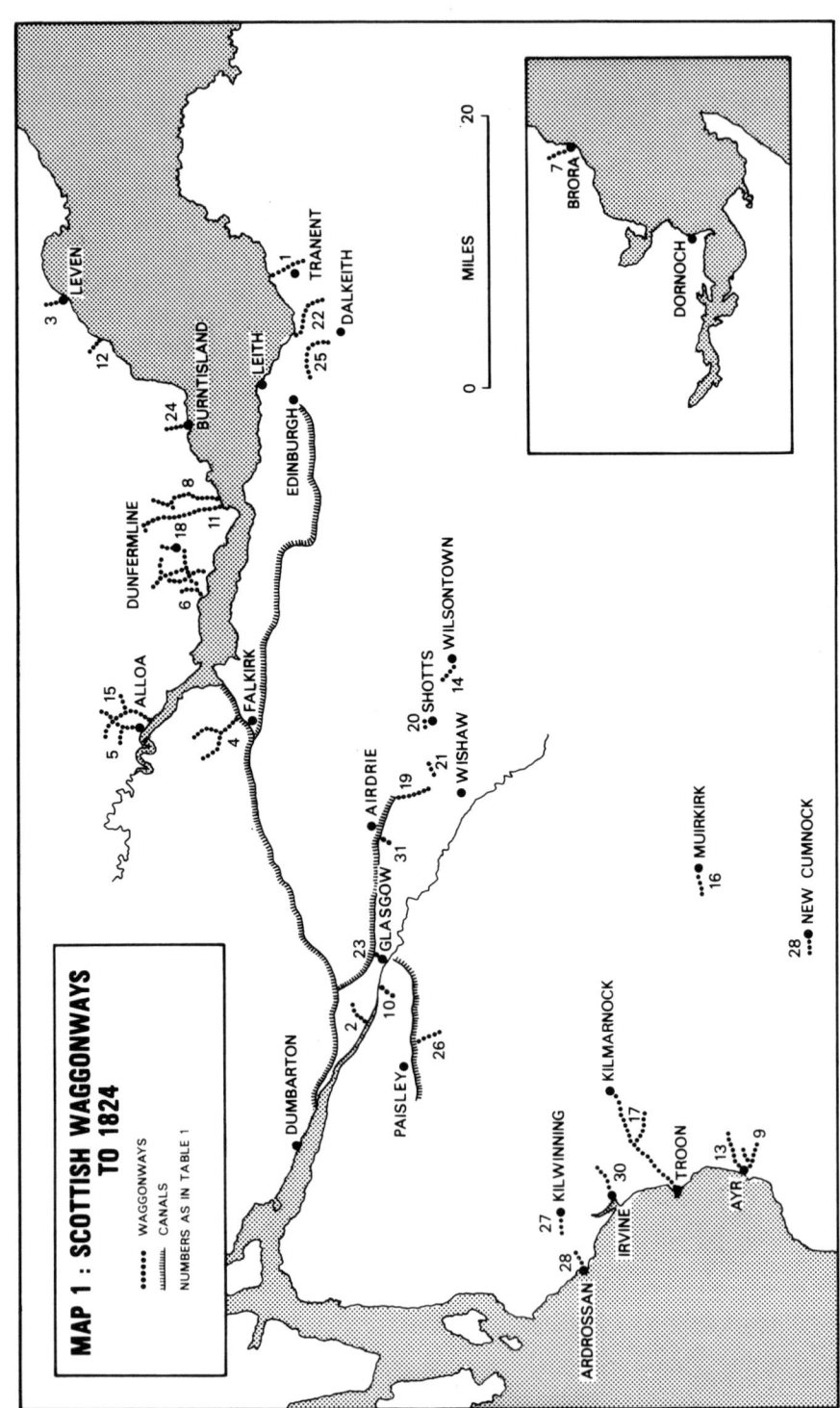

ments at Tranent. By 1779 the Cadell brothers, who had wide experience of coalmining in the Lothians and Fife as well as being involved in the founding of Carron Company, were willing to convert their tenancy into ownership when the pits and the waggonway were put up for auction.[33] But the Cadells may not have been particularly concerned with the waggonway, and they certainly did not send all their coal by rail and sea. The minister of Tranent, in his report for the *Old Statistical Account*, said nothing at all of the line, although he expatiated at length on the coal miners, and complained that:

> The great number of carts that daily resort to the different collieries, are extremely destructive of the roads; so that, in many places, the cross roads in this parish are almost impassable.[34]

In 1815 the Cadells relaid the way with cast-iron rails, retaining a single line with passing places, although shortly afterwards it was noted that railways were 'generally made double, one for going and the other for returning'.[35] They did not follow up the mysterious experiments conducted on the line by a Mr Ruthven, who in 1818 endeavoured 'to apply the principle of the crank so successfully used in his Printing Press, to propel waggons up an inclined plane'.[36] Cadell influence was also presumably responsible for the early ventures of the Carron Company into waggonway construction, the first way at Carron being built perhaps only a year after the company's foundation.[37]

Early English waggonway practice divided broadly into northern and southern patterns. On Tyneside relatively large waggons were hauled along tracks of substantial gauges, including the $4'8\frac{1}{2}''$ gauge that George Stephenson took to the Stockton & Darlington. According to Outram in 1799, most lines in South Wales were built to a gauge of $4'2''$. Along the Severn even smaller waggons and narrower tracks were common.[38] It was the latter pattern which the York Buildings Company had brought to Tranent, and which was copied in the Alloa waggonway built by John Francis Erskine of Mar. Erskine was co-author of the Alloa report in the *Old Statistical Account*, and not surprisingly he was willing to sing the praises of his innovation, while succinctly summarising the incentives for a coalowner to invest in a waggonway:

> It has often been asserted, that there have been more estates lost than made (especially in Scotland) by working coal mines. There probably has been some foundation for such an assertion. The expences of mining and keeping up a colliery are considerable, and the commodity will not bear a great price; so that it is only a large quantity, that can produce a profit adequate to the expence. While the coals of the barony of Alloa were brought to the shore in small carts by the tenants, the quantity was uncertain, and often not very considerable. In 1768 a waggon way was made to the Alloa pits, which proved to be so great an advantage, that it induced the proprietor to extend it to the Collyland, in 1771. The sales were by these means increased, from 10,000 or 11,000 chalders to 15,000 or 16,000.

Although the way cost at least 10/- per yard to lay, 'the proprietor has been long ago reimbursed, and is a considerable gainer'.[39]

In spite of the early introduction of narrow-gauge waggonways at Tranent and Alloa, most early Scottish ways conformed more closely to Tyneside practice, at least on the question of gauge. This was hardly surprising. North-east England was geographically close, and a large number of Tyneside colliery viewers, foremen and engineers came to Scotland on visits or contracts. Some settled permanently, like John Dixon of Sunderland, who bought Dumbarton Glassworks and supplied them with fuel via a new waggonway and the Clyde.[40] His son William rose from manager of Govan Colliery to be the 'mighty Zeus of Scottish coal and iron masters', owning a series of collieries and the Govan, Wilsontown and Calder Iron Works. In the 1770s he built a coal-to-water waggonway from Govan to the Clyde, parts of which were later incorporated in his son's Polloc & Govan Railway.[41] Under the influence of such men, gauges of four feet or more became common. The second edition of the *Encyclopaedia Britannica* in 1778 described four feet as 'the common gauge':[42] in fact, no one gauge ever seems to have been generally accepted, but the selection of four feet for, among others, the Middleton Colliery Railway, the Tanfield Waggonway, the Surrey Iron Railway and the Kilmarnock & Troon did mean that it was used on some seminal lines in the development of the railway.

While the gauge of Scottish waggonways varied considerably, there was even less consensus on the size of waggons. Tyneside-style large waggons, of about three tons' capacity, were noted about the turn of the century at Fordell, and on the Elgin waggonway, where the capacity increased from 50 to 60 cwt between 1784 and 1796. But the narrower lines kept to smaller waggons, of 40 cwt capacity at Tranent, and 30 at Alloa and Pinkie. Sometimes either evidence is contradictory or practice on a particular waggonway changed dramatically over time. Thus, Lewis suggests 60 cwt waggons at Ayr, but in 1811 Buchanan stated that a horse there normally pulled five waggons each of one ton net. Erskine of Mar specifically defended the use of three small waggons rather than one large one, as being easier to load and unload, less damaging to the track, and easier for the horses.[43] Over time, there was a gradual though not unanimous move to smaller waggons. The celebrated mining engineer Robert Bald recommended moderate-sized waggons to all coalowners in 1816, and noted in the following year that Fordell waggons now held only 48 cwt — surprisingly, they still required two horses per waggon. A few years later Robert Stevenson, who thought that large waggons destroyed the track, noted that in general waggons in Scotland were smaller than they had been before.[44] Again then, Tyneside practice was not automatically copied, but individual owners calculated, sometimes by trial and error, what would best suit their own circumstances (see Table 3).

The difficulties in tracing the physical development of a waggonway culminate in the changing patterns of the Earl of Elgin's lines in the Dunfermline area. During half a century of alterations and extensions, both the places of origin of the traffic and the harbour to which it was sent were changed. Baxter, indeed, tries to unravel the problem by treating it as a matter of two distinct waggonways.[45]

The Earls of Elgin had interests in lime and, later, in coal. It may have been local patriotism which stimulated the claim that their limeworks were 'the most extensive ... even in Britain, belonging to any particular person', but an annual production of 80,000 to 90,000 tons of limestone was certainly substantial. By the late 1790s the seventh earl also owned 900 acres of coalfield, with an annual production of up to 90,000 tons.[46] Limestone rather than coal was the initial catalyst for a waggonway built, according to the *Old Statistical Account*, by the fifth earl in 1777-8 to convey stone 'from the quarry to the kilnheads' at Limekilns:[47] since the fifth earl in fact died in 1771[48] it is reasonable to identify this line, as most authors have done, with one built about 1768 from Berrylaw to Limekilns.[49]

Although the minister of Dunfermline might be unsure of dates twenty or thirty years before, he presumably knew what was happening as he was writing in 1794. The seventh earl, he reported, finding himself short of coal to fuel the lime kilns, had recently bought the extensive coalfields of West and Mid Baldridge, Clune, Luscar and Rosebank: 'from these coal mines, his Lordship is making a waggon-way of 4 miles extent, to his lime-works'. Geographically, it seems possible that this could have been done by relaying the original line and extending it for a mile to Baldridge, but in fact it appears to have been a completely new route.[50]

Table 3. Waggon capacity on Scottish waggonways

Waggonway	Date	Waggon capacity cwt
Elgin	1784	50
Alloa	1795	30
Halbeath	1795	40–42
Fordell	1795	48
Elgin	1796	60
Fordell	early 19th c.	60
Ayr	early 19th c.	60 (?)
Ayr	1811	20 (?)
Tranent & Cockenzie	1819	40
Pinkie	1824	30

Sources: *Old Statistical Account* VIII, 617; X 507; XV 270. A Scott in R. Stevenson (ed.), 'Essays on rail-roads', *Trans. Highland Soc.*, 6 (1824), 24. R. Buchanan, *Report relative to the Proposed Rail-way from Dumfries to Sanquhar*, 14. C. MacLaren, 'Railways compared to canals and common roads', *Pamphleteer* 26 (1826), 60. M.J.T. Lewis, *Early Wooden Railways*, 189-90.

Meanwhile, there are complexities at the other end of the line. The earls owned three almost adjacent harbours, at Brucehaven, Limekilns, and the newest and best one built by the seventh earl at Charlestown (a village built by and named after the fifth earl). Lewis suggests that a branch had reached lime works at Charlestown by 1792, before the harbour was built, and that a connection from Pitfirrane pits to the new harbour was added by 1801.[51] Certainly at some point the emphasis of the line changed, so that the main flow of traffic was from the collieries to Charlestown. Duckham, apparently following Baxter, suggests this was when 'the iron Elgin Railway proper was opened in 1812' from Wellwood and Rosebank collieries to Charlestown.[52] The major part of this change was probably relaying the old line with iron-edge rails and extending it at least to Rosebank; but Dott adds to the confusion by suggesting that Wellwood and perhaps even Baldridge were not brought on to the line until about 1841, which might coincide with Chalmers' statements that the railway 'has recently been greatly improved'.[53]

Chalmers also speaks of 'a change in the line of the rail-road in 1821' substan-

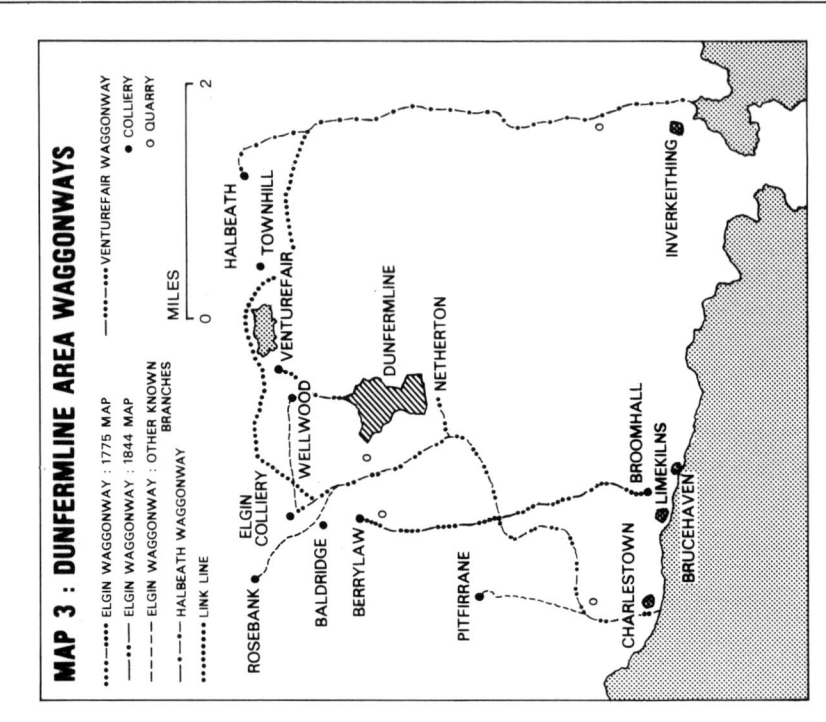

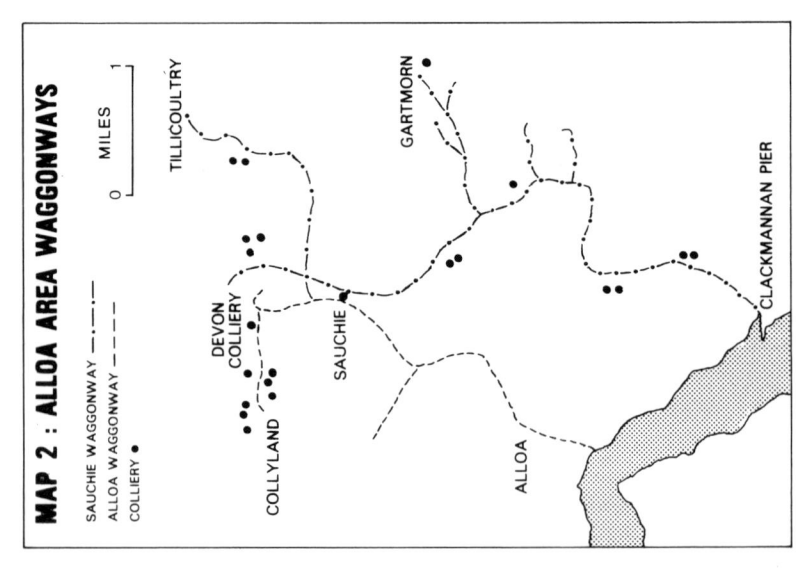

tial enough to involve the creation of two inclined planes. The two tracks between the inclines were graded separately and in opposite directions, so that a waggon might run under gravity in either direction from the head of one incline to the foot of the other. These were engineered by 'the ingenious Mr Landale of Dundee' and may be seen as a trial run for the later eccentricities of his Dundee & Newtyle.[54] A branch into Dunfermline was added in 1834, by which time the railway was known as the Dunfermline & Charlestown; the branch enabled the line to carry not only minerals, but also general goods traffic and passengers to and from the river steamers, which they had to reach in small open boats. Between 1838 and 1843 an average of 23,000 passengers per year passed through Charlestown harbour, most of whom travelled on the railway.[55]

The final main line of the Dunfermline & Charlestown, then, ran from Wellwood and Baldridge to Charlestown harbour, a distance of some six miles. The processes by which this was attained are sufficiently difficult to trace; there is even less information available on the dates at which various branches were abandoned. It is worth noting, however, though as weak evidence only, that the undated map in Chalmers' 1844 volume shows only the line from the Baldridge area to Charlestown and the Dunfermline branch, and omits any connections to Limekilns, Pitfirrane, Berrylaw or Wellwood.[56] Not far away, the Halbeath railway showed a similar, though less complex, pattern of altering tracks as conditions of coal supply altered; by 1847 the colliery which gave the line its name was no longer linked to it. In the 1850s a link line was built between the Elgin and Halbeath railways.[57]

III. The first railway?: Kilmarnock & Troon

In the early nineteenth century there are signs of greater ambition in the projectors of waggonways. Those built were in general more substantially constructed, and in some cases longer. Those planned but not built extended the range of conceptual possibility to lines of over a hundred miles, and to the servicing of agricultural areas. Where the eighteenth-century waggonways had been designed, so far as is known, by employees at the relevant collieries, engineers of national repute were now commissioned to survey and design cross-country lines. Alongside the creation of more short coal-to-water waggonways, there were now also plans for lines which would replace water transport rather than merely be an adjunct to it. The fact that these ambitious lines were not built showed that they were ahead of their time, and that the cheerful forecasts of 10% or 12% net profit were too hypothetical to extract capital from landowners who were concerned first with high wartime costs and later with low demand in the post-war depression: but the fact that the plans were made shows that the nature of the waggonway was changing.

There is no easy way of deciding when the waggonway becomes the railway. To insist on the presence of the steam locomotive seems unnecessary: the Stockton & Darlington did not become a railway simply because the directors were persuaded by George Stephenson to try his engine, and the Edinburgh & Dalkeith was certainly a railway although it was pulled by horses through the 1830s and after the mania. If a line must be drawn, it may be best to draw it on organisational criteria. When a line is established as a public railway by Act of Parliament, with its control thereby vested in the full panoply of directors and shareholders, and its charges and behaviour subject at least potentially to parliamentary interference, it has attained a status beyond that of the humble waggonway. If this criterion is accepted, the traditional claims of the Surrey Iron Railway and the Kilmarnock & Troon to be the first 'proper' railways in their respective countries can be justified.[58] Waggonways were no longer to be simply the private property of coal-owners with short-distance transport problems.

The Kilmarnock & Troon Railway, although longer than the others, was fundamentally just another coal-to-water waggonway. In one respect at least it was an old-fashioned one, in that it was constructed as a plateway at a time when other Scottish lines used the edge rail.[59] The plateway, or tram-road, with flanged L-section rails on which tram-engines ran with unflanged wheels, was popular in southern England in the eighteenth century, but never caught on in Scotland. Apart from Kilmarnock, a 'concave iron-track' was used at Ardrossan, and some form of plateway, possibly merely unflanged flat plates, was used at Shotts and inside the Carron works.[60] Dyos and Aldcroft refer to the pioneering use of wrought iron at Alloa in 1785 as a plateway:[61] this appears to be a misunderstanding of a common type of early rail in which iron plates were fastened on top of wooden rails. Baxter claims that tram-engines were used in 1831 on the Monkland & Kirkintilloch, but this was not a plateway.[62] On Kilmarnock & Troon plateway construction became particularly incongruous when the railway's only branch was laid with edge rails and to a different gauge.[63]

The problems of transport had exercised the minds of coalowners in the Kilmarnock area for some time. Much Ayrshire coal was exported to Ireland, and this trade had already been the reason for Messrs. Taylor's waggonways at Ayr.[64] Kilmarnock coal had to be carted eight miles to Irvine, at a cost of 5/6d to 6/8d per ton: even so, by 1790 40% of an annual coal production of 8,000 tons was travelling this way. In 1791 the minister of Kilmarnock noted a plan 'some time ago' for a canal to Troon which would be 'certainly one of the most desirable that can be made in Scotland'. He even hoped that it might be extended to Glasgow, which would have supplied Ayrshire coal with another very large potential market.[65] This may be the same canal plan 'originally proposed by Colonel Fullarton' which the Marquis of Titchfield noted in 1806.[66] Fullarton had been the chief opponent of Titchfield's long-planned project for a major harbour at Troon, although he had belatedly switched to approval.[67] Titchfield, from 1809 fourth Duke of Portland and the largest landowner in Ayrshire, proposed instead of a canal a railway to be promoted by all the landowners on the route as joint proprietors, with shareholdings in proportion to the amount of land required from them, preferring this to a canal in order to reduce both cost of construction and the disturbance to land.[68]

Titchfield's immediate success was limited to an agreement to make the line, for which an Act was passed on 27 May 1808. His neighbours were less willing to put up money, or to take more than a nominal interest in the project. Of the eighty £500 shares in the company, Titchfield had to take 74 himself, while the other three landowners were quite content to restrict themselves to one share each, and three remained unallocated.[69] It is not entirely clear why Titchfield went to the trouble of establishing his railway by Act of Parliament. The most probable explanation is that, given his previous experience with Colonel Fullarton's opposition to his harbour plans and the less than enthusiastic support he was receiving from his fellow-proprietors, he wished to safeguard the line against any future changes of mind. His investment went much further than the railway alone. On Fullarton's death, he had bought the colonel's estate, and was now building Troon harbour on it at an ultimate cost of over £100,000.[70] One contemporary at least appreciated the duke's efforts:

> It is no very common thing for an individual proprietor, to contract at one time for improvements of a public nature, which will probably cost upwards of £100,000 sterling, and to carry them into execution with all possible dispatch.[71]

The engineer of the Kilmarnock & Troon, William Jessop, who has a tenuous claim to be the inventor of the edge rail, had already built the Surrey Iron Railway and had worked with Telford on the Caledonian Canal.[72] His line was built to stronger specifications than previous Scottish waggonways, although it was suggested that too much haste had led to bad drainage on Shoalton Moss.[73] In spite of the problems of plateway construction, it paid its proprietors well. Whether it ever reached Titchfield's optimistic forecast of a 20% return[74] is doubtful, but in 1841 it was observed that although 'the principle [of a plateway] is bad, and it is standing in need of constant repair ... from the quantity of coal conveyed, it still continues, we believe, a very profitable speculation'.[75] Two years earlier, 130,500 tons of coal were carried on the line, of which 70,000 tons were from Portland's own mines. There were also about 70,000 tons of non-coal traffic, showing a development largely unknown to earlier waggonways.[76] From early years the line prospered: although by 1814 construction costs were £59,849 or £6,300 per mile, against Jessop's estimate of £38,167, in 1817 the company paid its first 5% dividend. It was to do as well or better in almost every year for the rest of the century, on top of the benefits it created in the way of increased trade and coal sales. W. McAdam's view that the general public derived little benefit from the Kilmarnock & Troon before it was taken over by the Glasgow Paisley Kilmarnock & Ayr in 1846 seems a very harsh judgement.[77]

The railway also became the first in Scotland to carry passengers, although its Act of Incorporation, unlike that of the abortive Berwick & Kelso, did not authorise such activity.[78] 'A regular system of travelling on Railways,' said Robert Stevenson in 1819, 'or the conveyance of passengers, has not been attempted, excepting, perhaps, from Kilmarnock to Troon.' Stevenson's qualifying 'perhaps' reflects the erratic and unscheduled nature of the service, which was run not by the company but by William Wright of Kilmarnock.[79] The company merely levied its tolls for so many tons of passengers, a practice that still prevailed in 1839.[80] In 1818 the French engineer Charles Dupin made his famous observations on:

> des diligences établies sur la route en fer de Kilmarnock à Troon Bay; elles donnent l'idée d'une

voiture nomade énorme, et, pourtant, trainée sans effort par un seul cheval.[81]

In 1829 the service was given credit for an increase in tourism at Troon, 'which has become a fashionable sea-bathing town'.[82]

The Kilmarnock & Troon saw the first trial on a Scottish public railway of a steam locomotive. In 1813 one of Portland's agents inquired of the Kenton waggonway in Northumberland about 'the New Mode of Leading our Coals by Means of Steam Engines instead of Horses'. The reply was encouraging:

> Should the length of Lead from his Lordship's Concerns in Ayrshire be considerable, I have no doubt but a considerable Saving will be made by adopting Mr. Blenkinsop's new Method.[83]

The trial, using an engine supplied by George Stephenson, took place in 1816 or 1817. Hard facts on the event are not easy to come by, but the artist John Kelso Hunter, writing fifty years later, left a graphic if not necessarily accurate recollection of a childhood experience:

> Early in 1816, Robert Stephenson, brother of the inventor, came to Kilmarnock with the first locomotive engine that ever appeared in Scotland. It was set down on the Duke of Portland's tram road about 400 yards below Kilmarnock House . . . As the steam got up the people stood further back. The liability to burst at the start had been much speculated on, and a strong desire that it should was fearlessly expressed. I stood in the Lower Woods Park beside Geordie Pettigrew, who had a heavy interest in the auld horse . . . When the engine had passed through the cut, he gave the final sentence: — 'To the tanyard every living beast: flesh and blood cannot stand against that!'
> . . . the engineer opened the safety-valve with a grand burst, which struck the air and the ears of the crowd at the same moment. It seemed to me as if the whole mass had been blown to fragments. The crowd instinctively swelled to such a size as burst the boundaries of the hedges on both sides of the road, and down a gentle declivity of about four feet there sprawled the mass . . . All sorts of murder shouts arose from the group. I was petrified, and held a death grip for a time, quite uncertain whether the people were killed, or if I were still alive . . .[84]

But the Kilmarnock & Troon's track and the locomotive were not made for each other. The minister of Kilmarnock blamed the engine, which had had to be altered for a plateway: 'from its defective construction and ill adaptation to flat rails, it only drew ten tons at the rate of five miles an hour'.[85] Hunter's recollection suggested that the problem was the height of the horse-path between the rails, which caught on the underside of the engine.[86] In fact the trouble was that Jessop's cast-iron plateway, supported as was usual on stone blocks, was not strong enough to take the engine's weight. George Buchanan, writing fifteen years after the event, confirmed that 'the locomotive engine had . . . succeeded well; but was given up, on account of its destructive effect on the cast-iron rails, although its weight was only five tons'.[87] The Kilmarnock & Troon remained a horse-drawn line until the Glasgow Paisley Kilmarnock & Ayr relaid it with edge rails from 1841.[88] Legend, however, persistently claims that Stephenson's loco, or possibly another one, was fitted with wooden wheels to reduce track damage and continued to run until 1848.[89]

The Kilmarnock & Troon, then, deserves its prominence in Scottish railway history, even apart from its pioneering act. It was the scene of interesting, if abortive, experiments in steam locomotion. More importantly the fact that it ran not merely to collieries but to an important manufacturing town (grandiloquently stated by Aiton to be 'in Ayrshire, as Manchester is in the county of Lancaster')[90] enabled it to mark the transition from a coal waggonway by developing an important trade in general merchandise and, gradually, in passengers. A few of the Scottish projectors of the early nineteenth century were beginning to take passengers rather more seriously. Robertson Buchanan, hoping they might bring substantial revenue to his Dumfries & Sanquhar plan, referred to the Welsh Sirhowy line where passengers were carried at six or seven miles per hour, 'in a manner more pleasant and easy than can well be conceived by those who have not experienced it'.[91] The Kilmarnock & Troon demonstrated that the provision of a passenger service met a demand. The horizons of Scottish promoters were widening.

IV. Coal, construction and costs

In spite of the lengthy list of waggonway projects, only about sixty miles of line had been built by the end of the Napoleonic wars, a figure which compared very badly with either Tyneside or South Wales, in spite of the generosity normally shown by Scottish landowners in granting wayleaves.[92] About 1819 Scott of Ormiston observed that 'in Scotland, railways are also employed for the conveyance of goods at all the collieries, and other works of any extent'.[93] This suggests that there may have been a good number of now unrecorded ways, generally limited to the confines of the works and, in the case of collieries, often underground. And, although occasionally Scottish projectors built small canals where waggonways would have seemed more logical (at, for instance, Campbeltown or Burnturk), there was no great development of canal transport either.[94] It is no part of this book to consider the structure and marketing techniques of the Scottish coal industry, which have in any case been admirably and comprehensively covered by Duckham, but it is desirable to say something about the extent to which the waggonways achieved their purpose.

The waggonways were, of course, built for economic reasons. In some cases, particularly those sponsored by iron works, the aim was to reduce the costs of manufacture of a firm's product. For these lines, detailed information on the costs of construction, the costs of transport, or the amount of coal moved, has not survived (and in some cases may never have been gathered), and we can only assume on logical grounds that they did in fact show advantages over road cartage. But most of the waggonways were sponsored by coalowners, and were meant to reduce the costs of marketing. That so few were built must raise questions as to whether even an efficient waggonway could enable a colliery to compete either with suppliers closer to the market or with coal brought in by water.

In practical terms, large markets for Scottish coal were limited. Apart from those mines which supplied large iron, lime, salt or other industrial enterprises, the possibilities were Glasgow, Edinburgh, or export by sea. The last was restricted, particularly on the east coast, by the efficiency and well-established trading network of the Tyneside coalfield, so that the 6,000 tons of coal exported from West Wemyss to Holland at the time of the *Old Statistical Account* stands as an exception rather than a rule.⁹⁵ Fife coal was also good for steamship use; by the time of the *New Statistical Account* it was claimed that Fordell 'has gained an ascendancy over most of the Scotch and English coal for that purpose', and that 'the steam-boats plying between Paris and Rouen are almost entirely supplied' from Charlestown.⁹⁶ But the main seaborne trade ran to Ireland from Ayrshire; the competition from West Cumberland was not overpowering, and the possibilities of the trade led to the waggonways at Ayr, Ardrossan and Troon. Scottish coal exports to Ireland doubled between 1780 and 1800, and by 1819 had reached over 175,000 tons, or a quarter of all Irish coal imports.⁹⁷

To the city markets, supplies were delivered by cart and, in the case of Glasgow after 1793, also by the Monkland Canal. The relative efficiency of carting frequently led to inter-city comparisons unfavourable to Edinburgh. In spite of attempts at cartelisation by the suppliers, the price of coal in Glasgow stood in the early 1790s at only 5/10d to 7/6d per ton, and rose in the latter part of the war to a peak of 11/8d.⁹⁸ In Edinburgh the cost in both 1793 and 1808 was £1 per ton; in 1813 Hugh Baird stated it at 15–21/- depending on quality; and by 1818 it had fallen to 14/- or 15/-. Baird claimed that in Edinburgh prices were one-third higher than in Glasgow, and three times those in the coalfield on his proposed Union Canal.⁹⁹ Some of the difference might be accounted for by the Edinburgh insistence (ridiculed by Bald) on 'great coal', and by the fact that the pits supplying the capital were on average one and a half miles further away than those supplying Glasgow. These factors could not, however, outweigh the difference in the cost of hiring a cart and horse — in Edinburgh 5/- per day, in Glasgow 9/- to 12/-.¹⁰⁰ The question was indeed one of carting efficiency. Glasgow's well-organised system, often run by the coalowners themselves, with sturdy carts of 24 cwt capacity averaging three journeys to the pit per day, contrasted well with the single daily trip of an inefficient 12 cwt Edinburgh cart, pulled by a horse worth perhaps one-tenth of its Glasgow rival.¹⁰¹ Bald estimated that cartage cost 8¾d per ton-mile in Glasgow, and 1/5d in Edinburgh, though Dixon at Govan cut the labour cost of transport to 2½d by employing his own carters and building his own waggonway.¹⁰² The efficiency of Glasgow carting, the possibility of supplies reaching the city centre by river, and the development of the Monkland Canal, seem to have dissuaded most coalowners in the Glasgow area from trying to improve their competitive position by the marginal help of waggonways.

In Edinburgh the opportunities were clearly better, and until the Union Canal opened in 1822, waterborne suppliers could come no nearer central Edinburgh than the docks at Leith. Here, given the quality of Lothians carting, was a chance to compete by waggonway. It is not then surprising that over half Scottish waggonway mileage was in Fife and Clackmannan, whence it was a short water journey to Leith. While Edinburgh continued to import coal from Newcastle — particularly specialist coals not found in Fife or Midlothian — any coalowner who could get his coal cheaply to the Forth had a chance in the Edinburgh market. The export figures given in the *Old Statistical Account* indicate the difference between waggonway harbours and the others — Alloa sent out about 56,000 tons per year, Charlestown-Limekilns 60,000 tons and Inverkeithing 25,000 tons, whereas

The Century of the Waggonways

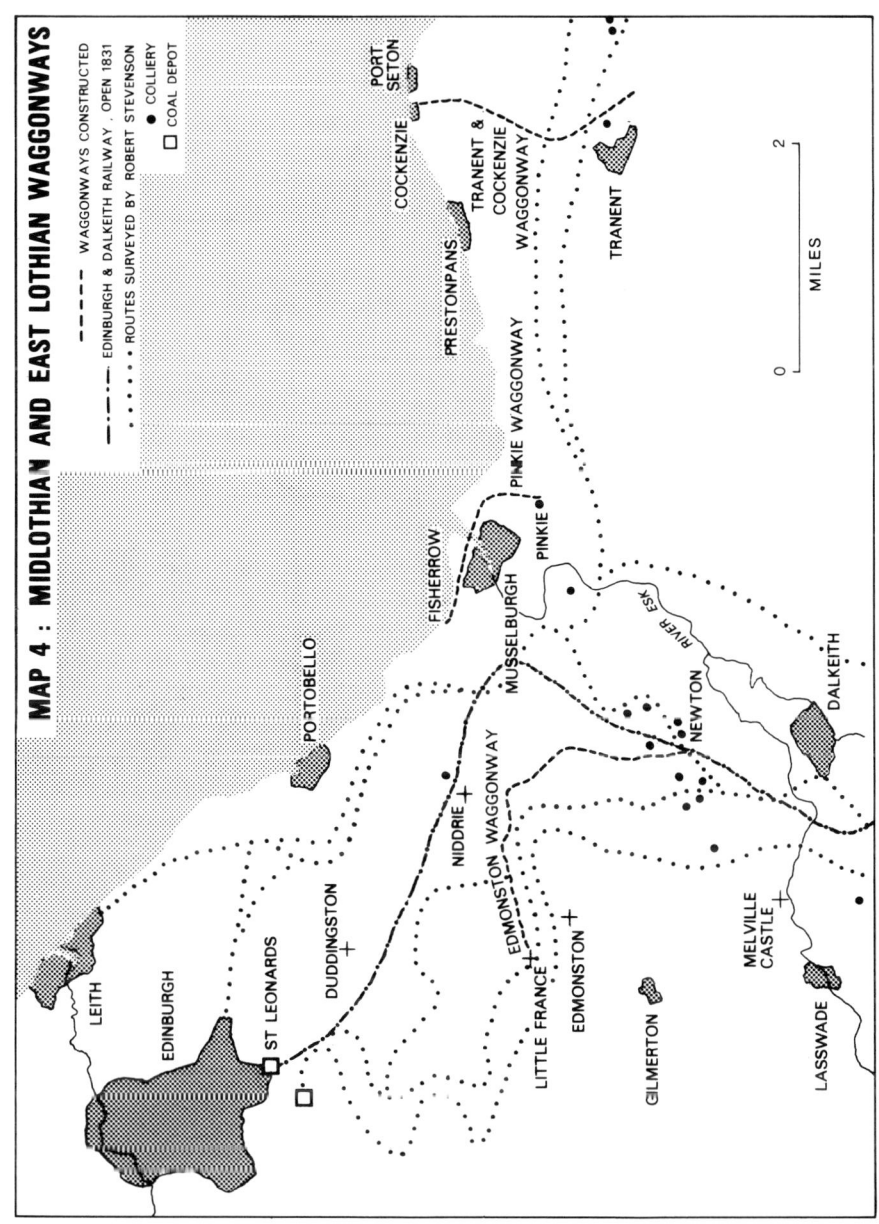

Dysart only achieved 4584 tons and Kirkcaldy a mere 600.[103] Bald noted that the Edinburgh sea-coal supply came chiefly from 'Halbeath, St. David's and Wemyss, upon the opposite coast, or from those up the river, the farthest of which is Alloa, distant about twenty-four miles', and that the price of this great coal was 10/- or 11/- on ship, plus 4/- freightage. Baird stated that coal from Fife, Alloa, and Newcastle was of sufficient quality to rate a premium of about 3/- per ton, and therefore cost 18/- to 21/- in 1813.[104] The shortage of Lothians coal at Leith even led to ships trading to the port unloading there, going to Charlestown or Alloa to bunker, and returning to Leith for an outward cargo.[105] Alloa coal, priced at about 6/- per ton at the harbour in 1783, rising to 10/- in 1810, had obvious attractions. Another advantage was noted in Dunfermline:

> Coals have been sent to the Forth, for exportation, on cast iron railways; and this mode of conveyance now saves the labour of not fewer than one hundred horses.[106]

After 1820, more coal came from Tyneside, and the Forth coal trade suffered some decline.[107]

The coalowners of the Lothians were not unaware of the need to improve their transport, although they tended to move after, and perhaps in response to, developments in Fife. In 1793 Sir Archibald Hope was declared by a rival to have 'almost a monopoly of the Edinburgh market' by the excellence of his coal and its convenient location. Twenty years later his son Sir John built the Pinkie waggonway in face of growing competition.[108] Plans for a general Midlothian coal waggonway, proposed by Henry Seton Steuart early in the century, reconsidered by Hope and John Clerk of Eldin in 1812, and surveyed by John Farey for the Duke of Buccleuch in 1816 and by Robert Stevenson in 1818, did not come to anything,[109] perhaps because of the mutual suspicion of various proprietors which surfaced again over the formation of the Edinburgh & Dalkeith Railway; the proprietors of the Newton pits did, however, build the Edmonston line for their own coal. The needs of Edinburgh spawned not only Stevenson's plan, based on his view that Midlothian coal would last the city for precisely 2,581 years, but also Baird's West Lothian Railway, based on the belief that Midlothian coal was almost exhausted.[110]

Apart from estimating the increased possibilities of his market, the potential waggonway proprietor had to consider his costs. In several cases, figures for construction costs are available, even if they are best regarded in most instances as orders of magnitude rather than accurate figures (see Table 4). The low figure for the Fordell was paid for in high maintenance costs and poor track: the high one for the Kilmarnock & Troon reflects its status as something more than a mere waggonway. It also indicates a problem which was to afflict almost all railways — underestimation of costs: Jessop's original estimate had been just over £4,000 per mile.[111] That a figure of about £1000 per mile was considered reasonable for a simple waggonway is confirmed by Henry Seton Steuart's two calculations, based on allowances of £837-10-0 and £1200, for his proposed twelve-mile link between Edinburgh and the major Midlothian collieries.[112] As waggonways became ambitious after 1810, estimated costs inevitably grew. But even if an estimate of over £3000 per mile was made, the cost would still only be about one-third of that of a canal.[113]

There were also the costs of running and maintenance to consider. On running costs, the Carron Company made a simple calculation. Before the waggonways, carriage had cost £1200 per week: the waggonways had allowed three-quarters of the horses to be disposed of, and had thus cut costs to £300 per week, and saved £10,800 per year.[114] Elsewhere the most cost-conscious line appears to have been, reasonably enough, Fordell. In 1791 Sir John Henderson was spending £138 on upkeep and £402 on running costs in the year, figures which agree well with mining engineer John Morrison's 1798 figure for working expenses of £560. The wooden rails and sleepers were easily damaged by horses, cast-iron wheels and sprag brakes, and Fordell horses hauled a smaller weight than on any comparable line.[115] Morrison noted that the neighbouring iron Halbeath waggonway had only half the running costs and one-eighth of the maintenance costs of Fordell, even though the minister of Inverkeithing stated that the Halbeath was 'kept in good repair at a great expence'.[116]

Table 4. Waggonway construction costs

	Date open	Total £	per mile £
Tranent & Cockenzie	1722	3,500	1,400
Fordell	?1752	—	450
Alloa (wood and iron strip)	1785	—	880 (min.)
Kilmarnock & Troon[a]	1812	59,849[b]	6,138
Pinkie	1814	—	800-1200
Edmonston	1818	— 4,000	800-1200 1,067

[a]: double track [b]: Figure from Highet. NSA V (Ayrshire), 554, says cost over £50,000.

Sources: OSA VIII, 618. J. Fraser in R. Stevenson (ed.), 'Essays on rail-roads', Trans. Highland Soc. VI (1824), 124. D. Murray, The York Buildings Company, 65. C. Highet, The Glasgow and South Western Railway, 7. J.C. & F. Inglis, The Fordell Railway, 13. G. Dott, Early Scottish Colliery Waggonways, 23.

The Fordell, in fact, was faced with a decision which came to all waggonways — when and how far to replace and modernise the track, and in particular when to move from wooden to iron rails. At Fordell, even although usage had fallen to 22 waggons by 1805, change was slow in coming. Bald's proposal in 1816 to place a stationary steam engine at the steepest incline was rejected, as, on grounds of expense, were two plans by leading Scottish engineers — the rising partnership of Grainger and Miller in 1828, and the experienced Stevenson in 1830. Not until the mid-1830s was a plan by Fordell colliery manager William Gofton, much altered by Robert Hawthorn of Newcastle, adopted — it cut costs by two-thirds and doubled the traffic.[117] The most detailed description of changes in track was gathered for the Alloa waggonway by the Rev. Peter Brotherston[118] and is summarised in Table 5. Most of the major older waggonways would go through several of these stages, although often, as on the Elgin, improved track could be combined with a planned change of route.

Table 5. Alloa waggonway track

	Date	Track material	No. of waggons per horse	Net load per horse (cwt)
1.	Pre-1768	Road	1	6
2.	1768	Wood: fir	1	30
3.	?	Wood: double fir[a]	1	30
4.	?	Wood: beech on fir.[b]	1	30
5.	?	Cast-iron plates on wood[c]	1	30
6.	?1785	Malleable iron plates on double wood[d]	3	90
7.	c.1810	Cast-iron edge rail	8	160
8.	c.1840	Malleable iron-edge rail[e]	8	160

Notes:
[a] Two fir rails one on top of the other; made repairs simpler.
[b] Top beech rail gave a smooth polished surface; reduced horse's work.
[c] 'Quite abortive'; rails broke.
[d] Good, but expensive once repairs were needed.
[e] 'If necessary, ten tons will be as easily drawn along the malleable iron railway as eight tons along the wavy surface of the cast iron railway.'

Source: *New Statistical Account* VIII (Clackmannanshire), 29-31.

The fundamental economy created by the waggonway was, of course, that it allowed a horse to pull a greater load, and therefore either increased the amount of coal transported or allowed a reduction in the number of horses and attendants employed. There is available from early nineteenth-century sources a number of scattered references to the amounts hauled by a single horse: unfortunately they sometimes omit to state whether they refer to net or gross weights, which, since waggons might themselves weigh up to a ton, is a serious omission. Their lack of internal consistency also underlines the varying quality of waggonway construction (see Table 6).

The advantages of iron rails over road haulage were clear. Canals, on the other hand, offered a much greater haulage per horse, but each barge required two men and a boy in attendance, whereas it was even possible on a waggonway for one man to look after two horses and their loads.[119] More important, though, was the advantage of the waggonway over the canal in terms of construction costs and of flexibility — the latter a major point for coalowners who might be opening new shafts and closing old ones at reasonably regular intervals. The problem is why more coalowners' waggonways were not constructed. The solution must be that, in so far as their decisions were taken on rational economic grounds, they felt that the extra competitive advantage they would gain would not be sufficient to offset the competition of better-placed collieries, the Monkland Canal, and/or imports from Tyneside.

V. Growing ambition: long-distance projects

By about 1810, the potential of waggonways for projects more ambitious than simply taking coal to the nearest waterway or ironworks was becoming clear, and

Table 6. Loads pulled by single horse

Place	Date	Type or condition of surface	Gradient	Load (cwts, net)	Source
Roads					
General	18th c.	—	—	2½ (packhorse)	8
Alloa area	pre 1766	—	—	6	13
Edinburgh	1808	—	—	12-15	5
Glasgow	1808	generally good	—	24	5
Sanquhar area	1811	poor	—	9½	4
Edinburgh area	1818	—	—	15	6
Glasgow	c.1818	—	—	24	7
General	c.1818	average	—	15-18 (gross)	7
,,	—	good	—	30-32 (gross)	7
Dalkeith-Edinburgh	1824	—	—	20	8
General	1825	—	—	30	10
Canals					
General	c.1819	water	level	600	6
,,	1825	water	level	900	10
Waggonways, observed					
Alloa	c.1770	Wood	very slight	30	1
Halbeath	1793	,,	—	24-48	13
Fordell	1790s	,,	—	30-60 (44-88 gross)	1
Fordell	1817	,,	—	24	14
Alloa	1785	Iron plate on wood	very slight	90 (150 gross)	13
Port Dundas	c.1818	Unflanged iron plate on road	1 in 15	60 (upwards) (69 gross)	7
Kilmarnock & Troon	c.1818	Plateway	1 in 660	93 (gross, down)	7
Alloa	post 1810	Iron edge	very slight	160-200	13
Ayr	1811	,,	level	100	4
Ayr	c.1818	,,	level	160-200 (gross)	7
Ardrossan	c.1818	,,	level	60-120 (gross)	7
Edmonston	c.1818	,,	varying	176-220 (gross)	7
Pinkie	c.1820	,,	varying	120 (168 gross)	11
Pinkie	1825	,,	varying	150 (210-220 gross)	9, 14
Pinkie	1828	,,	varying	210	12
Elgin	1828	,,	—	200 (240 gross)	12
Halbeath	1830	,,	—	120	14
Waggonways, estimated					
General	c.1818	Wood	level	48	7
,,	c.1818	Plateway	level	60	7
Kilmarnock & Troon	1811	,,	1 in 660	160-220 up, 200-240 down	3
General	1810	Iron edge	—	100	2
Dumfries & Sanquhar	1811	,,	1 in 528	100 up, 140 down	4
General	c.1818	,,	level	120-220	7
,,	1819	,,	level	160-200	6
,,	c.1820	,,	—	200 (267 gross)	11
,,	1824	,,	—	160	8
,,	1825	,,	level	up to 300	10
,,	1828	,,	level	100 at 4 mph, 200 at 2 mph	12

Sources for Table 6:
1. *Old Statistical Account* 2. J. Rennie, *Report respecting the Proposed Railway from Kelso to Berwick* 3. W. Aiton, *General View of the Agriculture of the County of Ayr* 4. R. Buchanan, *Report relative to the Proposed Rail-way from Dumfries to Sanquhar* 5. R. Bald, *A General View of the Coal Trade of Scotland* 6. R. Stevenson, *Report relative to Various Lines of Railway from the Coalfield of Midlothian to the City of Edinburgh and Port of Leith* 7. R. Stevenson (ed.), 'Essays on railroads', *Trans. Highland Soc.* VI (1824) 8. J. Grieve & J. McLaren, *Report on the Utility of a Bar-Iron Railway from the City of Edinburgh to Dalkeith* 9. *Quarterly Review* 31 (1825) 10. C. McLaren, 'Railways compared to canals and common roads', *Pamphleteer* 26 11. J. Rickman (ed.), *Life of Thomas Telford written by himself* 12. D. Rankine, *A Popular Exposition of the Effects of Forces Applied to Draught* 13. *New Statistical Account* 14. G. Dott, *Early Scottish Colliery Waggonways*.

much more elaborate schemes were proposed. William Jessop's engineering in Scotland was limited to the Kilmarnock & Troon and its semi-detached branch, but he was also asked for a second opinion on the most ambitious scheme of all. The idea of a waggonway across the country from Glasgow to Berwick, primarily to move Lanarkshire coal and lime eastward and Borders grain westward, had been mooted by, among others, Martin Dalrymple of Fordel, and a committee under Sir James Stewart Denham of Coltness commissioned Thomas Telford to examine the possibilities. Telford's proposed line ran through the Monkland coalfield, Carluke, Peebles, Melrose and central Berwickshire, to Tweedmouth. The estimated cost was £365,700, the revenue £55,559 per year, and the working expenses a round and undetailed £10,000, giving a 12% return on capital. The line would give to Lanarkshire the supplying of coal to the Borders, which had at the time to cart it for twenty to forty miles from Dalkeith, and it promised to reduce the price to 15/6 per ton in Peebles and 23/6 in Kelso. Jessop approved the scheme with only minor criticisms, and the committee agreed to arrange public meetings and issue shares of £100 each: but the project sank without trace.[120]

It helped, however, to establish the idea of the long-distance waggonway. Telford, whose long and distinguished career in road, bridge and canal building both made his views entitled to respect and ensured that he would not be an instant convert to waggonways, was now prepared to admit their value. The turnpike, he was sure, was outdated:

> In the best improved countries, they are now chiefly employed for the passage of travellers, and for articles of traffic, which are no great value in proportion to their bulk.[121]

He retained his affection for canals as 'the most perfect means which have hitherto been discovered', but admitted that they were not always practicable, and that a railway with easy gradients 'for facility and cheapness... nearly rivals a canal'. The railway could operate better in rugged country, if the traffic was downhill, if the weight of the articles carried was high in proportion to their bulk, or, of course, if water was in short supply. Therefore, he reported in 1803:

> In the instances where I have taken the liberty of mentioning that canals are much wanted, I beg leave to be understood to include iron rail-ways, and I strongly recommend, that in all future surveys it may be an instruction to the engineers, that they do examine the tracts of country with a view of introducing iron rail-ways, wherever difficulties may occur with regard to the making navigable canals.[122]

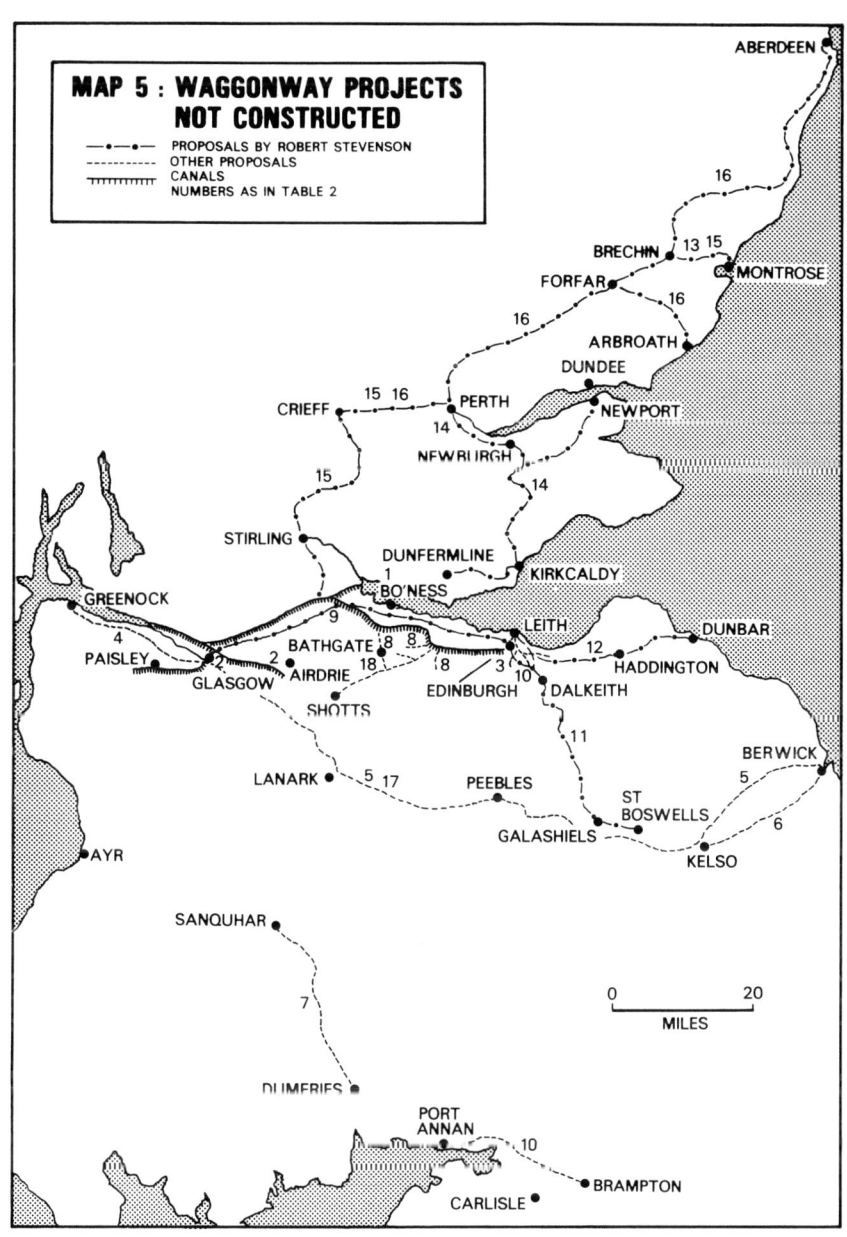

The country between Glasgow and Berwick, which was sufficiently hilly to require at least two inclined planes on the waggonway plan, clearly fitted his criteria.

Another project, which paralleled part of Telford's route, got as far as parliamentary authorisation in the following year. Again a prime motivating force was coal, but unlike the earlier waggonways the initiative came from consumers rather than producers. The projectors of the 22-mile Berwick & Kelso Railway estimated that their area required 16,000 tons of lime and 15,000 tons of coal each year, which could be brought in by sea and rail, and which would far outweigh the 6000 tons of grain exported.[123] They hired another distinguished engineer, John Rennie, and obtained an act which incidentally contained the first parliamentary authorisation of passenger transport by rail, but they failed to raise the necessary £90,000 capital.[124] The company lingered on in a moribund condition until it flickered into life again in the boom period of 1824-25. This time progress was halted by disagreement over whether to keep to Rennie's route south of the Tweed to Tweedmouth, or whether, as Berwick supporters wanted, to run north of the river into Berwick itself: the southern route, as far as they were concerned, would be 'objected to and impeded as much as possible'.[125] In 1836 a further meeting was held after a sub-committee of the Border Association for the Encouragement of Agriculture had suggested reviving the project, but no action followed. The company was finally dissolved in 1838.[126] Another plan worthy of notice was that published in 1811 by the engineer Robertson Buchanan, by which a 30-mile waggonway would both give the isolated Sanquhar coalfield an extended market and divert the coal supply of Dumfries away from English sources. Buchanan even visualised his line as eventually forming part of a through line from Glasgow to Carlisle, almost exactly along the route later taken by the Glasgow & South Western.[127]

By the end of the Napoleonic wars, a group of engineers was appearing whose commitment was to the railway as the transport of the future. In Scotland the engineer who was most totally convinced by the waggonway was the Glaswegian Robert Stevenson, who is yet another indicator of the truth that early nineteenth-century civil engineers were not restricted to particular specialities. He was the stepson of a lighthouse engineer, and his own most famous project was the Bell Rock lighthouse, built slowly in extremely difficult conditions during the first twelve years of the century. He was also responsible for much of the Edinburgh New Town, including the section on and around the Calton Hill.[128] His work on roads and early investigations into waggonways seems to have convinced him of the possibilities of rails. In a report to the Edinburghshire Road Trustees in 1812, Stevenson asked them to consider:

> how far *cast-iron cart-tracks* might not be advantageously laid upon the roads. Some years since the reporter got two or three yards' length of these iron tracks brought from the Shotts iron works, where they have been used for years with much advantage, and, it is believed, with economy. These cart-tracks would cost about £2000 per statute-mile, including upholding by the iron-founders for one year. It would be interesting to have also a trial made of these in some very public road . . .[129]

This idea involved simply the insertion of flat iron plates into an ordinary road.

This device was successfully tried on the road between Glasgow and the canal at Port Dundas in 1816, though the idea came directly from John Baird of Shotts Ironworks rather than from Stevenson, who by then was concerned more with waggonways using the edge rail.[130]

Stevenson has a claim to be a major populariser of the malleable iron rail which was an essential ingredient of the eventual rail network. George Stephenson certainly allowed him much of the credit for Birkinshaw's rails:

> Sir, — with this you will receive three copies of a specification of a patent malleable iron rail invented by John Birkinshaw of Bedlington, near Morpeth. The hints were got from your Report on Railways, which you were so kind as to send me . . . Your reference to Tindal Fell Railway led the inventor to make some experiments on malleable iron bars, the results of which convinced him of the superiority of the malleable over the cast iron — so much so, that he took out a patent. Those rails are so much liked in this neighbourhood, that I think in a short time they will do away with the cast-iron railways . . . I know you have been at more trouble than any man I know of in searching into the utility of railways . . .[131]

Birkinshaw confirmed that his attention was drawn to the subject by Stevenson's report.[132]

By 1820 Stevenson was recognised as 'the great authority on railways in Scotland', being for instance called in by the Stockton & Darlington promoters to consider Overton's original plan for their line, and chosen by the Highland Society to edit their series of prize essays on railways.[133] Yet the only waggonway designed by him which was actually built was a short coal line — the Edmonston or Newton waggonway, running from Newton Colliery to the Edinburgh-Dalkeith road, whence the coal was carted to the city.[134] It appears later to have been linked to, and was certainly eventually superseded by, the Edinburgh & Dalkeith Railway. It is unclear who commissioned Stevenson to build the line. The owner of the land was John Wauchope of Edmonston, who had leased his coal to Laing of Shawfair and Stenhouse of Whitehill, and the line is referred to variously as Wauchope's and Laing's. During negotiations with the Edinburgh & Dalkeith both Laing and Stenhouse talked as if they owned the line, with Stenhouse claiming a clause in the Edinburgh & Dalkeith's act to compensate him for damage to his private railway.[135] Either a partnership, or one or more sales of the line, are possible explanations. The line, like Sir John Hope's one at Pinkie, can be seen as fulfilling part of Henry Seton Steuart's earlier wish for a waggonway from Edinburgh to almost all the Midlothian collieries.

Stevenson's ideas were, however, much more elaborate than mere colliery lines: he was fortunate enough to find several groups of landowners willing to commission large-scale surveys from him, though less fortunate in that they never raised the money to put his ideas into practice. The routes which he surveyed between 1817 and 1827 later became, with some modifications, the main railways of eastern Scotland, but they had to await the proven success of the locomotive and in many cases the speculative atmosphere of the mania before they were built.

His plans show both the constraints he inherited from the colliery waggonways, and ideas which foreshadow the greater ambitions and capacities of the railways. He was opposed to the use of heavy rolling stock as being too damaging to the light track he intended to use. For most of his surveys he was not considering the possibility of locomotive traction, and even in 1817, when it had to be at least contemplated, he felt that in areas of limited traffic it would be more economical to use horses than to spend more money on strengthening the line.[136] Because he was planning for horses, his great quest was for the level line: if gradients were unavoidable, the line should be level as far as possible, with inclined planes worked by stationary steam engines to cope with necessary changes in elevation. The concept clearly owes much to canals.

On this basis, Stevenson three times surveyed the valley of Strathmore, in 1817 for a canal and in 1820 and 1827 for a railway, and emerged with a plan for a level line from Crieff to Aberdeen, with inclined planes taking branches down to Perth, Arbroath and Montrose.[137] In the Borders, however, the link from the Dalkeith coalfield to Galashiels would have to descend by the Gala Valley; Stevenson had to admit that the level route, by Gifford, Cockburnspath and Duns, would be 'a tedious and expensive line for the Vale of the Tweed'.[138] He was also enthusiastic when consulted by Sir John Sinclair about the most grandiose plan of all. James Watt had earlier suggested a canal from London to Edinburgh: Stevenson's view was that 'an iron road would not only be much more practicable but more commodious and useful for general intercourse than a canal', and that 'in almost every case, it is better to construct a Railway than a Small Canal'.[139] Other people were also thinking about connecting the two capitals — William Bell of Edinburgh, for instance, in 1824[140] — but their vision had to wait over twenty years for fulfilment.

The main purpose of the planned lines was still to carry heavy goods, and particularly coal. This was most obviously true for the network of possible routes which Stevenson surveyed between Edinburgh and the Midlothian coalfield, and for Hugh Baird's West Lothian Railway, authorised from Shotts to the Union Canal in 1825.[141] Stevenson had contemplated a line up the Esk Valley as far back as 1812, but his survey in the area had to await a commission in 1817 from a committee of coalowners chaired by John Clerk of Eldin and including the Duke of Buccleuch.[142] In spite of their engineer's reminder that Edinburgh:

> from the expence of land carriage, notwithstanding its local advantages . . . is under the necessity of being supplied, not only from the counties of Fife and Clackmannan, but even from England,[143]

the Midlothian coalowners had done nothing to improve matters, feeling no doubt that their 3/- per ton price advantage would still ensure sales. But now they were threatened by the advent of the Union Canal, bringing West Lothian and Lanarkshire coal into central Edinburgh by water. The threat, however, does not seem to have been sufficient, at least before the canal opened, to produce agreement among projectors some of whom were bound to be more favoured than others by whichever route was chosen — the fact that Stevenson surveyed four routes may have made things even more difficult to decide. Apart from individual enterprise by Wauchope and Hope, the area had to await the Edinburgh & Dalkeith Railway for improved links to the city.

The Midlothian scheme was projected as a rival to the canal, and as such was the first in Scotland to consider competing with a canal, if not along the same route, at least in taking the same product to the same market. The later West Lothian, not surprisingly since they shared the same engineer, was intended to cooperate with the Union Canal, supplying it with a traffic in coal and, via branches to Bathgate and Silvermines, in lime.[144] These were clearly coal and lime lines. But coal and lime were also the main motives for planning railways into agricultural areas. In Strathmore a chronic shortage of both 'operates nearly as a prohibition to the improvement of this part of the country', and the sponsors of the Roxburgh & Selkirk plan (from Dalkeith to the Borders) also made it clear that coal and lime were their first priorities.[145] In both cases, and unlike previous waggonways, the initiative came from the prospective consumers and the projectors lived at the agricultural end of the route. One promoter of the Glasgow & Berwick had earlier also put an optimistic view of the future for Borders industry:

> Were the Rail Road in Existence, the Woolen Manufactory would shortly be established on the banks of the Tweed, and the Farmer instead of sending the produce of the soil in Wool, to be manufactured in Yorkshire, whence it is again returned to this Country for Sale, would have it worked up upon the spot; thereby saving the double Carriage, and finding employment for an increased Population, in a district at present but thinly peopled and where from the high price of coals, there can be no inducement to establish manufactures of any sort.[146]

The possibility of other traffic was not ignored. In the Borders, for instance, Stevenson anticipated 2000 tons of general goods per year alongside 12,000 tons of coal and 10,000 tons of lime; and even passengers were possible, although 'we would not now calculate upon much revenue from this source'.[147] In the Midlothian survey, he saw the chance of combining profit with public health, by removing for agricultural use the city's copious production of manure, which 'has become a great nuisance, from its accumulation at the approaches to the City'.[148] And of course agricultural produce would flow in the opposite direction to urban markets.

The finances of railway building were treated usually with optimism, and often in very general terms. Sometimes the first problem could come as early as the survey stage, as some landowners were more eager to have a plan than to pay for it. The law firm of Gibson and Oliphant, agents for the Midlothian project, sent a memorial to the Edinburgh magistrates: after some judiciously flattering references to:

> The active spirit of Improvement, by which the City is at present distinguished — the liberal Patronage afforded by its Magistracy to every useful and ornamented work; and ... the circumstance of this undertaking being conducive, in so many ways, to the advantage of the City at large, and particularly of its Public Revenue,

they pointed out that the town council had backed the Union Canal on public grounds, claimed that they should do no less for the railway, and suggested they might start by paying for the survey.[149]

In 1820 the original projectors of the Roxburgh & Selkirk, a closely knit group of Borders landowners including Sir Walter Scott, thought up a much more elaborate way of using public rather than private money:

> Relying upon the present disposition of His Majesty's Ministers, the subscribers propose borrowing from them the requisite sum for completing the whole undertaking ...[150]

In this case, the railway could be opened in three years. Interest due to the government would be paid before opening from a levy on local heritors, farmers and towns: after opening, the tolls would repay this levy, continue to pay the government, maintain the line, and even allow for a sinking fund to liquidate the capital. The glowing picture went further still — once the capital was repaid, 'which must be the case in a few years', tolls would be required for maintenance only, and 'coal and lime shall then pass *toll free,* or nearly so'.[151] All of course depended on the government, since the projectors showed no immediate sign of raising the necessary £63,632 themselves. Their optimism was presumably based on previous state help with such projects as the Holyhead road, the Caledonian Canal and eighteenth-century roads in the Highlands, and perhaps also on the prominent positions held by some of their number in the local Tory party. In fact ministries would only help projects of national, and usually of military, importance: when Stevenson reported to an enlarged and more aristocratic committee in the following year, there was no reference to government money. Nor, unusually, was there a forecast rate of return — although the line would of course pay, the real advantage was to be in supplies of coal and lime.[152]

If the financial burden could not be passed on, it became important to have estimates of the rate of return on the proposed investment. For the coal waggonways, built privately by a single proprietor, the important calculation was the gain that they would bring to the enterprise as a whole. The waggonways were not being operated as separate commercial propositions, and in most cases were only carrying the goods of their owners — and though Henderson of Fordell, for instance, might bring in some extra income by allowing his neighbours to use his waggonway,[153] what was important to him was the reduction in costs and increase in output afforded to his own pits. For the newer and larger projects, the generalised benefit anticipated for the whole area of the line had to be at least to some extent separated from the financial returns to the investors in the company. Although, as one engineer remarked, 'Railways or undertakings of a similar nature, ought obviously not to be viewed in the light of mercantile speculation *merely*',[154] investment considerations were becoming increasingly important.

Unfortunately, estimating construction costs and traffic receipts was not an exact science. Both estimates were normally left to the engineer conducting the survey. His previous experience would certainly give him the ability to calculate construction costs in detail, and estimates for earthworks, bridges and so on were usually presented with an air of absolute precision and costed to the last penny. Since the lines were not built, we cannot know whether this confidence was justified, though the escalation of the Kilmarnock & Troon's construction costs from Jessop's estimate of £38,167 to an actual figure of £59,849 by 1814 (as well as the history of most later railways) may allow us to have doubts.[155] The usual allowance of 10% on the estimate for contingencies seldom covered the multifarious escalations of expense which could happen in railway construction.

Overall, the estimates in Table 7 appear reasonably consistent with each other, given the differences in the railways to be built and the territory they were to traverse — costs on the Midlothian line, for instance, were likely to be high because of the need to build in urban areas. The main exception is Rennie's apparently high figure, and it is interesting, if confusing, to note that an anonymous pamphlet in 1809 estimated the cost of his line at only £1532 per mile.[156] Figures for earthworks, masonry and the track itself could be estimated reasonably accurately, barring unforeseen circumstances such as the discovery of previously unknown bogs on the route or sudden changes in the prices of raw materials (see Table 8). The biggest unknown factor was the price which would be demanded for land, and here the estimates were generally optimistic. Buchanan, allowing only £1500 maximum for land for a 30-mile line, expected many owners to supply land free in return for the ensuing benefits.[157] Stevenson also felt that, even for non-coalowning proprietors, 'the advantages to agriculture, from this measure, must greatly compensate for any inconveniences'.[158] But even while Baird thought that land for the West Lothian should be cheap, as it was of mediocre quality and would be much improved, at least one landowner's agent was already calculating how high the claim for intersectional and amenity damages could be pitched.[159]

Tables 9 and 10 show the estimates for traffic on some of these lines. Here the engineer was on more difficult ground, having often no obvious qualification for making the calculation and little useful previous experience to work from. Telford, though not specifying the amount of general goods traffic he anticipated on the Glasgow & Berwick, was thinking in terms of a total annual traffic of about 125,000 tons. One of the supporters, however, was prepared to anticipate 200,000 tons, even though he thought one-fifth of that amount would give a 12% return.[160] And Telford himself in private claimed that 'the prospect of remuneration when fairly stated, will far exceed what has been held out by the warmest advocate of this useful project'.[161] Sometimes figures for the total possible amount of a particular traffic were available. Stevenson had statistics available to him for the coal demand and the refuse supply of Edinburgh. But he could only be making an educated guess in claiming that the Midlothian Railway would supply one-third of Edinburgh's coal plus all that of Leith (to which the Union Canal did not go), and that it would take away half of the manure. In the Borders he had to assume further that the presence of the railway and the decline of peat burning would raise coal consumption per head to an arbitrary figure of three tons per year (against $6\frac{1}{4}$ tons in the cities).[162] Other figures in the various estimates reflect rough calculations of, for instance, the demand for lime in the eastern borders (Telford) or the potential production of the Hopetoun limeworks in West Lothian (Baird).[163] Only Buchanan was willing to assume, without going into details, that building a railway would automatically increase the general traffic of the area: he added one-third for this to his estimate of the situation in 1811, thus starting the process by which later enquirers assumed a railway would double the previous traffic.[164]

Other figures in the reports are much more arbitrary (see Table 11). Working expenses are not detailed, but presented as a convenient round number of about 15-20% of receipts. Presumably this bore some relation to what the engineers knew of earlier waggonway experience, though nowhere do they say so. Nor does there seem to be much consistency between the different estimates. The figure for the rate to be charged seems almost random; virtually anything might be considered,

Table 7. Construction cost estimates for waggonway projects, 1810–27

Line	Date	Length m.	Track	Engineer	Total cost £	Cost per mile £
Glasgow & Berwick	1810	125	double	Telford	365,700	2926
Berwick & Kelso	1811	22	single, double works	Rennie	90,000	4091
Dumfries & Sanquhar	1811	30	single	Buchanan	44,242	1475
Midlothian	1819	11	double	Stevenson	52,600	4782
East Lothian	c.1820	27½	single	Stevenson	95,082	3457
Roxburgh & Selkirk	1821	44	single, double works	Stevenson	63,632	1449
West Lothian	1824	22½	double main line, single branches	Baird	54,892	2426
Strathmore	1827	84	single, double works	Stevenson	370,000	4400

Sources: Engineers' reports on the various lines.

Table 8. Details of construction cost estimates for waggonway projects, 1810–21

Line	Date	Length m.	Land total	Land p.mile	Earthworks total	Earthworks p.mile	Masonry & bridges total	Masonry & bridges p.mile	Roadway total	Roadway p.mile	Rails total	Rails p.mile	Misc.	Total cost	Total p.mile
Glasgow & Berwick[a]	1810	125	27,240	217.9[e]	41,244	330.0	36,478	291.8	55,650	445.2	193,296	1,546.4	10,900	365,700	2,926
Berwick & Kelso	1811	22	5,680	258.2	25,970	1180.5	21,135	960.7	included in works		26,660	1,211.8	11,355	90,000[d]	4,091
Dumfries & Sanquhar	1811	30	1,500[c]	50.0	4,000	133.3	7,000	233.3	11,880	396.0	15,840	528.0	4,022	44,242	1,475
Midlothian[a]	1818	11	8,950	813.6[e]	4,842	440.1	9,807	891.6	10,269	933.5	12,733	1,157.5	6,000	52,600	4,782
Roxburgh & Selkirk[b] (main line)	1821	35	2,650	75.7[e]	3,161	90.3	5,269	150.5	14,284	408.1	18,115	517.6	7,109	50,588	1,445

a: double track.
b: single track, double works.
c: expecting much to be given free
d: estimate to Parliament £88,709 by Rennie (HLRO 1811 51 Geo.III c.133).
e: allowances for land per acre were £60 on the Glasgow & Berwick, £50 on the Roxburgh & Selkirk, and £260–360 on the Midlothian with the highest price for land in Edinburgh.

Sources: Engineers' reports on the various projects.

Table 9. Estimates of traffic on waggonway projects, 1810–24

Line	Date	Volume of traffic (tons)				Rates p.ton–mile (d)	Returns on traffic (£)					Total
		Coal	Lime	Agricult'l produce	Other goods		Coal	Lime	Agricult'l produce	Other goods	Passengers	
Glasgow & Berwick	1810	39,054	51,200	16,000	unstated	$1\frac{1}{2}$	16,109	19,200	9,000	11,250	—	55,559
Berwick & Kelso	1811	15,000	16,000	6,000+	3,000+	2–3[a]	1,757	2,333	c.1,170	c.468	—	6,637
						3–4[a]	2,651	3,500	c.1,560	c.624	—	8,335
Dumfries & Sanquhar	1811	25,651	5,000	1,000	5,200	2	3,675	333	133	935	—	5,077+33% for railway = 6,769
Midlothian	1818	75,313	5,000	6,000	25,271	4+waggon dues	8,032	531	638	2,197	possible	11,367
Roxburgh & Selkirk	1821	12,000	10,000	3–4,000	2,000	2+waggon dues	3,046	2,413	644	1,150	some	7,252
West Lothian	1824											
— main line		76,000	10,000	28,000		1	3,650	250	738		150	4,788
— branches		41,000	36,000	12,000		1	1,100	900	350		—	2,350

a: Rennie made calculations on two different sets of rates. In each case some goods were charged more than others.

Sources: Engineers' reports on the various projects.

Table 10. Estimates of traffic on waggonway projects, 1810–24: percentages

Line	Date	Coal	Volume of Traffic			Value of Traffic			Passengers
			Lime	Agricultural produce	Other goods	Coal	Lime	Agricultural produce	Other goods	
Glasgow & Berwick	1810	—	—	—	—	29.0	34.6	16.2	20.2	—
Berwick & Kelso	1811	37.5	40.0	15.0	7.5	31.8[a]	42.0[a]	18.7[a]	7.5[a]	—
Dumfries & Sanquhar	1811	69.6	13.6	2.7	14.1	72.4	6.6	2.6	18.4	—
Midlothian	1818	67.5	4.5	5.4	22.7	70.4	4.7	5.6	19.3	?
Roxburgh & Selkirk	1821	43.6	36.4	12.7	7.3	42.0	33.3	8.9	15.9	?
West Lothian	1824	57.6	22.7	13.8		66.5	16.1	15.2		2.1

a: Using Rennie's higher estimates (see Table 9).

Sources: Engineers' reports on the various lines.

Table 11. Estimates of finances of waggonway projects, 1810–24

Line	Date	Length	Construction cost		Gross revenue		Working expenditure		% of gross revenue	Net revenue		Rate of return
			Total £	p.mile £	Total £	p.mile £	Total £	p.mile £		Total £	p.mile £	
Glasgow & Berwick	1810	125	365,700	2,926	55,559	444.5	10,000	80.0	18.0	45,559	364.5	12.5
Berwick & Kelso	1811	22	80,000[a]	3,636	8,335[b]	378.9	1,000	45.5	12.0	7,335	333.4	9.4
Dumfries & Sanquhar	1811	30	44,242	1,475	6,769	225.6	1,000	33.3	14.8	5,769	192.3	13.0
Midlothian	1818	11	52,600	4,782	11,367	1033.4	3,000	272.7	26.4	8,367	760.6	15.9
Roxburgh & Selkirk	1821	44	63,632	1,449	7,252	164.8	2,000	45.5	27.6	5,252	119.4	8.3
West Lothian	1824	22½	54,892	2,426	6,998	309.3	1,490	65.9	21.3	5,508	243.4	10.0

a: Excluding £10,000 which it was hoped would be contributed by local authorities for the Tweed Bridge.
b: Using Rennie's higher estimates (see Table 9).

Sources: Engineers' reports on the various lines.

provided that it undercut the rivals by a sufficient margin. Stevenson's suggestion of 5d per ton-mile in the Borders was simply based on the 8d charged by road carters from Dalkeith. In East Lothian he proposed a scale running from 6d for general goods, through 4d for coal and agricultural produce, to 2½d for lime and stone, and 2d for manure. Telford calculated on the basis of 1½d per ton-mile, but then remarked that 3½d would still divert the supply of Borders coal away from Dalkeith and into Lanarkshire via his railway.[165] It is difficult to avoid the conclusion that in some cases the desired rate of return (consistently about 12%) was established first: and the ton-mileage rate depended on how optimistic estimates of the volume of traffic had been.

The most obvious fact about these long-distance waggonway plans is that they were not constructed.[166] The years around 1820 mark a transitional period in the development of the railway, when promotional enthusiasm was running ahead of both technological ability and financial prudence. The railway had not yet captured the public imagination, although an increasing number of people were becoming enthusiastic. The waggonways had proved their competence in moving coal over short distances, often merely as subsidiary parts of a transport chain based on water. They had not yet convinced investors, or indeed many landowners, of their suitability over greater distances, and, unlike a short waggonway, a long line needed wide support. It could not be built by one man backing his belief and consulting his interests alone — it required a company structure, general support from landowners and a sharing of the financial burden of getting it built. In the difficult years of the Napoleonic wars and the post-war depression, risk capital was not to be attracted into such speculative ventures, and even men who could see personal advantage for themselves in a scheme might well still be unwilling or unable to help finance it. The only Scottish railway to be built by Act of Parliament before 1824 was short enough to be effectively a one-man operation, and the grudging nature of the support given to Titchfield by his Ayrshire neighbours did not prevent his going ahead. But over a longer distance than that from Kilmarnock to Troon, even the wealth of the Portlands would have been stretched to undertake a virtually single-handed operation.

Economic recovery in the 1820s was one precondition for the next stage of railway development. Another factor was technological change. To assert superiority over the canals, the railway required efficient locomotive traction, and the locomotive in turn required among other things malleable iron rails. The 1820s provided the rails, and locomotives of dubious though increasing efficiency. But until the power of the locomotive was established, the future of railways had to remain in doubt. After all, Stevenson was still searching for level lines in Strathmore in 1827, two years after the Stockton & Darlington opened; and shortly afterwards he appears to have lost faith to such an extent that he left railway planning and returned to his work on harbours, bridges and river navigations.[167] Until the *Rocket* and its fellows made out their case in the early 1830s, railway schemes in Scotland showed little real change from the patterns laid down by the Kilmarnock & Troon. But the 1820s saw not only an increase in the number of lines actually built, but also a great debate on the future of the railway.

2
The Coal Railways

I. The idea of the railway

THE 1820s and early 1830s mark a transition in the development of Scottish railways. The change is not primarily in the nature of the lines planned; there is, for example, no difference of principle between Stevenson's Midlothian scheme and its successor the Edinburgh & Dalkeith. Almost without exception, the new lines are short, and follow the waggonway principle of taking minerals to water or to the cities. In this respect, only the Garnkirk & Glasgow extended the concept of the railway, being the first Scottish line to compete directly with water transport for the same traffic and along much the same route. On the other hand, they differed from the waggonways by virtue of the joint-stock organisation conveyed by their parliamentary authorisation, by their wider range of ownership, and increasingly by their extension into traffic other than the coal and iron for which most of them were intended. And unlike the waggonways, most of them were to become integral and often important parts of the overall railway system.

There was also a substantial debate on the future of railways, initiated largely but not entirely by the locomotive engine. Such a debate could clearly not be conducted in purely Scottish terms. Discussions centred on technological achievement and potential inevitably used evidence and opinions from the rest of Britain and on occasion from overseas. But it would be equally unfair to deny the existence of a Scottish dimension to the debate, and to assume that the Scots were merely observing and copying events south of the border. A complex and recurrent theme of Scottish railway history is the impact of English influences. For much of the time, it may seem that Scotland follows developments in England, often imperfectly and almost always after an appreciable time lag. Thus, as noted already, English influences on the much smaller Scottish waggonway system were considerable: and, after the success of the Liverpool & Manchester in the early 1830s, England went ahead with large-scale plans which the Scots could not match until a decade later (and even then only with substantial help from English ideas, English engineers and above all English capital). But in the 1820s the two countries were moving ahead in parallel, under the same influences of rapid urbanisation, the growth of heavy industry, and the highest rate of economic growth on record. In both countries the demand for coal was of prime importance for the provision of transport, and in Scotland this demand was intensified from the end of the 1820s by Neilson's hot blast process and the development of an iron industry based on it. Hence, although the attention of historians has focused on English pioneers like George Stephenson, Thomas Gray and William James, there was sufficient stimulus for Scottish entrepreneurs to take action on their own account.

Two Scottish contributions to the debate were of particular importance in marking the transition from the horse waggonway to the apparently unlimited potential of the steam railway. In 1818 the Highland Society (a body whose influence and interests were very wide-ranging indeed) offered 'a piece of Plate, of Fifty Guineas value, . . . for the best and approved [sic] Essay on the construction of Rail-roads for the conveyance of ordinary commodities'. A year or two later the society repeated the competition, and on both occasions the prize was divided among several contributors. Robert Stevenson was then commissioned to edit and introduce selections from the essays, and the result was published in the society's *Transactions* in 1824.[1] It stands as the final analysis of the achievements and capabilities of the horse railway.

In spite of the apparent enthusiasm of the contributors for new developments, their analyses for the most part suggest only minor modifications to a well-established pattern, and most of the more radical ideas propounded were in fact to lead down blind alleys. The railways they propose are still drawn by horses, carrying coal, and either connecting to water or being constructed only because a canal is not feasible in the particular location. The emphasis of the essays is technological, as indeed was required by the terms of the competition, but the locomotive earns only a passing reference, by Alexander Scott of Ormiston, to George Stephenson and the Middleton Colliery Railway near Leeds.[2] Otherwise the horse is taken for granted, and discussion centres on how to increase its pulling power. The authors concentrated in particular on the type of track and on a series of ingenious and elaborate ways of overcoming gradients.

The argument about track was principally between the support of Stevenson and George Robertson for the new malleable iron-edge rail, and that of Scott for the plateway as less damaging to wheels. An eccentric variation was provided by Edinburgh mathematician George Douglas, who proposed U-shaped stone rails, taking ordinary waggon wheels in the central channel. Since the society had asked for consideration of 'whether rail-roads, or the wheels of carriages, may be so constructed as to be applicable to ordinary roads, as well as to rail-roads', the idea had possible attractions.[3] The general opinion, however, was that a combined road-rail waggon was not sufficiently desirable to outweigh the advantages of the edge rail: Stevenson considered that the only real problem was the distribution of coal to customers in the cities, and that, in this case, 'the removal of the body of the waggon with its load, by means of a crane, from the railway-carriage to the common-cart' would be the best solution. For road vehicles he wanted an extension of a practice already existing in Aberdeenshire and planned for Linlithgow High Street, whereby long flat stones were inserted longitudinally into the road at the gauge of a standard cart: a similar idea to reduce friction by the use of iron plates was offered by John Baird of Shotts Ironworks.[4]

The problem of overcoming gradients allowed for solutions ranging from the reasonable to the undeniably bizarre (and also permits some curiosity about the offerings which Stevenson did not consider worthy of inclusion in his edited collection). Stevenson could not find unanimous support for his advocacy of the level line: Scott qualified his approval by fearing that it might be hard on the horses, a point amplified by Robertson, who wanted an undulating line to allow the horse to vary the muscles in use. When a substantial change in level was unavoidable, suggestions ranged from a perpendicular hoist floating in water (clearly based on canal locks) to a rack-railway incline with haulage by stationary steam engine.[5] No one as yet realised that the locomotive was, if not to end the concern about gradients, at least to alter radically the dimensions of the problem. Perhaps, at the time when the essays were written, it was not surprising that the locomotive played little part in the authors' plans.

But between the offering of the Highland Society's prize and the publication of the essays six years later, much had changed. Admittedly no public railway using steam haulage had yet been opened, but already George Stephenson had demonstrated his engine to the Quaker entrepreneurs of Teesside. And a new breed of propagandist was appearing, with ideas based on the locomotive, and plans for covering the country with large-scale lines. In the English context, most attention has been given to two men — the railway monomaniac Thomas Gray, whose 1820 *Observations on a General Iron Rail-Way* proposed to relieve both postwar unemployment and the suffering of coach-horses by constructing a system of rack-railways over the entire country south of the Forth; and William James, whose early nineteenth-century projects and surveys included lines to link London to Brighton, Birmingham, Manchester and Liverpool.[6] Scotland also had her share of visionaries; Stevenson's planned network of lines in the triangle between Berwick, Glasgow and Aberdeen was not lacking in imagination, even if he had not intended them for the locomotive. The principal Scottish advocate of the new traction was Charles MacLaren, who was editor of the *Scotsman* for thirty years from its foundation in 1817. In December 1824, at the end of the year which had opened with the Highland Society's *Essays,* he wrote a series of editorial articles on railways which proved so popular that they were reprinted as a pamphlet before the final part had even appeared in the newspaper. This pamphlet, which was subsequently translated into most European languages, offers one of the most persuasive advocacies of the railway in its formative years.[7]

MacLaren appears in print as the cool rational advocate of progress, persuading by logic and reason rather than the enthusiastic fanaticism of a Gray. He had no practical engineering knowledge, and for technical detail relied largely on the published work of men like Nicholas Wood, Blenkinsop and Professor Leslie.[8] But, he complained, he had gathered most of his notes a year earlier and waited in vain for someone of more scientific background to take the subject up: 'writers of science . . . generally travel in beaten tracks',[9] and were, in his view, unwilling to give adequate consideration to the possibilities of the railway. His pamphlet covers some of the familiar ground of waggonway history, and of track types and gradients, but for the most part looks forward to the prospects opened up by the potential of the high-pressure engine. After discussing both coal and passenger traffic, and analysing the advantages to be gained over canals, MacLaren concluded that 'the Railroad, by its form, breadth, strength, and other qualities, should be adapted for . . . an extended and general system of commerce';[10] to this end he proposed a comprehensive network of Scottish railways, noting that it would be well to build them all to the same gauge — a point the neglect of which was to cause a little trouble in Scotland, and a great deal in England. Unlike Gray, he foresaw railways extending to the Highlands, if only a cheaper form of durable rail, perhaps made of wood, could be developed.[11] MacLaren's work was a major early analysis of the possibilities of the steam railway; he was also fortunate that he was in a position to ensure it wide publicity.

The key change in the discussion was that the proponents of railways were now advocating them as replacements for older forms of transport, rather than as auxiliaries. The *Quarterly Review* found such enthusiasm excessive:

> As to those persons who speculate on making rail-ways general throughout the kingdom, and superseding all the canals, all the waggons, mail and stage-coaches, post-chaises, and, in short, every other mode of conveyance by land and by water, we deem them and their visionary schemes unworthy of notice . . . The gross exaggerations of the powers of the locomotive steam-engine, or, to speak plain English, the *steam-carriage*, may delude for a time, but must end in the mortification of those concerned.[12]

The supporters of the locomotive were not in fact suggesting such a holocaust of alternative transport, but they were now convinced that the railway was superior at least for the transportation on busy routes of passengers and of most types of goods. As the *Glasgow Herald* observed:

> Although the locomotive engine is a late invention and not generally understood, it seems to be nearly perfect in construction, and it is efficient almost beyond belief in operation.[13]

Even the waggonway had demonstrated the advantage of rails over the uneven surface of the common road; in spite of the work of McAdam and Telford in road construction, and the rapid extension of the turnpike system from the late eighteenth century, railway proponents still felt that they had little to fear from carts and coaches. The real challenge they saw as coming from the canals.

Already Telford and others had catalogued some possible advantages of waggonways over canals, but these, in their view, had not been sufficient to make a railway more desirable than a canal in a location which was physically suitable for either. Rennie had added an economic consideration to the geographical ones when he observed that, although canals were the cheapest form of transport to operate, one in the Borders might prove expensive overall, 'there being no large towns to occasion a great consumption at one place'.[14] Canals fared best when carrying large quantities of heavy goods from the minimum number of sources to a few destinations or even a solitary market. Hence the conveyance of coal to a city was an ideal traffic, while there was never much profit to be had from a canal in an agricultural area. In the 1820s the Monkland Canal was a much better bet than that newly opened white elephant, the Caledonian Canal.

MacLaren reiterated the old arguments for the railway, on versatility and relative immunity to drought or frost, while emphasising that railways cost only one third as much as canals to build:

> Railways, partly from their comparative cheapness, and still more because they are practicable in all situations and on inclined as well as level ground, may be ramified over a whole country, and become the universal medium of communication. Not only every town and village, but every considerable farm, may have its branch.[15]

Stevenson considered that the costs of construction had already given the railway an advantage over the canal in Scotland:

> The wealth of England enables her to stand unrivalled in the formation of her Water-ways, or numerous Canals. By these the horse-load has been much extended, and the conveyance of merchandise greatly facilitated. In Scotland and Wales, her less wealthy neighbours have endeavoured to supply this want, by the construction of numerous Rail-ways, which are perhaps better adapted than Canals to the undulating surface of their respective countries.[16]

This may well have applied to Wales; and it is also true that Scotland had few canals. But, leaving aside the improbable chance that a large number of waggonways have vanished without trace, Scotland had also lagged behind in waggonway construction (though Stevenson himself had done his best, at least on paper, to remedy the deficiency).

The newest and most important advantage of the railway was the speed offered by the locomotive. As long as the movement of heavy goods was restricted to the two or three miles per hour of the horse-drawn barge or waggon, so long might the railway's cheapness of construction seem less important than the greater pulling power of a canal horse, and so long might flexibility continue to be obtained by running short waggonways down to the canals. MacLaren was prepared to argue that the running costs of even a horse-drawn railway was not dissimilar to those of a canal, since a horse could pull fifteen tons at two miles per hour on a railway under the supervision of one man, whereas it required three men to help it while pulling 45 tons at a similar speed on a canal. Two years later Stevenson confirmed this ratio, although his horses appear to have been only two-thirds as powerful in each case. If they were right, the horse-drawn railway was in fact competitive, but it was by no means proven to be so.[17]

At low speeds, the horse was cheaper to run than the locomotive, even on railways. An enthusiastic clerical amateur, the Reverend James Adamson of Cupar, based his calculations on the performance of locomotives at Killingworth and Hetton collieries in Northumberland, where an engine did the work of three or four horses at much the same speed, and concluded that:

> The estimates of the expence of the employment of steam-power upon rail-roads, do not seem in its favour, when compared with horses moving at the velocity most favourable to them, providing the cost of coals continues to bear the same ratio to the expence of supporting the horses as it does at present . . .

However, he continued:

> For all velocities beyond this, the advantage of the Rail-Road augments in a high ratio.[18]

This was the real point. MacLaren, using the evidence of Nicholas Wood, believed that the locomotive might have a cost advantage at all speeds, but agreed that it was at velocities of over four miles per hour that the railway demonstrated 'astonishing superiority' over the canal. Even the great canal engineer Telford, at least according to the editor of his autobiography, accepted the idea of railways for the rapid movement of passengers and was 'merely against them as a rival conveyance of heavy goods usually carried on canals, and in this he has been fully justified by experience'.[19] An anonymous pamphleteer summarised the argument:

> It is by the substitution of elemental for animal power that the important advantages promised to the nation are to be realised. The whole arcana of the business must be sought for in the Locomotive Engine . . . [which] cannot be used for canal conveyance with the same advantage that it possesses on railways; nor, as far as we know, can it be employed at all, or at least to any useful purpose, on turnpike roads.[20]

In other words, the steam engine allowed for the first time a combination of speed and substantial pulling power, and it was peculiarly suited to the railway. In spite of improved road-making techniques, horses on roads could not even compete with horses on rails; and in spite of the gallant pioneering ventures of Goldsworthy Gurney, Walter Hancock and others in the 1830s, it was to be a long time before mechanical vehicles were to be common on the highways. On canals, steamboats were certainly practicable, but at speeds greater than about four miles per hour they caused excessive damage to the banks.[21] Even apart from this, the rapid increase in water resistance with increasing speed put the canals at a disadvantage. On the other hand, it was widely believed that there was no limit to the speed at which a railway train might eventually travel, once technological problems had been solved. Charles Sylvester, for example, reporting to the projectors of the Liverpool & Manchester, appeared to assume that once initial friction had been overcome there was no further resistance to the progress of a locomotive. MacLaren entered an observant caveat, at a time when locomotives travelled at no more than about ten miles per hour:

> The resistance of the air, which in vulgar apprehension passes for nothing, comes to be the greatest impediment to the motion of the vehicles, and may in some cases absorb five parts in six of the whole power.

But even without his proposed primitive streamlining, the locomotive had, for many purposes, given clear superiority to the railway.[22]

The most obvious of these purposes was the conveyance of passengers. Up to this time only the Kilmarnock & Troon had had any semblance of a regular passenger service; it had also, of course, made an early experiment with locomotive traction. Now some of the proponents of the railway saw much of the future in terms of passenger traffic. Adamson journeyed to Stockton to see for himself, and concluded that:

> Of all the advantages of the Rail-way, the remarkable velocity of motion which it admits of, and which fits it so admirably for the conveyance of passengers, light goods, and mails, is the most striking and the most valuable, in a country where these objects are becoming every day of more and more importance.[23]

MacLaren also spent some time in detailing the advantages for passenger transport, and designed an early type of corridor coach, which was even to have dining facilities.[24]

In fact, prospective railway entrepreneurs, as against the propagandists, were less willing to emphasise possible passenger traffic. It had been proved that the carriage of coal was profitable: it had not been shown that a large demand for passenger services existed, and the idea the railway might create such a demand merely by providing the services was hardly considered. The most obvious passenger route, between Edinburgh and Glasgow, already had plentiful services by road and canal, and it was by no means clear that canal opposition could be overcome. The canals might not compete in speed, but they could in cost and comfort. Passenger boats had run on the Forth & Clyde Canal since 1783; in 1809 fast boats had been introduced, running from Glasgow to Falkirk in $3\frac{1}{2}$ hours and providing such creature comforts as food, drink and newspapers. By 1812 they carried 44,000 passengers per year, who paid over £3450. From 1828 the canal even offered a steamboat service, run by Thomas Grahame's *Cupid* at a gentle three miles per hour. Grahame at least was convinced that his vessel would fend off any forthcoming railway challenge, and the canal was indeed to offer the railways severe competition until traffic-sharing agreements were negotiated in the 1850s.[25]

Besides which, the railway locomotives of the 1820s were not reliable. They frequently broke down, and they occasionally blew up. Sometimes they even had to be pulled back to base by the not-yet-superseded horses. The possibility of a boiler explosion was unnerving, and the general public was not likely to be entirely reassured by Sylvester's view that there could only be danger through gross neglect or through a boiler being made of cast iron, which would fragment if it burst.[26] MacLaren's opinion that the high-pressure engine 'can be used with perfect safety, because it may be easily placed in a car by itself, a few feet before the vehicle in which the passengers are' must have seemed equally facile.[27] Nor were people to be persuaded by having the undeniable dangers of road travel pointed out to them. And, besides, the locomotive was only to go at perhaps eight miles per hour. Was the risk of death by explosion or derailment worth it? The railway builders of the 1820s were probably wise to prove the capabilities of the locomotive on a further series of coal lines.

Public doubts were encouraged by opposition to the locomotive from several sources. Some, of course, were simply opposed to all change, and particularly that created by industrial technology. Others had their own interests to protect: canal proprietors, turnpike trustees, innkeepers, foxhunters and privacy-loving landowners were vocal groups who disliked the railway. But among the more disinterested, two recurrent doubts appeared — that the locomotive was incapable of the proposed speeds, or that, if it could travel so fast, it would certainly be unsafe. The two most famous examples of these arguments may bear repetition. The first came from Nicholas Wood, whose views had to be respected:

> It is far from my wish to promulgate to the world that the ridiculous expectations, or rather professions, of the enthusiastic speculator will be realised, and that we shall see engines travelling at the rate of twelve, sixteen, eighteen, or twenty miles an hour. Nothing could do more harm towards their general adoption than the promulgation of such nonsense.[28]

The second, from the commentator of the *Quarterly Review* (who was by no means altogether unfriendly to railways), was aimed at the planned London & Woolwich Railway:

> It is certainly some consolation to those who are to be whirled at the rate of eighteen or twenty miles an hour, by means of a high pressure engine, to be told that they are in no danger of being sick while on shore; that they are not to be scalded to death nor drowned by the bursting of the

boiler; and that they need not mind being shot by the scattered fragments, or dashed in pieces by the flying off, or breaking of a wheel. But with all these assurances, we should as soon expect the people of Woolwich to offer themselves to be fired off on one of Congreve's *ricochet* rockets, as trust themselves to the mercy of such a machine, going at such a rate.[29]

Given the state of locomotive development in 1825, these views were entirely rational. The propagandists of a grand railway expansion were still visionaries, even if they were visionaries backed by some sort of evidence. The *Quarterly*, looking at things as they stood, wanted a parliamentary maximum speed of nine miles per hour. Even MacLaren accepted that twenty miles per hour was not yet practicable and should not be attempted, but added the crucial rider that 'no complex invention can be perfect at the moment of its birth'.[30] The contribution of the visionaries was to provoke debate and analysis on the future of the railway, and to realise that, with the probable development of the locomotive into a much safer and more powerful machine, the apparently unthinkable might become commonplace. Adamson summed up with a mixture of caution and optimism:

> The system of water conveyance we must look upon as nearly perfect; and the other [railways] as yet offering many chances of improvement; and from its applicability in some of its many forms to all imaginable situations, and its success in those wherein it has been attempted, we must esteem it eminently worthy of having its properties more accurately investigated.[31]

Meanwhile, however, the views of the *Quarterly* ('We scout the idea of a *general* rail-road, as altogether impracticable')[32] were those of the majority. It was to be a long time before the visions of MacLaren or Gray were to be realised; in the meantime, another series of small lines gave more evidence of what the railway could do.

II. The Monklands lines

The cyclical boom of 1824-25 contained the first signs of an outbreak of speculative formation of railway companies; since the processes of planning, surveying, and applying to Parliament took at least a year, the authorising acts continued into 1826. Of the major schemes, only the Liverpool & Manchester was authorised, while lines to link Birmingham to both London and Liverpool were among those rejected. Altogether, two acts were passed in 1824 for new British lines, and nine in each of the two succeeding years. Of these twenty acts, six were Scottish; it may be that a slight Scottish time lag behind England is already apparent, since four of the Scottish acts came in 1826. Among the projects which remained in embryo was a link between Edinburgh and Glasgow, and, in fact, all the nine Scottish projects authorised between 1824 and 1830 were recognisably successors of the waggonways (see Table 12). All were short: only two were over twenty miles long, and of these the construction of one was never started and the other was only partially completed. All were primarily concerned with the carriage of heavy goods, and seven of the nine were clearly coal lines. Of these seven, five were engaged in helping the coal on its way to water transport, and a sixth was covering the same ground as Stevenson's Midlothian project. They were all still designed mainly or entirely for horse traction. Although there might be a substan-

52   The Origins of the Scottish Railway System

Table 12. Railway companies authorised 1824–35

Company[b]	Acts to 1844	Date of Act	Route	Miles	Date open	Engineer	Gauge	Capital authorised (£) Share	Loan	Total	Estimate	Construction Costs (£) p.mile	Actual	p.mile
Monkland & Kirkintilloch	5 Geo.IV c.49	17.5.24	Old Monkland–Kirkintilloch (Forth & Clyde Canal)	10.03	1.10.26	T. Grainger	4'6"	32,000	10,000[a]	42,000	24,953	2,315	43,750	4,058
			br: Kipps	0.75										
	3&4 Will.IV c.114	24.7.33	br: Dundyvan					20,000	—	20,000				
	2&3 Vict. c.70	4.7.39						62,000	—	62,000				
	6&7 Vict. c.79	28.7.43						86,000	28,667	114,667				
Ballochney	7 Geo.IV c.48	19.5.26	Kipps (Monkland & Kirkintiloch)–Ballochney–Arbuckle	3.75	8.8.28	T. Grainger	4'6"	18,432	10,000	28,432	18,070	3,614		
			br: Airdrie											
			br: New Monkland											
			br: Clerkston											
			br: Stanrig & Whiterig	1.25										
	5&6 Will.IV c.97	21.8.35	br: Blackrig					10,000[a]	—	10,000				
	2&3 Vict. c.59	1.7.39						41,568	—	41,568				
	6&7 Vict. c.50	27.6.43						40,000	—	40,000				
Edinburgh & Dalkeith	7 Geo.IV c.98	25.5.26	St Leonard's, Edinburgh–Eskbank	8.50	4.7.31	J. Jardine	4'6"	70,125	—	70,125	50,000	5,882	c.130,000– c.150,000	c.15,294– c.17,647
	10 Geo.IV c.122	4.6.29	extn at Eskbank Niddrie–Leith	0.63 3.93	3.35			25,700 54,875	10,000 —	35,700 54,875	7,815 29,628	12,404 7,539		
	4&5 Will.IV c.71	27.6.34	br: Fisherrow br: Musselburgh					8,053	—	8,053				
Dundee & Newtyle	7 Geo.IV c. 101	26.5.26	Dundee–Newtyle	10.50	16.12.31 part 3.4.32 all	C. Landale	4'6½"	30,000	10,000	40,000	27,600	2,629	c.100,000	c.9,525
	11 Geo.IV & 1 Will.IV c.60	29.5.30						10,000	20,000	30,000				
	6&7 Wil.IV c.102	4.7.36						100,000	—	100,000				
	—	—	br: Dundee harbour (built privately)		7.34	J. Leslie								

The Coal Railways

Company[b]	Acts to 844	Date of Act	Route	Miles	Date open	Engineer	Gauge	Capital authorised (£) Share	Loan	Total	Estimate	Construction Costs (£) p.mile	Actual	p.mile
Garnkirk & Glasgow (from 1844, Glasgow Garnkirk & Coatbridge)	7 Geo.IV c.103	26.5.26	St Rollox, Glasgow–Gartsherrie (Monkland & Kirkintilloch)	8.18	5.31 goods 27.9.31 all	T. Grainger & J. Miller	4'6"	28,497	—	28,497	28,497	3,484	98,519	12,044
	8 Geo.IV c.88	14.6.27						9,350	—	9,350				
	11 Geo.IV & 1 Will.IV c.125	7.6.30						—	21,150	21,150				
	1&2 Vic. c.38	4.7.38						89,198	—	89,198				
	7&8 Vic. c.87	19.7.44						—	—	—				
Ardrossan	7&8 Geo.IV c.87	14.6.27	Ardrossan–Johnstone (Ardrossan–Kilwinning built)	22.50 (5.5 + 4 brs built)	1831	J. Jardine	4'6"	95,658	—	95,658	94,093	4,182		
	3&4 Vict. c.104	23.7.40						51,700	—	51,700				
Wishaw & Coltness	10 Geo.IV c.107	21.6.29	Coatbridge (Monkland & Kirkintilloch)–Wishaw–Chapel br: Cleland	11.03	1833 part 21.3.34 to Jerviston 1841 branch 9.3.44 all	T. Grainger & J. Miller	4'6"	60,000	20,000	80,000	53,000 +7,000 brs	5,893		
	4&5 Wil.IV c.41	16.6.34						—	—	—				
	7 Will.IV & 1 Vict. c.100	12.7.37						—	—	—				
	2&3 Vict. c.58	1.7.39						60,000	20,000	80,000				
	4&5 Vict. c.11	10.5.41						120,000	40,000	160,000				
	7&8 Vict. c.98	19.7.44						—	—	—				
Polloc & Govan	11 Geo.IV & 1 Will.IV c.62	29.5.30	Broomielaw, Glasgow–Tradeston, Glasgow br: Port Eglington (Glasgow Paisley & Johnstone canal) br: Polloc estate	0.85 / 0.34	8.40	T. Grainger & J. Miller		10,000	5,000	15,000	8,000	9,412		
											950	5,000		
	1&2 Wil.IV c.58	23.8.31						36,000	15,000	51,000	750			
	7 Will.IV & 1 Vict. c.118	15.7.37						—	—	—				
Rutherglen	1&2 Wil.IV c.35	2.8.31	Tradeston (Polloc & Govan)–Rutherglen		Not built			15,000	5,000	20,000				

Company[b]	Acts to 1844	Date of Act	Route	Miles	Date open	Engineer	Gauge	Capital authorised (£)			Estimate	Construction Costs (£)		p.mile
								Share	Loan	Total		p.mile	Actual	
Slamannan	5&6 Will.IV c.155	3.7.35	Arbuckle (Ballochney)—Causewayend (Forth & Clyde canal)	12.49	31.8.40	J. Macneill	4'6"	86,000	20,000	106,000	65,769	5,266	c.125,000	c.10,000
	7 Will.IV & 1 Vict. c.94	3.7.37						29,000	—	29,000				
	2&3 Vict. c.57	1.7.39						54,000	36,333	90,333				
Paisley & Renfrew	5&6 Will.IV c.85	21.7.35	Paisley–Renfrew (R. Clyde)	3.00	3.4.37	T. Grainger & J. Miller	4'6"	23,000	10,000	33,000	17,288	5,763	c.30,000	c.10,000
Newtyle & Coupar Angus	5&6 Will.IV c.86	21.7.35	Newtyle (Dundee & Newtyle)–Coupar Angus	4.82	2.37	W. Blackadder	4'6¼"	15,200	5,000	20,200	11,437	2,373		
Newtyle & Giammis	5&6 Will.IV c.92	30.7.35	Newtyle (Dundee & Newtyle)–Eassie–Glamis	6.61	4.6.38	W. Blackadder	4'6¼"	20,000	6,600	26,600	14,689	2,222		

a: Might be raised as either share or loan capital.
b: for West Lothian project, 1825, see Table 2.

Sources: HLRO, HC Deposited Plans of the various railways; PP1844(159)XLI, *Return of Moneys raised Railways Acts 1826–43*; J. Priestley, *Historical Account of the Navigable Rivers, Canals and Railways, of Great Britain*; F. Whishaw, *The Railways of Great Britain and Ireland*; *New Statistical Account*; G. Buchanan, *An Account of the Glasgow and Garnkirk Railway and other Railways in Lanarkshire*; H. G. Lewin, *Early British Railways*; GUEH, 'Monkland and Kirkintilloch Railway. Return required by the Railway Commissioners ... 1848'; GUEH, 'Monkland Amalgamation Bill. Brief for the Promoters, 1848'; GUEH, [Robert Dodds], 'General Remarks on the Wishaw & Coltness Railway ...'; S. G. E. Lythe & C. E. Lee, 'The Dundee and Newtyle Railway', *Railway Mag. 97 (1951)*; engineers' reports on the various railways.

The Coal Railways

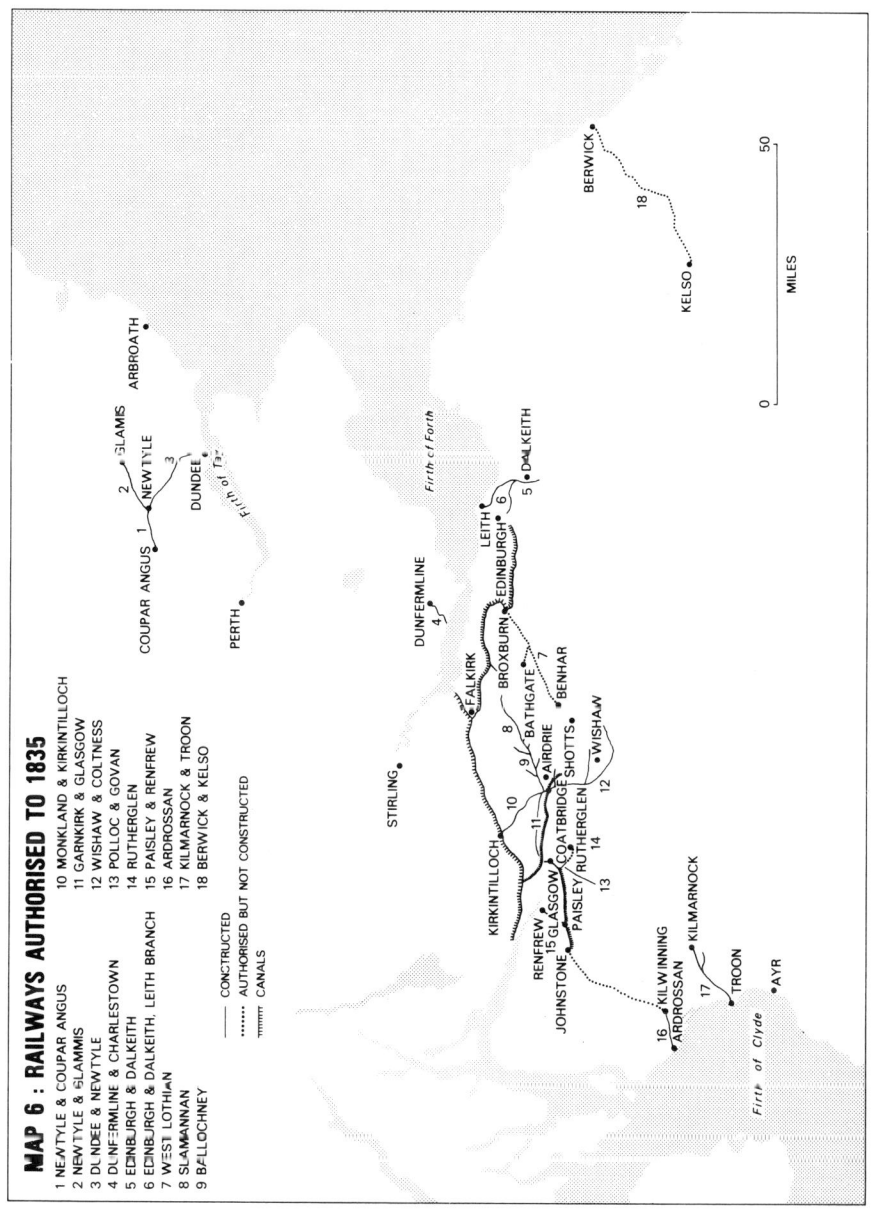

tial and imaginative debate on the future railway development of Scotland, the 1820s saw little sign of change on the ground. And indeed the early 1830s were no different. Of five authorisations between 1831 and 1835, two were coal lines (one linking to a canal), two were short feeders to an existing railway, and one connected to the Clyde. The longest ran for $12\frac{1}{2}$ miles, and all were designed for horses.

The underlying incentives to transport improvement were generally the same in kind as in the previous half-century, though they might have intensified in degree. On the supply side, coalowners were concerned about the dangers of overproduction, not so much in terms of their theoretical markets but in terms of the markets they could actually reach at a competitive price. Substantial total demand certainly existed, as long as excessive transport costs did not render it ineffective. The great cities were expanding rapidly: fuel was needed not only for domestic purposes, but for a rapidly growing industrial and commercial sector, including that fuel-hungry new development, the steamship.[33] By 1820 Glasgow was using half a million tons of coal a year, almost all of it subject to the high charges on the Monkland Canal.[34] In agricultural areas, which the waggonways had never reached in spite of all the planning, there was still a need for lime, coal and access to markets. Canal proprietors, in their search for new traffic, still regarded the short railway as an adjunct rather than a rival — and since the opening of the Union Canal in 1822 there were three substantial canals with interests in the coal area between Edinburgh and Glasgow. Any or all of them might gain an advantage by encouraging railway construction.

There was also the iron industry. The late eighteenth century had witnessed substantial expansion, mostly on the north Lanarkshire coalfield. Output rose by 1800 to 22,800 tons of pig-iron (9% of British production), but then stagnated due to high costs and poor quality. The later explosive growth and established prosperity of the industry is justifiably linked to Neilson's discovery of the hot blast process in 1828, and its use in connection with the local black-band ironstone. But for present purposes it is important to note that recovery had started earlier, spurred by the prosperity and high prices of 1825. New works were established in north Lanarkshire at Chapelhall (1825), Monkland (1826) and Gartsherrie (1828), and output rose from 25,000 tons in 1825 to 39,500 in 1830 — small compared to the 564,000 tons of 1848, but enough to strain existing transport arrangements.[35] As far as the iron industry was concerned, pre-hot-blast expansion was enough to generate railways; the startling growth of the industry thereafter offered these railways a chance of prosperity.

The first semblance, then, of a system of Scottish railways appears in north Lanarkshire, in the area known as the Monklands, from the mid-1820s. The early projectors included coalowners, ironmasters, landowners, canal proprietors and merchants: several qualified under more than one heading. Shareholders of the Forth & Clyde Canal were prominent at an 1823 meeting, held symbolically enough at Gartsherrie Cottage five years before the Bairds moved into the area, which led to the authorisation of the Monkland & Kirkintilloch in 1824 and its opening two years later.[36] The canal company's minutes later recorded that:

> A few of the proprietors of the Forth & Clyde Canal, have already by much exertion, and considerable advance of money, found means to set on foot and establish two railways, termed the Monkland & Kirkintilloch and the Ballochney Railway . . .[37]

One body which would certainly not have approved was the Monkland Canal Company, since the railway was an attempt to break its near-monopoly of the Glasgow coal trade, and the canal, along with the local road trustees and two small landholders, made up the dissentients to the railway bill in Parliament.[38] As its rivals observed, the Monkland Canal 'has for many years yielded a Dividend of Cent. per Cent . . . arising solely on its Tolls on coal'.[39] On the other hand, the Forth & Clyde dividend had risen from 5% in 1800 steadily to 25% in 1816-19. It had then fallen to 20% in 1820-25, indicating one reason for the attempt to infiltrate the rival's territory. In 1824 the Forth & Clyde ordered two new boats to deal with the anticipated increase in traffic from the railway, in 1826 the 25% dividend was restored, and by 1828 the Monkland & Kirkintilloch was delivering 1600 tons per week to the canal.[40] The railway ran from Cairnhill Colliery, in the heart of the coal and iron area, to the canal ten miles away at Kirkintilloch, and offered a route to Glasgow which, though longer than the Monkland Canal, avoided the locks at Blackhill. It also opened up the possibility of sending Monklands coal cheaply by the Forth & Clyde and Union Canals to Edinburgh. And it was of course of great help to customers who were on the railway but on neither canal; the cost of taking iron rails from the Forth & Clyde to Gargill Colliery was reduced from 6/11d to 9d per ton.[41]

The line also marks the debut as a railway engineer of Thomas Grainger who, with his partner John Miller, was to design much of the early Scottish railway network. According to a fellow engineer, looking back from the more advanced knowledge of 1832, the design betrayed a prentice hand:

> This railway having been executed at a time when the construction and operation of railways were but indifferently understood, it is much less perfect in some respects than those which have since been made.

However, he admitted, it 'has fully answered the expectations of its proprietors'. Before the Monkland & Kirkintilloch was opened, Grainger had also designed 'an excellent and well-finished road' for the proprietors of the Ballochney Railway.[42] As noted above, the membership of this group overlapped with that of the original railway's proprietors, and their line was an extension of the Monkland & Kirkintilloch's Kipps branch into an untapped mineral field. The Forth & Clyde Canal even subscribed money in its corporate capacity at a time when funds were hard to raise on the Glasgow market.[43] At much the same time a survey was made for another line south into Lanarkshire's Middle Ward: a variant of this line was projected in 1828 as the Garion & Garturk, and authorised in 1829 as the Wishaw & Coltness. Again, this was later said to have been projected 'principally by shareholders in the Monkland & Kirkintilloch Railway',[44] and again it was engineered by Grainger as a single-track line. His estimate of £50,000 for an 11.03-mile line seemed cheap, but excluded the costs of land, workshops and rolling stock — the sort of estimate that was to get engineers a bad name.[45] Shortage of funds meant slow progress in construction, and the company had to

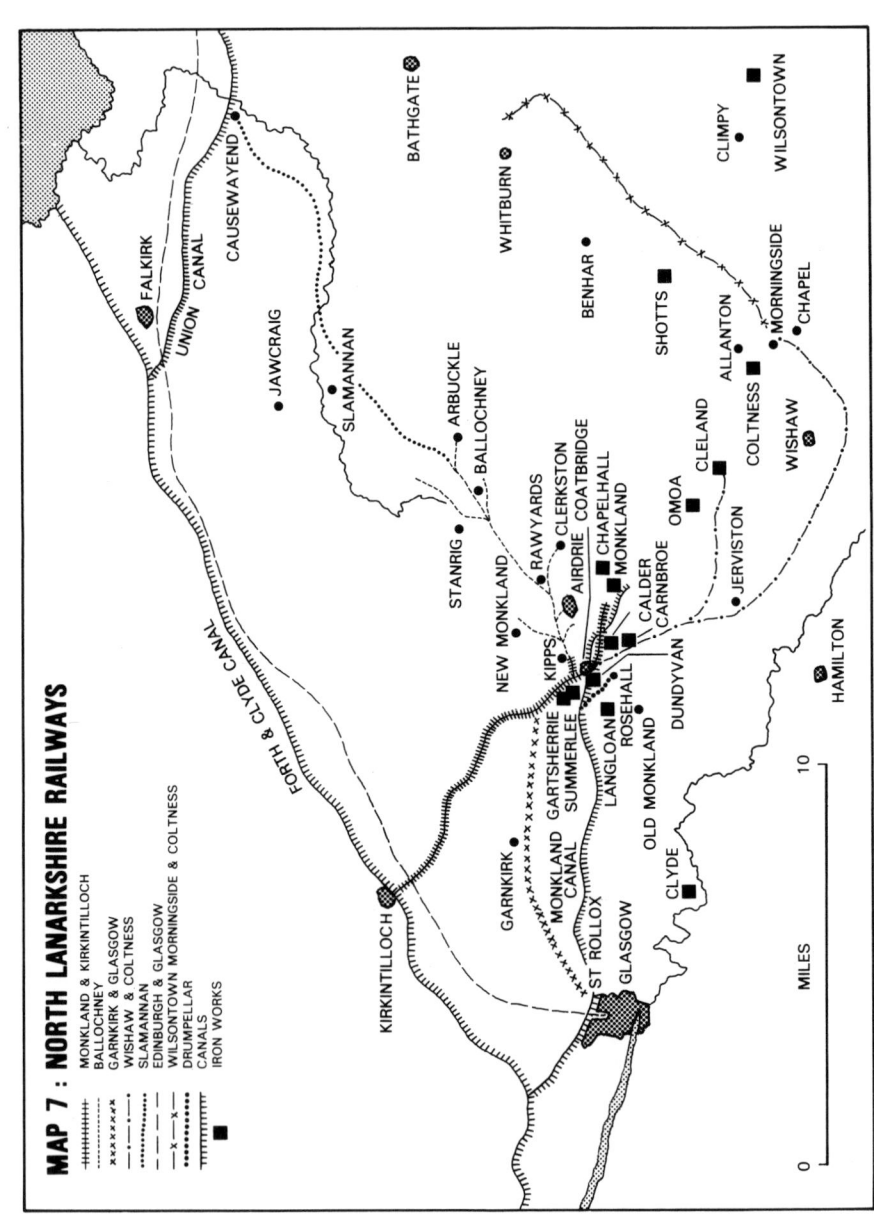

return to Parliament in both 1834 and 1837 for a three-year extension of the time allowed for completion of the works.[46] In 1839 the minister of Cambusnethan (a parish which included both Wishaw and Coltness) was still looking forward:

> Should the railway come through the parish, as is expected, it will open up the coalfields in various places, where there is at present no demand; and will add greatly to the wealth and improvement of the district.[47]

The company's share capital was doubled in 1839 and redoubled in 1841, in which year the line was finally opened throughout.[48] The line opened up coalfields on fourteen landed estates, whose owners included some prominent names: Lord Belhaven at Wishaw, General Sir James Steuart Denham at Coltness, and the almost inevitable William Dixon at Garturk. It also passed Dixon's Calder Ironworks, and sent out a branch to Cleland and the works at Omoa.[49]

Unlike the Ballochney, however, the Wishaw & Coltness traffic did not represent pure gain for the proprietors of the Monkland & Kirkintilloch and the Forth & Clyde Canal. By the time it was projected, an alternative route to Glasgow was authorised and under construction. The Garnkirk & Glasgow Railway — the unexpected order of names in the title presumably representing the direction of traffic flow — was the first Scottish railway to issue a direct challenge to the canals. Like the Monkland & Kirkintilloch, it was an attack on the traffic of the Monkland Canal, but in this case was to carry the coal by rail directly into Glasgow. It would thus also bypass the Forth & Clyde Canal, and all but a connecting couple of miles of the Monkland & Kirkintilloch, for coal coming from the Wishaw line. Although the railway was promoted by a group of Monkland & Kirkintilloch shareholders, any inference that it was therefore supposed to be for the benefit of the latter railway would be incorrect. The promoters were above all interested in getting coal to Glasgow by the most effective means, and they invested in any project which promised to help. The main initiative came from the Glasgow end of the line, led by the great coal merchant James Merry, and Charles Tennant of the St Rollox chemical works, which at the time used 30,000 tons of coal per year.[50] According to Buchanan, the reason for promoting the line was clear:

> It was obvious to the most careless observer, that if the coal could be sent eight miles along the Monkland & Kirkintilloch Railway, shipped on the Forth & Clyde Canal, conveyed nine miles along that Canal, subject to all the charges on it for dues, haulage, &c., and delivered at Port Dundas, and this with any chance of profit; much more would it be of advantage, both to the proprietor and the public, to send them at once by a railway, if this were practicable, the direct distance not exceeding nine miles in all, and no transference of the coal from one vehicle to another between the pits and the place of delivery; which for soft coal in particular is of very decided advantage.[51]

The line, as amended by a second act, ran roughly parallel to but at a moderate distance from the Monkland Canal, to a depot at St Rollox. Here it was close to a major industrial area, including Tennant's works and the city gasworks which required an annual input of 16,000 tons of coal. A short extension linked to the Forth & Clyde Canal, allowing for the possibility of sending coal onwards to the Clyde. The line from the Monklands was either level or downhill all the way, yet the depot was high enough to allow downhill cartage into central Glasgow.[52]

The mention of cartage, however, leads on to the unsolved Glasgow problem of the unsatisfactory links between the communications nexus developing beside the industrial area in the north of the city, and the harbour at the Broomielaw. Carting coal for just over a mile on roads with gradients of up to 1 in 12 through the centre of the city was 'exceedingly imperfect and expensive', costing about the same as ten miles carriage on a railway and also involving a double transshipment of goods. The carts were a recognised danger in the streets, and were claimed to cause more damage to the road surface than all other traffic put together. The Statute Labour Trustees had even considered a special toll on them. With the growing trade in coal, iron and industrial products, the matter was becoming increasingly serious.[53] The two canals had been planned and authorised before Golborne's improvements to the Clyde allowed 100-ton ships up to the Broomielaw, and long before the major river works of the early nineteenth century raised this figure to about 400 tons, otherwise either or both might have linked to the harbour rather than keeping to the north of the city. As it was, Rennie had planned a canal link from the Monkland Canal to the river in 1797, but the costs of 160 feet of lockage and the waste of water involved were too great. A Stevenson survey for a canal had also come to nothing, and although the Monkland Canal Company had in 1824 bought land for a connecting railway, that scheme too had foundered.[54]

In 1830 Grainger and Miller were commissioned to survey a line from the Garnkirk & Glasgow to Edinburgh and Leith.[55] To complete the link between the main east and west coast harbours, and to provide the final piece in the opening up of the Monklands, they also proposed a line from the Broomielaw, passing in a tunnel under Blythswood, and branching to link to both canals and the Garnkirk railway. While above ground, the railway would keep to undeveloped ground in industrial areas, thus avoiding the objections to a line through the streets which had scuppered the 1824 proposal. The plan received the imprimatur of George Stephenson, who described it as 'a plain practicable work unattended by danger or annoyance to the houses or other property under which it may pass'.[56] A powerful committee was set up, chaired by Tennant, containing among others Lord Belhaven, Dixon, Archibald Speirs of Elderslie and the Lord Provost, but it was unable to push the plan through. Apart from the fears of Blythswood householders about the effects of the tunnel on their property, shareholders of both the Monkland & Kirkintilloch and the Forth & Clyde Canal suspected a Garnkirk plan to appropriate their traffic. A tunnel advocate denied this — three-quarters of the shares, he said, were 'held by gentlemen of high respectability and intelligence who have no connection whatsoever with that Company' (which leaves a question hanging over the other quarter) — and announced that the railway might carry a ton of goods for about 4d against the carters' price of 1/2d to 1/5d downhill and 2/- to 3/- up.[57] The extent of opposition led to the rejection of the tunnel bill in committee. Of some importance was the ambiguous position of the key landowner, Archibald Campbell of Blythswood, MP for Glasgow Burghs; although he attended the meeting which approved Grainger and Miller's project (and did not apparently voice any objections), in the following year his opposition was taken as established:

> All the hope which the projectors now entertain is that Mr Campbell of Blythswood will, out of pique towards a great Canal proprietor [Kirkman Finlay], who lately contested the Burghs with him, assist in carrying a measure to which he was utterly opposed upon principle last session.[58]

In the event, the much-needed link between the railways and the harbour had to await the mania.

Table 13. Interlocking Directorships in the North Lanarkshire Railway Companies, 1826-29

	Monkland & Kirkintilloch 1826	Ballochney 1827	Garnkirk & Glasgow 1827	Wishaw & Coltness 1829
William Dixon	X		X	X
Robert Grahame	X	X		X
Thomas Grahame[a]	(a)	X		X
James Merry	X		X	
George M. Nisbett[b]	X		X	
Charles Tennant	X	X	X	X

a: Thomas Grahame was clerk to the Monkland & Kirkintilloch.
b: Nisbett, the owner of Cairnhill Colliery, did apparently find a conflict of interest, and resigned from the Kirkintilloch board after joining the Garnkirk.

Source: G. Buchanan, *An Account of the Glasgow and Garnkirk and Other Railways in Lanarkshire*, 5-7, 11.

Even without a harbour link, the Garnkirk line meant a substantial reduction in the cost of Glasgow's coal. At the opening, Buchanan estimated a reduction of perhaps 1/ per ton and thus a saving to the city of £35,000 on an estimated consumption which had by then risen to 700,000 tons.[59] He apparently assumed that the company would become both a monopoly supplier and a benevolent one: if the former were achieved, the latter was certainly not impossible, since the interest of most of the proprietors would be primarily to get cheap coal to their business concerns or their markets rather than to exact the maximum profit as railway owners. This common emphasis on the coal rather than on the railways as profitable enterprises in themselves may explain the lack of open rivalry, at least in the earlier years, between competing lines. Men such as Tennant, Dixon and Merry found it perfectly proper and congenial to belong to the boards of both the Monkland & Kirkintilloch and the Garnkirk & Glasgow (see Table 13), and most Garnkirk shareholders also had money in the original railway.[60] But conflict between the rival routes is apparent by the 1840s. In 1844 the Garnkirk obtained an act to make a direct link to the Wishaw & Coltness, bypassing the intervening two miles of the Monkland & Kirkintilloch, whose right to charge an extra short-distance toll had become contentious. The move brought into the open a division among the Wishaw proprietors, who had already set up a committee of inquiry into the company's lack of success. This committee noted that proxy voting forms for company meetings were apparently being issued only to members who were also shareholders in the Monkland & Kirkintilloch — a statement unconvincingly denied by a director, and used by Tennant's pro-Garnkirk faction to force the resignation of company secretary James Mitchell, who was also secretary of the Kirkintilloch company.[61] The idea of solving the problem by amalgamation or traffic-sharing, first floated by Lord Belhaven in 1842 with the suggestion that the traffic of the three lines:

> should be conducted as a whole by an arrangement among the three companies under a Board of Management constituted by members of each company upon the principle of the charge to the traders being such as to pay five per cent on the capital employed and the expenses of running the establishment[62]

also came to nothing. The future of the Wishaw & Coltness and the Garnkirk & Glasgow lay in absorption into the Caledonian at the time of the mania: shortly afterwards, the other lines amalgamated into the Monklands Railways and eventually, in 1865, joined the North British empire.

The Monkland & Kirkintilloch did not, of course, lose all its traffic to the incoming Garnkirk, although both it and the canals had to reconsider their charges: passenger rates on the Monkland Canal, for instance, were reduced by one-third when the new railway opened.[63] And the Kirkintilloch and Ballochney lines were also developing a growing interest in markets to the east. The view towards Edinburgh was responsible for the last link in the early Monklands railways. The Slamannan Railway, authorised in 1835 and opened piecemeal over the following five years, made very little sense if viewed in isolation. Its termini hardly even rated as villages, while the propectus remarked accurately that the intervening country 'may be said, from the total want of roads, to be as yet hermetically sealed up'.[64] The line was essentially a link in the coal chain between the Ballochney at one end and the Union Canal at the other. Its $12\frac{1}{2}$-mile length allowed a reduction of 25 miles in the journey to Edinburgh from the collieries on the Ballochney, and bypassed the staircase of locks at Falkirk (the Union Canal was level from the railway connection at Causewayend to Port Hopetoun in Edinburgh). A saving of 2/- per ton on the price of coal in the capital was anticipated.[65] The line would also open up some new sections of coalfield. In many ways it was a compromise. All things being equal, the canal interests and many of the Monklands coalowners would have been happier without the railway — the Slamannan, after all, both reduced the distance travelled, and therefore the toll paid, on the canal, and increased the possibility of Lothians coal travelling westward to compete in the Glasgow market (a fear which had created much opposition to earlier proposals to extend the Ballochney eastwards).[66] But the probability of a through line from Edinburgh to Glasgow being authorised was increasing every year, and the Slamannan was a despairing effort by the Union Canal interest in particular to retain its share of the inter-city traffic. The Union Canal Committee of Management agreed to contribute to surveying costs and to pay up to £250 towards parliamentary expenses: according to a disgruntled canal shareholder:

> This scheme was considered by its projectors to be so utterly worthless, that they refused to be at the expense of applying for the Act, and the money was chiefly, if not altogether, paid out of your funds.[67]

The Slamannan suffered during construction from overspending, 'considerable difficulty' in obtaining land, procrastinating contractors, high material costs, and problems of money raising in the deepening depression of the late 1830s.[68] Although for a brief period after opening it was part of the fastest passenger route between Glasgow and Edinburgh, and in spite of the claim that it was 'executed on a scale of rigid economy', the company was never successful. Its future depended on supplying coal to Edinburgh in particular, and it could not overcome the disadvantage of transshipment to the canal, of lack of proximity to important pits, or of the smaller transport costs of West Lothian and Midlothian coal, even apart from the directors' 'disappointment in not finding ironstone'.[69] When it eventually joined the Monklands amalgamation, it was very much as the junior partner.

III. Other early railways: Dalkeith, Newtyle and the Clyde

In 1825, as far as the coalowners of Midlothian were concerned, the problems of supplying Edinburgh were getting worse. Not only had Stevenson's Midlothian project failed, but the Union Canal had opened in 1822, bringing West Lothian and even Lanarkshire coal into the capital, and the Monkland & Kirkintilloch promised to intensify the competition further. Hugh Baird, the canal's engineer, had originally proposed short waggonways to bring the West Lothian coal and lime to the water; these had not been built, but in 1825 an act was passed for his West Lothian Railway. Its $14\frac{1}{2}$-mile main line would, he estimated, carry 76,000 tons of coal each year, mostly to the canal for Edinburgh.[70] Against this threat of canal/railway cooperation, and against the continuing seaborne trade from Fife and Tyneside, the Midlothian interest had to act. To encourage them they had the rising trend of Edinburgh coal consumption, from 200,000 tons at the beginning of the century to 350,000 tons about 1830.[71]

In 1824 a report by John Grieve, factor to the Duke of Buccleuch at Dalkeith, sounded a very positive alarm. The duke's coal, he stated, was selling in Edinburgh for 3/- per ton more than canal-brought coal thanks to its superior quality; but the opening of the Monkland & Kirkintilloch would bring coal of equal quality to the capital; and, furthermore, the best Midlothian coal was, at least in existing pits, almost exhausted. Rejecting a canal because it might flood the mines, he proposed a railway into the city and a branch to Fisherrow harbour, with various sections powered by locomotives, horses, stationary engines, and even possibly gravity, at a total estimated cost of £22,471. A few months later he raised his estimate to £36,863, to allow for a short extension at the Dalkeith end and the use of locomotives throughout. At this time the cost of carting coal from Dalkeith by road averaged 4/- per ton; his railway would supply it for 2/2d, of which 11d would be for carting from the depot at St Leonard's to central Edinburgh. As a final optimistic touch, he suggested that a further reduction of 6d could be achieved if the railway proprietors limited their dividend to 15% — and indeed a higher dividend 'would not be judicious'.[72]

The Edinburgh & Dalkeith Railway, as built under an act of 1826, was engineered not by Grieve but by the more celebrated James Jardine. His line followed Grieve's closely, but was designed solely for horse haulage, apart from a gravity-worked plane at St Leonard's.[73] The opening in 1831 at last gave the coalfield cheap, and slightly speedier, access to Edinburgh, although the benefits were restricted at the supply end, for the first decade at least, to the five coalowners whose property was either intersected by the line or, in the case of Stenhouse of Whitehill, linked to it by the old Edmonston waggonway. Initially the main beneficiaries were Sir John Hope and the Marquis of Lothian, with about three-quarters of the traffic between them, but a smaller owner, Dundas of Arniston, was enabled to increase his shipments from under 200 tons per year by road to over 7,000 by railway. The fifth supplier was the Duke of Buccleuch, who built a private extension from his own coal in 1838. By late 1839 the breakdown of total coal traffic was: Hope 39.57%, Lothian 33.75%, Stenhouse 11.76%, Dundas 10.01% and Buccleuch 4.90%.[74]

Not surprisingly, Hope (£2500), Lothian (£2500), John Grieve (£2500, presumably held for Buccleuch), and two members of the Dundas family (a combined holding of £3000) were among the largest subscribers to the railway in 1826, only the Earl of Wemyss (also £2500) having a comparable holding. An 1834 shareholders' list, which accounts for £57,700 of the possible £70,125 share capital, is headed by Robert Dundas (£4250), Hope (£2750), Buccleuch (£2550) and Lothian (£2500).[75] These large subscribers had an obvious interest in the railway, identical to that which had spurred the coalowners of the Monklands. However, their combined holding was only 25.2% of the original capital listed in the 1834 document, and 20.7% of the authorised original capital, which had presumably all been issued since a second successful share issue had already taken place. The line attracted fairly widespread interest even before authorisation, with 87 original subscribers against the 12 of the Garnkirk & Glasgow.[76] The subscribers had an aristocratic flavour, being led by a duke, a marquis, two earls, a viscount and five baronets, but they also contained several of the Edinburgh lawyers and bankers who were to be prominent in Scottish railway financing.[77] The capital sources of the Leith branch, which was floated as a separate company, were similar: this branch was designed not only to supply Leith itself, but also to divert the bunkering trade away from the Fife ports and to create an export trade which would reduce freight rates by supplying outward cargoes.[78] The projectors may even have considered attacking west of Scotland markets via the Forth & Clyde Canal. An undated but early shareholders' list accounts for 1028 £50 shares, of which 119 were held by Buccleuch, 100 by Lothian, and no more than 60 by any other individual.[79] It may also be noted that the Town Councils of Edinburgh and Musselburgh took shares in the main line, and that of Leith in the branch.

The Edinburgh & Dalkeith, more than any other early line, entered popular folklore. Its familiar and affectionate soubriquet of the 'Innocent Railway' was not due, unless inaccurately, to the legend that no one was ever killed on it, but rather to an air of old-fashioned unreality which stood by the leisurely horse-drawn tradition long after it had been abandoned elsewhere. Robert Chalmers, who coined the nickname, gently enjoyed himself at its expense:

> By the Innocent Railway you never feel in the least jeopardy; your journey is one of incident and adventure; you can examine the crops as you go along; you have time to hear the news from your companions; and the by-play of the officials is a source of never-failing amusement.[80]

Its eccentricities reached parliamentary notice when Lord Seymour's Select Committee of 1839, which had already perused the company's byelaws which *inter alia* forbade drivers to graze their horses while pulling trains, heard this reply from manager David Rankine:

> How do you take your tickets on the Dalkeith Railway? — We do not use them, there are so many different places for lifting passengers; it is a very populous country; there are a great many villages; and we have always found that many persons would not tell, or did not make up their minds, where they were going, which causes great confusion in using tickets.[81]

In spite of, or perhaps even because of, its eccentricities, the minister of Dalkeith in 1844 was prepared to affirm that 'few undertakings have contributed more to the commerce, convenience and health of the surrounding neighbourhood'.[82]

Elsewhere, there was one line in the 1820s which was not predominantly concerned with coal. In spite of the canal and waggonway surveys of Stevenson and his successors, improved transport had not reached the valley of Strathmore, where fertile agricultural land lay in urgent need of supplies of lime and access to markets. The nearest port and major town was Dundee, which in turn needed Strathmore's food and stone, but in between was the barrier of the Sidlaw Hills. There was also an expanding water-powered linen industry in the valley towns, which might take in flax or yarn through Dundee and sent back cloth for export. Not surprisingly, an 1817 proposal to build a canal over the hills was not followed up, and it seemed that Strathmore's trade would develop through Arbroath, Montrose or Perth. It was to prevent this that Dundee Town Council took the first initiative towards promoting a railway at the beginning of 1825:[83] the initiative was soon taken over by the local landowners, particularly the two largest, so that, according to one pioneer:

> It was chiefly, I may almost say entirely, through the spirit of Lord Airly & Lord Wharncliffe, supported by other landowners and the tenants around, that the Railway between Dundee and Newtyle was formed.[84]

Wharncliffe's influence was indicated in a newspaper report of a company meeting:

> A considerable disposition to grumble existed at the meeting; but Lord Wharncliffe had been got to take the chair: he strongly recommended unanimity, and there was unanimity.[85]

The line's engineer, Charles Landale, although described by the Board of Trade's General Pasley as an apothecary, was sufficiently well recognised to have been in charge of the transformation of the Elgin waggonway into the Dunfermline & Charlestown Railway.[86] But with the Dundee & Newtyle, he stood in some danger of becoming to Scottish railways what his fellow-townsman William McGonagall became to Scottish poetry. He first determined to run his line to Newtyle, where there was not even a village, rather than to Forfar, since Newtyle was in his view a better place from which to extend along the valley.[87] He then took the railway straight over the Sidlaws, with three inclined planes of between 700 yards and a mile each, as well as a tunnel, in a total distance of just over ten miles. Pasley was astonished by:

> a great many injudicious curves... I was informed that the projector of this railway had a favourite plan of making the wheels of locomotive engines travel around the outside rails of curves on their rims, and that his desire of proving the efficiency of this mode was partly the cause of his adopting such sharp curves... The projector of this railway seems to have set all ordinary rules at defiance.[88]

A separate branch through the streets of Dundee to the harbour was later added by some of the shareholders, without prior parliamentary approval; in spite of steep gradients and sharp curves, it was agreed by Grainger and Miller to be suitable for traffic at under three miles per hour and with no more than two waggons per horse. A train conveyed from Newtyle to the harbour, a distance of under eleven miles, was powered at various times by three stationary engines, two locomotives and a horse.[89] The Committee of Management, examining their affairs in 1836, described their line in terms which seemed to emphasise the poverty of its prospects from the beginning:

> It is believed that there has never before or since been a more adventurous railway than this originally was! There was nothing to look for its support but the general trade of the country. There was no coal, or lime, or iron in the district. There was only Dundee and its port, then beginning to rise in importance, at the one end, and the valley of Strathmore at the other. At Dundee, the railway was not allowed to extend onward to the Harbour; and therefore there was an expensive loading and unloading, and carriage between the Depot and the port, and consequent injury to the commodities conveyed, and loss of time; and at the north end, the line literally terminated in an arable field... Then, as to passengers, now the great support of the Railway, there was not one public conveyance on the road, except the coach between Dundee and Blairgowrie, which, during the summer months, travelled twice and sometimes three times a-week...[90]

In these circumstances, the conveyance of 31,264 passengers and 24,393 tons of goods in 1832-33 (the first full year of operation), and increases of 82.8% and 79.7% respectively in these amounts three years later, were not insignificant achievements.[91]

The line is a first-class example of a project which benefited its customers considerably, but did nothing for its shareholders. As Lythe concludes, 'the Committee's acceptance of the basic principles enunciated by Landale doomed the line to commercial failure'. Landale's estimate of £25,600 excluding land was hopelessly low, and the £40,000 authorisation of capital in the original act was almost finished by 1830, with the line far from complete.[92] A new act allowed completion but could not create viability. Although the volume of both goods and passenger traffic was well above expectation, the costs involved in using five separate power sources for each train journey (six if running to or from the harbour) ensured that there would not be a profit. Even before opening, financial constraints led to internal wrangling. Landale's position as engineer included the post of superintendent of works: in 1829 the treasurer, David Dobson, resigned, complaining that because of Landale's incompetence he himself was increasingly at risk under clause 83 of the company's act, which declared that debts not paid by the company could be recovered 'by the Distress and Sale of the Goods and Chattels of the Treasurer for the Time being'. Although at the meeting at which his resignation was accepted 'Mr Dobson left in bad humour' and 'not a word was said regarding Landale's dismissal', within two months the engineer had been eased out of the superintending part of his duties.[93]

The financial problems did not go away. In 1831 Dundee Town Council refused further help, and the company was within a week of having to stop work. In the following year the Dundee Union Bank, who held a mortgage for £24,000, threatened to foreclose. Nicholas Wood, called in to advise, could only suggest contracting out the management of the line (presumably in the hope that someone else might do it better) and extending to tap a larger area of Strathmore, since the cost of carting to Newtyle was still a deterrent to most of the towns.[94] Since there was no way in which the Dundee & Newtyle could have financed extensions, they were carried out by two miniscule companies, the Newtyle & Coupar Angus and the Newtyle & Glammis, to a construction standard reflected in Pasley's view that the latter was impracticable even for horses.[95] They still failed to reach either of Strathmore's principal towns, Forfar and Blairgowrie.

In 1836, with interest charges running at £1200 per annum and its rails still unpaid for, the company took the emergency step of trebling its authorised share capital, although it then had to offer the new shares at 10% discount, and even so could only get three-quarters of them taken up, mainly by creditors.[96] By 1839 the creditors suggested foreclosure, and were not opposed by a dispirited board. At this point the value of the line as a stimulus to the local economy was made clear by a group of gentry who wanted it kept open and were therefore prepared to advance £40,000 in exchange for valueless stock. In investment terms, the chairman admitted, 'it is a very bad concern; it has never paid a farthing': a view echoed a few years later (and with remarkable honesty, considering that he was trying to negotiate a lease of the line) by the clerk: 'it is true that there has been no dividend, and it is most likely there never will be one'.[97]

The minutes of the company in the early 1840s record a board wrestling in an increasingly dispirited manner with intractable financial problems. Shareholders lost interest, and in March 1842 a quorum could only be found for the company's statutory half-yearly meeting at the fourth attempt: a year later only the company's clerk turned up, but the meeting was quorate since he also held Lord Wharncliffe's proxy.[98] The company had taken a lease of the Coupar Angus line, but in 1841 the directors noted firstly 'that it is not in their power to pay any part of the Coupar Railway Rent this week', and two months later that they would pay when they could — 'to expect more is out of the question'. By the end of the year the lease had been terminated by mutual agreement, but the back rent was not paid until May 1844 and even then amounted to only about half of the sum claimed.[99] Other pressing creditors included the collectors of the government's passenger duty, but above all the Union Bank, which in 1841 was applying so much pressure that the directors resigned, and in 1842 went to law to force the company to sell the railway to anyone who would take it. The directors were quite willing to dispose of the line but wished, unlike the bank, to ensure that any new owner kept it open for traffic. By 1844 both sides had agreed to try and negotiate a lease;[100] their eventual success in 1846, when the Dundee & Perth agreed to lease it for a guaranteed £1400 per year, owed more to that railway's desire to keep the Scottish Midland Junction out of Dundee than to any intrinsic value attached to the Newtyle. But in terms of opening up trade to and from Dundee the line was a considerable success. And, as the minister of Newtyle said in his only recorded comment on the railway, 'the undertaking was at least a bold one'.[101]

Of the other lines of the 1820s, perhaps the most interesting is the one known variously as the Glasgow, Paisley & Ardrossan, the Ardrossan & Johnstone, or simply the Ardrossan. At the beginning of the century, at much the same time as Titchfield was contemplating the development of Troon, the Earl of Eglinton had promoted an ambitious scheme for Ardrossan, with the intention of making it the principal port for Glasgow. He spent over £100,000 on creating the best harbour on the Firth, and in 1806 obtained an act for a canal linking it to the city; his fellow shareholders included Dixon of Govan Colliery, who hoped to export coal via the canal. There were also to be branches to Clyde Ironworks and to Hurlet colliery.[102] But of the authorised share capital of £140,000, only £44,342 was raised: by the time the canal was built from Glasgow to Johnstone, this had been spent, and besides the company was £71,209 in debt.[103] And the dredging work being carried out on the Clyde had both challenged the canal's principal reason for existence and ensured that it would not receive a government grant as a work of major national importance.

In the 1820s the canal proprietors returned to the question of their missing link, and obtained an act to fill the gap with a railway. This vested the management of the railway and the canal in separate sub-committees of the company, with the outstanding debts entailed on the canal alone. Thus the whole of the £95,638 not previously raised was available for the railway, which tallied almost suspiciously well with Jardine's estimate of £94,093 for constructing the line. This time the line was started from Ardrossan, probably in order to give Eglinton's collieries a link to the sea at the earliest possible date. Capital, which would have been totally unobtainable if saddled with the old canal debt, was still difficult to find in the harsh economic conditions of 1827-28; only £28,950 could be raised by subscription and £20,000 borrowed, and the line petered out at Kilwinning.[104] After all, it was not clear how much traffic would be carried — sizeable ships could now reach Glasgow harbour, and the extra distance they had to travel was less of a handicap for goods traffic than the double transshipment from canal to railway to ship involved in the Ardrossan route.

In 1830 the company proposed to obviate this problem by turning the canal into a railway and extending it from Johnstone to Kilwinning. The scheme was, however, allowed to fold, perhaps because of money problems, and perhaps because of a hostile report (made to rival projectors) by Grainger and Miller who, considering the sharp curves of the canal totally unsuitable for a railway, refused to believe 'that there is any serious intention of persevering in this scheme'.[105] By the mid-1830s the proprietors accepted that they would not succeed in linking their two enterprises, and in 1840 the Ardrossan Railway Company was separated from the canal, taking with it a debt of £29,150. New share capital of £10,250 was issued, the line was doubled and altered to standard gauge, most of the curves and gradients were rebuilt in order to introduce locomotives, and the railway found its future as an offshoot of the Glasgow Paisley Kilmarnock & Ayr.[106]

Elsewhere, the Polloc & Govan Railway was in effect an updating and extension by William Dixon of his Govan waggonway, designed to link Govan with the Broomielaw, the Glasgow Paisley & Johnstone Canal, and the mineral deposits on the Pollok Estate of the Stirling-Maxwells.[107] In 1831 Dixon proposed to extend the line into the main Lanarkshire coalfield, linking with and indeed crossing the Wishaw & Coltness, and eventually connecting with his Calder ironworks. This provoked a petition of protest from the Trustees of Hutcheson's Hospital, who were also coalowners and through whose land the line would run. Dixon's initial act, in the previous year, had made the line terminate on the Hospital's ground 'whereby the fair advantage which the measure was calculated to produce might be secured to the institution': now they accused him of attempting to 'violate entirely the conditions upon which alone the other parties interested gave their consent', of cutting the Hospital's coalmines off from both suppliers and markets, of lowering the value of their land and of attempting to gain a local monopoly.[108] The planned extension was rejected, but an extension of the Polloc & Govan to Wellshot, near Cambuslang, by the Rutherglen Railway Company, was authorised. This, however, was never built, while plans to extend the line across the Wishaw & Coltness to the Coatbridge area in 1837 and 1839 were abandoned before submission to Parliament.[109] The link between the Polloc & Govan and the Lanarkshire railways had to await the mania and the construction of the Clydesdale Junction.

Finally, the Paisley & Renfrew was built as both a rival and an auxiliary to water transport. It was intended to bypass the tedious and inadequate navigation of the River Cart and to take passengers and light goods to connect with the river boats on the Clyde. It was not a line with any prospects of success. The canal made it unlikely to be of much interest as part of a route between Paisley and Glasgow, and before it even opened an act had been passed for the Glasgow Paisley & Greenock, which was to appropriate the traffic in the opposite direction. In 1841 the directors sadly announced that there would be no dividend, which there would have been:

> had the trade of the Company not been so materially injured by the opening of the Greenock Railway on the 30th March last, since which period, the falling off in Passengers had been more than one-third from the corresponding period of last year.[110]

Whishaw used it as a peg on which to hang the observation that 'short lines of railway are far less profitable to their proprietors than long lines', but there were few early lines of any length which were so essentially misconceived as the Paisley & Renfrew. Even its later attempts to be taken over were beset with difficulties. In 1846 it agreed to sell out to the Paisley Barrhead & Hurlet for £11-2-6d per share, plus its accumulated debt, but this mania company never even managed to build its own line. In the following year it was sold to the Glasgow Paisley Kilmarnock & Ayr, but it had no physical link to its parent company until it was rebuilt to standard gauge in 1866.[111]

IV. The impact of the locomotive

One crucial decision facing all these railways was what to do about the locomotive. Were the extra speed and the extra capacity of steam-hauled trains desirable for the amount and type of traffic anticipated, or would it be more profitable to save on construction costs and build a road suitable only for horses? At this point there appears to have been no clear agreement as to whether locomotives or horses were cheaper to maintain, although Grainger and Miller were convinced by 1830 that:

> The decided superiority which Railways, combined with Locomotive Engines, possess over every other communication, seems to be admitted by every unprejudiced man who has paid the least attention to the subject.[112]

The responses of the companies varied, and not always in the most logical fashion. Locomotive speed might be expected to be useful to lines which had a substantial passenger traffic into the cities, and by 1831 Grainger and Miller could state that locomotives were safe, economical, and usually run at 16 to 24 miles per hour;[113] but, although the Garnkirk & Glasgow was a locomotive line from the start, the increasing number of passengers on the Edinburgh & Dalkeith did not persuade it to abandon the horse. The line remained a horse railway until absorbed by the North British in 1846, as did the Ardrossan until linked to the Glasgow Paisley Kilmarnock & Ayr in 1840. In both cases the anti-locomotive decision was presumably made easier by the flimsy nature of the track, and in both cases the line had to be strengthened and relaid to standard gauge before locomotives were used.[114]

One major problem was the existence of any impediment on the line which would prevent the running of locomotives from end to end. The Monkland & Kirkintilloch was the first British railway to include clearly in its act of incorporation the power to run locomotives, but it was designed and started out as a horse-drawn line.[115] When the decision to introduce locomotives was taken, the Bedlay tunnel, which was unsuitable for steam traffic, had to be converted into a cutting — otherwise the company would have had the additional costs of having to use two locomotives and intermediate horse haulage on every train.[116] The original locomotives, built by Murdoch and Aitken of Glasgow following Stephenson's Killingworth design, were the first to be constructed in Scotland.[117] The Ballochney had an even greater problem to overcome, since in the middle of its line were two inclined planes. On a horse line, the inclined plane was a sensible and desirable way of overcoming gradients, allowing the horse to do most of its work on a level or near-level line, and bringing stationary steam-engines or the force of gravity into play for short sharp ascents and descents. On a locomotive line, an inclined plane was simply a nuisance, involving duplication of haulage facilities on either side of it, but usually necessitating extremely expensive construction work to get rid of it. An inclined plane might be tolerated at the end of a line if there was no other way of reaching a desirable station site in a city centre or at a major harbour, and such inclines were to be features of both the London & Birmingham at Euston and the Edinburgh & Glasgow between Cowlairs and Queen Street.

But, in spite of the powers for steam haulage in the Ballochney act, the existence of the double plane in the centre of the line shows that it was planned for horses. When it was being strengthened for the introduction of locomotives in 1840, the work was done only as far as the foot of the inclines.[118] The upper Ballochney, and the Slamannan beyond it, were for the time being left to horses.

On the other hand, the Dundee & Newtyle was not deterred by its three inclines. Landale had remarked in his original report that:

> There is good reason for believing that such machines [locomotives] will work at much less expense than horses ... yet I do not recommend that they should be employed in the infancy of the establishment

and Nicholas Wood had agreed, shortly after the line opened, that the traffic was not yet sufficient for locomotive power.[119] But by 1834 locomotives were running on the two upper levels of the railway, while the sharp curves and street crossings of the harbour branch restricted it to horses.[120] The two small railways branching from Newtyle were substantially horse lines, though locomotives were tried on both, and the Coupar Angus company even experimented with wind-powered trains.[121] The costs of all the various types of haulage had much to do with the fact that working expenditure on the Dundee & Newtyle was steadily over 80% of gross revenue.[122]

Another type of impediment was created by agreements with landowners that locomotives would not be run across their land. The Wishaw & Coltness had had to accept such a clause, as well as paying an exorbitant price for land, to overcome the opposition of Drysdale of Jerviston to a line which, he claimed,

> would enable the Landed Proprietors to the south of Major Drysdale to enhance the value of their own Estates at the expense and ruin of his.[123]

His daughter later charged £1000, of which Thomas Houldsworth of Coltness Ironworks agreed to pay half, to rescind the restriction.[124] A similar provision applied to the Prestonfield estate of Sir Robert Keith Dick on the Edinburgh & Dalkeith: this did not disturb the company, who had no desire to change the arrangement, but it hampered an 1830-31 plan to make the line part of a through route to Glasgow.[125]

The two railways which adopted the locomotive from the start were the Garnkirk & Glasgow and the Paisley & Renfrew. These two were the only lines in direct competition with water transport, and had to make the most of their potential advantage in speed. This was particularly important for passenger traffic, where higher fares than by canal or river might be acceptable if a faster service was supplied. In March 1842 the Paisley & Renfrew had to revert to horses for reasons of economy; by that time the Greenock railway had in any case taken away many of the anticipated passengers. Four months later, the Renfrew company succeeded in leasing their line for £1001 per year, but before the three year contract was completed they had to accept a reduction to £801.[126] The Garnkirk, although primarily a coal line, developed considerable passenger traffic with the help of locomotives, having started the service on 1 June 1831, almost three months before the formal opening of the railway.[127] Before the end of the year, the company's engineers considered that steam had made its point:

> From what has already been done upon the Manchester and Liverpool Railway, we are persuaded that it will be admitted, on all hands, that steam, combined with railways, infinitely surpass every other mode of conveyance for *passengers* and *light goods;* and upon the Garnkirk and Glasgow Railway, under our own eye and immediate direction, it has been ascertained and proved, that they are equally superior, in point of economy, to every other communication for the conveyance of *coal* and *heavy goods*.[128]

An incident in 1833, however, suggests both that the Garnkirk was not fully committed to locomotives and that all was not completely peaceful between the Lanarkshire railways. The occasion was a bill promoted by the Monkland & Kirkintilloch to allow the doubling of their track, into which the Garnkirk & Glasgow interest persuaded a House of Commons committee to insert an apparently harmless clause giving their company running powers over the section linking their line to the Wishaw & Coltness, and insisting that traffic in opposite directions should be kept on different tracks. Before the full House, it was explained for the promoters of the bill that their aim was to segregate locomotives and horses, because the Garnkirk refused to run locomotives on the disputed section:

> When the proprietors of the Kirkintilloch railroad found that in proceeding upon this one mile and a half of road, they were compelled to accommodate their pace to that of the [horse] trams, . . . much hostility existed between the men employed by the respective parties; and . . . it frequently occurred that the drivers of the trams would let the horses loose, for the purpose of leaving the trams in the way of the locomotive engines, so that the latter must run foul of the trams . . .

If the Garnkirk were allowed to gain their clause, the Monkland & Kirkintilloch would have to build a third track. The case convinced the House, and the clause was rescinded.[129] In any case, by 1838 the Garnkirk was running locomotives over both the Monkland & Kirkintilloch and the Wishaw & Coltness.[130]

The advent of the locomotive dictated a tightening up of organisation, with closer control by the management of all that happened on the line. At first, most railway proprietors regarded themselves as running more of a toll road than a carrying business. Rates were quoted carefully broken down into separate sums for the use of the road, for hire of motive power, for hire of waggons, and possibly also for wharfage and terminal charges; and on mineral lines it was hoped that customers would supply their own waggons and horses. In 1831 the Ballochney was not entirely happy to note that none of its customers was yet in a position to do this as cheaply as the company could; however, they observed:

> This carrying business will in a short time be in a great measure given up as the Kirkintilloch Railway Company are about to introduce a system of draggage by Locomotive Engines along their road.

At that point the company could sell off most of its horses, buy more waggons to hire out, and virtually give up carrying on its own account.[131] It is not clear whether carrying was simply seen as an improper activity for railway proprietors, or whether it was regarded as too risky. On the other hand, opinion later swung towards companies carrying for themselves. In 1842, for example, the Wishaw & Coltness bought 323 waggons from traders to ease the traffic flow by reducing the number of operators on the line and 'to keep down the complaints of Traders as to the Engines damaging their Waggons'.[132] The company may not have received much of a bargain; by the end of the decade, when the Caledonian was taking over the Wishaw & Coltness, Joseph Locke observed:

> That was one of the things that became essential to the Caledonian Company, to remove the old plant from the line. The coal-waggons were without springs; ... in order to adapt them to the ordinary traffic of a great public Railway, new waggons, with springs, and entire new engines, have been put on, and the thing altogether has been really re-modelled.[133]

The toll-road principle had been inherited by the railways from canal practice, but even on a horse railway it involved problems of management which did not affect canals. Trains could not overtake each other, or on single-track lines even pass in opposite directions, without a system of passing-places or sidings not required by barges. Still, a fairly haphazard system of management was possible; its evolution can be traced in the byelaws of the various companies. They had already progressed well beyond the Kilmarnock & Troon regulation of 1813 which stated that:

> No person be allowed to drive cattle along the Railway ... excepting the proprietors, the members of the Committee of Management, and the officers of the Company.[134]

The byelaws of the Edinburgh & Dalkeith laid out a series of rulings on who should back when trains met between passing-places (not unlike modern informal practice on Highland single-track roads).[135] The Ballochney gave priority to loaded over empty waggons, to descending over ascending traffic, and to locomotives over horses, but failed to say what happened when an ascending locomotive met a descending horse.[136] But increasing traffic volumes and in particular increasing locomotive speed meant stricter rules had to apply. In 1840 the Garnkirk & Glasgow drew up new byelaws which indicated some of the problems of intermingling locomotives and horses. Passenger trains in both directions were to share one set of rails with eastbound horse trains, leaving the other set free for the slow heavy-laden westbound waggons. Waggons were always to give way to passenger trains and to locomotives, and were not to start on a journey unless certain of reaching a siding before being overtaken. Decisions were largely left up to the drivers, except for a blanket clause declaring that:

> Waggoners and others shall, on all occasions, obey the directions of the company's servants, to prevent their impeding or coming into contact with other carriages.[137]

Once a company had decided to adopt the locomotive, and particularly to run a regular and frequent passenger service, much stricter timetabling had to be imposed. The position of horse traffic became increasingly difficult, and gradually it was eased away from through lines. In 1839 the Wishaw & Coltness, having decided to adopt the locomotive, concluded that its powers should be general and the multiplicity of horse traders on the line should be reduced, in order 'to do away with the collisions which are daily taking place between the drivers'.[138] In general, the presence of outside carriers on the lines became increasingly inconvenient.

All lines authorised after 1836 conducted the whole of their own carrying trade from the start, and most of the older ones gradually reduced the use of outside carriers. By 1843 only six companies, or seven if the Leith branch of the Edinburgh & Dalkeith is counted separately, reported outside carriers on their lines (see Table 15).[139] The Leith branch, indeed, reported 100% outside carrying, though this probably included a few trains run by the parent company. Even among the six companies, practices varied. The Edinburgh & Dalkeith, which was clearly reluctant to do any carrying at all on its own account, carried none of the coal which made up over 90% by volume, and over 95% by value, of its trade. The Garnkirk & Glasgow, on the other hand, was by the first half of 1845 carrying all its coal traffic, while outside carriers conveyed 18.96% of other goods.[140]

Table 14. Dates of Introduction of Locomotives on Scottish Railways

Company	Date
Monkland & Kirkintilloch	1831
Garnkirk & Glasgow	1831
Dundee & Newtyle	1833
Paisley & Renfrew	1837
Newtyle & Coupar Angus	1837, abandoned by 1842: reintroduced 1846
Ardrossan	1840
Ballochney	1840
Edinburgh & Dalkeith	1846

Sources: H.G. Lewin, *Early British Railways;* G. Buchanan, *An Account of the Glasgow and Garnkirk Railway* . . .; S.G.E. Lythe, 'The Dundee & Newtyle Railway', *Railway Magazine* 97 (1951); E.H. Ahrons, *The British Steam Locomotive 1825-1925;* F. Whishaw, *The Railways of Great Britain and Ireland;* GUEH, Draft Report of Committee of Management of Ballochney Railway Company, 1 Feb 1831; GUEH, Bye-laws of Ballochney Railway Company.

By the 1840s, outside carrying was limited to coalfield lines: it was presumably only worth while for an individual or firm to own a substantial number of railway waggons and horses when also sure of sending regular large amounts of traffic down the railway. Admittedly the horses and drivers could also in theory be used for non-railway purposes, and waggons could be hired from the railway company; but in practice waggon rentals did not account for a large part of railway revenue — 5.6% of traffic dues on the Monkland & Kirkintilloch between 1836 and 1839, for instance, or 8.4% during the first seven years' trading by the Ballochney.[141] But after the earliest years small users, such as Angus farmers, did not hire waggons to use with their own horses, and such trade was carried by the railway companies. The major waggon owners were the great coal and iron masters of Lanarkshire and the Lothians, who effectively kept their own carrying establishments.[142]

Outside carrying continued in places for a long time; Finnie, the lessee of the Duke of Portland's coal, was still carrying on the Kilmarnock & Troon in 1852.[143] But there is a clear correlation between the increased use of locomotives and the reduction of outside carriers. And although the statistics issued by the companies do not distinguish between main line and branch traffic, it is probable that the segregation was even more substantial than overall figures suggest, with locomotives kept to main lines, and carrier traffic largely using the proliferating coalfield branches. Hence, for instance, the increase in tonnage carried by outside

Table 15. *Goods Traffic — Companies allowing other carriers on their lines, 1843-1845*

Company	Total traffic in two years tons	Carried by company %	Carried by others %
Edinburgh & Dalkeith[a]	215,763	3.3	96.7
Ballochney	715,062	21.7	78.3
Monkland & Kirkintilloch	2120,344	37.8	62.2
Slamannan	124,944	51.0	49.0
Wishaw & Coltness	1057,431	61.2	38.8
Garnkirk & Glasgow	511,753	89.4	10.6

a: excluding Leith branch.

Source: PP 1846 [698] XXXIX, *Report of Railway Department for 1844-45*, appendix II.

carriers on the Ballochney from 71.5% in the first half of 1841 to 85.8% in the corresponding period of 1845 may correlate with the railway's expanding system of small branches, while the limited and falling outside carrier trade on the Garnkirk may reflect a company which concentrated on a locomotive-hauled main line.[144]

Another variation on the theme came in 1843, when opposition from traders to a Monkland & Kirkintilloch bill was bought off by a promise to allow private locomotives on the line. Powers for this were taken by inserting a clause in the Wishaw & Coltness Act of 1844 — another example of the possible close cooperation between the Lanarkshire railways.[145] There is no evidence that private locomotives were widely used on the line. The Ballochney fought against a similar request from traders in 1843 on the grounds that private locomotives would cause too much danger, particularly on a line with inclined planes.[146] And in 1842 the Wishaw & Coltness secretary was told by Lord Belhaven:

> What I now beg to call the attention of the Committee to is whether they permit Mr Houldsworth and myself to make our own haulages, we claim the right to do so and it would be better to avoid if possible all dispute on the subject.

Such a request from the company chairman and one of the largest traders, who made it clear that they wished to run their own engines all the way to Glasgow, set the reluctant committee a problem. They solved it by promising to introduce company locomotives on the southern half of the line, which was still restricted to horses, and to make the intervening Calder bridge safe for their passage.[147]

V. Financing the coal railways

Even in the 1820s, some of the perpetual problems which were to face railway projectors were becoming clear. Like most projects involving large-scale capital investment, railways tended to be proposed during the optimistic upswing phase of the business cycle, when money was available and expectations of return were high. There was, however, a long time lag between promising the capital and having to produce it. Preliminary and parliamentary processes took at least a year;

then, after the act was passed, capital was raised by a series of calls on the shareholders levied as the money was required for construction. Hence the time lag between promising to subscribe and actually paying the bulk of the calls could be three, five or more years, and by this time the economic climate could have changed for the worse. Railways originally projected in the good years around 1824, 1834 or 1844, and authorised one or two years later, all found themselves having to raise money in the troughs of the business cycle.

This had not been such a problem for the waggonways. In the first place, the time lags were shorter in almost all cases since parliamentary approval was not required. And since a waggonway was financed normally by one man or one company, as an auxiliary service to major interests in coal or iron, the question of whether the waggonway taken in isolation made a profit was usually incidental. But of the railways under discussion in this chapter, only William Dixon's Polloc & Govan was effectively an updated waggonway of this type. In the others, shareholding was not limited to a small group of people with a personal interest in sending traffic along the line. Even the Ballochney, which might, from its geographical position and its sole function as a mineral carrier, have been expected to be owned by a small group of coalowners and major coal users, had 50 original subscribers, and 48 shareholders in 1831. No individual then held more than £1000 of the £17,000 issued stock.[148] Among the shareholders were nine women — and women generally held shares for the sake of investment income, rather than either capital appreciation or outside business interests. Similar diversification of ownership applied elsewhere. The bulk of capital certainly came from such sources as the aristocratic landowners along the Edinburgh & Dalkeith or the Glasgow merchants and Lanarkshire coalowners with a personal interest in the Wishaw & Coltness, and, as already noted, in 1839 the landowners of Angus did rally round the Dundee & Newtyle in a crisis, investing extra money on which they could hardly hope to see a dividend rather than risk the closure of the railway. But the companies could not rely on being bailed out in this manner, nor indeed in most cases did the potential direct beneficiaries of the line put up all the money in the first place. Capital was coming more and more from sources whose concern was purely with investment, and whose money might be moved elsewhere if returns or prospects were unsatisfactory. Some of this capital had not even ties of local loyalty to the line — as early as 1826, £3,300 came from England to the Ballochney, and by 1844 £22,500 of the Wishart & Coltness's capital was English money.[149] The companies were having to give primary consideration to their profitability and to their rate of return, even though the first interest of most of the directors might be in the volume of minerals transported.

How then did these railways stand up to scrutiny? Much of the evidence suggests that at best they grew only slowly to prosperity, and that in many cases they never rose above marginal financial viability. Even for the mineral lines, where the volume of traffic increased steadily and where, for the lines authorised in the 1820s at least, the demand from the iron industry boom was an unanticipated bonus, profitability was not always easily achieved. The failure of the Dundee & Newtyle ever to pay a dividend, and the almost equally unfortunate state of the Paisley & Renfrew, have already been mentioned. The Wishaw & Coltness had only paid dividends in three half-years out of a possible 21 when it set up a committee of inquiry in 1844 to look for economies.[150] Between its foundation and the amalgamation of 1848, the Slamannan paid only a single dividend, in 1841 when it was briefly part of the fastest passenger route between Glasgow and Edinburgh: in 1842 it was unable to raise the capital to build a 6½-mile extension to Bo'ness, even although it was expected to increase traffic by £18,000 per year for an outlay of £46,000.[151] In 1836 it was noted of the Garnkirk & Glasgow that 'it has not yet begun to pay, and it cannot be conjectured when it will do so'. The same reverend commentator remarked that the Monkland & Kirkintilloch, which had then been open for ten years, was only beginning to be profitable.[152] In fact both it and the Ballochney, which were the most successful of the mineral railways, grew gradually to prosperity and by 1838-42 were paying respectively 10-12% and 14-16% dividends. Even by 1836, the Ballochney had been profitable enough to provide half the capital for the Slamannan.[153] They were not so successful in the mid-1840s, but at the amalgamation of 1848 the Monkland & Kirkintilloch had averaged a 5½% dividend over 22 years' operations, and the Ballochney 7¼% over 20 years.[154] These might not be the returns of 10% or even 20% forecast by engineers in the 1820s, but they were much better than the performance of rival lines.

The dividend was the result of an equation between construction costs, working expenditure and revenue from traffic, and it is not unfair to say that in their forecasts the projectors got them all wrong. Invariably they underestimated: it was fortunate that their underestimation of traffic to some extent made up for the extra expenditure required on the other items. Even so, by 1840, ten of the twelve schemes which had actually been built had had to return to Parliament for authorisation for more capital (see Table 16).

In two-thirds of the cases, the original capital had to be doubled, and in most it was required at least in part to complete the original main line. The relative restraint of the Ardrossan's extra requirements was of course achieved by not building three-quarters of the planned line. Apart from the Paisley & Renfrew and the Newtyle & Glammis, whose virtue in keeping within construction estimates was in both cases counterbalanced by a sad lack of traffic, only the Monkland & Kirkintilloch, the Ballochney and the Newtyle & Coupar Angus managed even to open before going back for a second act. One apparent problem, therefore, was that any surplus revenue available for dividends was going to be spread over a much greater share capital than originally intended.

It was not always so straightforward, however. Where the money was needed, as on the Dundee & Newtyle, simply in order to complete the original scheme, it was not creating extra traffic sources. But much of the extra capital raised by the coal lines was used to push out branches and extensions to more and more pits or iron works. By the mid-1840s, the four-mile main line of the Ballochney had been added to by nine separate branches and innumerable sidings.[155] And a project like the Leith branch of the Edinburgh & Dalkeith was an entirely rational use of extra capital: the large demand of the port for coal was, after all, to be an important part

Table 16. Capital authorised, 1824–40

Company	Original authorised share and loan capital £	Share and loan capital authorised in subsequent acts to 1840 £	Increase in authorised capital %
Garnkirk & Glasgow	28,497	119,698	420.0
Monkland & Kirkintilloch	32,000	123,333	384.4
Polloc & Govan	15,000	51,000	340.0
Dundee & Newtyle	40,000	130,000	325.0
Ballochney	28,431	51,569	181.4
Edinburgh & Dalkeith[a]	70,125	98,628	140.6
Slamannan	106,000	119,333	112.6
Wishaw & Coltness[b]	80,000	80,000	100.0
Newtyle & Coupar Angus	20,200	15,000	74.3
Ardrossan	95,658	51,700	54.1
Paisley & Renfrew	33,000	—	—
Newtyle & Glammis	26,600	—	—

a: If the Leith branch is excluded, the figures become respectively £70,125: £62,982: 89.7%.
b: A further £160,000 was authorised in 1841.

Source: PP 1844 (159) XLI, *Return of Moneys raised under Railway Acts 1826–43*.

of the calculations even of the Edinburgh & Glasgow pioneers. It was perhaps unfortunate that in the act which authorised £35,700 for the Leith branch and £7815 for a short extension at Eskbank, the Edinburgh & Dalkeith also had to raise £47,060 to complete their main line.[156] But it is always necessary to look closely at a company's plans to see how far extra capital is a result of previous miscalculation and how far part of legitimate future planning.

Again, extra capital could be required simply because a line had been more successful than anticipated. The boom in the iron industry after the opening of the Gartsherrie works in 1828 found the railways unprepared for the enormous ensuing traffic. On the Monkland & Kirkintilloch and the Ballochney, trade changed from carrying limited amounts of coal to the canal; now there were increasingly large movements of coal and ironstone to the works, and of pig iron as well as coal in the other direction.[157] As early as 1829, the Monkland & Kirkintilloch had to start doubling its main line to cope with demand,[158] and other companies followed over the next decade. Improvements had to be made when a company found itself carrying extra traffic from a new railway opened beyond it, as with Slamannan traffic on the Ballochney. For most companies also, the introduction of locomotives meant not only the cost of the engines themselves, but also relaying the track with heavier rails. In 1844 Robert Dodds, manager of the Wishaw & Coltness, drew up a comparison of Grainger and Miller's original estimate and the company's actual capital expenditure, and demonstrated at least to his own satisfaction that the extra amounts had been either unavoidable or desirable[159] (see Table 17).

The extra costs came mainly under three headings. Firstly, Grainger and Miller had deliberately not estimated for some essential items, including land, rolling stock and sheds. Secondly, there were items connected with increasing traffic, such as doubling the line (which included doubling a tunnel), or with locomotives, such as the heavier rails. The third smaller but no doubt infuriating group were costs forced on the company from outside, such as the £2435 cutting necessary to pass under the main road at Motherwell when the trustees refused to allow the company to raise the road for a mere £206.[160] Once all this was taken into account, Dodds claimed that the line had been completed for as near to its estimate as any in Great Britain. On the other hand, one landowner, admittedly as part of a legal case against the company, claimed that the estimates for the southern section of the line had been at least 40% too low.[161]

If the decision to issue extra capital was in many instances not simply due to original miscalculations, but was a deliberate decision in order to increase the revenue by attracting extra traffic, it should be possible to judge whether these decisions were justified by the outcome. Unfortunately this is a case where it becomes possible to prove almost anything by judicious (or injudicious) use of the available statistics. In the first place, we do not have complete runs of figures for the various companies: in the second, the results obtained from the limited figures available depend entirely on the years selected for comparison. To take as our example the two most successful of the early lines, neither the Monkland & Kirkintilloch nor the Ballochney increased its called-up capital between 1837 and

1839, or in 1842-3, but both increased it by almost 150% between these periods. Whether the statistics indicate that this extra investment was repaid by proportionately increased profits depends almost entirely on how the figures are manipulated (see Table 18).

Table 17. Wishaw & Coltness Main Line Costs

Original estimate by Grainger and Miller	£	Subsequent costs to 1844	£
Parliamentary and preliminary	1,500	Land	26,971
Earthworks	15,000	Locomotives	3,958
Masonry, bridges	9,300	Waggons	11,833
Iron work, rails	11,200	Carriages	1,125
Laying rails	1,000	Sheds	628
Stone blocks	1,800	Station houses	200
Ballast and horsepath	2,000	Doubling and new rails	30,700
Fences and gates	1,200	Doubling extension	32,000
Waggons etc. during construction	1,000	Heavier rails	17,300
Salaries, surveys, etc.	2,000	Bridges and level crossings	10,000
Cleland branch	6,000	Cuttings for bridges etc.	12,000
Rosehall branch	500	Depressions	365
Law Branch	500	Coltness branch	205
Contingencies	4,000	Cleland branch extension	2,040
		Other branches	463
	57,000		149,788

Sources: GUEH, Robert Dodds, 'General Remarks on the Wishaw and Coltness Railway from its Commencement to the present state', 1844; T. Grainger and J. Miller, *Report relative to the Proposed Railway to connect the Clydesdale, or Upper Coal Field of Lanarkshire, with the City of Glasgow, and the East and West Country Markets* (1828).

Table 18. Monkland & Kirkintilloch and Ballochney Railways: Capital and Revenue 1837-43

	1837-42 %	1839-43 %	1837/9-1842/3 %
Monkland & Kirkintilloch:			
Increase in capital called up	142.5	142.5	142.5
Increase in gross revenue	245.1	18.3	93.0
Increase in net revenue	274.1	1.8	88.5
Ballochney:			
Increase in capital called up	146.3	146.3	146.3
Increase in gross revenue	190.3	33.5	96.3
Increase in net revenue	197.4	7.9	81.5

Sources: Calculated from GUEH, 'Monkland & Kirkintilloch Railway: Return required by the Railway Commissioners ... in reference to the Monklands Railways Amalgamation Bill 1848'; GUEH, 'Explanatory notes and details of several statements and calculations contained in the Remarks by the W. & C. Ry. Directors on the Interim-report by the Committee of Enquiry' [1844]

It may be that the main fact reflected in these figures is that 1843 was a bad year for both companies — the start indeed of a period of decline from the years of double-figure dividends, which was to push the companies first into an attempt to amalgamate with the Edinburgh & Glasgow, and then into amalgamation with each other and with the Slamannan. There are also other considerations which the figures cannot bring out. In many cases there would be a time lag between useful investment and the extra revenue resulting from it: if a company bought new waggons to expedite traffic already held up by a shortage, the effects would be immediate, but if it extended a branch into an untapped mineral area, it might have to wait until new pits were sunk. There is frequently no way of saying that a particular increase in the gross revenue figures was caused by a particular item of previous investment.

Alternatively, investment which did not in itself seem to have any obvious financial benefit may still have been necessary. For the Garnkirk & Glasgow to adopt the locomotive was logical, since it was in direct competition with the Monkland Canal and one of its main advantages was speed. For the Wishaw & Coltness to do so, when it had little interest in non-mineral traffic and no competitors in its area, was less obviously necessary, unless it would clearly increase profits to an extent greater than the costs of further strengthening the tracks. But, even apart from the pressures exerted by Belhaven and Houldsworth,[162] there was the point that, although Parliament was always inclined to respect the territorial rights of existing companies, it might be inclined to sanction a competitor if the incumbent was not running an efficient operation — and by 1840 locomotives were part of efficiency. There were to be plenty of envious eyes cast at the Lanarkshire coal and iron traffic — in the early 1830s for instance the Ballochney was seriously concerned about an Edinburgh & Glasgow scheme surveyed through its area, more than once an extension of the Polloc & Govan was contemplated, and in 1841 the Wilsontown Morningside & Coltness was authorised to open a further section of coalfield[163] — and the existing railways had to satisfy their customers, their shareholders and Parliament of their right to retain their local monopolies.

Even if capital was authorised, it was not always easy to raise. The boom conditions of 1825-26, when the opening of the Stockton & Darlington and the promotion of the Liverpool & Manchester first made railway investment popular, allowed subscription lists to be filled up and shares to be fully allotted. But in the more depressed years at the end of the decade, calls on shares might not be forthcoming — defaulters on the Ballochney even included two of the Baird brothers of Gartsherrie[164] — and extra finance for lines with apparently bleak prospects was hard to come by. In 1839 the Slamannan issued a nominal capital of £140,000, but the lack of enthusiasm shown by investors necessitated issuing some of the shares at a discount, and the company received only £114,980.[165] Admittedly share capital was not the only source of money for construction. Companies were also authorised by their acts to borrow money once half of their share capital was paid up, and invariably they used this facility. Unlike later and larger companies, these railways tended simply to run up an overdraft with their bankers; between 1836 and 1843 the debt of the Monkland & Kirkintilloch to the Royal Bank rose from

£9956 to £54,906, and that of the Ballochney from £13,843 to £21,119. In both cases some of the share capital raised in the intervening period had been applied to debt reduction, and the Ballochney had even cleared its entire debt in 1840, but the bank does not seem to have exerted heavy pressure on the companies.[166]

Other companies, whose prospects were less satisfactory, found life more difficult. The troubles of the Dundee & Newtyle have already been noted: as early as 1828 a report to the Committee was looking for desperate expedients to raise money or reduce expenditure. These included feuing rather than buying land, persuading shareholders to lend to the company or to give their personal security against loans from the bank, prosecuting shareholders who failed to pay their calls on time, and not paying the company's lawyers, engineers and salaried officials until the line was open, when the money could come from traffic receipts.[167] The Wishaw & Coltness, whose difficulties in raising money were primarily responsible for the eleven years taken to complete the line, found in 1835 that they were unable to borrow a further £20,000 unless the directors gave their personal security for the payment of at least the interest due. Since none of the directors were willing to give such a guarantee, construction had to be stopped at Jerviston with only the northern half of the line completed.[168] This in turn led in the following year to a legal case which could have had incalculable repercussions for all railways. Sir Henry Steuart of Allanton, a landowner in the coal area at the end of the proposed line, had subscribed for £3000 in 1829: now he accused the company of having taken 'not one single step . . . to carry forward the line of the railway towards Allanton', which was seven miles from the point at which construction had stopped, and of having no intention of ever completing the line. Grainger's estimates, he declared, were clearly inadequate, and even if all the authorised capital were raised, it would all be spent long before the line reached his land (in this Steuart was entirely accurate, though it would have been fairer to accuse the company of miscalculation rather than bad faith). In these circumstances he refused to pay calls on his shares, claiming that he was legally entitled to do so until the railway reached Allanton, when he would pay up with pleasure.[169]

Two surviving documents from the beginning of the company give some weight to Steuart's contention that there had originally been such an understanding. One is a copy of a letter in September 1828 to Sir James Steuart Denham of Coltness: the author is unidentified, but the presence of the letter in the records of the Wishaw & Coltness suggests that he had some standing in the company. As part of his encouragement to landowners to support the railway, he states:

> there might be a condition attached [?] to your subscriptions that instead of being liable to the calls for the payment of stock as the work proceeded, you should only pay by agreed-on percentages according to the Sale of your Minerals, — 10, 20 or 100 p.cent on the net profits of the minerals sold as the Proprietors respectively agreed being applied in reducing or paying your Stock subscribed for . . . The Act of Parliament (similarly to what is done in Turnpike Acts for receiving subscriptions out of the Lands or Rents) would of course contain a clause for securing payment of such subscriptions out of the Minerals of the Proprietors subscribing in all the Estates along the Line. There would thus be no difficulty in borrowing at once from a Bank the whole sums required to pay the contractors of the work on assignment or Mortgage of the Subscriptions, and the Contracts might be obtained on the most advantageous terms, and no inconvenience be felt by any subscriber.[170]

It was a skilful idea for painless financing, though it made a large assumption about the degree of cooperation to be expected from the banks. A similar proposal is contained in a manuscript draft of 'Proposals for raising Funds for making a Railway from Allanton Coltness Cleland &ca to join the Glasgow Railways and Canals', which proposed that the subscriptions should be a burden on the minerals to the extent of a third or a half of their value, and makes it clear that no call would be made until the line had reached the relevant estate.[171]

The railway company, apart from protesting their good faith and intention to build, could clearly emphasise the illogicality of being promised the payment of money only after they completed the work for which the money was necessary. Since Steuart's attitude was also being taken by three other landowners — Lord Belhaven, Sir James Stewart and William Dixon — with a combined subscription of £15,950 or 18½% of the entire share capital, and since the banks neither would nor legally could lend all the necessary funds, there was no way in which the company could manage without the money. Describing Steuart as 'one of the chief bars to the progress of the work', they pointed out that he was not entitled to special consideration as a landowner, but must be treated like any other subscriber. Steuart had invested in the company because he thought it would pay and because it would help his lands:

> His name & the amount of his subscription may have induced many to become subscribers who had only the first of these motives in view — & tho' it may turn out an unprofitable speculation they have no right & do not attempt to withdraw.[172]

Another claim by Steuart, that the company was not allowed in law to apply revenue from the part of the line already opened to capital works on the remainder was correctly rejected by the company.[173] In fact there was nothing to stop this so long as a company could persuade shareholders that this was a better use of the money than issuing it in dividends.

It turned out that legally the company were on extremely shaky ground. In the lower court Lord Jeffrey found that they had no claim on Steuart until the line reached Allanton, and his verdict was upheld on appeal by the Inner House of the Court of Session. The company was preparing a further appeal to the House of Lords when in 1837 Steuart died. His heirs — Robert Steuart of Carfin, Laurence Hill and Charles Tennant — were all strongly committed to the success of the railway: a compromise was reached by which the subscription calls were paid up, though without any interest charge for late payment, and the company undertook the legal costs of both sides.[174] Fortunately for railways in general, no one else followed up the principles of Steuart's claim, which again suggests that a formal agreement may have been made with him in 1828-29 on which the company were attempting to renegue. The finances of the Wishaw & Coltness also improved; all but 35 of the 1200 new shares created by the 1839 act were taken up by existing shareholders.[175]

VI. The lines in operation

The most arbitrary figure in most railway estimates, if indeed one was given at all, was that for working expenditure. Even in prospectuses, it was neither broken down nor justified, and it was normally either a round sum itself or a figure which would give a convenient round number for the total estimate of net revenue. It was frequently set at about 10% of gross revenue, and experience was soon to show that this figure was unreasonably low. Table 19 gives statistics for the four main Lanarkshire coal lines.

Table 19. *Working expenditure, Lanarkshire railways, 1835–43 (average per year)*

Company	Years	Gross revenue £	Working expenditure £	Net revenue £	Working expenditure as % of gross revenue
Monkland & Kirkintilloch	1835–38	7,882	2,783	5,099	35.3
Ballochney	1835–38	4,846	1,884	2,963	38.9
Monkland & Kirkintilloch	1839–43	18,749	7,489	11,260	40.0
Ballochney	1839–43	13,441	5,050	8,390	37.6
Garnkirk & Glasgow	1839–43	15,337	7,863	7,474	51.3
Wishaw & Coltness	1839–43	11,125	4,105	7,020	36.9

Source: GUEH, 'Explanatory note and details of several statements and calculations contained in the Remarks by the W. & C. Ry Directors on the Interim-report by the Committee of Enquiry'.

Working expenditure was clearly too high, whether in good or bad periods of the trade cycle. Traffic and revenue increased on all these lines until 1841 or even 1842, and, although 1843 was a bad year in all cases, omitting it from the reckoning would make a difference to the percentage of working expenditure of less than $1\frac{1}{2}$ points in all cases except the Wishaw & Coltness. According to the 1844 Committee of Inquiry, working expenditure on the Wishaw & Coltness was then running at 50% when other companies averaged 37%. In their defence, the directors claimed this was 6% less than the Garnkirk.[176] From 1845 to 1848 the working expenditure of the three railways which became the Monklands amalgamation averaged 55%.[177]

Companies clearly had to find some explanation for such expenditure even if only to fend off accusations of incompetence from dividendless shareholders. Sometimes the explanation was obvious enough: if a train from Dundee harbour to Newtyle was pulled at various stages by a horse, three separate stationary engines and two separate locomotives, its running costs would hardly fail to be high. On the coal lines a number of factors were involved. In 1848 George Knight, secretary and general manager of the three Monklands companies as they moved towards amalgamation and James Wright, a mining engineer employed by the Duke of Buccleuch at Dalkeith, indicated the problems of a system of short main lines, numerous branches and sidings, and short-haul traffic: the Monklands complex consisted of 36 miles of railway proper and 12 miles of sidings, and had connected with it another 48 miles of private railways built by the various extractive and industrial interests. Although a through journey of 25 miles was possible on the system — from the eastern end of the Slamannan to the Kirkintilloch canal basin — 30% of all traffic travelled less than a mile, and half of it less than $2\frac{1}{2}$ miles. Hence locomotives were involved in a ceaseless pattern of stopping and shunting, and averaged only 24 miles per day against the 90 miles normal on the Edinburgh & Glasgow.[178]

Over and above this there was a suggestion that traders were getting unreasonably favourable treatment, not surprisingly since Knight's main purpose was to argue for generous maximum rates in the amalgamation act. The sidings were expensive to work, and even private sidings required main line points which had to be renewed every three or four years at the company's expense; these numerous points also meant the employment of a large number of men to supervise them. Traders could also benefit from using company waggons, and were not charged for their use on sidings and private lines. Knight also complained of excessive damage to waggons when under the control of mineowners' employees 'by whom they are much abused', and of long detention of waggons by the traders which meant that they averaged only $5\frac{1}{4}$ miles per day against 23 miles on the Edinburgh & Glasgow.[179] In 1846 Edward Woods, an outside engineer called in to give evidence on a Slamannan bill, confirmed that:

> The time of the actual travel is as nothing compared with the time of waiting to load and discharge and assort and shunt the waggons into sidings and private lines.[180]

Knight appeared envious of not only the Edinburgh & Glasgow, but also of the Garnkirk & Glasgow, since it too collected coal trains which had been made up on other railways and transmitted them over a single through line to the market. Yet the Garnkirk's working expenses were proportionately even higher than those of its rival: the conclusion in this case may be not that the expenses were simply too high, but that the gross revenue was too low and that canal competition was forcing the company to carry at uneconomic rates.[181]

If working expenditure was too high a proportion of gross revenue, there were various possible lines of action. Clearly one was to look for reductions in outgoings, and companies with a long record of not paying a dividend often found themselves being investigated by committees of shareholders intent on cost reduction. The great days for such committees, however, came in the aftermath of the mania in the late 1840s: on the small lines in the 1830s and early 1840s, the emphasis was more on trying to increase income. Naturally all extra costs incurred, whether the capital expenditure involved in building branches or the revenue costs of running more trains, were undertaken in the anticipation that their returns would help to reduce the proportion of working expenditure. But there might also be ways of increasing revenue with no extra costs at all. The most obvious was to raise the rates charged.

All railway acts of Parliament laid down the maximum rates the company were permitted to charge, but these were usually well above the plausible level which the traffic would bear. The room for manoeuvre available to the companies varied considerably, depending partly on the availability of alternative transport and partly on the price at which it would simply not be worth sending goods to market. But it was almost predictable that any application for an act by a company would be opposed by groups of traders who were not in fact averse to the purposes of the bill but who wanted to force a reduction of rates; and it was equally likely that a company faced by inadequate revenue would be tempted to raise charges without close calculation of demand elasticities and the overall effect on receipts. Arguments over rates were an inevitable and recurring feature of railway management.

The evidence presented by the Slamannan in support of its 1846 bill for branches to Bathgate and Jawcraig shows a typical situation. The existing tolls on the line were 3d per ton for the first mile and $\frac{3}{4}$d per mile therafter: on top of this, both locomotive haulage and waggon hire were charged at $\frac{1}{2}$d per ton-mile. The net effect was a rate of about 2d per ton-mile for goods travelling along the whole line, and slightly more for the six-mile trip made by much of the traffic. In response to traders' petitions for lower rates, the company, through consulting engineer James Thomson, stated that their proposed $2\frac{1}{2}$d per ton-mile maximum was in effect also a minimum, and that 'witness considers the demands of certain traders unreasonable and impossible to be complied with'. The company had paid no dividend since 1841, and was substantially in debt. Not only was a general rate increase necessary to keep the line alive, but the short-distance clause, allowing higher charges to cover the increased proportion of handling costs involved in traffic which moved only a short way, should be extended to cover journeys of up to three miles. Experiments with lower rates in the previous year had not increased the traffic, and there was no reason to believe that they would do so now.[182] Scenes like this were conducted before parliamentary committees with monotonous regularity. In 1848 the whole Monklands group was claiming the need for higher rates for short-distance traffic: this, they said, would allow a reduction in the high charges made for haulage and waggon hire, which in any case failed to catch traders using their own horses and equipment.[183]

If companies frequently miscalculated construction costs and working expenditures, they also fortunately often under-estimated the volume of traffic. There was no way in which traffic estimation could be an exact science, although gallant efforts were made, particularly in parliamentary estimates, to make it seem so. Some facts could be established with more or less certainty. The number of passengers using existing public transport by road or water could usually be ascertained, as indeed could some idea of the amount of goods moved by existing carriers. If public statistics or the carriers themselves could not or would not divulge the information, it was common for railway projectors to conduct their own censuses of road traffic. Again, some estimates of existing demand were available, such as the coal consumption of Glasgow or Edinburgh, and on the corresponding supply side mine-owners and ironworks proprietors could furnish figures of output and sales.

Beyond this were much greyer areas. How much of this existing traffic would actually transfer to a railway? And how much more traffic might be created by the existence of a railway, either because it opened up an area previously without transport, or because it offered advantages not possessed by its rivals? Would the known if unquantified demand for lime and coal in Strathmore bring prosperity to the Dundee & Newtyle? Would the landowners of the Monklands sink pits because the Slamannan was made? Would they be willing to pay more for railway speed and the absence of transshipment on the Garnkirk & Glasgow, or would they stay with the canal? Although the views of potential traders and customers could be canvassed, the final estimates were often little more than educated guesses. And the degree of optimism or pessimism in these estimates depended in turn on prevailing economic conditions and the view taken of future prospects. With all these possibilities for error, it is not surprising that only too often the companies got it wrong.

With hindsight, it appears the projectors of the earliest railways tended to be cautious, though they did assume that they would receive all the traffic previously carried over the same route by road. They were less sure of an advantage over canals, although they usually claimed that they would take most of their light and much of their heavy traffic. But for many it took time to realise that the mere existence of a railway might generate traffic which had not formerly existed. Although this idea had been specifically enunciated by, for example, Robertson Buchanan in 1811,[184] it was not until the 1830s that it became a commonplace assumption. The railways of the 1820s clearly expected more minerals to be moved after they were built than before, since that was the reason for their construction, and allowance might easily be made for resources such as lime or ironstone on the line. But it was more difficult to visualise the expanding trade in general goods and passengers, and these tended to appear in estimates in terms of the already existing demand on the route. As more railways, and particularly more passenger-carrying railways, were opened, estimates for traffic on new projects were adjusted upwards to reflect the unexpectedly large demand that appeared.

All these railways were primarily constructed for the transportation of goods, and in almost all cases this remained the major part of their traffic: some indeed long remained almost solely mineral lines. But to the outside observer who was not conveying or receiving goods traffic, a very conspicuous development of the early railways was in the carriage of passengers. Of the lines authorised in the 1820s, only the Garnkirk & Glasgow and the Dundee & Newtyle carried passengers from opening, and only the latter anticipated that they would be an important part of their revenue[185] — perhaps because it was not so obvious as on a mineral line where the other traffic was to come from. In their report to the Wishaw & Coltness projectors in 1828, Grainger and Miller had been almost diffident in suggesting that £400 per year, 'which may by some be thought large', might come from passengers in a total revenue of £5054. But the Stockton & Darlington had convinced them that 'railways are found to be remarkably suited for the cheap and expeditious conveyance of passengers';[186] the Liverpool & Manchester was about to convince a great many other people.

The initiative for passenger services on the horse-drawn Edinburgh & Dalkeith came not from the board but from 'an enterprising individual connected with their establishment' in May 1832. By the end of the year *Chambers' Edinburgh Journal* hailed a success:

> The great utility of railways, and their productiveness in a pecuniary point of view, have just been exemplified at Edinburgh, where a railroad, formed for the purpose of introducing coal to the city from pits a few miles distant, has been covered with vehicles for the conveyance of passengers to all parts of the adjacent country. A similar degree of success has attended a railway in the west of Scotland...[187]

In the first year of service, the Edinburgh & Dalkeith carried 150,000 passengers for a revenue of almost £4000.[188] The 'railway in the west' was the Garnkirk, where passengers had been carried before the official opening, and where not only did a regular passenger train leave Coatbridge for Glasgow every morning and return in the evening, but:

> Each time, also, the engine starts with a load of coals from the upper part of the line, or with empty waggons returning, a small passenger waggon is attached, not being regulated by any hour; and a considerable number of stragglers find their way in this manner along the line.[189]

Table 20. Non-road Passenger Traffic between Glasgow and Destinations to the East, 1830-35

Date	Monkland Canal		Forth & Clyde Canal		Garnkirk & Glasgow Railway		Total
	No.	%	No.	%	No.	%	No.
1830	20,923	21.3	77,449	78.7	—	—	98,372
1831	25,129	22.1	88,571	77.9	—	—	113,700
1832	23,589	14.8	72,845	45.8	62,605	39.4	159,039
1833	28,332	11.7	117,086	48.5	96,003	39.8	241,421
1834	31,632	10.0	166,077	52.6	117,743	37.3	315,452
1835	27,605	7.8	189,664	53.6	136,724	38.6	353,993

Source: PP HL 1837-8(185) XX, *Select Committee on Edinburgh & Glasgow Bill*, evidence, pp. 9-10.

By 1834 the company was running four passenger trains per day, with round trip fares of 9d and 6d for 16 miles.[190] Table 20 indicates that from the start the Garnkirk line attracted almost two-fifths of non-road passenger traffic on the east side of Glasgow. There is no evidence as to how much of this was attracted from the roads rather than from the canals: the total growth of demand, however, was such that, except in the railway's first year, the canals suffered no absolute loss of passenger traffic.

The other Monklands lines realistically did not attempt to compete with the Garnkirk — for passengers, a roundabout route including a change on to a canal boat could not be made attractive, whatever might be the situation for minerals — and neither the Monkland & Kirkintilloch nor the Ballochney ran passenger trains on their own account. Indeed, to discourage private enterprise by traders on the line, the Ballochney byelaws of 1836 stated that passenger traffic would be allowed only if written permission had been received.[191] By the 1840s, however, both lines were receiving some passenger income as trains belonging to the Slamannan and the Garnkirk & Glasgow passed over parts of their lines. Doubts elsewhere were typified by the Ardrossan, where passenger trains started in 1834, 'but such was the uncertainty of its succeeding by running at regular hours, that it was kept to be hired for pleasure trips'.[192]

It is not always easy to determine the importance of this traffic to the companies, particularly in those cases where it was clearly a relatively minor part of the whole. The records kept by the companies in the 1830s have for the most part not survived, and even when statistical returns were required by the Board of Trade from 1841 onwards, small companies not infrequently failed to provide them. For all the evidence offered by the Newtyle & Glammis or the Polloc & Govan to the Board, they might never have existed. Other companies were erratic in their returns: the Wishaw & Coltness, for instance, after recording that 6.9% of its revenue came from passenger traffic in the eighteen months from July 1842, then ceased to make any returns of passengers at all. The Ballochney sometimes made a separate note of its receipts from Slamannan passenger traffic and sometimes did not. In the year from July 1841 to June 1842, it even appeared that no Scottish company in the second half of the alphabet subscribed returns to the Board, but this is presumably an omission by the Board's compiler.[193] And where more than one record of particular traffic exists, the figures are not always in agreement, although they are seldom unreasonably far apart. The keeping of absolutely precise records was not a matter to which companies operating with very small managerial staffs always paid full attention, and there was no attempt by the Board of Trade to enforce tighter standards. Given these caveats, Table 21 gives some evidence on the relative importance of goods and passenger traffic.

The figures indicate the extent to which passenger traffic became important even on these short mineral railways, particularly if they were part of a direct route to a city. Admittedly only on the Newtyle & Coupar Angus, the Paisley & Renfrew and, rather surprisingly, the Edinburgh & Dalkeith did passengers ever amount to more than half of the total revenue, but at the other end of the scale only the lines extending from the Forth & Clyde canal and the Dunfermline & Charlestown — the updated Elgin waggonway — had less than a quarter of their income coming from passengers. The fact that, of the three most passenger-orientated lines, the Edinburgh & Dalkeith and Newtyle & Coupar Angus were always horse-drawn and the Paisley & Renfrew's flirtation with mechanical power was brief, cannot be taken as a condemnation of locomotives: the Renfrew line had only to compete with the unsatisfactory River Cart, and the Dalkeith, though scarcely faster than road transport, offered a smoother and, at a fare of $\frac{3}{4}$d per mile, a competitively priced ride. On locomotive lines, average speeds of passenger trains were broadly related to the relative importance of passenger traffic (see Table 22).

Traffic of course fluctuated for various reasons. Some variations were due to the development of the railway system itself. The Slamannan's *annus mirabilis* of 1841 came to an abrupt halt in February 1842 when the Edinburgh & Glasgow opened and took away the through traffic.[194] The Paisley & Renfrew was severely hit by the opening of the Glasgow Paisley & Greenock. On the other hand, the growth of passenger traffic on the Ardrossan dates from its link to the Glasgow Paisley Kilmarnock & Ayr, and the associated development of steamer services to Ireland and Fleetwood.

More general variations came from changing economic circumstances and fluctuations in industrial activity. Although all forms of traffic would be affected by depression, the effects would be most severe on goods traffic, as firms reduced their demands for transport or even closed down altogether. The closure of many ironworks in the early 1840s was a severe blow to railways largely dependent on moving coal and iron. The Ardrossan directors, explaining the absence of a dividend after the exceptionally depressed year of 1842, blamed it primarily on 'the universal stagnation of trade and commerce which has prevailed throughout the country for the last twelve months', which in their case had closed the ironworks at Dalry and Kilbirnie. Similar sentiments were echoed in directors' reports across the country, although the Ardrossan was also faced with another unique cause of falling revenue:

> In consequence of the almost universal stoppage of the Irish distilleries and breweries, resulting from the abstinent habits lately cultivated to so great an extent in Ireland, the demand for coals for these works, which used to constitute the principal portion of your coal traffic, has all but ceased.[195]

In such depressed circumstances, the less variable passenger traffic would assume a proportionately greater importance.

Table 21. Traffic on early railways, 1842–44

Company	Passengers		Goods		Total	Passenger receipts as
	No.	£	Tons	£	£	% of total
Ardrossan	193,653	5,217	140,685	7,340	12,557	41.5
Ballochney	n.a.	806	1,075,896	31,085	31,891	2.5
Dundee & Newtyle	167,825	8,038	96,418	10,812	18,850	42.6
Dunfermline & Charlestown	37,488	874	84,011	9,302	10,176	8.6
Edinburgh & Dalkeith	807,779	16,438	321,104	14,706	31,544	52.1
Edinburgh & Dalkeith Leith br.	207,625	2,629	74,522	2,461	5,090	51.7
Garnkirk & Glasgow	641,434	14,699	724,234	27,920	42,619	34.5
Monkland & Kirkintilloch	—	—	3,185,020	44,153	44,153	—
Newtyle & Coupar Angus[a]	53,736	1,274	19,456	1,194	2,468	51.6
Slamannan[a]	35,502	1,740	130,771	4,425	6,165	28.2
Wishaw & Coltness[a]	117,413	3,581	1,184,970	30,164	33,745	10.6
Paisley & Renfrew[b]	38,715	645	n.a.	384	1,029	62.7

a: July 1842 to Dec 1844 only
b: July 1842 to June 1843 only

Sources: PP 1843 [440] XLVII, *Third Report of the Railway Department*, 28 Feb 1843, appendix II; PP 1844 [551] XLI, *Report of the Railway Department for 1843*, appendix II; PP 1846 [698] XXXIX, *Report of the Railway Department for 1844–45*, appendix II.

For most lines, however, the concentration of effort was on increasing and encouraging mineral traffic. This led to complicated negotiations between companies and prospective traders: both clearly had much to gain from an expansion of rail-borne traffic, but the balance of advantage would depend on the rates and conditions under which it was carried. Negotiations were often given added piquancy by the fact that a director in his private capacity might find himself bargaining with the rest of the board of which he was a member. Matters to be settled included both the actual construction of railway facilities, and the terms on which they might be used.

Table 22. Average Train Speeds, including stops, 1845

Company	Passenger trains m.p.h.	Goods trains m.p.h.
Ardrossan	22	6
Garnkirk & Glasgow	20	8
Wishaw & Coltness	15	8
Slamannan	14	14
Ballochney	12	5
Edinburgh & Dalkeith[a]	10	4
Dundee & Newtyle	9 [b]	6
Newtyle & Coupar Angus[a]	8	$2\frac{1}{2}$
Dunfermline & Charlestown[a]	7	3
Monkland & Kirkintilloch	n.a.	$5\frac{1}{2}$

a: horse-drawn
b: 24 m.p.h. excluding inclined planes

Source: PP 1846 [698] XXXIX, *Report of the Railway Department for 1844-45*, appendix II.

One device resorted to on occasion by early railways was the wayleave — an annual payment, varying with and often directly proportional to the amount of traffic, made to a landowner for the right to cross his land. The system could have short-term attractions for a financially pressed company, since although they were burdening their revenue account with the wayleave payment into the indefinite future, the sum which they had to pay out immediately for the land would be much reduced. It also might be the only condition on which a landowner such as Sir Robert Keith Dick of Prestonfield, 'whose land is not likely to be feued, or to be dug for coal or quarries . . . or to have its value enhanced by the vicinity of the Railway', would surrender the necessary land. The wayleave negotiated by Sir Robert with the Edinburgh & Dalkeith varied from £490 per year if the traffic carried was under 500 tons per day to a maximum of £990 for over 900 tons per day. The company also granted four landowners the right to free use of the railway on their own land; two of these also received a wayleave payment of $\frac{1}{2}$d per ton on all traffic.[196]

By the 1840s, however, wayleaves were unpopular with the companies. On a short line, with only one or two adversely affected landowners to be placated, the payments might not be too great a drain on company revenue, although the steady increase in traffic made them increasingly burdensome. On a longer line, and particularly on one where many landowners were not looking to the exploitation of minerals, numerous wayleaves would be impracticable. No company aiming at substantial passenger traffic ever considered them, and the mineral railways became disillusioned. Thus the Wishaw & Coltness had negotiated a wayleave with Campbell of Islay, which had cost them only £181 from the opening of the line until November 1836, but had cost £161 in the following year as traffic increased. In 1840 they bought it out for £12,000 as it otherwise 'would have continued a serious and permanent burden upon the trade of the Railway'.[197] By 1846 the Monkland & Kirkintilloch was indignant in its refusal of a proposal by the hard-bargaining firm of Bairds of Gartsherrie that the projected Chapelhall branch should pay a wayleave of, at the Bairds' option, either £100 per year or 1d per ton, and rejected the whole concept of a wayleave 'in such a district as this'. Indignation was fuelled when Bairds also attempted to deny the use of the branch to rival coal and ironmasters: the company:

> regret to find that the Petitioners [Bairds], who have been more benefitted than any others by the formation of the present line and its connections through the properties of other Mineral Owners should attempt to throw impediments in the way of other Coal and Iron Masters connecting themselves with the Monkland & Kirkintilloch Railway ... Their demand is unprecedented, and if such a principle were recognised it would be impossible for the Company to carry on the system of connecting the neighbouring Mineral Fields with the extensive Iron Works on the line, and with Glasgow.[198]

Another area of negotiation concerned the terms on which a company would build a branch to the property of an individual mineral proprietor. Such a man might well build a private railway, for which parliamentary authorisation was not required, to join a public line which crossed his own land. He was, however, unlikely to build beyond his own land; even if the permission of neighbouring landowners were obtained (and since they would almost certainly be rival mineral owners it might well be refused), a substantial wayleave might be charged — and of course the costs of maintenance and operation would increase with the length of the line. Hence it was in his interest to persuade a public railway company to extend a branch to his property. Equally naturally, the company wanted reasonable assurance that the forthcoming traffic would pay them a fair return on their investment. Sometimes, particularly with main lines, they would build into undeveloped mineral areas on the general expectation that they were bound to gain a large traffic; on other occasions they tried to make their returns more certain.

In 1831, for instance, William Dixon was opening up a new ironstone field at Whiterig. The Ballochney company gave him a branch to it only on condition that he agreed:

> to do business on the Branch to the full extent of the outlay within a given period. The arrangement is such as to make this part of the road productive of a larger income to the company compared with the outlay than any other portion of the line.[199]

By 1836 the company had made this a general principle — branches would be made:

> on the usual understanding that the principal sums and interest laid out in their construction shall be reimbursed by the traders within a certain space from tonnage dues.[200]

In a similar way, the completion of the original main line of the Wishaw & Coltness, after the company had run out of money halfway, was prompted by the purchase of the Coltness estate in 1836 by Manchester entrepreneur Thomas Houldsworth. In order to allow the development of his minerals, Houldsworth agreed that if the company extended their line to Coltness, for which they would require to borrow £20,000 and to get an act allowing an extension of time for construction, he would personally guarantee the interest on the loan and pay £300 towards the act. He also agreed to give the necessary land on the Coltness estate free (apart from compensation to tenants), and to send all his goods by the railway if the company charged him the prevailing rate. Between them the two parties also bought the right to run locomotives over Jerviston estate. In view of this evidence of the value put on the railway by a member of one of the great ironfounding families, the next general meeting of the company agreed to extend the line 'to Coltness or as far as their funds would allow them'.[201]

Once the physical link had been provided, there remained the question of the rates to be charged. Within the maxima authorised by their acts, railway boards had freedom to set charges at the levels they thought the traffic would bear, though they could not discriminate between equal customers. As might be expected, the traders conducted a steady campaign, at varying levels of activity, for rate reduction in any case where the railway was not clearly unprofitable. The Ballochney, for instance, faced campaigns to have the rates reduced by the traders on the line in 1843, by John Wilson of Dundyvan Ironworks in 1846, and by a group of major ironmasters in 1848. On each occasion the campaign arose because of the opportunity to oppose in Parliament a bill prompted by the company.[202] The boards occasionally adjusted the charges in a more or less trial-and-error manner to maximise their revenue. But particular attention had to be paid to major customers, and the surviving records of the companies are full of usually incomplete references to negotiations with the great coalowners over rebates and guarantees.

The Wishaw & Coltness provides some typical examples. In 1841 the manager, Robert Dodds, reported on the traffic of the Castlehill ironworks, which was two miles from a section of railway about to be opened, and which was sending iron to Glasgow by road for 7/6d per ton. By rail it could not be done for less than 6/11d, of which only 2/1¾d would come to the Wishaw & Coltness (even including waggon hire), while 3/- was required for carting to the railway and in Glasgow, and the rest would go to other railway companies on the route. A sevenpence advantage was unlikely to make up for the inconvenience of twice loading and unloading the goods: yet Dodds rejected the idea of offering a 6d reduction in the rate, since it would also have to be given to the Coltness proprietors who were content with existing proposals.[203]

In 1844 came a move by the great firm of Bairds to play off the railway against the canal. If rates were cut to 1d per ton mile, they would abandon the idea of making a line from their works to the Monkland Canal, or of supporting a branch canal. This brought into question the tapering scale of charges, which companies used to encourage traffic from the end of their lines farthest from the markets. Dodds' conclusions about the effects of three possible rating systems are indicated in Figure 1.

The second system would give the same rates as existed for the substantial amount of traffic which travelled for four miles, including that from the Summerlee and Carnbroe ironworks, but would give much higher charges for traders at the Coltness end of the line, including company chairman Lord Belhaven. The third would satisfy Baird's demands, without offending the longer-distance users.

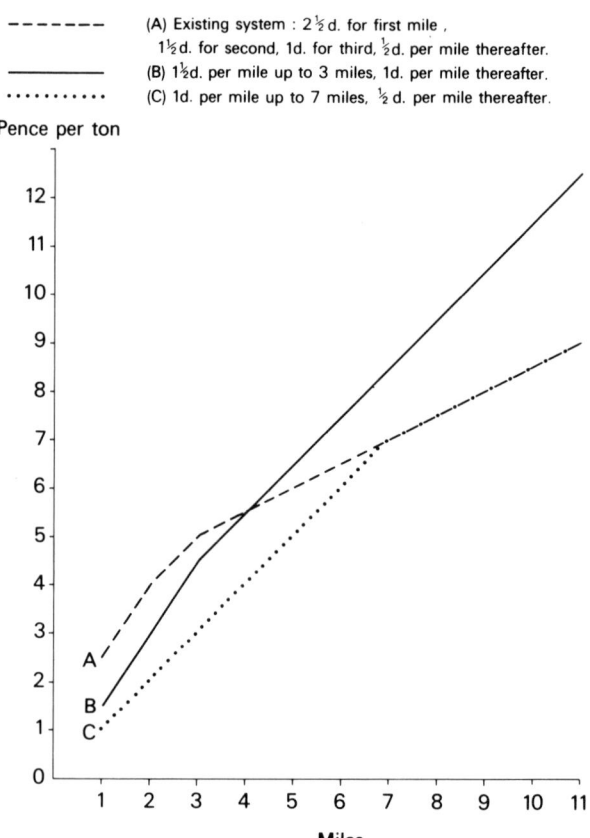

Figure 1. Wishaw & Coltness Railway: Effects of Possible Rating Systems, 1843

Rating system	Income on 1843 traffic volume £	Loss if change made £
A	6490	—
B	3908	2582
C	2582	2943

Source: GUEH, Robert Dodds to Wishaw & Coltness Directors, 16 Jan 1844.

But both involved a loss in revenue greater than any losses that could come even if all Baird's needs were satisfied by canal.[204] Too much of the line's traffic was short-distance for concessions to be made on it, as is indicated by Figure 2.

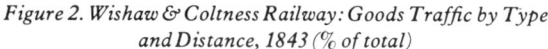

Figure 2. Wishaw & Coltness Railway: Goods Traffic by Type and Distance, 1843 (% of total)

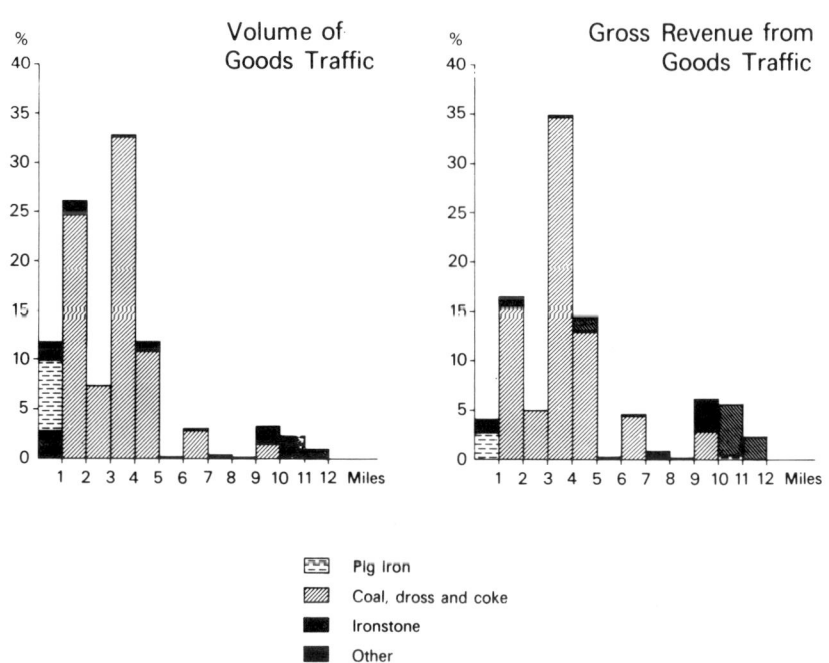

Sources: PP 1844[551]XLI, *Report of the Officers of the Railway Department for 1843*, appx. II; PP 1846[698]XXXIX, *Report of the Railway Department for 1844-45*, appx. II.

In 1845 a more complex suggestion had to be referred to the company's lawyers for their opinion. Another major ironmaster, Wilson of Dundyvan, had offered to develop a new coalfield and send at least 50,000 tons per year along seven miles of the line, or pay the toll for any deficiency, if he was guaranteed fixed rates for 19 years. The offer was obviously extremely attractive, being equal to over 10% of the total tonnage carried in 1844 and travelling over almost two-thirds of the line, but could the company accept it without having to give similar concessions to other traders? There was a clause in their original act allowing them to lease the tolls and dues of the line for up to 19 years: could this be used to lease only Wilson's own traffic to him? The lawyers were discouraging: the company could only lease the whole dues on any particular article carried, whoever sent it, and any contract to convey a particular trader's traffic at given rates would have to be extended to his rivals. Opening up the issue might even bring the tapering scale into question and lead to much trouble. Another interesting idea had to be passed over.[205] Instead a general increase in rates was applied.

These railways were now at a stage of transition in their relations with their customers. The early waggonways, as still exemplified at this time by the Polloc & Govan, were in general answerable to and controlled by a single owner. At the next stage, companies like the Edinburgh & Dalkeith, or the Ballochney in its early years, were faced by a few major traders, who might well be on the board or have their representatives there. Concerted pressure from these traders would be difficult to resist, and certainly any concession given to one would have to be given to the others. But the figures for the Wishaw & Coltness in Table 23 indicate a growing problem. The five largest traders still account for over half the traffic, and the top ten for almost two-thirds, but there are now a large number of smaller operators each with less than 1% of the total. Although the company could hardly afford to offend the Bairds, who had almost 16% of the traffic and who could change their source of supply, they could not afford either to make concessions if these would also have to be given to all other users — hence the elaborate search for legal loopholes in the acts. At the next stage, the larger inter-urban lines of the 1840s would not be nearly so open to pressure from individual customers simply on the grounds that their trade was too important to lose.

Table 23. Examples of dependence on particular traders

Edinburgh & Dalkeith 7–12.1839	Tons	%	Wishaw & Coltness 7–12.1843	Tons	%
Sir John Hope	47,971	39.57	William Baird & Co.	35,299	15.92
Marquis of Lothian	40,908	33.75	John MacAndrew & Co.	24,494	11.05
A. & M. Stenhouse	14,258	11.76	James Merry	20,732	9.35
Executors of R. Dundas	12,139	10.01	Stevenson Coal Co.	20,152	9.09
Duke of Buccleuch	5,944	4.90	Summerlee Iron Co.	13,459	6.07
			John Watson	11,270	5.08
			Robert Steuart	7,585	3.42
			Carnbroe Iron Co.	7,413	3.34
			Lord Belhaven	2,466	1.11
			George Inglis	2,348	1.06
			Others	78,875	35.57
	121,219	100.0		221,743	100.0

Sources: SRO, GD 224/554, Buccleuch Muniments, Bundle 'Edinburgh & Dalkeith Railway 1834 to 1841', Minute of Edinburgh & Dalkeith meeting 5 Aug 1840; GUEH, 'Table showing the Abatement of Tolls to the Traders upon the Wishaw & Coltness Railway, during 1843'.

3
From Town to Town: Planning and Authorisation

I. Expansion in the 1830s

When the Liverpool & Manchester Railway opened in 1830, the political and economic prospects were uncertain. But in the first half of the decade the fears of revolution spreading from France, the agricultural distress of 1830 and the consequent riots, and the clamour over the Reform Bill all were appeased or died away. Instead the full prosperity of maturing industrialisation was becoming apparent. The growth of industrial output was more rapid than at any time in the century, trade was buoyant, the harvest — that crucial influence on prosperity even long after the Industrial Revolution — was good for several years in succession, and a reforming Whig government was tackling a range of issues from the slave trade to local government reform, from an embryonic factory inspectorate to the new Chadwickian poor law. Confidence was rising, profits were being made, and government stock conversions in 1830 and 1834 which made the return on consols an unexciting $3\frac{1}{2}\%$ encouraged reinvestment in commerce and industry.[1] The success of the Liverpool & Manchester, indicating the potential for railway traffic in general goods and passengers as well as minerals, helped to divert some of this money to railways.

The first and lesser British railway mania appeared in 1835, about the peak of the cyclical boom. The greatest number of acts authorising new lines came in 1836 (see Table 24), the summer of which year also produced the greatest number of new prospectuses for lines many of which were bubbles while others were overtaken by financial pressures and investors' caution as the cycle turned down.[2] But railway growth was not simply a product of high prosperity and the top of the trade cycle. The planning of a substantial number of major English lines had started very early in the decade, in spite of sceptics like J.R. McCulloch, who as early as 1832 doubted 'whether there be many more situations in the kingdom where it would be prudent to establish one'.[3] The inherent lags in the authorisation process — the time needed for planning, surveying, finding subscribers, and in some cases the loss of a year by failing to pass Parliament at the first application — meant that the first major acts, for the London & Birmingham and the Grand Junction, came in 1833, followed by the London & Southampton in 1834, the

Great Western (after three years of effort) in 1835, and in 1836 the South Eastern, the Eastern Counties, the Manchester & Leeds and a number of future components of Hudson's Midland. Pollins considers that these lines 'were put forward at an inappropriate time', before political and financial conditions were fully prepared for them;[4] if so, their success with both investors and Parliament would be one more stimulus to rapid expansion when conditions were more appropriate.

Table 24. Railway Acts 1835-37

Date	England & Wales		Scotland		Ireland		U.K.	
	Acts	For new lines	Acts	For new lines	Acts	For new lines	Acts	For new lines
1835	14	12	5	5	1	0	20	17
1836	29	24	4	3	2	2	35	29
1837	32	13	6	2	4	3	42	18

Source: PP 1867 [3844] XXXVIII, *Select Committee on Railways*, appendix EK.

The English lines of 1835-37 were developing not only on the proven success of the Liverpool & Manchester but also on the confidence engendered by these large-scale schemes. By the time they were completed, the main framework of the English railway system, with the principal exception of the Great Northern, was in place. As early as November 1835, the *Circular to Bankers* could suggest, while indicating the main geographical inspiration of the railway expansion, that railways were already a settled and familiar part of the economic system:

> The Rail-way system advanced and became established in the public confidence almost wholly without the assistance of the Stock Exchange. The support afforded to it was derived almost exclusively from the capitalists and men of thrift and opulence in the mining and manufacturing districts of the north of England.[5]

As a result of the first mania, railway construction was by the end of the 1830s among the most important areas for investment in England, and Parliament was already concerned about:

> the difficulties that must arise from an extended inter-communication throughout the country, solely maintained by Companies acting for their private interests, unchecked by competition, and uncontrolled by authority.[6]

The Scottish experience was different. No new company was authorised between 1832 and 1834, and in 1835 there came only four insignificant extensions of the existing small system. In terms of acts of Parliament, the Scottish main lines start at the earliest in 1836, with the first inter-urban lines in Angus, and the major early link — the Edinburgh & Glasgow — was only authorised in 1838. Nor did anything like a Scottish network appear, the companies being limited to the area between Edinburgh and the Firth of Clyde and the outlying small group in Angus. It is not unreasonable to say that Scotland had to wait until the great mania of the mid-1840s to achieve what England had managed a decade before. Scotland had not been backward in the 1820s in the use of railways to open up mineral fields, but there is at first sight a reluctance to extend the concept in the 1830s to inter-urban services aimed at general goods and passengers.

It may be, as Francis Whishaw thought in 1840, that traditional Scottish caution played a part:

> The people of North Britain are proverbially cautious, but they are also sagacious and persevering. A long time elapsed before they would venture on so large a project as a grand trunk railway; but having carefully watched the movements of their brethern of the south in respect to these important results likely to accrue from their general introduction in Scotland, they have commenced operations in earnest . . .[7]

It was more to the point that Scotland was a poorer country than England, and that there were other pressing claims on such investment capital as was available. Between 1825 and 1841, for instance, eleven new ironworks opened in the west of Scotland, largely to exploit the new possibilities of the hot-blast process: even in the depressed years from 1837 to 1843, almost sixty new furnaces were started. On the Clyde the steamship boom was under way, with production rising from 17,600 tons in the 1830s to 81,400 tons in the following decade (a rise also from 24% to 66% of total British production). The ideas associated with James Smith of Deanston were leading to widespread and expensive drainage schemes in agriculture. With all this Scottish resources could not also finance large-scale railway building, and the attractive prospects being claimed by English companies (the Eastern Counties, for instance, forecast a dividend of 22%) made it more difficult to attract English capital into the relatively unfamiliar country north of the border.[8]

The demand for railway accommodation was also less in Scotland. As the *Scotsman* observed, 'Scotland generally is an unfavourable country for Railways. Its surface is too uneven, its population too thin, and its large towns too distant from one another'.[9] The booming industrial sector was limited to the central belt, and largely to its western half, and for the most part required little more than an extension of the Monklands type of railway provision. Over most of the country, including everywhere north of the Highland line, there was no apparent way in which a railway could generate enough traffic to make a profit, particularly as it was feared that the nature of the terrain would inevitably mean high construction costs. The railway requirements of the Scottish economy in the mid-1830s were, firstly and above all, a link between the major centres of Edinburgh and Glasgow; secondly, more lines to carry minerals, industrial products and, increasingly, passengers in the expanding west; thirdly a link to English markets (as long as it was cheaper and more reliable than transport by sea); and fourthly an east-coast extension to Dundee and Aberdeen — a need felt less by the industrial centre than by the northern towns, which were not entirely satisfied either with east-coast shipping services or with the often dilatory connection to Glasgow provided by the Forth & Clyde Canal.

II. The essential link: Edinburgh to Glasgow

The route between Glasgow and Edinburgh was almost as obvious a candidate for railway construction as that between Liverpool and Manchester. In 1831 the two great Scottish cities, one the focus of expanding industrialisation, the other the centre of government and the professions, contained respectively 202,000 and 162,000 inhabitants, or between them 15.4% of the Scottish population. The intervening counties — Lanarkshire, Stirlingshire, West- and Midlothian — contained a further 632,000 people,[10] uncalculated resources of coal, extensive lime works, and a booming iron industry. The influence of a railway, particularly if it could link Leith and the Broomielaw, would extend beyond its termini to overseas trade. If taken in a straight line, the route was only moderately hilly, and by curving to the north an almost level line could be achieved. Trade along the route was inevitably substantial, in spite of the delays and the expense of both road services and the Forth & Clyde Canal. According to one group of railway projectors (who had of course an interest in denigrating existing services), the road carriers took anything from 18 to 96 hours over the journey, and charged from 16/- to 40/- per ton. Waterborne trade from eastern Scotland to western England or Ireland either went by sea round the north of Scotland 'with a sea-risk actually greater than to the West Indies', or used the canal 'at an expense more than equivalent to the risk and expense of the more circuitous route'.[11] George Stephenson, advising the projectors, added his condemnation of 'the dilatory, uncertain, and expensive mode of transit, afforded by the existing monopolist of the carrying trade, the Forth and Clyde Canal'.[12]

The first survey for a railway, after Stevenson's very undetailed examination in 1817, was conducted by James Jardine in 1825-6, at the time of the promotional boom following the repeal of the 1720 Bubble Act — perhaps 'the first cyclical boom of the modern sort' — which *inter alia* almost doubled the number of British railway companies in two years.[13] In 1827 one commentator estimated that a rail link between the cities could pay on coal traffic alone. A line through the northern part of the coalfield, charging $\frac{1}{2}$d per ton-mile, could deliver top-quality coal from Benhar to Edinburgh for 7/- per ton, against current prices varying with quality from 11/- to 17/-, and could expect to appropriate at least two-thirds of the 200,000 tons used annually in the city. Similarly Glasgow used 600,000 tons annually, as well as exporting substantial amounts to Ireland, and the railway might anticipate a traffic of 300,000 tons. From these sources alone the return would be £15,750; thus the line could cost £315,000 to build and still give a 5% return on capital. And the Stockton & Darlington had cost £150,000 for 23 miles plus branches. The case, he suggested, was proved, even without any allowance for lime, ironstone, agricultural produce or passengers. Even although he had also failed to allow for the costs of running the line, he could plausibly claim that:

> It is only, we have no doubt, owing to the late melancholy stagnation in trade, that the work, as well as others of a similar kind . . . have not already been commenced.[14]

In fact, although the downturn from 1826 might well have discouraged the projectors, it was not clear that some of them really wanted a railway at all. Jardine's survey was commissioned by a committee under the Duke of Hamilton which included a number of men closely associated with the Forth & Clyde Canal. There was no apparent reason why these men should promote a project which threatened

Table 25. Railway Companies Authorised 1836–43

Company	Act to 1844	Date of act	Route	Miles	Date open	Engineer	Gauge	Capital authorised (£) Share	Loan	Total	Estimate	Construction Costs (£) p.mile	Actual	p.mile
Dundee & Arbroath	6 & 7 Will.IV c.22	19.5.36	Dundee–Carnoustie–Arbroath br: to Almericloss, Arbroath (Arbroath & Forfar)	16.75	6.10.38	T. Grainger & J. Miller*	5'6"	100,000	40,000	140,000	87,790 + br. 12,054	5,364	153,100	9,140
	5 & 6 Vict. c.83	18.6.42						50,000	10,000	60,000				
Arbroath & Forfar	6 & 7 Will.IV c.34	19.5.36	Arbroath–Forfar br: link to Dundee & Arbroath at Almericloss	15.25	24.11.38 part 3.1.39 all	T. Grainger* & J. Miller	5'6"	70,000	35,000	105,000	62,750 + br. 5,709	4,489	136,700	8,964
	3 & 4 Vict. c.14	3.4.40						50,000	5,000	55,000				
Edinburgh Leith & Newhaven (from 1844, Edinburgh Leith & Granton)	6 & 7 Will.IV c.131	13.8.36	Canal St., Edinburgh–Canonmills–Newhaven br: Heriothill–Leith harbour br: Newhaven–Newhaven harbour br: Newhaven–Trinity Chain Pier br: Newhaven harbour–Trinity Chain Pier	2.50 c.2.5	31.8.42 Scotland St.–Trinity 19.2.46 to Granton 10.5.46 to Leith 17.5.47 tunnel	T. Grainger* & J. Miller (by 1840, J. Macneill)	4'8½"	100,000	40,000	140,000		c.20,000 max	310,000	62,000
	2 & 3 Vict. c.51	1.7.39	abandoned: brs. to Leith and Newhaven harbour	(–c.2)										
	7 & 8 Vict. c.81	19.7.44	br: Trinity–Granton harbour br: Water of Leith bridge–Leith harbour	2.08				73,300	24,433	97,733				
Glasgow Paisley & Greenock	7 Will.IV & 1 Vict. c.116	15.7.37	Tradeston, Glasgow–Paisley–Greenock brs: Bogle St. and Virginia St., Greenock–Greenock harbour br: Port Glasgow harbour	22.56	15.7.40 joint line to Paisley 3.41 all	J. Locke & J. E. Errington (original survey by T. Grainger)	4'8½"	400,000	133,333	533,333	383,000 + brs. 10,000	20,684	780,287[a]	41,068[a]
	3 & 4 Vict. c.107	23.7.40	abandon all branches in 1837 act. br: Mansion House–Custom House, Greenock br: Chapel St. Greenock–E. India quay br: Port Glasgow harbour (realigned)											

Railway	Act	Date of Act	Plan date	Description	Mileage	Opening	Engineer	Gauge	Capital				Cost	
Glasgow Paisley Kilmarnock & Ayr	4&5 Vict. c.5 6&7 Vict. c.49	6.4.41 27.6.43	15.7.37	lease to West Ferry (to Dumbarton) Tradeston, Glasgow–Paisley–Dalry–Ayr br: Dalry–Kilmarnock br: Cochran Mill–Over Johnstone br: Woodside–Byres (Ardrossan) br: Fullarton–Drybridge (Kilmarnock & Troon) br: Fullarton–Irvine harbour br: Drybridge–Troon harbour	40.25 10.25 c.4.5	5.8.39 Ayr–Irvine 15.7.40 joint line 21.7.40 main line 4.4.43 Kilmarnock br.	T. Grainger & J. Miller*	4'8½"	100,000 150,000 625,000	33,333 50,000 208,300	133,333 200,000 833,300	583,900 + brs. 39,389	11,230 1,029,663[a]	21,908[a]
	3&4 Vict. c.53	4.6.40		abandoned: Over Johnstone br. br: Woodside–High Monkcurr (link between Ayr and Kilmarnock branches)										
	5&6 Vict. c.29	13.5.42							312,500	104,100	416,600			
Edinburgh & Glasgow	1&2 Vict. c.58	4.7.38	44.75	Haymarket, Edinburgh–Falkirk–Queen St., Glasgow br: at Falkirk		21.2.42	T. Grainger & J. Miller*	4'8½"	900,000	300,000	1,200,000	850,000	18,994 1,029,000	22,370
	3&4 Vict. c.108 5&6 Vict. c.12 7&8 Vict. c.58 7&8 Vict. c.70	23.7.40 29.4.42 4.7.44 4.7.44		extn. XX: Haymarket–North Bridge br: Myrehead–Causewayend (Slamannan) (promoted as Slamannan Junction Company)	1.25 1.00	1.8.46 28.8.47			225,000 281,250 12,000	75,000 93,750 —	300,000 375,000 12,000	8,000		
Wishawtown Morningside & Coltness	4&5 Vict. c.43	21.6.41	8.65	Chapel (Wishaw & Coltness)–Morningside–Knowton–Crofthead–Wilsontown–Longridge, nr. Whitburn		2.6.45	J. Macneill		70,000	23,000	93,000	63,572	7,349 81,381	9,408
Drumpellar	6&7 Vict. c.63	4.7.43	1.82 (0.86 built)	Cuilhill (Monkland canal)–Rosehall, Old Monkland			N. Robson		26,000	8,666	34,666	24,865	13,662	

a: including half of costs and mileage of Glasgow and Paisley joint line
*: indicates whether Grainger or Miller was the engineer actually responsible

Sources: *Local and Personal Acts* (as listed); PP 1844 (318) XI, *Select Committee on Railways, Fifth Report*, appx. 2; HLRO, HC Deposited Plans of the various railways; H. G. lewin, *Early British Railways*; H. G. Lewin, *The Railway Mania and its Aftermath*; F. Whishaw, *The Railways of Great Britain and Ireland*; S. Salt, *Railway and Commercial Information*.

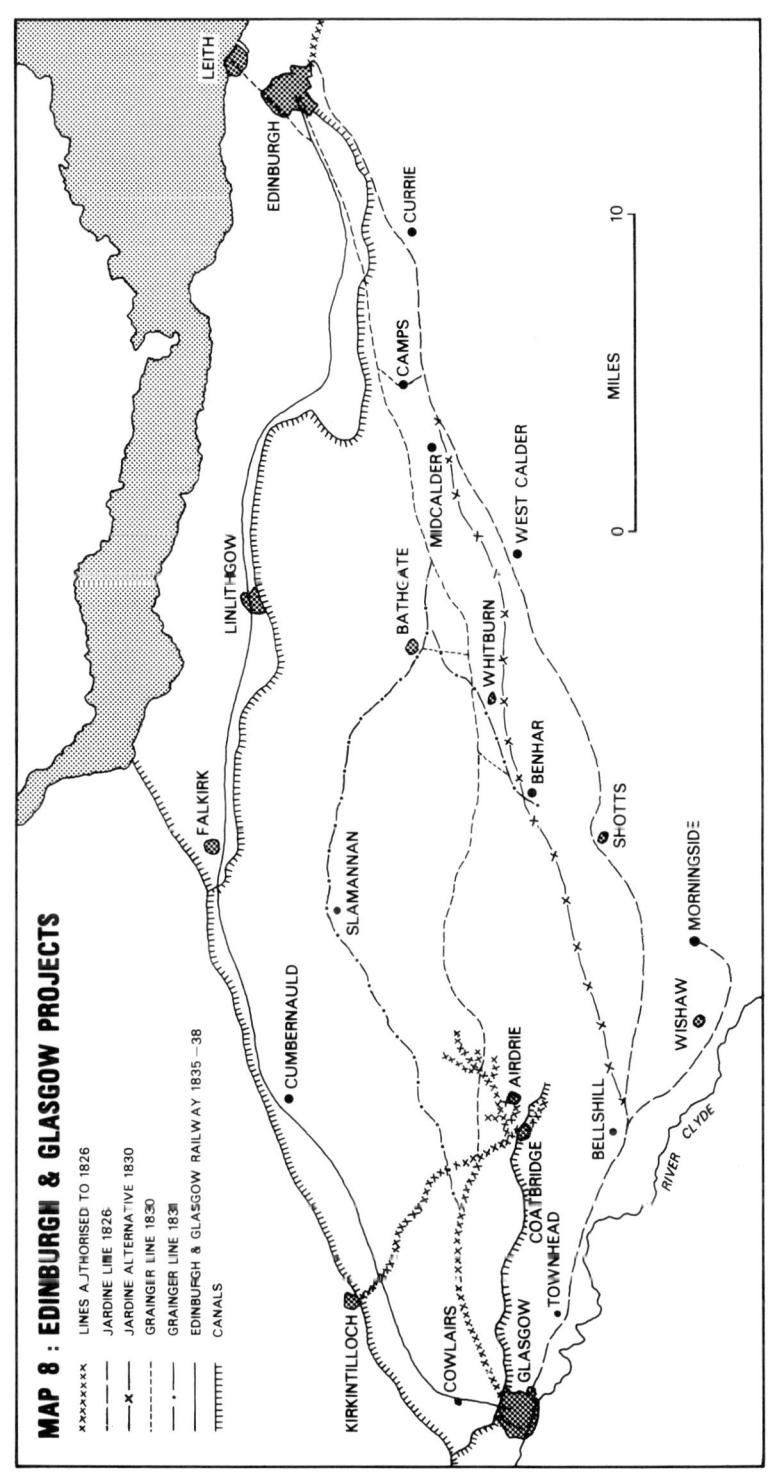

to break the canal monopoly, and in some cases might also harm their interests in the Monkland railways. A suspicion grew that their initiative was designed to pre-empt any move by other possible projectors by letting it be known that an eminent engineer was already undertaking a survey. This suspicion was increased when the result of the survey was not published until a rival committee was established in 1830, and when it was then discovered that the projectors had neither formed a company nor raised any funds towards promoting the line.[15] The survey report also showed that Jardine, after making a general examination of the Forth-Clyde valley, had been instructed by his committee to make his detailed survey of a line through the middle of the coalfield by Wishaw and Benhar, about as far away from the canal as possible. It was in effect another coal railway to supply the two cities, and almost incidentally to make a link between them.[16] Meanwhile an 1829 plan for a line via Airdrie had been abandoned when its projectors learned of the existence of the Jardine survey.[17] Another threat to the canal interest came from the opening in 1829 of the Newcastle & Carlisle railway, offering an alternative route for traffic from the west of the country to the east, and on to Europe. This was a particular danger to the port of Leith, since it was likely that, as an anti-canal pamphleteer stated, 'the delay of three days in going by our only present conveyance (the Forth and Clyde Canal) will not be submitted to, when a Railway is made by which two thirds of the time may be saved'.[18]

In 1830 a number of tentative proposals were made, but the only one to become a serious proposition was also based on an existing interest. Of the five men who initiated the scheme, three — Charles Tennant, coal- and ironmaster Mark Sprot and landowner John Lang — were directors of the Garnkirk & Glasgow, and their chosen engineer was Thomas Grainger. Two other Garnkirk directors joined the committee of management, as did two from the Wishaw & Coltness, William Dixon and the chairman of the new committee, Archibald Spiers.[19] To the Garnkirk there were obvious advantages in becoming part of a through line between the cities, while for the coalowners along the Wishaw & Coltness it offered a quicker and cheaper route to eastern markets. Grainger had worked under Jardine on the 1825 survey, but he and Miller now had ideas of their own. Their route ran from the still incomplete Garnkirk, by Slamannan, Bathgate and Midcalder, to Edinburgh, and was much more obviously a competitor to the canal.[20] Several of the new committee had been supporters of the Jardine line five years before, but now, as they informed the chairman of the Jardine subscribers:

> We certainly had considered the above association [the Jardine line committee] as abandoned or under the control of those who did not seriously desire any Railway, or whose interest either pledged them to preserve existing monopolies or to adopt some particular line or mineral field . . . The association of 1825 is now no more than a Shadow, made to walk forth when it is expedient to alarm the public.[21]

In reply to a proposal from the new committee, the Jardine group refused to let the question of the best route be decided by neutral engineers, and demanded a payment of £1500 before they would let anyone see their plans. Spiers reaffirmed to his committee that:

> from these circumstances, and from the whole conduct of the old or alleged Company, during the last six years, that they were **under an influence hostile to the execution of any railway from**

Edinburgh to Glasgow whatever, and which merely kept alive the name and machinery of an undertaking proposed in the dark era of 1825, to impede the execution, if possible, of any other.[22]

As the new committee was established, the Jardine line announced that it was postponing further surveying in view of the depression and in order to observe the progress of the Liverpool & Manchester, but that it would apply for an act in the following year.[23] Although it might be true that they would prefer to see no railway at all, they would certainly fight for their own southern route rather than concede the inter-city traffic to a rival.

The battle for the route between Edinburgh and Glasgow was the first in Scotland between rival railway promoters; earlier railways had been opposed by existing canal and road interests, and in the case of the Garnkirk by some of those connected with existing railways, but in no case had two groups disputed the right to make a new line. The aim of each committee was firstly to satisfy the conditions imposed by Parliament in the standing orders of the two houses, but then more essentially to persuade Parliament that their line was superior in terms of local support, engineering, cost of construction and anticipated traffic. For the first time in Scotland, a project had to be shown to be not only viable and desirable in itself, but also to be a positive improvement on an alternative scheme.

Also relevant was the greater scale of the project. Jardine's line was estimated to cost £520,130 to construct, and Grainger's in its final form £410,000 (see Table 26), whereas the highest previous estimate had been £60,000 for the Wishaw & Coltness and the highest authorised share capital £70,125 for the Edinburgh & Dalkeith. It was unlikely that such substantial sums could be found simply from the resources of the projectors and their contacts, or even from the area of the line alone. The information in the prospectus and the names of the provisional committee had not only to impress members of Parliament but had also to attract money from investors outside the area of the line.

The provisional committees of the two rivals were interestingly contrasted. The Jardine list, of 27 names, leant heavily on the landed interest (the only identifiable industrialist being Jacob Dixon of Dumbarton Glassworks), and was headed by a duke, two barons, two baronets and four knights, most of whom owned minerals on the proposed route.[24] The Grainger line's links with Debrett were restricted to one coalowning baronet, while of the 21 committeemen fourteen were described as merchants. Of these, ten came from Glasgow and only one from Edinburgh; two Liverpudlians and a Londoner were co-opted 'in respect of the large body of subscribers in England'. The remaining seven included Charles Tennant and a group of mineral owners on either the Garnkirk or the new line.[25] The Grainger line's committee was much closer to the way in which such bodies developed, at least for railways in industrial Scotland. Rural railways might continue to fill their provisional committees with gentry, culminating in the phalanx of peers, clan chiefs and high military officers which graced the Highland in the 1860s, but elsewhere the money and the directorial talent which created the railways came predominantly from the merchant interest (and above all from Glasgow merchants), with support from the proprietors of coal and iron.

106 The Origins of the Scottish Railway System

Table 26. Edinburgh and Glasgow Projects 1830–31: Estimates of Construction Costs[a]

	Mileage	Land			Earthworks			Rails & roadway			Masonry & tunnels			Other		Contingencies		Total	
		£	p.mile	% total	£	p.mile	% total	£	p.mile	% total	£	% total		£	% total	£	% total	£	p.mile
Jardine, Oct. 1830																				
main line	43.59	92,755	2,128	20.5	90,057	2,066	19.9	147,800	3,391	32.7	63,322	14.0		16,617	3.7	41,060	9.1	451,611	10,360
branches[c]	5.60	39,000	6,964	56.9	9,936	1,774	14.5	9,273	1,656	13.5	5,500	8.0		4,400	6.4	—		68,519	12,236
total	49.34	132,155	2,679	25.4	99,993	2,027	19.2	157,073	3,183	30.2	68,822	13.2		21,017	4.0	41,060	7.9	520,130[a]	10,542
Grainger, Dec. 1830																				
main line to Leith	37.82[b]	44,949	1,188	12.9	120,079	3,175	34.4	99,389	2,628	28.5	84,307	24.2		375	0.1	—		349,100	9,231
branches[d]	8.32	10,009	1,203	25.0	12,701	1,527	31.7	12,906	1,551	32.3	4,369	10.9		30	0.1	—		40,015	4,809
total	46.14	54,958	1,191	14.1	132,780	2,878	34.1	112,295	2,434	28.9	88,676	22.8		405	0.1	—		389,115	8,433
Grainger, Nov. 1831																				
main line to Haymarket	37.33[b]	23,472	629	7.3	102,601	2,748	32.1	110,026	2,947	34.4	63,655	19.9		940	0.3	19,306	6.0	320,000	8,572
branches[e]	12.51	25,228	2,017	28.0	14,671	1,173	16.3	21,957	1,755	24.4	21,531	23.9		200	0.2	6,413	7.1	90,000	7,194
total	49.84	48,700	977	11.9	117,272	2,353	28.6	131,983	2,648	32.2	85,186	20.8		1,140	0.3	25,719	6.3	410,000	8,226

a: All estimates for double main line with single branches. Jardine estimated that the total cost would be £438,019 if the main line were single with double earth-works, £507,130 if the Glasgow harbour branch ran along the streets, and £425,019 if both economies were made.
b: Grainger planned to use 6.41 miles of the Garnkirk line to St. Rollox, and a further 2.49 miles of the proposed tunnel line to reach the Broomielaw.
c: To Glasgow harbour (1.5 miles), Leith docks (3.4) and Camps lime works (0.7).
d: To Princes Street, Edinburgh (1.66 miles), Bathgate (1.85), Benhar (3.79) and Camps (1.02).
e: To Leith docks (3.66 miles), Benhar and Shotts (7.81) and Camps (1.04).

Sources: James Jardine, *Report to the Subscribers for a Survey and Plan of a Railway from Edinburgh and Leith to Glasgow*, 15–19; *Edinburgh Glasgow and Leith Railway, Reports by Messrs. Grainger and Miller of Edinburgh, and Mr. George Stephenson of Liverpool, Civil Engineers* (Jan. 1831), 3–11; *Edinburgh & Glasgow Railway, Reports by Mr. George Stephenson of Liverpool, and Messrs. Grainger and Miller of Edinburgh, Civil Engineers* (Dec. 1831), 25–27.

By 1830 a comparison of the proposed routes probably favoured Grainger. This was not entirely Jardine's fault: apart from the constraints imposed on him by his committee, much had happened since his survey. The Wishaw & Coltness had been authorised to open up the central Lanarkshire coalfield; the increasing potential of the locomotive made the two inclined planes on his line appear inconvenient and old-fashioned; and the Liverpool & Manchester had revealed the possible importance of passenger traffic on an inter-urban line. One of Jardine's opponents claimed that events since 1826:

> render the speculations of engineers at so early a date . . . comparatively valueless. And to this circumstance we are inclined mainly to attribute many of the errors, as we regard them, into which so eminent an engineer as Mr. Jardine has fallen . . . The transit of passengers seems, by common consent, to take precedence of the carriage of goods by this new means of conveyance.[26]

George Stephenson also saw the main purpose of an Edinburgh-Glasgow railway as 'the expeditious, safe, and cheap conveyance of passengers'.[27] If this view was accepted, and the provision of a rapid direct route between the cities was more important than connecting them to new sections of coalfield, the Grainger committee were in a strong position. Even for coal traffic, the growing network of lines in Lanarkshire and Grainger's proposed branch to Benhar and Shotts made it less necessary for the main line actually to traverse undeveloped mineral fields. Grainger's route also allowed a summit level of 635 feet against Jardine's 804 feet, important on lines to be worked by primitive locomotives or even possibly horses. Jardine's counter-suggestion that his summit could be reduced to 620 feet by inserting a $3\frac{1}{2}$-mile tunnel near Shotts only served to emphasise his route's disadvantages, and was answered by a realignment of Grainger's route to achieve a 495-feet summit, even although this added £30,000 to the estimated cost.[28]

For both projects, terminal arrangements in the two cities, and particularly the crucial accesses to the harbours, were a problem. Since Grainger was approaching Glasgow over five miles of the Garnkirk line, 'through a cluster of Vitriol and Soap-works, Foundries and Dung-depots . . . the very reverse of unexceptionable',[29] he would use that company's station, and was dependent on the ratification of his Blythswood tunnel plan for access to the harbour. Opponents pointed out the disadvantages of the St Rollox station — 'the depot would be perched upon a height which every day's experience proves to be almost inaccessible to loaded carriages'[30] — although it could be argued that this at least meant that the distribution of goods from the depot throughout the city would be downhill. Jardine approached Glasgow along the Clyde valley to a station at Barrowfield, east of Glasgow Green, from where access to the harbour would be either by a tramway along the streets (which was likely to cause chaos at the junctions with the bridges), or by a line through the back gardens of some fairly influential citizens. In 1830 Jardine produced the surprising compromise suggestion that, if the Blythswood tunnel were authorised, his line might be altered to run into the Garnkirk; more importantly, he revised his plan to replace the street tramway by a bridge at Rutherglen and a connection via the Polloc & Govan to the south side of Glasgow harbour.[31] After the rejection of the Blythswood tunnel, the advantage here seemed to rest with Jardine.

In Edinburgh, Grainger's line followed the route eventually used at the end of the decade to Lothian Road, or in the revised version only to Haymarket, with an apparently straighforward engineering line down the Water of Leith to the docks,[32] although opponents claimed that this branch would cost the company £16,000 in property damages alone.[33] Jardine, on the other hand, excited the scorn of opponents by planning to curve round the south side of Edinburgh to join the Edinburgh & Dalkeith at St Leonard's, and to reach Leith in the first plan by a tunnel through the Calton Hill and a street tramway, and in the revised plan by the circuitous Leith branch of the Dalkeith line:

> Mr. Jardine, as is well known, was the surveyor of the Edinburgh and Dalkeith Railway. That he should feel a paternal affection for his offspring is most natural, and everyone knows that by a wise provision of nature, the more unpromising the offspring, the stronger is the parental tenderness.

Not only did this plan entail a peripheral Edinburgh station and a devious route to Leith, but the terms of Sir Robert Keith Dick's wayleave prohibited the use of locomotives on his land.[34] Leith was recognised as an important part of both committees' planning, since not only were substantial industrial exports from Glasgow to Europe anticipated, but ships unloading at Leith were anxious for back cargoes of coal. It was still common practice, as in the waggonway days, to unload at Leith, go on to collect coal at a Fife port, and sometimes return to Leith for the rest of the cargo.[35]

Although the Edinburgh Glasgow & Leith committee had commissioned Grainger's survey, they called in the highest available engineering authority to adjudicate between it and the Jardine line.[36] George Stephenson, having approved Grainger's 1830 plan, enthused about the improved proposals of 1831, where the lower summit level was achieved at the cost of 3.7 extra miles of construction, mostly on longer branches. The new line also had better gradients (a maximum of 1 in 180, against Jardine's 1 in 100),[37] and the realignment took the route away from the area of the Ballochney company, which would otherwise certainly have opposed the scheme.[38] The plans were also approved by three other engineers, James Walker of London, Nicholas Wood of Newcastle, and John Gibb of Aberdeen.[39] The employment of Stephenson, Walker and Wood emphasises the deference in Scotland to English engineering skills; not only were the great English engineers recognised as the top of their profession, but their names gave confidence to prospective English investors. Throughout the early years of Scottish railways, highly skilled native engineers like Grainger, Miller, Neil Robson and Joseph Mitchell had to submit to having their work reviewed by consulting engineers imported from the south.

Estimates of traffic and construction costs were scrutinised by both opponents and potential supporters quite as keenly as the proposed routes. In 1830 the Jardine committee issued his 1826 estimates with minor variations, while Grainger and Miller, after first producing some provisional 'observations', published their estimates two months later: at the end of the following year appeared their estimates for their revised line (see Tables 26 and 27). Inevitably the figures were

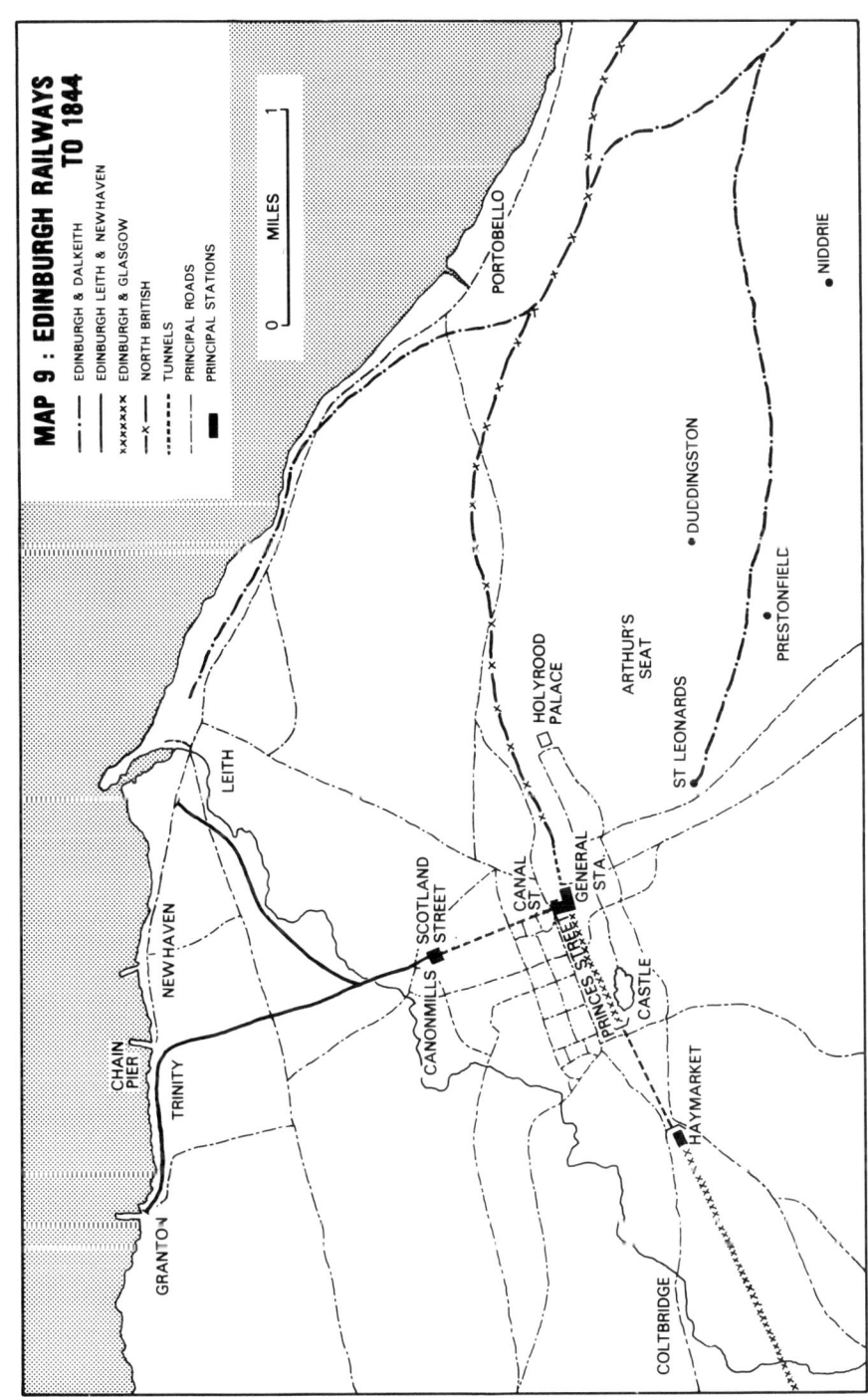

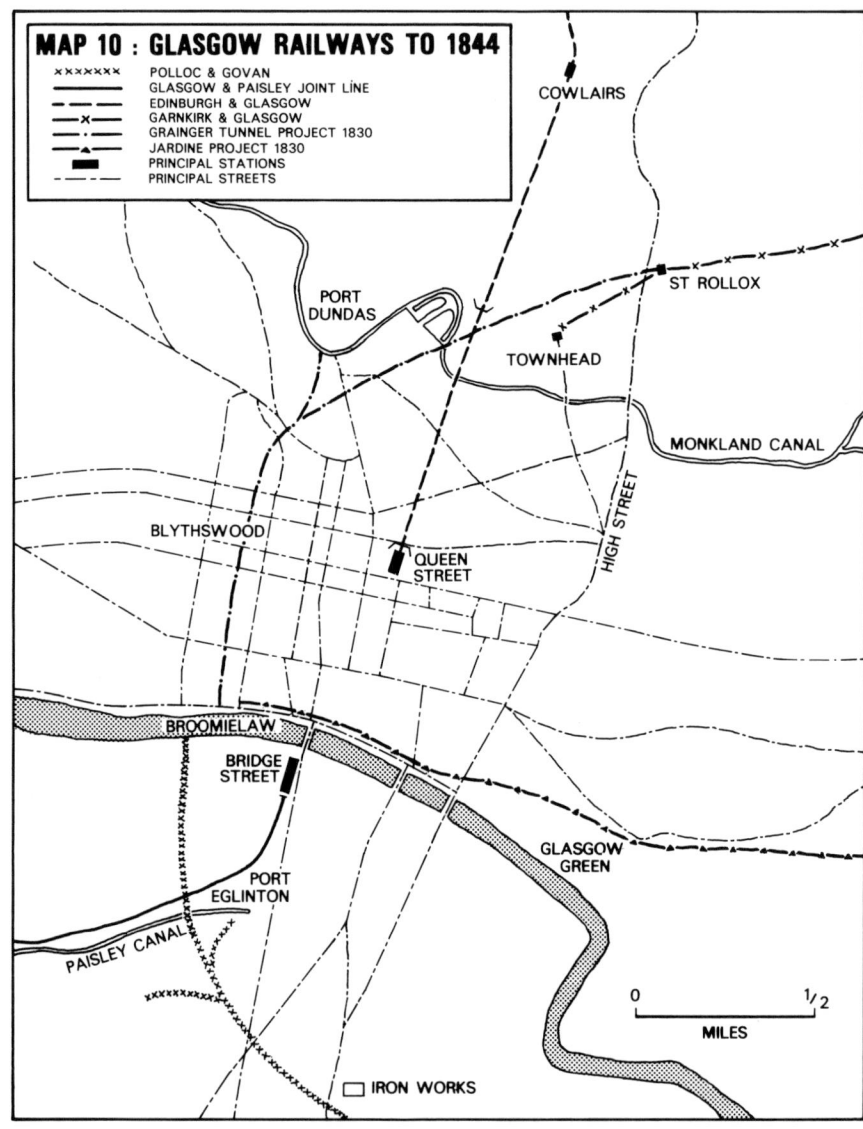

Table 27. Edinburgh and Glasgow Projects 1830–31: Estimates of Traffic

	Coal	Iron and ironstone	Freestone	Lime	Agricultural goods & manure	Other goods	Total goods	Passengers	Total revenue	Working expenditure	Net revenue	Return on capital
Jardine, Oct. 1830												
tons	243,000	11,300	66,000	241,000	—	25,900	604,700+	93,900[a]	—	—	—	—
revenue (£)	17,150	2,809	2,750	2,425	6,550	11,500	43,185	17,606	60,791	9,500	51,291	
% of total revenue	28.2	4.6	4.5	4.0	10.8	18.9	71.0	29.0	100.0	15.6	84.4	9.86%
Grainger, Dec. 1830												
tons	60,000	6,000	4,000	10,000	—	50,000	130,300	—	—	—	—	—
revenue (£)	11,250	900	500	1,388	3,000	18,750	35,788	48,000	83,788	—	—	—
% of total revenue	13.4	1.1	0.6	1.7	3.6	22.4	42.7	57.3	100.0			
Grainger, Nov. 1831												
tons	100,000	6,000	—	16,000	30,000	30,000	182,000	—	—	—	—	—
revenue (£)	16,000	1,500	—	2,400	16,125	19,500	55,525	45,000	100,525	42,525	58,000	
% of total revenue	15.9	1.5		2.4	16.0	19.4	55.2	44.8	100.0	42.3	57.7	14.15%

a: Number of passengers

Sources: James Jardine, Report to the Subscribers for a Survey and Plan of a Railway from Edinburgh and Leith to Glasgow, 6–9; Edinburgh Glasgow & Leith Railway, Reports by Messrs. Grainger and Miller of Edinburgh, and George Stephenson of Liverpool, Civil Engineers (Jan. 1831), 12; Edinburgh & Glasgow Railway, Reports by Mr. George Stephenson of Liverpool, and Messrs. Grainger and Miller of Edinburgh, Civil Engineers (Dec. 1831), 8–9.

subject to close analysis in a series of newspaper comments and usually pseudonymous pamphlets.

The most obvious discrepancy in the traffic estimates was that, while Jardine anticipated 29% of revenue from passengers, Grainger at the end of 1830 allowed for twice that proportion (and for 2.7 times as much money). These could be explained at least in part by the nature of the two lines, and by the probability that Jardine's figure was still the one he had originally estimated in 1826. It was at first sight more surprising that Grainger's estimates of passenger traffic should fall steadily from £53,040 to £48,000 to £45,000 (and more dramatically as a proportion of total income), in the three successive estimates. In fact the first figure included income to and from Glasgow, some of which would belong to the Garnkirk company, while the third sum was inserted into the report by the directors rather than the engineer, and was simply the amount paid by passengers on all existing public transport between the cities. It was assumed that the reduced fares would be balanced by an increase in usage. Since the railway expected to halve the coach journey time of $4\frac{1}{2}$ to 5 hours, and to charge fares of 7/6 and 5/- against the ordinary coach fares of 12/- and 8/-, it was reasonable to expect a total transfer of road passengers to rail.[40] It was however not proven that reduced fares would create the necessary increase in volume, though Grainger and Miller pointed out that between 1817 and 1830 coach fares on the Glasgow to Paisley road had been halved and that 26 coaches were running where there had originally been only two. Still, claimed their opponents, 'speculative anticipation of increased numbers is not warranted by the relations existing between the cities'.[41]

To their employers Grainger and Miller admitted that the effects of rate reduction might be uncertain. On the Monkland Canal, a rate reduction from 1/5d to 1/- per ton, to compete with the Garnkirk railway, had increased the volume of goods traffic from 210,999 tons to 222,254 tons — an increase in volume of 5.4%, but if all else was equal a reduction in revenue of 25.6%. On the other hand, passenger traffic had proved more price-elastic, and a 1d cut in fares had increased numbers by 20.1%.[42]

More surprising was the increasing optimism of Grainger on goods traffic, as his estimates of income rose from £28,720 to £35,788 to £55,525. Traffic estimation could not be an exact science, and much speculation was involved in guessing how much goods traffic might be attracted from the canal, or how many new sources of minerals might be opened up simply by the provision of a railway. It may even be that Grainger's escalating figures are a reflection of the views of the rival line, since only his final one was higher than Jardine's £43,185. What is surprising is that, although neither side state their proposed rates per ton-mile or the average expected length of goods journeys, Jardine allows for an income of only 1/5d per ton against Grainger's 6/1d. Even on coal traffic, Jardine's 1/5d is met by a rival figure of 3/2d. It is unlikely that rates per ton-mile would have diverged to this extent, and it must therefore be assumed that Jardine's coal traffic was travelling for much shorter distances — in fact, that he was concentrating on supplying Glasgow from central Lanarkshire, while the much smaller volume of coal expected by Grainger was mostly to travel from Benhar, Shotts and the

Monklands to Edinburgh. It is much more difficult to understand why Jardine expected to transport fifteen times as much lime as the rival line, but only to receive the same income from it. Both sides also tended to imply that their estimates were conservative; according to Grainger and Miller, for instance:

> There are many sources of revenue besides those above stated, and benefits to be derived from the improved communication, to which we have not averted.[43]

The speculative nature of many of the estimates left them open to attack, and the third party to the debate, the canal companies, joined in. A pamphlet issued by the directors of the Union Canal both made much of the discrepancies between the rival railways, and challenged many of their assumptions, particularly the transfer of all passenger traffic or of any goods traffic at all from the canal. A horse, they claimed, could draw about six times as much on a canal as on a railway, and 'if steam, or any other power, is substituted in both cases, the result will probably be in a similar proportion in favour of canals'.[44] The Union Canal was sponsoring its own railway project, the Edinburgh & Glasgow Union Railway, under the chairmanship of J.H. Colt of Gartsherrie, to build a line from the Garnkirk via Cumbernauld to the end of their canal, with a branch into Falkirk and a connection to a proposed line to Stirling.[45] The plan was similar to the later Slamannan, but, as that company discovered, had no hope of competing as an inter-city route with a through railway. But for the moment the canals appeared reasonably confident of fending off their rivals: to their optimism might be opposed the view of a railway supporter that 'when the Railway is made, the Canals will have no passengers, for which they are not at all fitted'.[46]

Estimates of construction costs should logically have been less speculative, since engineers were dealing with matters closer to their previous experience, and were not being required to guess the reactions of entrepreneurs and prospective passengers to the arrival of a railway. But, as the canal company were willing to remind investors even at the cost of a little self-criticism, estimates had a reputation for understatement. Hugh Baird, with the broad agreement of Telford, had estimated the cost of the Union Canal at £246,322 against an annual revenue of £52,728. By 1827-8 the actual figures were respectively £448,956 and £15,538, and the anticipated 20% dividend had been reduced to 2.7% (and even this was retained by the company as part of an emergency capital levy of £46 per share). Turning to railways, Jardine's own estimate for the Edinburgh & Dalkeith had been doubled before the line even opened, and even the great Liverpool & Manchester was already 60% above its original allowance of £510,000.[47] Against this sort of evidence, the claims of a railway supporter for the accuracy of their estimates as against those of canals seemed facile.

> The whole expense of a Canal consists in what is technically called *formation*; a deep trench, *water tight*, must be formed, and supplied with water; an operation, the details of which, can never be previously calculated with any degree of certainty, for there is no article in its construction which bears a definite price. But a Railway is totally different — the *formation* is the least part of the expense . . . The rest of the expense consists in the price of the ground, which can be very nearly ascertained — the cost of blocks, chairs, keys, malleable iron rails and laying, masonry, fences, and fieldgates. Every one of these articles can be estimated to a nicety, at least the trifling variation of the market price (in iron, for example) makes all the difference. Now it

must be obvious that where the expense of formation, which is indefinite, is thus confined to one-fifth part of the whole cost, the possibility of error in the estimate is also diminished in the same proportion; hence any material deviation in the expense is precluded by the very nature of the work.[48]

Nor did the claim that the different engineers were calculating 'on precisely the same principles' and reaching the same costs per mile tally with the figures; Jardine's estimates of costs per mile were in fact 28% higher than Grainger's final version.

Since Grainger and Miller intended to reach the Broomielaw over some six miles of the Garnkirk and $2\frac{1}{2}$ of the Blythswood tunnel project, they required only some 37 miles of new main line against Jardine's $43\frac{1}{2}$ miles. However, their more northerly route required a branch of almost eight miles to reach the important coal and iron area around Benhar and Shotts,[49] so that the total construction required by the two schemes was almost identical at between 49 and 50 miles. For a double main line and single branches, Jardine allowed £520,130 against Grainger's £410,000. Although Jardine's hillier route might have been expected to lead to higher costs for earthworks, his willingness to accept severer gradients meant that in fact this part of his estimate was lower than his rival's. He allowed £500 per mile more for rails and roadmaking, which may in part reflect the higher cost of iron when he originally drew up his figures in 1826, but the main difference between the estimate was his allowance of £2,679 per mile for land against the £977 of Grainger.

Not surprisingly, Grainger's estimate for land came under heavy fire, particularly as he gave no reasons for his optimism. For urban land the estimates were similar — for the respective Leith branches (admittedly not on the same route) Jardine's £6,142 per mile was only £33 higher. But for the eighteen miles from Bathgate to Edinburgh, including some good suburban building land, Grainger allowed only £17,371, or less than the Garnkirk had paid for eight miles 'in one of the most barren and bleak districts in Scotland'. Similarly, though with less direct comparability, it was claimed that the £6,101 allowed for fifteen miles between Bathgate and Gartsherrie was less than the Garnkirk had paid for its first mile out of Glasgow.[50] Possibly Grainger's previous experience in building railways for projecting mineral owners who were willing to grant their land cheaply because of the profits which the railway would bring to their coal and iron interests had made him over-sanguine. Though the failure of both schemes in 1830-31 meant that the estimates were not tested, some support for Grainger is given by the Slamannan, built along much the same route as the western part of his project in the later 1830s, which even as late as 1848 had only spent £821 per mile for land.[51] This, however, was for a single-track railway to carry minerals; nor did it include any of the more expensive land near Edinburgh. In 1831 doubts about the figures, coupled with the known underestimating of the costs of earlier railways, may have given investors pause.

Even so, Grainger and Miller's revised report of November 1831 was a confident document, both about the future of railways in general and about their own project in particular. Since their previous report a year before, the Liverpool &

Manchester had triumphantly completed its first year, and the Edinburgh Glasgow & Leith committee observed that:

> the unparalleled returns already made by the Liverpool & Manchester Railway afford the best proof of the Revenue likely to be drawn by a Similar Mode of Conveyance . . . between the two great Cities of Edinburgh and Glasgow.[52]

Locally, steam locomotives were demonstrating their abilities on the newly opened Garnkirk. With their route approved by the great Stephenson, their backing by a substantial and wealthy group of Glaswegians, and the inevitable doubts of the seriousness of the intention to build of the Jardine line committee, they felt confident of success. Yet parliamentary authorisation for the Edinburgh & Glasgow railway was held back until 1838, and was then for a railway along a completely different route.

The failure of the 1831 project was due both to the vigour of the opposition and, in the end, to insufficient confidence among potential supporters. The interests opposed to the line, though not all individually powerful, were cumulatively formidable. The Jardine committee had claimed in 1830 that the bill for the Grainger line was bound to be rejected because of the degree of interference with existing transport interests, in that apart from threatening the traffic on the Ballochney, the turnpikes, and the Forth & Clyde, Union and Monkland Canals, it also planned physically to cross the first four of these.[53] Foremost among the opponents was the Union Canal, which at least in goods traffic probably had most to lose. Although the management clearly feared competition, their case against 'the delusion in regard to a Glasgow railway' was based on the claim that the railway would not in fact take away their traffic and could therefore never pay more than a minimum return.[54] The trustees of the main inter-city turnpike, on the other hand, implicitly granted the railway's claim that it would take most or all of the traffic, and pleaded for the protection of existing interests — a plea intensified by the fact that, of the trust's £60,000 debt, £9,600 was held on the personal security of the trustees. Where alternative transport already existed, they stated:

> the Trustees and Creditors humbly apprehend, that the Legislature will not sanction *an uncalled for experiment*, such as the proposed Railway, whereby a capital, much larger than what is anticipated, will be embarked in an undertaking *not required by the public*, and which, while it is hazardous to the proprietors, will be injurious, if not ruinous, to many other establishments (Roads, Canals, and Railways) of tried utility . . . [55]

More substantial opposition came from the trustees of the Bathgate road. The price at which they were prepared to be bought off is unknown, but it was high enough for the committee to minute that 'owing to the opinion entertained by the Committee of the injustice of the Bathgate Road demands that the same should not be acceded to or any compromise made with them unless in an extreme necessity'.[56]

The reference to other railways was relevant. Apart from the rigid hostility of the Jardine promoters, the alliance against the Edinburgh Glasgow & Leith also included, to a greater or less extent, the Monkland & Kirkintilloch and the Ballochney, which were still closely allied to the Forth & Clyde Canal. Although the east end of the Ballochney was the nearest point of the Lanarkshire railways to Edinburgh, Grainger (even apart from the interests of his committee) was unwilling to link his scheme to a line unsuitable for passenger traffic, and therefore ran independently through the section of coalfield from which the Ballochney drew all its traffic.[57] In spite of a group of shareholders who had rather larger holdings in the new plan, the Ballochney came out in opposition. A section (later scored out) from a draft report to a company meeting shows both the strength of feelings, and the ambiguous position of Charles Tennant:

> From a report on the subject made by a sub-committee of the Monkland & Kirkintilloch railway... it will be seen that the rival railway is not adapted for any public object but is a manifest job calculated solely to promote the private interests of a few individuals, and particularly those of one of the Committee of Management of both the Monkland & Kirkintilloch and Ballochney railways.[58]

The amended line of 1831 removed the immediate threat to the Ballochney, and persuaded it into neutrality, but nothing could prevent the hostility of the other transport interests. And at the same time the rejection of the Blythswood tunnel line removed the essential link to the Broomielaw, giving the opposition a very strong argument indeed. Finally, there was a substantial number of hostile landowners, many of them connected with one or other of the rival concerns, others simply apprehensive about the interference with their estates and their privacy. 'Nothing,' remarked a doleful railway supporter, 'excites so much bad feeling and expensive opposition to a public measure as any inroad on the privacy of our gentry.'[59] It was above all the 'bitter and persistent' opposition of landowners to which Joseph Mitchell later attributed the failure of the project.[60]

Another possible influence on some potential investors (though probably not one of great influence) was the appearance of another rival form of transportation. The early 1830s was the one period in which the steam road carriage looked as if it might be a serious proposition. The experiments of Goldsworthy Gurney and Walter Hancock in the 1820s led to services by Sir Charles Dance between Gloucester and Cheltenham in 1831 (with a Gurney machine), and by Hancock in London in 1833.[61] Grainger and Miller dismissed the new contender; an experimental journey by road from Edinburgh to Glasgow had taken three or four days to complete, and 'it has not been thought expedient to make a second attempt'. British experience, they claimed:

> has proved, beyond doubt, that steam can never be employed with advantage as a moving power upon even the smoothest and most level of our turnpike roads, and that a smooth and impenetrable surface, such as that afforded by railway bars, is absolutely necessary.[62]

All the same, road steam gained the support of, for instance, Sir John Macneill and John Louden McAdam, and of the *Scotsman*, which, having been in the forefront of early railway propaganda, now approved the new technology:

> We feel therefore constrained to believe, that we shall soon have steam coaches plying on all our roads, the very uneven ones excepted, at 12 miles an hour, without waiting for the construction of Railways at an expense of many millions of pounds.[63]

John Scott Russell, later more famous as Brunel's partner in his shipbuilding ventures, started a regular, and for a time successful, road service between Glasgow and Paisley. But the promoters of road steam services failed in the face of a series of mechanical failures, inadequate supplies of coke and water, and the hostility of the turnpike trusts, which feared damage to their roads and demanded extortionate tolls. Russell's service was ended by order of the Court of Session in 1834 after a boiler explosion had resulted in five deaths.[64] Although the 1839 Select Committee on Turnpike Trusts spent a considerable amount of time discussing steam on roads, almost all the vehicles had already disappeared. One promoter saw it all as a railwaymen's plot:

> a very strong and powerful conspiracy, you may almost call it, among all the attorneys of England, and all the engineers and ironmongers, and all other mongers, to get up railroads, and whether the railroads succeeded or not, they filled their pockets with 60,000*l* or 70,000*l* in expenses and in getting the Bill passed through this House.[65]

But in the early part of the decade road steam may have marginally hindered railway promotion by diverting some of the more adventurous projectors, and by persuading others to wait and see which of the contending systems would triumph.

The evidence of the company's minutes is that the real difficulties, in spite of Joseph Mitchell's recollection, were in raising money and satisfying Parliament rather than in placating landowners. Some land was even given free, and the difficult land cases all seemed capable of solution, even if in one case it meant persuading a committee member's brother to buy a £20,000 estate whose owner would only sell all his land or none.[66] Raising subscriptions to an amount previously unknown in Scotland was much more problematic. The attempt to present the bill to Parliament in 1831 had to be abandoned when, according at least to disgruntled shareholders, only a third of the capital could be raised.[67] Subscription collecting for the revised line to go forward in 1832 was better organised, with a separate committee under Theodore Rathbone in Liverpool as well as the ones in Glasgow and Edinburgh. Even so, the committee complained that the agents of the Forth & Clyde Canal were persuading subscribers to withdraw by misrepresenting the project. The minutes also indicate their awareness that the subscription contract as presented to Parliament was not duly signed by all the subscribers, and that some who had signed now wished to withdraw.[68] In May 1832, as the bill approached the committee stage, this minority published their claim that the prospectus was delusive, half the subscription list false, and many of the names those of people who had approved the 1831 bill but not the present one.[69]

Trouble came, as with so many railway bills, before the committee of the Commons. Sometimes a solution could be achieved, as when powers to take over the Garnkirk & Glasgow (which had been approved by that company) were dropped when the committee objected to the possibility of a company not yet incorporated taking over one which was. But the opposition of the road trustees and the irregularity of the subscription contract led to the bill's rejection on standing orders. Although the promoters claimed that the vote was purely 'on a point of form' and:

> notwithstanding of the very decided Opinions not only of their English Counsel and Solicitors and almost all the Official and other Gentlemen in Parliament acquainted with such business that *the late vote of the Committee was decidedly wrong*

they withdrew the bill rather than going to appeal.[70] It was clear that they would be vigorously opposed on every point in both Houses by their share of 'the formidable opposition which *every* Railway Bill seems to meet with in Parliament' and which, they claimed, 'originates in narrow views of Self Interest and is not supported by principle'. Fighting on would be very expensive and, they must have realised, quite probably unsuccessful. The promoters determined to let other railways fight the battles of principle and set the precedents in 1832 and 1833, and reintroduce their own bill in 1834.[71]

It never happened. The project had cost £9,433, even although Grainger and Miller agreed to forego £500, and other professional advisers £600, of their fees. On the credit side, £4,289 had been raised by the original call on the subscriptions. For eighteen months after the withdrawal of the bill, the committee endeavoured to persuade most of their subscribers, including several former committee members, to pay up the 26/- per share which would cover the deficit, and meanwhile to fend off creditors who were increasingly consulting their lawyers. The situation in April 1833 was summed up in a laconic note in the minutes:

> The meeting at the same time instructed Mr Bannatyne [the company's lawyer] to mention to Mr Sprott, that it was unnecessary to discuss the amount of his claim at the moment, as there were not funds to pay it.

Rather than prosecute offenders themselves, they circularised their creditors, inviting them to undertake the prosecutions. Meanwhile only the most half-hearted attempts were made to prepare for the 1834 session, and, although the committee in November 1833 minuted its unenthusiastic determination to continue, no further meetings were recorded.[72] Once a project had lost the confidence of a significant number of its subscribers it had little hope of recovery. *Chambers' Edinburgh Journal* was left to complain that:

> In Scotland railways are as yet unknown, except in one or two instances, in which they have been constructed chiefly for the transfer of coal. Nothing continues to astonish us so much as the want of a railway betwixt Edinburgh and Glasgow.[73]

The recession of 1830-31 was not the best of times to promote a railway, and although the economy was turning up by 1832, the seeds of failure had already been sown. The next attempts to link Glasgow and Edinburgh came amid the general optimism and prosperity of 1835-36. They were made by a new set of projectors, although Tennant later became associated with the successful company, and Rathbone was the representative of the Liverpool interest on its board. The change of personnel among the promoters marks the transition from men who saw railways as essentially adjuncts to their other interests in mining and industry, to those who regarded the success of the railway itself as the main consideration. The scale of vision was widening rapidly, and the leaders of the new Edinburgh & Glasgow projectors were to go on to promote the Glasgow Paisley Kilmarnock & Ayr and the North British, and to lead Scotland into the railway mania of the mid-1840s. The majority were merchants: the chairman was Glaswegian John Leadbetter, later deputy chairman of the Ayrshire and, in the mania, chairman of the Glasgow Dumfries & Carlisle; his deputy was John Learmonth, coachbuilder, merchant and ex-provost of Edinburgh, later not only to take over the chair of the Edinburgh & Glasgow, but also those of the North British and the Edinburgh & Northern.[74] The difficulties and opposition besetting their project were similar to those faced by the earlier promoters, and their determination had to survive three annual applications to Parliament before the eventual success of 1838.

Grainger and Miller were again employed as engineers, with Miller by now being the principal partner in the work. Once again he had to submit to English supervision, since the prospectus promised that the choice of line would be left to 'one of the most eminent English Railway-Engineers, who has not hitherto been on the line'.[75] Those called in, to confirm rather than choose the line, included Rastrick, Vignolles and Locke.[76] The new plan bore no resemblance to that of 1830-32, but followed the valley north of the coalfield used by the canals and earlier rejected by Stephenson as being too circuitous. The overwhelming consideration was now the inter-city traffic; as the prospectus said:

> Since 1832, much knowledge has been acquired as to the best construction and probable cost of Railways... It has been proved that they are principally advantageous and profitable when applied to the conveyance of passengers, and light goods... the existing intercourse is always increased *more than threefold* by the new facilities thus afforded.[77]

The coalfield was being opened up by branch lines specifically designed for mineral traffic, including the new Slamannan railway, and coal traffic for the new main line should be brought by these adjuncts rather than by taking the line through the coal areas. The plan instead was for a high-speed route, almost level from Edinburgh to Cowlairs, with consequently elaborate engineering including two major viaducts, at an estimated maximum cost of £550,000. The main engineering problem was the steep approach from Cowlairs to the Glasgow station at Queen Street. The refusal of the Forth & Clyde Canal to sanction an overbridge forced the railway to design a steeply graded tunnel under Port Dundas, to be worked by a stationary engine.[78] The traffic estimates showed how much the emphasis had moved away from coal and towards passengers: Grainger and Miller forecast revenue from passengers of £75,503 (81.8%), from parcels £2,062 (2.2%), from mail £1,633 (1.8%) and from goods only £13,058 (14.2%). Even if working expenses of 42.85% were incurred, being the actual figure recorded on the Liverpool & Manchester, the net income of £52,719 would yield almost 10% on capital.

Opposition from established railways was not a serious problem, since the new route was no threat to the Monklands area lines. Opposition from the canals, however, was, if possible, intensified by the new railway route. Though the railway prospectus claimed to expect little diversion of traffic from the canal, opining that 'there is *enough of trade for both*',[79] by 1837 Leadbetter was more equivocal before a House of Lords committee:

> [Joy (counsel for the canals)]: you hope to take off a good deal of our Traffic, to leave us no further trouble about it? — [Leadbetter:] I do not know that; I do not think that we shall destroy you entirely.[80]

Joy's arguments that the canal was efficient, since speed was irrelevant in goods transport and the canal did not require transshipments, was undermined by the provost of Falkirk, who stated that not only was his town unanimously in favour of the railway, but that the neighbouring farmers preferred to send grain to Glasgow by carts rather than by the canal since 'they find it more regular and expeditious, and they travel in greater safety'.[81]

The anti-rail alliance of the canals was not without strain. The Union Canal revived its plan for a railway from its Falkirk end to the Garnkirk line, this time under the name of the Glasgow & Falkirk Junction, as the only hope of ever making its canal pay.[82] Even if the plan had succeeded, and had prevented authorisation of the Edinburgh & Glasgow, it would have severely damaged the Forth & Clyde Canal. But, as a malcontent Union Canal shareholder pointed out, if a canal plus a railway was better than two canals, logically a through railway would be an even greater improvement; and even if the Glasgow & Falkirk Junction was authorised, the canal would in effect be at the mercy of both it and the Garnkirk.[83] In fact control of the Falkirk Junction scheme slipped away from the canal interest, until early in 1837 the promoters agreed to abandon their plan in favour of the rival in return for an allocation of £50,000 of Edinburgh & Glasgow stock.[84] It is worth noting that as late as December 1838, five months after the Edinburgh & Glasgow act, the Union Canal was still trying to arrange a deal with the Lanarkshire railways to save some inter-city passenger traffic. A report by John Macneill, who had engineered the Slamannan, examined the possibilities of a $4\frac{1}{4}$-hour journey using the Garnkirk, Monkland & Kirkintilloch, Ballochney and Slamannan railways, with almost half of the distance and three-fifths of the time being spent on the canal. By some startling arithmetic, which assumed that the estimated 100,000 passengers per year would distribute themselves equally among the available trains, he calculated that the service could be run with only two locomotives and eight carriages in stock, and that it would yield 100% profit. The report made no mention of the Edinburgh & Glasgow line, and no financial allowance for the substantial works necessary to fit the Ballochney and the Monkland & Kirkintilloch for passenger trains. But in the event the canals had to submit to the loss of through passenger traffic.[85]

The change of route meant that the new company was dealing with previously unaffected landowners, and inevitably much time was spent placating the gentry. Leadbetter's report to his committee in April 1837 on his progress in London is almost entirely devoted to negotiations with hostile or greedy landowners. Among others, Sir James Gibson-Craig would only relinquish his land if he was allowed to select the 'neutral' arbiter who would assess its value, while Brown of Coy Mill had to be supplied with a stationary steam engine to make up for the diversion of water from his mill. And there was no way of satisfying either the Marquis of Hopetoun or J.M. Hog of Newliston, both of whom carried their opposition to Parliament, even although the latter was to become a director in the boardroom revolution of 1846. Those old adversaries, the Bathgate road trustees, who were now perhaps less directly affected since the route had changed, were conciliated by the railway's agreement to take over £17,500 of their debt.[86]

As the decade wore on, and more and more major railway schemes were authorised in England (to say nothing of a few in Scotland), it became increasingly difficult to claim that there was no justification for a line between Edinburgh and Glasgow. The canal opposition became more and more a question of delaying tactics, by creating the greatest possible amount of difficulty in Parliament on both points of substance and points of procedure. In 1836 they achieved the rejection of an inadequately and hastily prepared submission from the company; as the directors told Locke:

> from the hurried manner in which the line was laid down considerable alterations may yet be made upon it of which notice will afterwards be given.[87]

In 1837 the company was properly prepared and, according to the heavily biased *Railway Times:*

> A fair, honourable or well-reasoned opposition to the formation of a railway between Edinburgh and Glasgow, was soon found impossible to be sustained; and accordingly, the agents of the Canal Company had resource to the vexatious methods which are afforded so abundantly by the new Standing Orders of the House. In order to carry over the Bill to another session, the arguments of Counsel were purposely spun out on the most frivolous subjects, until the patience of members of the Committee was entirely exhausted . . .[88]

Much might depend on the stamina of a committee and the attitude of its chairman, but the canals were too open to the charge of maintaining their own monopoly to be able to hold off the railway indefinitely. But their delaying tactics were rewarded in 1837 when the death of William IV on 20 June brought the parliamentary session to a premature close with the bill still in committee. In 1838, however, all the arguments had been so thoroughly gone over in previous years that the railway had comparatively little difficulty in achieving authorisation.[89]

III. Landowners' initiatives: north of the Tay

The prolonged gestation of the Edinburgh & Glasgow meant that it was not the first of the new generation of inter-urban lines to achieve authorisation; indeed five other companies, of varying importance, started their progress later but completed their parliamentary odyssey in 1836 and 1837. The former year produced two lines in Angus and a small but potentially important one in Edinburgh; the latter saw two companies connecting Glasgow to the Firth of Clyde, one of which was to become the pivotal line of the future Glasgow & South-Western.

In Angus the Dundee & Newtyle, in spite of its eccentricities, had been useful to an admittedly limited part of Strathmore. As director Hugh Watson observed in 1838:

> though the Shareholders have never received any dividend whatever, yet all the Landowners and not a few of the Tenants acknowledge, that they have been great gainers. Their properties have been increased in value far beyond the money they have sunk in the Railway.[90]

There were two ways of extending railway facilities in the valley. One, encouraged by Dundee interests, was to extend the Dundee & Newtyle itself, and in 1835 the lines to Coupar Angus and Glamis were authorised; Watson indeed hoped to continue the former by Blairgowrie to Dunkeld.[91] But the inhabitants of Forfar and the east of Strathmore were unlikely to find this attractive; they looked towards links to nearer ports by lines unencumbered with inclined planes. Inevitably, they turned to the partnership of Grainger and Miller for practical advice.

Most anxious for a link to the coast were Brechin and Forfar. Stevenson had surveyed between Brechin and Montrose in the 1820s, and a few years later Grainger and Miller examined the ground again but without further results. In 1837

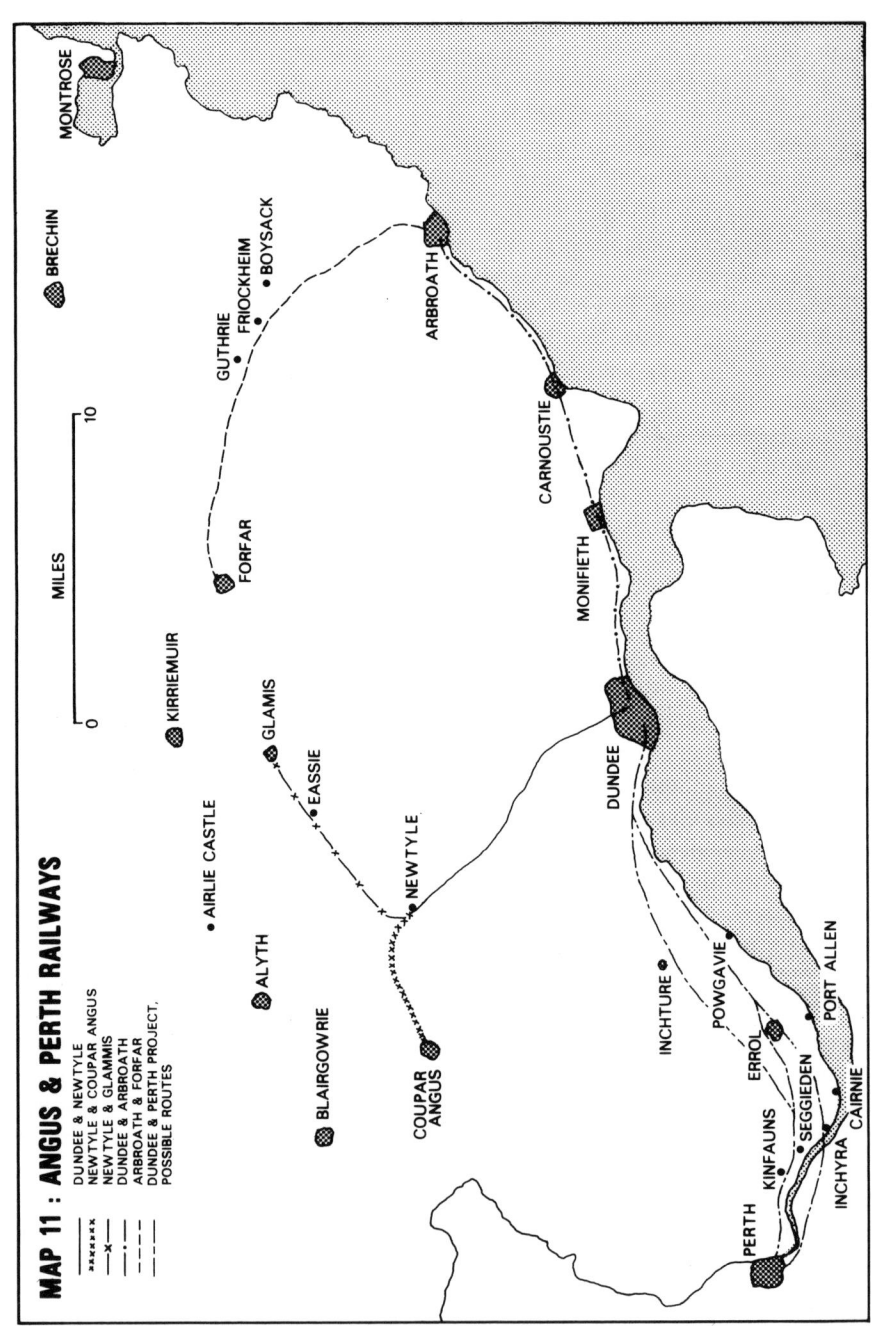

they were still trying to get paid for their work on 'this almost forgotten railway', while Brechin town council refused to contribute more than £50 towards total costs of £117, although they had previously agreed to share the costs with Montrose.[92] The two engineers had more luck when they were privately employed by local landowner and agricultural improver William Lindsay Carnegie of Spynie and Boysack to find a route between Forfar and Arbroath. Again the country had formerly been surveyed by Stevenson, with the help of William Blackadder, in 1826: Grainger and Miller for the most part followed Stevenson's route except where individual landowners had become hostile, though, since they were no longer concerned with the dead level line, they dispensed with Stevenson's inclined plane into Arbroath, as well as his final link to the harbour through the streets of the town which would be 'neither for the interest of the Public, nor of the Railway'.[93] The experience of the Dundee & Newtyle harbour branch (although in a location where no other route was possible), and of the Polloc & Govan, had swung opinion against running rails along public roads.

Almost all the driving force behind the creation of the Arbroath & Forfar Railway came from Lindsay Carnegie himself, and he remained the company's chairman throughout its independent existence until it was leased to the Aberdeen in 1848, and thereafter until his death in 1860. At the end of the century, Arbroath's historian had no doubt of his contribution:

> No service of equal value to that of Lindsay Carnegie's in originating and carrying out the introduction of the railway system had been previously or has since been rendered to the town of Arbroath by any single individual.[94]

The town council, equally convinced at the time, made him a freeman of Arbroath within ten days of the passage of the act of authorisation. His active support came mainly from landowners, including the influential Ogilvy family, one of whom joined the committee, and from Arbroath merchants.[95] Seven-eighths of the capital was locally subscribed. Support came more from the seaward end of the line at Arbroath than from Strathmore, even although the isolated inward areas might have appeared more obviously in need of improved transport. Admittedly, Arbroath, with 14,568 inhabitants in 1841, had almost twice the population of Forfar (7981), but it contributed almost three times as much to the initial subscription list (£23,625 against £7,400).[96] The people of Strathmore required in particular supplies of coal and lime, but, with the partial exception of the Dundee & Newtyle, no railway had yet been created by a group of potential consumers, and earlier conceptions such as the Roxburgh & Selkirk waggonway had ended in failure. They also needed larger markets for agricultural produce, but as yet the limited amount of such traffic and the difficulties of attracting it to the railway from a wide agricultural area had deterred the construction of railways aimed primarily at carrying foodstuffs. The initiative for railway building had come essentially from two groups — mineral owners, who required to move very large quantities of heavy materials from a small number of particular sources to a not much greater number of specific destinations, and merchants, by whom (as a generalisation) any innovation which increased the total volume of trade was looked on with favour. Mineral production was not important in Angus (although

Carnegie himself and some of his colleagues possessed stone quarries), but the merchants of the port of Arbroath were keenly interested in the development of their hinterland. Before the arrival of the railway, traffic along the route was so little developed that there was not even a regular coach between the two termini. There was little serious opposition to the line, apart from half a dozen of 'Lindsay Carnegie's friends and neighbours', who objected both as landowners and as road trustees.[97]

Grainger's original estimate of £57,243 for a $15\frac{1}{4}$-mile single line was revised to £68,459 before application to Parliament, mainly due to the rising costs of land in Arbroath and of iron, and to a decision to use heavier rails. The line was to be built to a gauge of 5'6", simply because the engineers thought it was a suitable one for such a railway.[98] Since they had built to 4'6" on Lanarkshire mineral railways, and were planning for $4'8\frac{1}{2}"$ between Edinburgh & Glasgow, Grainger and Miller clearly had as yet no concept even of a Scottish national network, let alone a connection to England. The originally proposed share capital of £60,000 had risen to £70,000 by the time of authorisation. When the line was opened in January 1839, still in an incomplete state, the costs were already some £15,000 or £20,000 over the estimate, and an act to allow a further £50,000 issue of share capital was necessary in the following year.[99] All the new stock carried a guaranteed dividend of 5%, as did the company's debentures; within a few years this heavy load of guarantees was to make it very difficult to pay any dividend on the original ordinary shares.[100] Even so, in the severely depressed conditions of 1840, which had hit the Angus linen industry particularly badly, it was not easy to dispose of the new shares. Even some of the chief promoters could not or would not help further. The company clerk, who was also the town clerk of Arbroath, wrote to one leading landowner:

> I regret that you entertain so unfavourable an opinion of the Railway ... The new stock has not been run upon by the partners as was anticipated. Almost all the principal shareholders have neglected to take up their proportion of the Stock ... Had you & Mr. L. Carnegie & the Towns of Arbroath & Forfar taken your proportions there would have been far more shares disposed of for a time at least.[101]

It may be assumed that in many cases, and certainly in that of Lindsay Carnegie, the problem was tightness of money rather than loss of faith in the railway, but it all served to emphasise the difficulties of constructing a railway when the economy was sliding into the most intense depression of the century.

Lindsay Carnegie was also concerned with the promotion of another line authorised on the same day and in the same area as his Arbroath & Forfar, but the key moving spirit behind the Dundee & Arbroath was William Ramsay Maule, first Lord Panmure. He it was who called the original meeting of promoters in October 1835, employed Grainger and Miller as engineers, contributed £2500 to the subscription list, and personally piloted the bill through the House of Lords in 1836.[102] Since he also owned most of the land through which the line ran, his offer to give it for a nominal feu duty, and even to remit that during his own lifetime, was of considerable value: Miller, who was in charge of the engineering while Grainger concentrated on the Arbroath & Forfar, was enabled to estimate the land costs of the main line at a mere £340 per mile. The total estimate of £99,844 included £12,054 for the company's share of a link line to the Arbroath & Forfar at Arbroath; the estimate of about £5500 per mile for the main line reflected the lack of engineering difficulties anticipated as well as Panmure's generosity.[103] Opposition was limited to one landowner and the inevitable road trustees,[104] and the parliamentary procedures were satisfied with little difficulty. To move from original meeting to authorisation in $7\frac{1}{2}$ months was an impressive achievement.

The Dundee & Arbroath, however, had problems, one of which was the nature of the competition. Early railways had generally had little difficulty in overcoming directly competing road transport. Both the Liverpool & Manchester and the Garnkirk line had demonstrated that railways could compete effectively with canals, even if the canals had not admitted defeat. But the Dundee & Arbroath, where over 80% of the line and all the towns served lay along the coast, had to compete not only with a road of good quality but also with estuarial shipping, which did not have to pay for expensive capital works (except through harbour dues) or government passenger duty, nor, unlike canal boats, to worry about the effects of increased speed on the banks. Like the London & Greenwich (1833), the London & Blackwell (1836) and the Glasgow Paisley & Greenock (1837), the Dundee & Arbroath was taking on a rival form of transport which might well be able to undercut its rates, and whose comparative lack of speed might be no deterrent to either third-class passengers or goods traffic.

The company was also geographically isolated. Grainger and Miller again built it to a gauge of 5'6", suggesting that the only link they had in mind was to the Arbroath & Forfar. But the two companies failed to build their proposed junction line round the town, and the 600-yard gap between the stations had to be unsatisfactorily bridged by a horse tramway through the streets, although the companies did co-ordinate their timetables in order to try to divert traffic between Dundee and Strathmore away from the Newtyle lines.[105] At the Dundee end of the railway, the line was opened to Craigie, $2\frac{1}{4}$ miles from the city centre, in October 1838, but it took another eighteen months to reach 'a wooden shed containing two lines of railway, with no waiting room and no conveniences for passengers' at Trades Lane, near the docks.[106] Since the dockside land which offered the only route westward was owned by the Harbour Trustees, who were unwilling either to sell or to lease in perpetuity or even long-term, no connection could be made to the Dundee & Newtyle or planned towards Perth — only the advent of the Dundee & Perth in the mania provided sufficient incentive for the Harbour Trustees and the Dundee & Arbroath to work out a tolerable compromise. Meanwhile, the railway had to be launched as a local line whose chances of becoming part of a national system were extremely uncertain.

Another Tayside landowner had rather less success than Panmure and Lindsay Carnegie. Lord Kinnaird had privately commissioned surveys and issued parliamentary notices for a line from Dundee to Perth, even before a meeting of landowners in November 1835 agreed to promote it and set up a provisional committee.[107] The landowners of the Carse of Gowrie were, however, far from unanimous and the project got bogged down in their disputes. Even leaving aside the few whose opposition was absolute, the rest could reach no agreement as to which of three possible routes should be followed. Hay of Seggieden, whose land at the western end of the Carse could hardly be avoided, demanded inordinate compensation, so that the committee in desperation considered bridging the Tay at Inchyra and approaching Perth by the south bank.[108] Meanwhile the road trustees set the price of their neutrality at £10,000, being the sum required to liquidate the accumulated debt on the road. The sub-committee appointed to negotiate with landowners reported their

> regret that they have not been met by some of those interested in the proposed measure with the cordiality which the advantages to be derived from it would have entitled them to expect.[109]

But in 1835 no railway had yet proved itself profitable in a strictly agricultural area, and many landowners were not convinced that it would make much difference to their estates. Either Dundee or Perth was never more than a dozen miles away, carting on the flat Carse roads was easy, or in some cases goods could be sent along the river from little piers such as Port Allen, Cairnie or Inchyra. The railway promoters offered no certain prospect of connections beyond its termini, although there was talk of a railway southwards to Stirling and beyond, so that for longer-distance trade the sea routes from Dundee and Perth were not yet faced with competition.

These doubts and difficulties discouraged potential subscribers, and prevented any submission to Parliament in 1836. In May, £105,000 of the proposed £160,000 share capital remained unsubscribed for, and no hope remained of further help from Perth. By June many English applicants were withdrawing, made apprehensive by the delays and by fears of oncoming recession. No attempt was made to revive the project for 1837, and it was formally wound up in the following year, when it was reported that £87,250 in shares had eventually been subscribed for, 72.3% of it locally. The holders lost 8/- per £25 share for their efforts.[110] There was no self-evident reason why the scheme should have failed when the two to the east of Dundee succeeded, since its prospects of traffic were certainly no worse, and its engineering difficulties no greater. It may be that Kinnaird was not as efficient or as single-minded an organiser as Panmure and Lindsay Carnegie, or that he was less adequately supported by his committee. It was relevant that, with few exceptions, he failed to interest the merchants and professional men of his terminal towns in the project, since the landed gentry alone were not a firm enough foundation. And the arguments over land which delayed progress in 1836 were crucial: by 1837 tighter money had lessened enthusiasm for isolated railway plans of uncertain potential. As in the mania a decade later, a year's difference was often all-important; schemes brought to Parliament in 1836 and 1845 in general went on to fruition, while those designed for 1837 and 1846, often equally valid on paper, all too often either collapsed before application or failed thereafter.

Even in its failure, the Dundee and Perth plan spawned subsidiary projects. A subscription was raised in 1835 for a survey between Perth and Dunkeld, but the project died in the face of the indifference of Perth Town Council and the hostility of the Duke of Atholl, who wanted to maintain the level of pontage on Dunkeld bridge.[111] In the following year a survey was proposed from Stirling to Newburgh, to connect to both the Dundee and Perth at the southern end of the Inchyra bridge and to a proposed line through Fife; connections from the southern end might be made to the Forth & Clyde Canal and the proposed Edinburgh & Glasgow, thus creating through routes between Glasgow, Edinburgh, Dundee and Perth. Another committee was examining the ground between Perth and Crieff,[112] and in 1837 a survey was made from Perth to Aberdeen,[113] but in no case was there any further activity. Although five distinct routes from Perth were under examination in the mid-1830s, no railway to the town was authorised until 1845.

IV. Problems and profligacy: Granton, Leith and the route to Fife

The optimists of the mania period were inclined to quote the success of companies like the Arbroath & Forfar and the Dundee & Arbroath to support their cases for new lines in often improbable places. Even the cautious officials of the Railway Department, after reviewing the progress of the two lines, added:

> These circumstances induce us to hesitate in recommending the rejection of lines proposed to open out new districts in Scotland, where the local advantage will be considerable, and where substantial parties are ready to undertake them; although we may feel doubts, looking at the existing amount of traffic, as to its sufficiency to yield a fair return on the required capital.[114]

The record of the third promotion of 1836 should have engendered more caution, since the Edinburgh Leith & Newhaven demonstrated in its short existence most of the things, short of actual bankruptcy, that could go wrong with a railway company. Yet the initial prospects were bright. There was no shortage of existing traffic between Edinburgh and the series of harbours which stretched along the Forth two or three miles away — the major east coast port of Leith, the fishing community of Newhaven, and the new facilities being developed at Trinity and (by the Duke of Buccleuch) at Granton. Apart from sea traffic, Leith was a substantial trade-generating town in its own right, with a population of about 25,000.[115] Its only existing rail link was the circuitous offshoot of the Edinburgh & Dalkeith, offering horse-drawn facilities to the inconvenient station at St Leonard's. The failed 1831 Grainger and Miller scheme for the Edinburgh to Glasgow link had proposed an almost equally devious connection between Leith and the centre of Edinburgh via the western suburbs, while the revised, and eventually successful, project ignored the port altogether. It was anticipated that all Edinburgh's seaborne trade, including the important coal traffic from Fife and Tyneside, as well as the passengers for the Fife ferries and for summer bathing, should be captured by a direct railway from the city centre to the harbours, and the arguments for making one appeared strong.

On turning from traffic to topography, the problems became more apparent. To reach a central Edinburgh station at Canal Street, near the east end of Princes Street, the railway had to tunnel for over 1000 yards on a gradient of 1 in $27\frac{1}{2}$ under the ridge of the New Town.[116] When it emerged northwards, it was for almost all of its length either in deep cuttings or, at the seaward end, on a substantial embankment. Most of the ground required was urban, with correspondingly high land prices and inevitable disturbance of private residences. The ensuing vigorous opposition was orchestrated by the determined figure of Dr Patrick Neill, printer, conservationist and chief founder of the Edinburgh Horticultural Society. The line as planned was calculated to cause him maximum offence, since it ran through his own notable private garden at Canonmills and ended at a station built on East Princes Street Gardens, which had been laid out in 1820 by a committee under his convenership.[117] Although he succeeded in forcing a deviation round his own property, he continued to attack the spoliation of Princes Street by the station and warehouses 'perhaps four or five storeys high', the danger of the tunnel to the houses above it, and the unpleasantness of the tunnel itself — 'This Edinburgh tunnel will be nearly of the dimensions of the *cloaca maxima* at Rome, and may perhaps vie with it in sweetness'. With the horrors of the tunnel, and with a threepenny fare over so short a distance, locomotives would not give the railway a sufficient advantage to attract passengers away from the horse-buses. The failure of the town council to oppose the railway he could only attribute to the fact that the company's agent was a bailie and the engineer (Grainger) a councillor.[118]

Neill was right to predict a troubled future for the company, although he touched on only some of the reasons. It was to be six years after the act of authorisation before any part of the line was open, almost ten years before it reached Leith, and almost eleven before the opening of Canal Street Station and the tunnel.[119] The work proceeded so slowly that the committee — a group who took little part in the development of any other Scottish railway company — were enabled to change their plans on more than one occasion, switching their attention from Newhaven first to the comparatively unsuccessful new harbour plan at Trinity and then to Granton and an uncomfortable relationship with the Duke of Buccleuch. A confused and unseemly boardroom wrangle in 1837, during which four directors attempted to disqualify themselves by selling their scrip but failed to do so when the new holders, reassessing the company's prospects, decided to cut their losses by not registering the change of ownership, led to a replacement of most of the board, and in 1839 to a new act reorganising the company.[120] This act also authorised a change of line to concentrate on Granton (an emphasis belatedly recognised in 1844 by a change of name to the Edinburgh Leith & Granton),[121] brought Dr Neill back into the fray as the realigned route was now back in his garden, and, according to another opponent, ensured that *'no portion whatever of the traffic* between Leith or Newhaven and Edinburgh is intended to pass on the present line'.[122]

The stimulus to change, according to an authority with the benefit of hindsight, was that 'the estimates were utterly fallacious, and . . . the engineering difficulties of the Leith branch were insurmountable'.[123] In 1840 a disgruntled English shareholder claimed that, even for the revised plan, 'these "detailed estimates" were so grossly inaccurate, that not the slightest reliance could be placed upon them', and work on the line had to be halted while more money was raised.[124] By mid-1841 95% of the capital had been called up, only $1\frac{1}{4}$ miles of track were anywhere near completion, work had yet to start on the tunnel, and the English directors had resigned.[125] Grainger, however, was still calculating that the line would pay 3% as it stood, 4% when the tunnel opened, and 7% if linked to a good harbour and to the Edinburgh & Glasgow railway.[126] In the following year chairman Richard Dawson made the startling claim that 'this *hitherto* much *calumniated* railway' would be completed within the parliamentary estimates, a statement which bordered on the fraudulent.[127]

The Edinburgh Leith & Newhaven's prospects were never good, but so long as it remained isolated from connecting transport they were catastrophic. Buccleuch's law agent John Gibson had in 1840 come to essentially the same conclusions as Grainger: 'the Newhaven railway can never pay except by the formation of a great shipping port at its terminus, & its communication directly with Glasgow'.[128] Though the company did eventually complete its branch to Leith, the main line's hopes for sea traffic were concentrated on the duke's new harbour at Granton. A rail link to the city centre would also of course be of benefit to the harbour, but in spite of this apparently symbiotic relationship the duke and the company were uneasy colleagues. In 1840 Buccleuch had vigorously rejected a proposed gift from shareholders in Bristol of fifty paid-up shares in the company; as their spokesman Henry Wait Hall admitted, 'It is hoped by your Grace's consenting to join this present Railway Company it will obtain a character in the estimation of the public which it does not at present possess'.[129] In 1844 he refused an offer by the company to let him appoint one of the directors.[130] The route of the railway was not the one the duke would have chosen, since on its way to Granton it passed by Newhaven and Trinity, giving these potential rivals a shorter connection into Edinburgh. More importantly, it offered as yet no link to the Edinburgh & Glasgow, and Buccleuch was primarily interested in access to western trade; throughout the 1840s he contemplated with enthusiasm a westward link from his harbour to the Glasgow railway west of Edinburgh, which would have cut the Edinburgh Leith & Newhaven out of the trade altogether.

It was thus essential for the railway to establish a link to the Edinburgh & Glasgow. Between their respective stations at Canal Street and Haymarket lay the level valley between the Old and New Towns which had formerly contained the Nor' Loch, and which, according to Provost Adam Black, 'was apparently intended by nature for the floor of a railway'.[131] The Edinburgh & Glasgow board was keen; not only would the plan give the company a connection to Leith without having to undertake a branch of its own, but several of the directors, led by John Learmonth, were already projecting the line eastwards to Dunbar and Berwick which became the North British. Their chosen station site was at North Bridge, contiguous to Canal Street, and a link to Haymarket would complete planned through routes from Glasgow and Leith to England. The disapproving Gibson noted that the North Bridge site happened to belong to Learmonth:

> He anticipates disposing of his property to great advantage for the benefit of both companies . . . I look on it as a most notorious job, the object of which is to save the Newhaven Railway from the ruin it deserves, & to further the views of some interested parties.[132]

The essential part of the plan was the connection from Haymarket to the North Bridge. But, although the link involved virtually no engineering problems and no demolition of buildings east of Lothian Road apart from some markets under the

North Bridge, it met with prolonged and bitter opposition, and was not authorised until 1844. It became the first major confrontation between the interests of railways and the interests of public and private amenity. In 1816 the proprietors of the houses in the greater part of Princes Street had formed themselves into an association, and obtained an act which among other provisions entitled them to lay out gardens in the valley on land leased from the town council, and forbade further building on the south side of the street. In the following years they extended their domain by leasing the slopes on the south side of the valley, hired Robert Stevenson to drain the Nor' Loch (which he did with qualified success — flooding remained a problem until further draining in 1854), and laid out Dr Neill's communal gardens, with entry restricted to the proprietors and a few other fortunate key-holders. The result was both an improvement to the city and a valuable amenity to their own houses.[133]

Almost from the beginning of railways, projectors eyed the valley with interest. In 1830 Grainger and Miller considered taking their Glasgow line not only to Lothian Road but along the south side of Princes Street on a level with the causeway, but decided that their project had enough opposition without offending the Princes Street Proprietors:

> The idea of carrying a Railway along that Street to the North Bridge or the Post Office may now by some be considered as somewhat visionary if not very absurd. We are aware that in the Country a prejudice exists against the Introduction of Railways into Cities, but we are convinced such will soon wear off.[134]

All that an extension to the North Bridge offered in 1830 was a station site which was more convenient for the population of the eastern half of Edinburgh. When the committee of the revived and rerouted Edinburgh & Glasgow returned to the matter late in 1836, it also offered a link to the Forth via the newly authorised Newhaven company, as well as encouragement to Learmonth's eastward ambitions. In October he unveiled the new plan, for a railway at the foot of the Gardens valley, screened from view by bushes and, if desired, by an embankment.[135]

The Princes Street Proprietors were furious. On their behalf, James Skene argued that they had spent some £7000 on creating an agreeable resort for their families and an attractive prospect for the rest of the citizens:

> It is plain that anything which would tend to interfere with the privacy, quiet and comfort would utterly defeat these objects. It is accordingly vain to suppose that any consideration could compensate for such a loss ... It is inconceivable to imagine how a railway in whatever manner masked and enclosed, along the line of which numerous steam carriages will be constantly passing ... can prove otherwise than a most serious annoyance.[136]

His fears were shared by that champion of conservation Lord Cockburn, who was pessimistic about the chances of thwarting the railway:

> The result will in time practically be, that the whole of that beautiful ground will be given up to railways, with their yards, depôts, counting-houses, and other abominations ... which will ruin the peculiarity of the valley between the old and new towns, and by rendering the preservation of the open space less important will possibly lead to building on the south side of Princes Street. Could a Judge agitate, I should raise the very stones against this project ... My prediction is that the nuisance will be introduced, though some parties say they are resolved to oppose it. But what chance has taste against a railway before a committee of the House of Commons?[137]

Indeed, while such committees did take questions of amenity very much into consideration, experience suggested that they were more inclined to respect the amenity of a great country house owned by a peer than that of an urban setting and a group of prosperous but untitled merchants and professional men. Other arguments against the railway were required.

Some of the arguments produced were bizarre, such as that reported by *Chambers' Edinburgh Journal:*

> that the locomotive engines, in passing through the Princes Street Gardens . . . would throw out sparks, and set fire to the powder-magazine on the top of the castle — a building, let us remark, situated fully four hundred feet above the place and described as bomb-proof.[138]

But there was one more serious engineering objection. The Gardens were (and are) divided into two sections by the Earthen Mound, a gigantic spoil heap from the New Town development, which had been dumped on the boggy remnants of the Nor' Loch and had not yet convinced everyone of its stability. Besides providing a convenient roadway from the Old Town to the New, the Mound carried in its interior the water main for the New Town, Leith and other northern districts, and on its surface the new, and in 1837 not yet complete, Royal Institution.[139] James Skene, wearing a different hat as secretary to the Board of Trustees for the Encouragement of Arts and Manufactures in Scotland, who occupied the Institution, informed the Treasury of the Trustees' fear for their £40,000 building and forwarded two hostile reports by experts. One was W.H. Playfair, the Institution's architect; the other, James Jardine, was the water company's engineer, who perhaps had no love for the railway which had displaced his own Edinburgh to Glasgow project. In reply the railway company, by offering to pay for it, persuaded the Treasury to sanction a further examination by 'a neutral English engineer of eminence, unconnected with the promoters or opponents of the Bill'. The consequent report by James Walker and William Cubitt concluded that a tunnel through the Mound was practicable, but that the company would do better to wait until the railways with which a connection was proposed at the North Bridge were built. The company, already faced by quite enough opposition, retreated, and settled for a station at Haymarket.[140]

Haymarket, however, left the Edinburgh & Glasgow with no link to the Forth and no real hope of diverting the goods trade between Leith and Glasgow away from the Forth & Clyde Canal, and the directors had no intention of giving up their link. In 1840 John Gibson reported that they were rumoured to be interested in purchasing the Newhaven company, thus strengthening their case for linking two parts of a joint operation:

> The point they now aim at is to get the Glasgow railway carried through the Princes Street Gardens, & to say the truth, I think it very likely they will succeed in this . . .[141]

Though this came to nothing, and indeed the construction troubles of the Newhaven company reduced the urgency of making a connection, the prospect of the North British line maintained the need for action. By early 1843 the Edinburgh & Glasgow board had voted to reapply for the Gardens link, though only by seven votes to five with the five Glaswegian directors in opposition. The decision was supported by a meeting of shareholders in Lancashire (where a large part of the stock was held), although they favoured merely obtaining parliamentary authorisation but leaving construction until the Newhaven line was complete.[142]

The opposition to the Gardens link had by the 1840s lost a little of its vigour, and perhaps the idea of the railway was becoming less unthinkable. A pamphlet by Alexander Douglas, for 35 years clerk and treasurer to the Princes Street Proprietors, repeated the old arguments about interference with 'the quiet, retirement, and amenity of these beautiful gardens', the lowering of property values and the danger to the Mound, and estimated the total cost of a mile of line at £110,000, including £40,000 for land costs and compensation.[143] Another opponent, attempting to raise hostility among shareholders, argued that the ten Scottish directors, most of whom were both merchants and relatively small shareholders, were more interested in helping their own businesses than in improving the prospects of the company.[144] But general economic arguments, the national growth of the railway system, and the clearly imminent authorisation of a line to Berwick were increasingly persuasive; and any alternative line to complete a connection from the west to the east of Edinburgh would be so circuitous as to be impracticable for passenger traffic. Even the Buccleuch interest was coming round; in 1840 the duke had commented of the Gardens link that 'I disapprove of it upon Public Grounds, tho' to myself personally it is a matter of total indifference', but by the beginning of 1844 John Gibson, despairing of a westward link from Granton harbour, had come round to grudging support of the Newhaven, and therefore necessarily of the link line: 'if the Glasgow Company's branch is not to be executed, it seems to be desirable to give the other Company every reasonable encouragement'.[145]

In these circumstances the Princes Street Proprietors, while maintaining their formal opposition on principle, concentrated on compensation, on landscaping to hide the railway, and vainly on an attempt to ensure 'that Locomotive engines should not be admitted into the ground opposite Princes Street',[146] thus restricting the line to horse-drawn trains. The 1844 act confirmed an agreement by which the company paid the proprietors £1909 compensation, £85 expenses, and the costs of restoring the gardens.[147] The amenity lobby had been either bought off or defeated by commercial pressures, for the first but far from the last time in Scottish railway history. Lord Cockburn, at least, remained obdurate:

> No one does, or can, believe, that though this ground had been refused, Edinburgh would have been without a sufficient railway. At any rate, if what has been done injures the beauty of the town, I listen to the plea of convenience nearly as if it were urged in recommendation of a crime.[148]

The proprietors took their money and the landscaping, supervised by Playfair, which attempted to preserve their view. But it is unlikely that many of them shared the view of one railway enthusiast that the line:

> cannot in any respect be injurious to that property, but rather an advantage, for the passing to and fro of the immense trains of merchandise and passengers will be a source of amusement to the promenaders in the Gardens.[149]

It also became clear that, although the through line was necessary, any further railway encroachment was not. A few years later the majority of the Town Council, helped by the influence of city MP and junior Treasury minister William

Gibson-Craig, defeated an attempt by the Edinburgh & Glasgow to take over the whole of East Princes Street Gardens for station extensions.[150]

Meanwhile the Newhaven line opened from Trinity to Scotland Street, at the north end of the tunnel, in August 1842. In 1844 it became the Edinburgh Leith & Granton at a time when it had yet to reach any one of these three destinations: Leith and Granton were achieved in 1846 and the tunnel opened in May 1847.[151] For eleven years the shareholders saw very little traffic, escalating costs and no hope of a dividend. When in 1847 the company ended its career of independent profligacy, it had spent at least £310,000 against an original estimate of £100,000: the works alone had cost £175,480 and land a further £81,606.[152] Grainger had once again underestimated seriously, although in his defence may be cited the shortcomings of his board and the unprecedented difficulties of the tunnel. Compared to the other major tunnel at Queen Street in Glasgow, the one in Edinburgh was a much greater proportion of the total works (so that errors in estimating its costs would have a correspondingly greater effect overall), its gradients were twice as steep, and the property above was of greater value and belonged to men who for the most part had little interest in the line and would push for the highest available compensation. Since the plan for the more comparable Blythswood tunnel in Glasgow had been defeated, Grainger and Miller had been denied that opportunity of discovering how accurate their estimates might be. It may not be entirely coincidental that the remainder of the Scottish railway network was built almost entirely without substantial tunnels, and that the 'up-and-over' principles championed by Joseph Locke overcame the tunnel-boring tradition inherited from George Stephenson.

The Edinburgh Leith & Granton was rescued from the prospect of a bleak future by the events of the railway mania. The link to the Edinburgh & Glasgow was essential, but even it did not assure profitability, let alone Grainger's predicted 7%. There were already rivals appearing for the goods traffic. In 1845 the North British took over the Edinburgh & Dalkeith, and started upgrading its Leith branch;[153] although this gave a circuitous route to central Edinburgh and on to Glasgow, goods traffic, unlike passengers, would not complain about a little extra mileage, and the close connections between Learmonth's two companies might well induce them to send traffic between Glasgow and Leith without touching the Granton company's lines. If Leith held off the challenge of Granton and remained the principal port for the Edinburgh area, the distribution of goods traffic was not easy to predict. For passengers, however, the Granton company offered a shorter route by rail, even if the unpleasantness of the tunnel might induce some to stay with the horse buses; and Granton, not Leith, was the pier for the Fife ferries. That ferry service enabled the Granton company, not to make a profit in its own right, but to get itself taken over by a larger company which did itself eventually come through to profitability.

The need was for an extension of the railway system into Fife from the north end of the ferry at Burntisland. Fife had been prominent in coal waggonway development, and plans for a more extensive railway had appeared early. In 1819 Robert Stevenson had projected an Edinburgh & Dundee Railway to link the

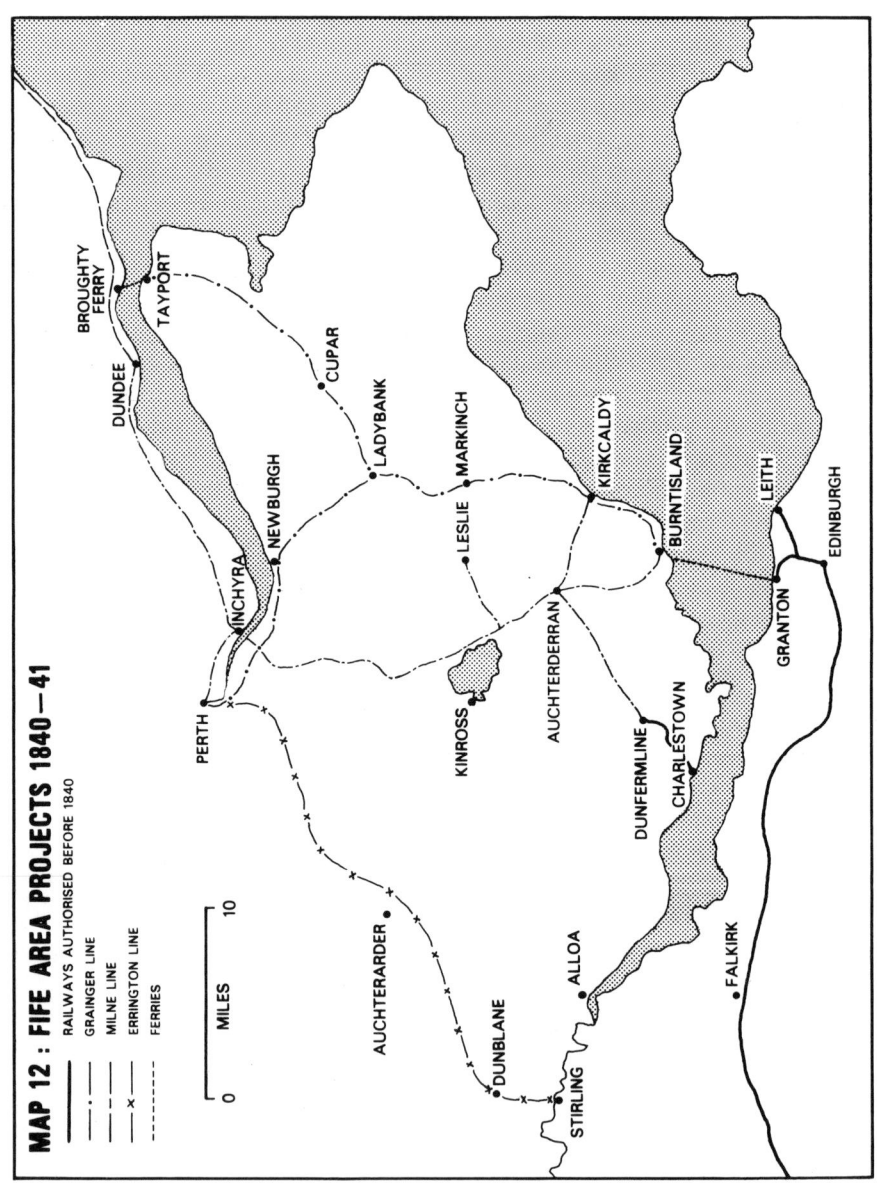

Forth and the Dunfermline coalfield to Perth and the Tay ferry at Newport.[154] Six years later a waggonway was proposed from the coal of Balbirnie and the lime of Forthar to Newburgh harbour;[155] since these minerals straddled the watershed north of Markinch, it would have been shorter in distance and probably easier in construction to run the line south to the Forth, so that the route was presumably chosen to give access to the coal-starved area round the Tay rather than compete with the numerous collieries which already had connections to the Forth. The midthirties boom brought plans from John Geddes in 1835, from Stevenson (making a brief return to railway engineering) in 1836, and from Grainger and Miller in the same year.[156] All proposed routes were similar to that actually built after the mania, from the ferry at Burntisland via Kirkcaldy to a point near Ladybank, whence the railway divided to Perth and Newport. In no case was work commenced. By now projectors were looking at more than local coal and lime, seeing their lines as connecting the central belt to the newly authorised railways in Angus, and even as parts of a future through route from London to Aberdeen.

The main problem was inevitably the need for two ferries between Edinburgh and Dundee, which might (and eventually did) mean that most goods and much passenger traffic would prefer a longer route by Stirling and Perth. In the early 1840s the situation was made worse since the absence of low-water piers at Granton and Burntisland meant that for much of the day the ferry could not operate, and Thomas Grainger, whose plan was again forward under the name of the Edinburgh Dundee & Northern, complained that:

> Viewing this communication as a great National Ferry, it is not creditable to the Government or the public spirit of the country, that it has remained in its present inconvenient, imperfect and dangerous state. Every one who has attended to public matters must be convinced, that, if a ferry of much less importance had existed on any such great line of communication in which Ireland had any interest, low water piers, if not splendid harbours, would, long ere this, have been erected at the sole expense of the nation.[157]

Help came, however, not from Peel's ministry, but from one member of it and the father of another, acting as private individuals. The Duke of Buccleuch was not only the owner of Granton harbour, which was overlooked by one of the smaller and less commodious of his dozen residences, Caroline Park; he was the largest landowner in Scotland, and the holder of a long list of public and political offices, including that of Lord Privy Seal. His interest in the ferry stemmed from his desire to improve and profit from his harbour, and he was willing to invest in order to create adequate facilities for a large traffic. By 1844 his expenditure was extravagant enough to alarm the cautious John Gibson, but he had provided Granton with a comprehensively equipped deepwater harbour.[158]

Across the Forth at Burntisland, Buccleuch worked in co-operation with another notable figure, John Gladstone of Fasque, later Sir John, father of the President of the Board of Trade and future Prime Minister. Gladstone, whose fortune had been made in Liverpool trade and West Indian development, including slaving, is assessed by his biographer as:

> an indomitable man... forming his own views of policy and persisting in their advocacy... assertive and quarrelsome, a man who bickered over matters great and small, who had an overpowering need to be vindicated...[159]

His interest in railways went back to the first provisional committee of the Liverpool & Manchester in 1824; by 1843 he had £169,700 invested in railways, of which over half was in the Grand Junction and the remainder sufficed to make him the biggest single shareholder in the Lanarkshire coal railways. He was also the largest shareholder, with £40,000, in the Forth & Clyde Canal, which he hoped to see joined in a monopolistic merger with the Edinburgh & Glasgow. In 1840 he was already aged 76, 42 years older than the duke.[160]

The alliance of these two strong-minded and disparate men — the landed aristocrat of long standing, always conscious of rank and obligation, and the self-made merchant — was forged by common interest, lubricated by Gibson's tact, and was, in the outcome, very successful. Together they constructed a low-water pier at Burntisland, and planned a steam ferry service at a cost of £15,000, which in spite of opposition from Leith opened in September 1844. An act of 1842 gave them a 25-year monopoly over both the ferry service and Burntisland pier.[161]

Gladstone's interest came from his position as one of the principal backers of Grainger's railway. This was now opposed by the project of an engineer called John Milne for a line from Burntisland to Perth via Lochgelly, Loch Leven and Glenfarg. Though Milne was dismissed by Grainger as having 'no experience whatsoever in the planning and execution of Railways', his line had the advantage of not requiring a Tay ferry, even if his bridge near Perth might well be opposed by the Admiralty and certainly meant a longer route to Dundee. In any case, Milne argued, Perth was a more important target than Dundee, since the latter was well served by boats.[162] Both sides paraded dubious statistics and more dubious estimates about the length of line required, the population served and the amount of coal traffic that might be expected; when Milne claimed to service twice as many people as Grainger, while Grainger claimed four times the population of his rival, potential investors may well have become chary of both of them. Grainger's prime consideration was to find the lowest possible summit level for his line, and for this he was willing to deny himself a line through the centre of the coalfield: the logic was in a direct line of descent from Stevenson's 'level line', and implied possible doubts about the continuing improvement of the locomotive.[163]

Planning railways in the severe depression of the early 1840s showed the faith of projectors in the future of the system, but made it difficult to attract sufficient support. Milne's Forth & Tay Railway disappeared from view, and although the Edinburgh Dundee & Northern went to Parliament in 1842 and 1843, it was without success. Gibson complained that 'the Fife proprietors have as usual been cold & indifferent to this proposed great improvement in their county'.[164] The projectors fought on: early in 1844 Gladstone was 'seriously engaged in promoting the making of a Rail Road across Fife', although on the same day he refused on the grounds of age to become chairman of the company (now named the Edinburgh & Northern).[165] Although he was still named as chairman in the prospectus issued three months later, the position went in fact to John Learmonth, who had been introduced to Gladstone by Buccleuch in 1842,[166] and whose ambitions envisaged an empire of lines radiating from Edinburgh. The creation of the Edinburgh & Northern came with the mania, and its arrival gave security to the Edinburgh

Leith & Granton, not so much because of the traffic generated but because it provided the Fife line with its only link to Edinburgh and the south. Indeed the Edinburgh & Northern could not afford to have this link under hostile influence, and after Learmonth had been forced to resign from the chairmanship of the Fife company in 1846 over a conflict of interests with his North British, there was a real fear that he might take over the Granton line. In a quest for security, the Edinburgh & Northern agreed in 1847 to amalgamate with the unfortunate and unprofitable Edinburgh Leith & Granton.[167]

V. From Glasgow to the Firth of Clyde

The two remaining companies to come out of the boom of the mid-1830s were authorised by consecutive acts in 1837. Both ran westwards from Glasgow and were to a greater or less extent in competition with the boats on the Clyde — the Glasgow Paisley & Greenock clearly so, since for much of its length it ran parallel to the river, the Glasgow Paisley Kilmarnock & Ayr perhaps more effectively as it reduced the journey time between Glasgow and the Ayrshire coast by four or five hours. Both on the other hand wished to co-operate with at least some shipping interests, since much of their anticipated trade was to be in conjunction with boats to western England, Ireland or beyond. And although they had different projectors, managers and engineers, the two companies managed to co-operate with each other to the extent that they promoted the railway between Glasgow and Paisley as a joint line, after learning that Parliament was unlikely to look kindly on two projects along more or less the same route.

This section of the route had seen some previous activity. In 1830, even before the Garnkirk had seen how well it could challenge the Monkland Canal, a company was formed to survey between Glasgow and Johnstone, along the territory of the canal which had originally set out for Ardrossan, but 'from the late season of the year at which the railway was started, and from other opposing obstacles, the undertaking was for the time abandoned'.[168] In the following year the committee, which included William Dixon and Charles Tennant, called upon Grainger and Miller for their survey. Their report was hopeful, estimating a revenue of £11,600 from passengers, £4600 from goods, working expenses of £4200, and a net revenue of £12,000, giving a return of more than 12% on the estimated £99,150 cost. Passenger services on the road between Glasgow and Paisley had increased from one coach per day in 1812 to 27 in 1830, while the canal carried 79,455 passengers in 1830-1. As later between Edinburgh and Glasgow, the engineers doubted whether the railway could challenge the canal for heavy goods, and they had left any consideration of mineral traffic out of their estimates:

> Within the last fifteen or twenty years, many railways have been formed both in England and Scotland for the transport of coal and other heavy commodities, but the advantages to be derived from this means of conveyance, as formerly understood, fall far short of those divulged by the experiments made on the Liverpool and Manchester Railway, in October 1829 . . . This

is . . . a railway, the revenue from which must depend principally on passengers and light goods, and where speed and safety are the great desiderata. In a locality, such as that now under consideration, the transport of minerals and of heavy commodities, must always be considered subservient to the more important objects in view.[169]

A counter-threat by the canal company to convert the canal itself into a railway was dismissed as impracticable due to numerous sharp curves. But in spite of the view of supporters that construction of the line would inevitably lead to its extension to Ardrossan, the project was started too late for submission to Parliament in 1831, and it became another casualty of the generally depressed condition of the economy.[170]

When the plan for the Glasgow Paisley Kilmarnock & Ayr (often known as the Glasgow & Ayrshire, or simply as the Ayrshire) emerged in the middle of the decade, conditions were more favourable. The dozen names on the original provisional committee included two from the Monklands railways, Tennant and George More Nisbett, but Tennant soon dropped out, and the leadership was taken by a group of Glasgow merchants led by James McCall of Daldowie. Grainger and Miller (in this case principally Miller) selected a route by Paisley and the Garnock valley which had a remarkably low 95' summit level and required very limited earthworks, and estimated the maximum cost to be £550,000 or a very reasonable £11,224 per mile, for what was already seen as a major project in itself but also as the possible first link in a route to England. Traffic estimates suggested an income from passengers of £62,713 and from goods of £80,399, working expenses of one-third gross revenue, and a net profit of £53,600 or almost 10% on capital. Once again the engineers had excluded any consideration of minerals, and on this occasion also of agricultural produce; significant amounts of both could certainly be expected, and the prospects were promising.[171]

The chosen route served Kilmarnock by an eleven-mile branch from Dalry, and thus gave the town a rail link to Glasgow twelve miles longer than by road. Kilmarnock Town Council commissioned Messrs Scott, Stephen and Gale (a firm which was not to make a mark on railway engineering) to find an alternative route. Their proposal cut five miles off the journey by crossing the high ridge south of Glasgow to go via Kilmarnock to Ayr: by a process of peculiar arithmetic based on the claim that 'there is no disadvantage in rising higher to redescend afterwards, provided the slopes are within certain limits', they even persuaded themselves that their gradients were favourable.[172] The Glasgow promoters settled the argument not only by trusting to the growing reputation of Grainger and Miller, but by bringing up the heavy artillery of George Stephenson. His report, seen as an arbitration by the Glasgow & Ayrshire projectors though not by the Kilmarnock group, was in favour of Miller's line 'even when viewed solely as a communication from Glasgow and Paisley to Kilmarnock', and comprehensively so as a link to the coast, preferable for its gradients, its probable working costs, its limited earthworks, its direct link to the north Ayrshire ports from Ardrossan to Troon, and its route along the foot of the Garnock valley, which allowed equally convenient access by traders on both sides. His calculation that the gradients on the direct line were equivalent to an extra thirteen miles on the level gave Miller an advantage of

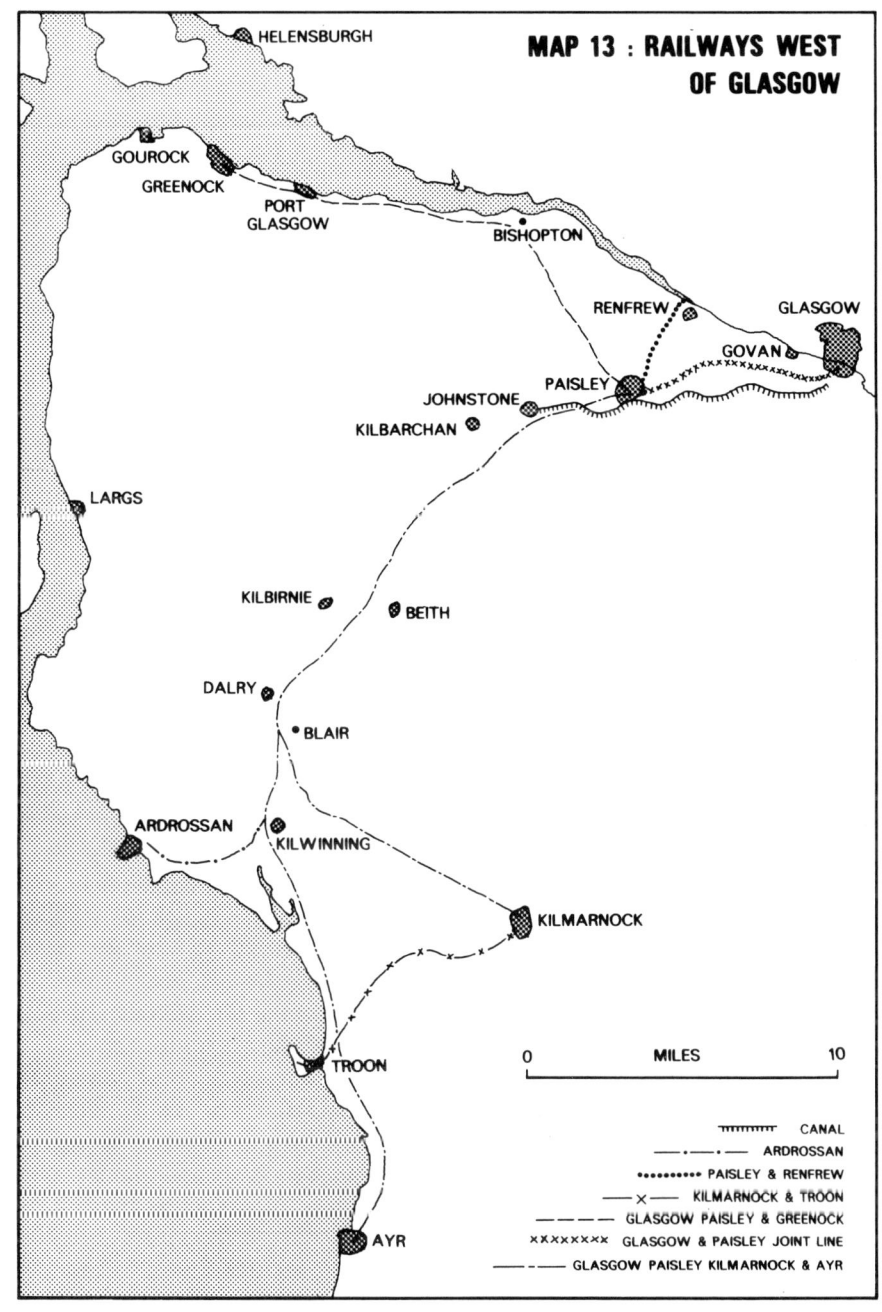

a hypothetical eight miles even to Kilmarnock.[173] Little more was heard of the Scott, Stephen and Gale plan, apart from an indignant pamphlet in which they complained that Stephenson had been employed by their opponents and had shown

> an extraordinary want of tact ... To have done the thing with apparent candour, Mr Stephenson ought really to have discovered some good points in our line, for which he should have pretended to give us credit.[174]

To do them justice, the lack of a more direct line from Kilmarnock to Glasgow remained a grievance until remedied in 1873, and they were also right to emphasise Kilmarnock rather than Ayr as the logical junction for a western route to England. But the prestige of their opponents, and in particular the still almost mystical deference given to the judgment of a great English engineer, ensured their defeat.

The Ayrshire was firmly founded on Glasgow merchant support. Of the twenty provisional directors named in the submission to Parliament, eleven were Glasgow merchants, five landowners on or near the line (including two earls), three coal- and ironmasters, and one a banker from Irvine.[175] The average subscription of the merchant directors was £8386, against the £2333 of their colleagues. Of the 85 individuals who subscribed £2000 or more to the project, 59 were merchants, 54 were Glaswegians, and 43 were both. The other significant group of large subscribers were the twelve Lancastrians, eleven of whom were merchants in Liverpool or Manchester.[176] Whereas the main impetus for the coal railways had come from the owners of the minerals, with the support of consumers in industry, the pressure for longer-range lines with more varied traffic was more and more from traders whose profit was made as middlemen. From the mid-1830s onwards, Glasgow merchants took the lead in Scottish railway promotion, and supplied a disproportionate amount of the capital; by the time of the Railway Mania, they were prominent in almost every one of the 170 or so Scottish mania schemes. In the case of the Ayrshire, they had an obvious interest in a route to the coast which both cut the time taken and challenged the monopoly of the river carriers; it was also expected that the railway would open up further coal and iron resources which, in the continuing absence of the Ardrossan canal, were too far from good transport to be worth developing.

The principal opposition to the project came from the proprietors of existing water transport, on both the Paisley Canal and the Clyde. Such an uncomplicated engineering route could hardly be said to be unsuitable for the construction of a railway, and all that opposition engineers could do was to impugn the estimates. This task was undertaken by two of the proponents of the defeated Kilmarnock line, William Gale (who was also the canal company's engineer), and Robert Scott, and Miller had some difficulty in explaining why he, on behalf of the Ayrshire, and Grainger, for the Greenock company, had produced substantially different surveys for a supposedly joint line between Glasgow and Paisley. Again the Ayrshire promoters called upon Stephenson, whose prestige and knowledge of parliamentary procedures had increased greatly since his unhappy experiences before the committee on the Liverpool & Manchester a dozen years before, and

Stephenson said all that they required.[177]

Argument therefore centred on traffic. Both sides were bound to agree that traffic existed and could be increased, although reputedly independent witnesses had to be called to assure the committee that this was not merely the general optimism of transport operators — the company called upon Alexander Guthrie, coal overseer to the Duke of Portland, Robert Farquharson, manufacturer and ex-provost of Paisley, and Councillor Hugh Miller of Ayr, harbour trustee and county commissioner of supply, who happened also to be the engineer's brother. The main claim of the opposition was that the railway could not compete on price — an argument not borne out by English experience, but more plausible if one assumed, as the canal company's counsel did, that the maximum rates proposed in the bill were the ones which would actually be charged. If passengers, for instance, were indeed to pay $3\frac{1}{2}$d per mile, the boat fare of 4/6d between Glasgow and Ayr would clearly be attractive however much longer the journey took, but in practice such a rate would have been abnormally high even for first class in a British context, and outrageous in a Scottish one. More likely were fares ranging from about 2d per mile for first class to 1d per mile for third, in which case the poor might save money as well as time by taking the train to the coast.[178]

The canal, which charged fares of 9d and 6d for the seven-mile journey between Glasgow and Paisley, could compete on cost but took up to $1\frac{1}{4}$ hours and had inconveniently located termini; Provost Drummond of Paisley considered that labourers would pay a little more for a considerable saving of time. Less attention was given to goods traffic, perhaps because Miller did not expect any diversion of heavy goods from the canal. Even so, ex-Provost Farquharson observed that, even although the canal charged 5/- per ton against the 6/6d–7/- of road carriers, many users including at least one canal proprietor preferred the carts as less damage was done to the goods and delivery could be door-to-door. The Paisley Gas Company, of which he was chairman, brought coal from Kilbirnie by cart for 7/- per ton, whereas the maximum railway rate allowed by the bill would be $3/1\frac{1}{2}$d.[179] Such a difference may explain why road interests did little to oppose the rail plan — in effect, road carriers across Britain now had little hope of claiming that they could offer as good a service as the railway over long distances, and were generally resigned to their fate. William Chaplin's evidence to the 1837 Select Committee on Railroad Communication was typical:

> 1080. You calculate that no coach travelling that can be planned can effectively compete with the railroad travelling? — In no shape or way; it is quite out of the question . . .
> 1100. As the system extends, you calculate upon a probability of a great diminution of the business of coach proprietors? — Annihilation is the best word to apply to it.

Since coaches charged on average $2\frac{1}{2}$d per mile outside and $4\frac{1}{2}$d – 5d inside, they were more expensive as well as much slower.[180] Goods carriers might find that sometimes lack of speed was less important, but they were still totally unable to compete on price.

The vigour of the opposition pushed the Ayrshire's parliamentary costs up to £20,000, but there was little real danger of the bill being rejected. Unlike the Edinburgh & Glasgow, the Ayrshire got its act in 1837 before the king's death curtailed the session, and also, as the directors observed, before a set of new and more stringent standing orders came into operation.[181] The differences over the Glasgow to Paisley joint line were ironed out, and the Ayrshire and Greenock companies agreed to divide the costs equally. The House of Lords committee unexpectedly insisted on inserting a clause to make the company spend money on construction equally at both ends of the line — a reminder that the line was among other things a successor to the ill-fated Ardrossan & Johnstone, and that powerful iron and coal interests in the Garnock valley wanted to ensure that the railway should provide their link to the coast as well as concentrating on the more immediately profitable Glasgow-Paisley line. The directors were unconcerned, calculating that once they could open from Glasgow to Paisley and from Ayr to Irvine there would be enough traffic for a 10% dividend.[182]

The Ayrshire company was primarily the creation of the Glasgow merchants. The Glasgow Paisley & Greenock, although authorised on the same day and sharing seven miles of joint line, was almost entirely ignored by them; in 1844 the company secretary could claim, even if not quite accurately, that the railway 'is not a Glasgow concern; we have not, I suppose, six shareholders in Glasgow'.[183] Glasgow trade had of course long been conducted by water, in the eighteenth century through Greenock and Port Glasgow, and then, as the Clyde was dredged and improved, increasingly from the city's own harbour. Merchants who had their interests and their money in Clyde river shipping would not support a rival railway along the river bank; those who wanted an alternative to the river put their capital and their effort into the Ayrshire railway, in spite of claims that the inferior harbours at Ardrossan and Ayr would not tempt shipping away from Greenock.[184] The main Glaswegian supporter of the project was not a merchant, but coal and iron entrepreneur William Dixon, who objected to the rates he was charged for moving his coal by water.[185]

The initiative, then, came from Greenock. This had been foreshadowed as early as 1802, when the magistrates of Glasgow showed no interest in a feeler from Greenock Town Council for the construction of a connecting waggonway.[186] When the idea resurfaced in 1835, the council, and Provost Macfie in particular, were again in the lead, supported by the MPs for the burgh and the county.[187] When the subscription list reached Parliament, it showed promises of £173,175 from Renfrewshire, against a mere £11,025 from the much greater resources of Glasgow.[188] The distinction was not surprising since the main purpose of the plan was to restore Greenock to its former place as the entrepot for the bulk of Glasgow trade. It was too much to hope that providing a railway would make all ships bound for the Broomielaw offload at Greenock instead, but some might prefer not to commit themselves to the tedious journey up the overcrowded river, where in 1838-39 there were 69 collisions.[189] Shipping charges of 5/- per ton and river dues of 1/6d made the river expensive, even if it avoided the need for an extra transshipment. John Poynter, a Glasgow merchant who also owned a mill at Greenock and was a partner in a steamboat company, told the House of Commons committee on the bill that 'I find it excessive to pay 7/- for carrying goods 22 miles which I can get conveyed to Dublin or Liverpool for 10/-'. He also anticipated an increase in traffic between Greenock and Paisley, where little went by road and the River Cart boats connecting with the Clyde were unsatisfactory. Although the Paisley & Renfrew had just opened, it had yet to reach agreement with the Clyde Shipping Company, whose boats passed its wharf during the night and refused to unload since the railway company had no one on duty.[190]. In the meantime, most goods from Greenock to Paisley went via Glasgow: for those which did use the river and the Paisley & Renfrew Railway, the rate was 7/- per ton, against an expected 3/9d (calculated at Liverpool & Manchester rates) on the new railway.[191]. In 1840, a year before the line opened throughout, it was expected to take 144,000 tons out of the 352,650 tons then carried between Greenock and Glasgow.[192]

Hopes of passenger traffic were even better. In the summer Glaswegians migrated en masse to the Firth of Clyde, and it was common ground between the Greenock and Ayrshire projectors that most of this traffic would come to the railways. The customers varied from impoverished day-trippers to 'some hundreds' of men like John Poynter, who lived in the city but also either owned a house near Greenock or rented one for their families for the whole of the summer. Poynter, who was in effect a summer commuter on the river, had kept a record of 150 round trips in 1835-36: the boats took an average of 2 hours 19 minutes downstream and 2 hours 31 minutes up, were frequently delayed by fog, and on occasion went aground.[193] The train, in contrast, would take about an hour. As the provost of Greenock remarked:

> Although the conveyances by river must be acknowledged to be cheap and comfortable, yet they are occasionally rather tedious ... The superior speed and certainty of the railway will induce, at all events, a considerable proportion of passengers to give it a preference.[194]

The railway promoters expected to divert half of the existing Glasgow-Greenock river passengers and a third of those travelling beyond Greenock; they also counted on half the cabin and a tenth of the steerage passengers on the canal. Already also they were looking to links beyond Greenock, and in 1839 came to an arrangement with the Loch Fyne boats by which a horse bus from their station to the pier would connect with a boat direct to Ardrishaig.[195] Overall the projectors' case sounded convincing, and the line was authorised over the strenuous antagonism of the shipowners.

The Greenock men who promoted their railway, though worthy local citizens, were not in possession of either the wealth or the connections of the great Glasgow merchants, nor did the line have wealthy and generous landed supporters like the Dundee & Arbroath's Lord Panmure. Since they could not depend on Glasgow money, they looked perhaps more than any previous Scottish company to attracting funds from the great centres of railway investment in England. To reassure the English they went one stage further than the other lines which had had their engineers' plans vetted by English experts: although Grainger had drawn up the original plans, he was not only checked by Joseph Locke, but was formally replaced by him as the company's engineer, in co-operation with his partner John Errington.[196] Locke, fresh from success with the Grand Junction, thus started the Scottish work which led on to the Caledonian and its adjuncts, and made him by far the most influential of the imported English engineers.

VI. A selection of failures

After the authorisation of the Edinburgh & Glasgow in 1838, there was a lull in Scottish, and indeed in British, railway promotion. No proposals for new Scottish companies reached Parliament in either 1839 or 1840, and, although several did appear in the next three years, only two small coal lines were authorised. A time for reflection, consolidation, and examination of the progress of the established companies may well have been desirable, and Matthews considers that the lack of activity was due to the prudence of projectors awaiting the results of the 1830s promotions rather than to any stagnation of business.[197] However, inactivity would have been encouraged by the depth of the depression into which the economy was subsiding; declining trade, falling prices, rising unemployment and increasingly idle resources meant that there was little obvious attraction in further railway expansion. Edinburgh stockbroker John Reid felt in 1841 that the great days of speculative railway construction were over:

> The days of speculation in new Railways are at an end; many heavy losses were sustained by speculating in some of them; these distressing facts have sobered down men's minds, and if any more Railways are brought forward to unite cities more closely, or to intersect the country, *they must be supported by the individuals interested in the localities where they are proposed to be established.*[198]

And Sir William Acworth, looking back in 1890, saw 1843 as the one year in which the railway system had briefly appeared to be in stable equilibrium.[199]

Nevertheless, projectors were still actively planning against the time when economic circumstances would be more propitious, and when, as they confidently expected, the achievements of existing railways would encourage investment in extending the system. Apart from the Fife railways discussed above, and the ambitious hopes for Anglo-Scottish lines examined in chapter 5, projectors contemplated ideas as different as further coalfield extensions in Lanarkshire and Fife on one hand, or opening up the Highlands on the other. Table 28 lists the bills which reached parliamentary consideration, some on more than one occasion, in the early 1840s.

Plans to extend the railway system to Aberdeen had been considered in the 1820s by Stevenson and others; by the end of the next decade not only had a further survey to the city been made, but the idea of extension to Inverness had been floated.[200] In 1840 at least 43 assorted meetings were held in Aberdeen, Inverness, Elgin and other northern towns to plan the advent of the railway; further action was, however, delayed, partly by disagreement as to whether a line to the south from Inverness should go by Aberdeen or by the Drumochter Pass, and partly by a chronic shortage of money.[201] A petition from Inverness in March 1841 requested government aid for a railway survey: in the following month the authorities in Aberdeen asked that the Commission appointed to examine possible trans-border routes should be instructed to extend its investigation as far as the

city.²⁰² The *Railway Times* regretted this tendency to turn to the state for help:
> We should like much . . . to see this spreading desire to participate in the benefits of the Railway System a little less tainted with a coincident desire to dip into the public purse . . . These Hill people have been too much and too long accustomed to draw on the public Treasury for the expenses of all their local improvements, (witness that great job, the Caledonian Canal) . . . There are, we believe, not a few districts of Scotland in which a railway would return a very fair rate of direct profit, but the main inducement must be a consideration of the indirect advantages which such undertakings, whether lucrative to the promoters or not, never fail to secure.

The journal also believed that any line to the north must go by Aberdeen: as for the Drumochter route from Perth to Inverness:
> excepting shepherds and sheep, with a few sportsmen and dogs in August, we know of no other living creatures betwixt the two towns whom it would accommodate.²⁰³

No government aid was in fact forthcoming, and the resources of the north were as yet unable or unwilling to finance lines of a length as yet unattempted in Scotland through sparsely populated areas, whatever might or might not be the indirect advantages. Before the mania, only the little line planned by some citizens of Elgin to give them access to the coast at Lossiemouth even got as far as a survey.²⁰⁴

Table 28. Bills Deposited for New Railway Companies in Scotland, 1841-43

Company	Year deposited			
Bathgate & Slamannan			1843	
Brechin	1841			
Caledonian (1 and 2)			1843	
Drumpellar			1843	authorised
Edinburgh Dundee & Northern		1842	1843	
Gartsherrie & Stanrig		1842	1843	
Glasgow & Dundyvan Junction	1841	1842		
Glasgow & Hamilton		1842	1843	
Glasgow & Hamilton Junction	1841	1842	1843	
Halbeath & Lochgelly			1843	
Inverkeithing Halbeath & Lochgelly	1841			
Leven Valley		1842	1843	
Lochgelly & Kirkcaldy			1843	
Shotts & Wilsontown (later Wilsontown Morningside & Coltness)	1841			authorised

Source: PP 1843(571)XLIV, *Return of Railway Bills and Acts, 1839-43*

A line as far as Aberdeen was a reasonable proposition in terms of both engineering and potential traffic, but it was also reasonable to wait at least until a link had been made from Glasgow to Perth. Aberdeen was well served by shipping services to the rest of the east coast, and a railway which reached only Perth or Dundee would be of limited value — the important connection was to the mineral and industrial area in the west. Lines into the Highlands were as yet visionary; even in the mania, when many such visions were promoted and some achieved authorisation, it was found impossible actually to bring any of them to fruition.

More surprising in the early 1840s was the failure to make much advance into the coalfields, where the early Monkland lines had demonstrated that profits could be handsome even in a depression. Yet between 1840 and 1843 expansion by existing companies was limited to a small branch of the Ballochney. Of the two companies authorised one, the Drumpellar, was little more than an old-fashioned waggonway attached to the Monkland Canal; it proposed a line of less than two miles and actually built less than one. The other, which entered Parliament as the Shotts & Wilsontown and left it as the Wilsontown Morningside & Coltness, extended from the Wishaw & Coltness towards an unexploited coal area and the isolated Wilsontown Ironworks.[205]

Other coal projects failed. Sometimes the failure was not surprising. Attempts to link the Lochgelly coalfield in Fife to the sea, either by extending and reconstructing the old Halbeath waggonway or by a new link to Kirkcaldy, made sense in isolation, but had to wait until the route of the main line through the county was fixed. Back in the Monklands, a proposed line from Gartsherrie to Stanrig was a blatant attempt to divert traffic from an area already developed by the Ballochney, and was in any case unlikely to pass a House of Commons which was inclined always to protect the rights of established concerns. But the existing complex of lines between the Monklands and Glasgow had one obvious fault — in spite of all efforts, there was still no rail connection to the north side of Glasgow harbour and, after the rejection of the Blythswood tunnel, it was difficult to see how one could be achieved. Hence proposals were made for connections between the coal and the south side of the harbour, using or replacing the old Polloc & Govan. One plan, the Glasgow & Dundyvan Junction, revived the plan of the abortive Rutherglen Railway of 1831, running from the Polloc & Govan across the Clyde to a junction with both the Monkland & Kirkintilloch and the Wishaw & Coltness near Coatbridge. In spite of Glasgow merchant backing and the name of Locke as engineer, the plan was quashed by the opposition of existing railways (particularly the Garnkirk, which had much to lose) and by financial stringencies.[206]

Another sign of the times was the failure of the £300,000 Glasgow & Hamilton, which issued a prospectus in 1841 and came before Parliament in the following two years. Its sponsors were distinguished. Unusually, the committee appointed a governor, perhaps to suggest reliability by an analogy with banking, but also because the Duke of Hamilton was willing to accept the position. The six provisional directors represented the cream of Glasgow commerce. They were led by Kirkman Finlay, perhaps the greatest of Glasgow merchant-manufacturers, and long-time chairman of the Forth & Clyde Canal, who had previously restricted his involvement with railways since he was unsure that they had advantages over canals. With him were John Leadbetter, chairman of the Edinburgh & Glasgow; James McCall, chairman of the Ayrshire; James Dunlop of Clyde Ironworks; and two leading merchants, Theodore Walrond and J.G. Hamilton. Their line would open up a new section of coalfield south of the Clyde, would not interfere with the territory of any other railway, and would of course have access to the harbour over the Polloc & Govan. They made out a persuasive case for an annual traffic of £76,000, which would allow a return of over 13% on the estimated cost even after deducting a substantial 40% for working expenses. Although they intended to charge $2\frac{1}{2}$d per ton-mile for minerals against the 2d common on the Monklands lines, they would still be much cheaper than rival road transport. They were even well placed to become part of an eventual Anglo-Scottish route. Yet applications to Parliament failed in both 1842 and 1843, partly perhaps because of the existence of a rival, if less impressively backed, Glasgow & Hamilton Junction scheme.[207]

VII. Sources of capital: an examination of subscription contracts

Since pre-mania share registers for the Scottish inter-urban railways of the 1830s have not survived, the subscription contracts which the companies had to submit to Parliament are the best available record for the sources of their finances. Even so, the contracts have for many reasons to be treated with caution. Successful applicants for shares in a company were informed in a letter of allotment of the number of shares which they had been allocated, and on payment of their deposit received either a scrip certificate which would be converted into shares on authorisation of the company, or a banker's receipt which could in turn be exchanged for the scrip certificate. The scripholder in turn signed the subscription contract, and undertook an obligation to pay the calls on his shares when they fell due. Scrip certificates were however made out to bearer and might be transferred, although the obligation to pay calls remained with the original signatory unless and until the new holder registered his claim to the shares with the company. A market developed not only in railway scrip, but even in letters of allotment and bankers' receipts, so that although companies anticipated that the signatories to the contract would actually pay the calls a year or two later, in practice the holding might have passed through several hands in between, with the railway being powerless to prevent it.[208]

Some subscribers might be expected to dispose of their holdings rapidly. In some cases the companies certainly connived at the process. They wished to impress Parliament both with the respectability of their subscribers and with the volume of local support for their line; individuals who came into one or both categories might well sign the contract on the understanding that they did not intend to be long-term holders. They might be discouraged from a lengthy commitment by the time that would elapse before they could hope for a return on their investment — even although their money would be extracted gradually as the line was constructed, it would be at least three or four years from subscription contract to dividend, and in most cases much longer. Also disheartening was the fact that until authorisation a company was still only a private partnership with unlimited liability, so that cautious investors might hold back from involvement until the act was secured.[209] In any case an ordinary scripholder might well be tempted by a rise in the market to sell for the capital gain rather than await the uncertainties of the dividend.

Companies were also faced by stock exchange speculators. As far as possible promoters would limit their lists to the respectable and the local, and if their project was oversubscribed they might be able to keep a close control on their allocation of scrip. But if subscriptions were difficult to find, they might have to accept applications from men of uncertain reputation who merely intended to gamble in the scrip, some of whom might well have no chance of ever paying calls if they were unable to dispose of their holdings before authorisation. A Scottish company might find it particularly difficult to check on English applicants; to directors in Greenock or Dundee, an unknown applicant describing himself as a 'gentleman' and writing from an address in the City of London might be almost anything. Even by 1837 the Select Committee on Subscription Contracts uncovered a labyrinth of malpractice and fraud by both promoters and subscribers, mostly on railway schemes in the south of England promoted in the London market. To some extent the limited London contacts of almost all pre-mania Scottish companies may increase the trust which can be placed in their subscription contracts. It is however quite possible that they contain examples of false names, forged names, bogus business descriptions, accommodation addresses (such as the use of domestic staff in large houses) to add respectability, applications from bankrupts, payments to subscribers, and the numerous other misdemeanours revealed by the Select Committee.[210]

Sometimes the contracts suggest that there may have been confusion rather than dishonesty. It is unlikely that the promoters of the Edinburgh & Glasgow, which was a well-subscribed company, would have deliberately falsified some of their largest subscriptions, where any malpractice would have been most likely to be noticed. But did two military men from the 29th Regiment, Assistant Surgeon John Hankey and Captain Christopher Humfrey, really empower promoter Charles Davidson to sign the Edinburgh section of the company's contracts on their behalf while they were stationed at Tralee, for sums of respectively £12,000 and £4,500, and then when they had moved to Plymouth sign the Liverpool section themselves for the same sums? Londoner John Shewell appears on both the company's Liverpool and Manchester lists for £10,000; did he really intend to take on a £20,000 commitment? Similarly London bookseller John Andrews signed for £11,300 on the Manchester list, while in Edinburgh Charles Davidson claimed that he was mandated to commit him for a further £10,300. It cannot be proved that any or all of these and similar cases are deliberate or unintentional duplication, but they do add to the doubts about the value of subscription contracts.[211]

As far as possible company promoters and officials preferred to make the allocations themselves, but the increasing involvement of English capital in Scottish schemes inevitably led to some dependence on English brokers. The Greenock company appointed their Liverpool broker specifically to vet the credentials of applicants in the area, although they also encouraged him to dispose of as many shares as possible.[212] In general the fact that the brokers were paid by commission, and had little real long-term interest in the prosperity of the company once it was established, might well encourage them to accept subscribers with only the most cursory check on their *bona fides*.[213] The choice of locations for brokers might also be expected to affect the geographical balance of subscriptions; certainly agents in Lancashire were an early priority for most companies, and the bulk of English railway capital in Scotland came from Lancashire. But the chain of causation is the other way round — the known capital resources and enthusiasm for railways of the Lancastrians induced the companies to concentrate their efforts in the county. Agents in other areas, such as those appointed by the Greenock company in Birmingham and Dublin, often showed little in the way of positive results.[214]

Subscribers, and in particular distant subscribers, also had to some extent to take on trust the credentials of a scheme's projectors; it was not unknown for respectable subscribers to pay the deposit on their subscriptions and then find that the promoters had no real intention of going through with their scheme or of repaying any money. In the 1830s, however, Scotland does not seem to have suffered from such projects, although by the time of the mania the country had its share of fly-by-night promotions. One piece of apparently sharp practice by promoters does appear in the closing days of the failed Edinburgh Glasgow & Leith project in 1833. As a committee member explained:

> There is a class of subscribers in Liverpool, who are mere "applicants", and who refuse to pay on the ground, that by the law and custom of England, they are not liable for either of the two calls, not having been informed, that shares were assigned them, or asked to sign the Parliamentary undertaking.

It appears that the committee was attempting to spread the financial burden of the project's failure by bringing in people to whom it had, for whatever reason, decided not to issue scrip at the time of the original allocation. 'Tho' confident that the applicants might be compelled to pay', the committee was not in fact prepared to take the issue to the courts.[215]

Parliamentary orders on the amount of capital to be covered by a subscription contract varied. In 1837, as the three main pre-mania lines sought authorisation, the House of Commons insisted that half of the proposed share capital should be subscribed before the petition for the bill could be presented, and two-thirds before the committee on the bill reported; the House of Lords increased the requirement to five-sixths before giving a third reading.[216] Thus contracts could be collected in sections, with extra evidence going to Parliament as required. Since the first names to be gathered would be those in the locality of the line and those elsewhere known personally or by reputation to the promoters and their agents, it may be that contracts submitted during the earlier stages of a bill's progress may overemphasise local enthusiasm. Certainly the Glasgow Paisley & Greenock, where two-fifths of the subscriptions eventually came from outside Scotland, depended almost entirely on its Scottish lists to satisfy the first procedural requirements of the Commons. Fortunately, of the surviving contracts in the House of Lords Record Office for the six companies under consideration in this chapter, only that for the Arbroath & Forfar covers less than 82% of the proposed capital, and other evidence indicates that its financing remained almost entirely local until authorisation.[217]

Subscription contracts are best regarded as indicators of the sources from which promoters expected to obtain their finance, rather than of the eventual balance of shareholding in practice. Reed's analysis of a number of English companies suggests that, among other examples, only 25% of the subscriptions to the London & Birmingham in 1833 were still held (as shares) by the same individuals in 1837, and only 24% of North Midland subscriptions in 1836 were retained until 1842; in the Great Western it even appears that 52% of the allotments made in 1835 were disposed of by the allottees within a year.[218] English companies looked largely to London and Lancashire for finance: the London market was notoriously volatile, and Lancashire subscribers to lines which were not directly connected to the county might also be looking for a quick profit on their investment. Pre-mania Scottish lines, with little connection with London financial resources and often with substantial local support, might well have a more stable shareholding pattern, but in the absence of share registers this cannot be proved.

With all these qualifications, the subscription contracts of the Scottish lines authorised in the late 1830s probably give a reasonable indication of the social status and location of their supporters. Those for the three lines which passed Parliament in 1836 are however sadly uninformative, recording only the names of subscribers and the sums promised. Table 29 breaks down these amounts by size of subscription, and indicates one apparent difference between the three relatively small companies, with the Arbroath & Forfar being heavily biased towards the small subscriber, the Edinburgh Leith & Newhaven depending more on substantial individual subscriptions, and the Dundee & Arbroath in an intermediate position. As already noted, the Arbroath & Forfar was very much a locally supported line, running between two towns of only moderate size and wealth through a farming area which, in view of its lack of lime, was also of modest prosperity.[219] It is therefore not surprising that, apart from the large subscriptions of Lindsay Carnegie, Panmure and the two town councils, the company depended on some moderate sums from Arbroath merchants and on small sums from numerous local supporters: 57.2% of subscribers took only one or two of the £25 shares. The table uses a surviving and presumably early subscription list covering less than two-thirds of the proposed share capital: by the time that the parliamentary committee on the bill reported, there were 459 subscribers offering a total of £56,175. Of the subscribers, 91.1% were local, 2.8% had trading links with the area, and only 6.1% were outsiders; as a proportion of the amount subscribed, the respective figures were 88.6%, 3.1% and 8.3%.[220]

The recognisable names on the Dundee & Arbroath list suggest that it too was essentially a locally supported line, but corroborative evidence has not survived to the same extent as for the Forfar line. A third of subscribers elected to offer £500 — a sum perhaps compatible with the status of a successful merchant in an important provincial centre — while the average subscription was almost $2\frac{1}{2}$ times that to the Arbroath & Forfar. Even so, small subscribers were still numerous, whereas on the Edinburgh Leith & Newhaven (in spite of the shares being issued

Table 29. Subscriptions to Scottish Railways, 1836, by amount

Subscription per head £	Arbroath & Forfar				Dundee & Arbroath				Edinburgh, Leith & Newhaven			
	Subscribers		Amount		Subscribers		Amount		Subscribers		Amount	
	No.	%	£	%	No.	%	£	%	No.	%	£	%
2,000+	4	1.0	11,000	24.4	3	0.8	7,500	7.8	11	5.4	24,600	23.3
1,000-1,975	1	0.3	1,000	2.2	15	4.2	15,700	16.3	25	12.2	25,900	24.5
750-975	1	0.3	750	1.7	1	0.3	875	0.9	10	4.9	8,300	7.9
500-725	9	2.3	4,500	10.0	65	18.4	32,550	33.9	24	11.7	13,500	12.8
400-475	4	1.0	1,650	3.7	6	1.7	2,400	2.5	41	20.0	16,400	15.5
300-375	6	1.5	1,800	4.0	25	7.1	7,550	7.9	10	4.9	3,000	2.8
200-275	17	4.4	3,700	8.2	82	23.2	17,200	17.9	47	22.9	9,400	8.9
100-175	105	27.1	12,300	27.3	106	29.9	11,475	11.9	47	22.9	4,700	4.4
25-75	231	59.5	9,225	20.5	54	15.3	2,400	2.5	—	—	—	—
Total	388		45,075		354		96,050		205		105,800	
Average holding			116				271				516	
Proposed share capital			70,000				100,000				100,000	
% covered by subscription contract			64.4				96.1				105.8	

Sources: HLRO, HC Deposited Plans of the various railways.

in £20 denominations) no offer of less than £100 was accepted, and almost half of the total amount, which strangely exceeded the proposed share capital, was raised in four-figure sums.

At first sight this pattern might appear reasonable for a line connecting a capital city to its port, where wealthy local merchants and professional men could be expected to contribute. In fact, however, the city's commercial and social leaders held back from the scheme; the most prominent Scottish subscriber was Sir Andrew Leith Hay, MP for Elgin Burghs and the first chairman of the company (although by 1838 the internal wranglings of the board had led to his resignation).[221] Instead the company was the first in Scotland to go to the English market for a large part of its subscription list. The list includes a number of prominent Lancastrians, including Richard Dawson (later to be company chairman during its troubled period of construction), William Houldsworth, Edward Braithwaite and three members of the Brancker family. From Leeds came Christopher Saltmarshe, later to be one of the most prominent Englishmen in Scottish mania promotions. The correspondence between Henry Wait Hall and the Duke of Buccleuch indicates that there was a group of supporters in Bristol, while by November 1837 there were at least 23 London subscribers who were so disgusted with the company that they petitioned for its dissolution.[222] According to another malcontent shareholder, publishing anonymously from Bristol, the subscription list was filled:

> chiefly ... by parties resident in Liverpool and Manchester, (for the Scotch interest appear, from the commencement, to have been too wide awake to embark their money, to any considerable extent, in such an undertaking) ... .[223]

Looking to English sources for finance was not of course new. English contributions to such lines as the Ballochney and the abortive Edinburgh & Glasgow scheme of 1830 have already been noted; Gourvish and Reed claim that, in the early 1830s, 31% of the Ballochney's shares were held by fourteen London residents, although their Londoners include such solidly Scottish individuals as the Earl of Breadalbane and Sir Gordon Drummond. Some apparently English money in this early period certainly came from exiled Scots, and John Gladstone's important holding in the Monklands lines was built up while he was still a resident of Liverpool.[224] Englishmen with no Scottish connections were also showing more interest in projects in the distant north; Lancastrians in particular, conscious of their importance in the commercial world, of the lead in railway experience given to them by the Liverpool & Manchester, and of the efforts already being made by the Grand Junction to promote links northward, were willing to contemplate Scottish investment. Even so, the Edinburgh Leith & Newhaven was the first Scottish company to be largely promoted by Englishmen (even if the original board of directors was composed of Scots), and to look for most of its subscriptions south of the border. Certainly Richard Dawson was the only Englishman to become chairman of a pre-mania Scottish railway.

The events of the mid-1830s period of railway promotion and speculation emphasised the extent to which shareholding was being undertaken purely in the hope of a good return on the investment. Promotion might still be undertaken primarily by men like the merchants of Glasgow, Edinburgh or Greenock, but the scale of the projects increasingly forced them to look for funds beyond their own localities, from investors whose other interests would be in no way affected by the existence of the railway; the two Angus lines of 1836 were the last Scottish railways of any size to be almost entirely financed from local resources. Fortunately the need for large inputs of outside money and the willingness of English speculators in particular to supply it arose together in the mid-1830s promotional enthusiasm. English money could however be more volatile than funds coming from local interests. Kinnaird's Dundee & Perth committee found in 1836 that English subscribers led the withdrawal which led to the collapse of the project. Although the committee laid part of the blame on a general downturn in the market, they made a specific complaint which suggested that the parliamentary efforts of James Morrison and his allies to apply governmental control to railways were having an effect:

> One principal cause was the dread entertained by Capitalists of the contemplated measures limiting the profits arising from these undertakings, or, at all events for reserving a power to the legislature to revise the several Acts at stated periods with the view, if necessary, of making a reduction in the rates of carriage.[225]

Potential users of the railway were more likely to remain loyal to the project, and it is reasonable to assume that, if all other factors were equal, promoters would wish to attract as many of their subscriptions as possible from the area of the line. Turning to England, or to Glasgow by companies in other parts of Scotland, was in itself an indication of the inadequacy of local resources.

English money came preponderantly from Lancashire, the great early centre of railway investment. The only other potentially important English source was London, the centre of financial activity and Stock Exchange activities, which had the mechanisms for channelling investment into joint-stock companies. In this respect Scottish companies were in much the same position as those in other parts of England, where in the late thirties companies as varied as the London & Birmingham, the Eastern Counties and the Stockton & Darlington all drew more than half of their share capital from Lancashire and Cheshire.[226] It is not obvious at this time how far Scotland should be regarded for railway purposes as a separate country, and how far simply as a region of a Britain dominated by the financial power of Lancashire and London. Certainly the structure of the capital market did not make it any more difficult for funds to flow from Lancashire to Scotland than to Bristol or Newcastle. On the other hand, there was clearly a tendency in England to regard Scottish railways as in some way set apart. The railway press was prepared to generalise about them as a group in a way that it would never have done about the railways of Yorkshire or East Anglia. Their pre-mania physical detachment from the English railways heightened this sense of distinctness, and even when transborder links were achieved there remained a very clear demarcation between the Scottish and English systems at Carlisle and Berwick. Relatively few English investors would have personal knowledge of Scotland, and investment there might involve more complex judgments than investment in more familiar territory to the south. Even so, English money flowed freely into almost all important Scottish lines from the mid-1830s onwards.[227]

The subscription contracts of the Edinburgh & Glasgow, the Ayrshire and the Glasgow Paisley & Greenock are much more informative, giving each subscriber's address and occupation or social status. The information is tabulated in Tables 30 and 31, with a more detailed breakdown in the Appendix. The geographical categories used are for the most part self-evident; Local Counties include all subscribers in the counties through which the line proposed to run, but not in Glasgow or Edinburgh, while Local Total includes Glasgow and where appropriate Edinburgh. Welsh subscriptions are included with Other rather than with England; they are not in any case of great significance, amounting to 1.1% of the total for the Edinburgh & Glasgow, 0.5% for the Ayrshire, and nothing at all at Greenock. The few subscribers who gave more than one address have been allocated to the one nearest to the area of the line; similarly, some Glasgow merchants who might subscribe either from their business addresses or from their residences outside the city have been allocated to Glasgow rather than to the surrounding counties. London has been taken to include all of Middlesex, and the parts of Kent, Surrey and Essex closest to the city. On the other hand, areas of Cheshire along the Lancashire border, in which businessmen from Liverpool and Manchester might well reside, have not been included in Lancashire because of the difficulty in identifying these individuals. Again any distortion of the figures is not severe, since subscriptions from the whole of Cheshire amounted to 1.2% of the total for the three railways.

Occupational groups are in many ways more arbitrary and require more explanation. Eleven headings are used:

> Non-occupational: includes aristocracy, baronets and knights, esquires, gentlemen, residenters, and MPs where no other occupation is known. Esquires, gentlemen and residenters form the most difficult groups of all. A 'gentleman' on a subscription list might be an individual generally recognised as of social standing appropriate to the description, but he might also be someone whose real occupation was so lowly as to make it unlikely that a committee of management would allot him any shares. According to a witness before a select committee in 1837, 'in railway offices, a man who has no business is necessarily a gentleman'.[228] Except where a subscriber is known from some other source to have a more specific occupation, it is necessary to accept at face value the description on the contract.
>
> Land: includes farmers, tenants, factors, etc. Most of the non-occupational subscribers would in fact be landowners, and the true involvement of the landed interest is probably not severely overstated by taking the two groups together.
>
> Trade: includes wholesalers, retailers, shipowners, innkeepers, brokers, agents, road and canal transport operators, etc. By far the greatest number described themselves as merchants, which in Scotland could mean either wholesale or retail traders. The brokers and agents, whose concern was to trade in the shares as commodities in their own right or sometimes to hold shares as a nominee for an individual who preferred to remain obscure, are few in number, although they were to become very prominent during the mania.
>
> Industry: includes manufacturing, mining, building, publishing and printing, etc. Borderline groups who both manufactured and retailed their goods — such as bakers or tailors — have generally been allocated to trade unless they are known to have been primarily manufacturers. In most cases they were small men with small subscriptions.

Law: a straightforward category.
Finance: includes banking, insurance.
Professions: includes doctors, clergymen, accountants, architects, educationalists, officers in the forces, the merchant navy and the East India Company service, and senior civil servants and local government officers.
Contracting: includes contractors, engineers and surveyors — groups which might hope to gain directly from employment on railway construction.
Miscellaneous: includes clerks, craftsmen, servants, and a motley collection of occupations from artist to hairdresser. A heterogeneous group, fortunately not numerous.
Women: includes all women, even where an alternative occupation is given. Almost all were either widows or spinsters; married women could not yet legally hold shares, but on the few occasions that a married woman's name appears on a contract, the subscription has been credited to her and not to her husband.[229]

Clearly there may be difficulties of classification at the margin of the groups, but overall they offer a tolerably coherent and logical coverage. Two categories — the Yorkshire location and the Contracting occupational group — may seem too insignificant to be treated separately; both were to become much more important, and are here distinguished so that comparisons with mania railways may be made in future. The North British, for instance, in 1844 drew 19.6% of its subscriptions from Yorkshire and 12.1% from Contracting.[230]

In the light of previous discussion in this chapter, the figures hold few surprises. The Ayrshire is confirmed as a creation of the Glasgow merchants, with 44.5% of the money coming from the Glasgow Trade category, while the next largest category (Lancashire Trade) contributed only 8.0%. Glaswegians contributed larger individual sums — an average of £2,216 against £829 for other Scots and £1,772 for English subscribers; one may however note here the massive £11,500 subscription of John Wilson of Dundyvan ironworks in Lanarkshire, who had little immediate chance of making much use of the line although he may have hoped for the success of one of the projects to link the Monklands with the railways on the south side of Glasgow harbour. Support along the line was substantial, with the 141 subscribers from Renfrewshire and Ayrshire actually being greater by eight than the number of Glaswegians. Scots contributed five-sixths of the total, while Lancastrian support was comparatively modest and that of the rest of England a mere 5.2%. It is not clear whether Lancashire was not asked to subscribe more, or whether railway interests in the county, already hoping for a rail link northward from the Grand Junction, were discouraged by the Ayrshire's apparent enthusiasm for co-operation with connecting steamships. The Ayrshire was always to be an essentially Glaswegian line with only limited English support. By the time that its extension to Carlisle and the English lines was authorised in 1846, the Grand Junction and its Lancastrian backers were firmly committed to the rival Caledonian route. In 1851 the company (by now the Glasgow & South-Western) was excluded from the Octuple and Sextuple Agreements which carved up Anglo-Scottish traffic between the east and west coast routes, and not until 1876, when the Midland reached Carlisle, could it find a useful English ally.[231]

The tables indicate how little the Greenock company owed to Glasgow. Over half of the money came from Renfrewshire; indeed, 45.4% of it and 52.8% of

Table 30. Subscriptions to Scottish Railways, 1837–38, by Location and Occupation of Subscriber

	% of amount subscribed					% of number of subscriptions			
	Glasgow Paisley Kilmarnock & Ayr 1837	Glasgow Paisley & Greenock 1837	Edinburgh & Glasgow 1838	Total	Glasgow Paisley Kilmarnock & Ayr 1837	Glasgow Paisley & Greenock 1837	Edinburgh & Glasgow 1838	Total
By location:									
Glasgow	55.8	3.3	16.7	26.1	36.8	3.2	16.4	16.0
Edinburgh	7.6	2.1	13.6	6.8	8.9	1.7	16.0	8.3
Local counties	17.3	52.3	0.4	15.6	39.0	62.6	1.4	35.9
Local total	73.1	55.7	30.7	48.5	75.9	62.6	33.7	57.4
Other Scotland	3.0	3.2	2.2	2.6	1.9	2.9	3.8	3.0
Total Scotland	83.7	60.9	32.9	53.9	86.7	70.3	37.6	63.2
London	2.7	4.1	11.6	7.4	2.8	1.7	5.7	3.3
Lancashire	9.8	29.6	42.1	29.8	7.8	22.1	48.1	27.4
Yorkshire	0.8	—	1.9	1.2	0.6	—	1.0	0.5
Other England	1.8	3.9	7.1	4.8	1.4	3.5	4.4	3.3
Total England	15.1	37.6	62.7	43.2	12.5	27.3	59.2	34.5
Other	1.3	1.5	4.1	2.7	0.9	2.3	2.8	2.1
Unknown	—	—	0.4	0.2	—	—	0.4	0.1

By occupation:								
Non-occupational	10.2	10.3	15.0	12.6	11.1	4.9	10.1	8.2
Land	0.4	1.5	0.1	0.5	0.8	2.2	0.2	1.2
Trade	62.5	55.9	53.3	56.6	48.5	54.7	54.5	53.1
Industry	5.9	9.5	7.1	7.2	6.6	11.6	9.9	9.8
Law	8.1	7.8	9.4	8.7	13.0	6.2	9.5	9.0
Finance	1.2	4.7	2.4	2.5	1.9	2.9	2.2	2.4
Professions	7.6	7.5	10.1	8.8	9.4	9.1	8.5	8.9
Contracting	0.7	0.1	0.4	0.4	0.6	0.7	0.6	0.6
Miscellaneous	0.4	1.5	1.2	1.0	1.7	4.4	2.4	3.0
Women	2.9	1.1	0.6	1.4	5.5	3.5	1.6	3.4
Unknown	0.2	—	0.4	0.3	0.3	—	0.4	0.2

Sources: HLRO, HC Deposited Plans of the Various Railways. Also see Appendix.

Table 31. Subscriptions to Scottish Railways, 1837–38, by amount

Subscription per head £	Glasgow, Paisley Kilmarnock & Ayr				Glasgow Paisley & Greenock				Edinburgh & Glasgow			
	Subscribers		Amount		Subscribers		Amount		Subscribers		Amount	
	No.	%	£	%	No.	%	£	%	No.	%	£	%
10,000+												
7,500–9,975	1	0.3	7,500	1.4	1	0.2	8,000	2.4	8	1.6	92,200	10.7
5,000–7,475	37	10.2	185,000	35.0	3	0.5	15,000	4.5	2	0.4	17,100	2.0
4,000–4,975					2	0.3	8,750	2.6	20	4.0	115,600	13.5
3,000–3,975	6	1.7	18,000	3.4	3	0.5	9,950	3.0	15	3.0	69,050	8.0
2,000–2,975	58	16.1	139,450	26.4	28	4.7	68,125	20.6	55	11.1	173,400	20.2
1,000–1,975	111	30.7	124,750	23.6	86	14.4	101,675	30.7	31	6.3	72,650	8.5
750–975	5	1.4	3,750	0.7	21	3.5	15,775	4.8	161	32.5	220,200	25.6
500–725	76	21.1	38,450	7.3	95	15.9	48,800	14.8	13	2.6	11,100	1.3
400–475					5	0.8	2,150	0.7	111	22.4	66,350	7.7
300–375	3	0.8	900	0.2	11	1.8	2,775	0.8	1	0.2	450	0.1
200–275	35	9.7	7,450	1.4	124	20.8	30,600	9.3	67	13.5	20,450	2.4
100–175	25	6.9	2,700	0.5	140	23.5	15,600	4.7	6	1.2	1,300	0.2
25–75	4	1.1	200	0.0	77	12.9	3,625	1.1	4	0.8	400	0.0
									1	0.2	50	0.0
Total	361		528,150		596		330,825		495		859,500	
Average holding			1,463				555				1,736	
Proposed share capital			625,000				400,000				900,000	
% covered by subscription contract			84.5				82.7				95.5	

Sources: HLRO, HC Deposited Plans of the various railways.

subscribers came from the town of Greenock itself. In the absence of Glaswegian support, the company turned to Lancashire for most of the rest of its money. The confidence-boosting appointment of Locke helped the fund-raising efforts led by Patrick Maxwell Stewart, MP for Lancaster and brother of Sir Michael Shaw Stewart, MP for and major landowner in Renfrewshire, and an enthusiastic supporter of the railway. Lancastrians subscribed for 30% of the company capital, with the bulk (21.7%) coming from merchants and traders in Liverpool. For a line promoted by the citizens of a relatively unimportant town and with its prospects of success doubtful in the face of river and canal competition, it was an achievement to organise so much southern support even at a time of speculative fever.

The Edinburgh & Glasgow, as an important line even by British standards, and apparently assured of plentiful traffic and satisfying dividends even in face of canal competition, was a more obviously attractive proposition to distant investors. Even by 1837, after the peak of speculative enthusiasm had passed, the committee could apparently be selective in their choice of scripholders, and accepted only eleven offers to take fewer than twelve shares (see Table 31). It might be expected that the committee would attempt, as was customary, to impress Parliament with the strength of its local support, and it is therefore surprising to find that less than one-third of the money was promised from Scotland. There were admittedly still alternative investment opportunities in Scotland, even if the economy was moving into recession. The continuing expansion of the iron industry, coupled with the adequacy for heavy goods purposes of the canals and the existing railways, may help to explain why the coal- and ironmasters of the Monklands ignored the project completely, while the 1837 slump in cotton, coupled with the violence of the spinners' strike, might in part account for the lack of interest from Glasgow manufacturers: optimists in textiles had also the alternative of investing in the spread of the power loom.[232] All the same, Glasgow was strangely lukewarm about the project, subscribing less than half the sum it had offered to the Ayrshire (£143,500 against £294,700); could it be that, in spite of all the complaints, the monopoly of the Forth & Clyde canal was seen as less offensive than those of the Paisley canal and the river steamships? Edinburgh, a smaller and less commercially minded city, subscribed even less, and had committed very little money to railways elsewhere, while the railway was almost entirely ignored by everyone living between the terminal cities.

The Edinburgh & Glasgow subscription contract was gathered in four sections, based on committees in Glasgow, Edinburgh, Liverpool and Manchester; names from other parts of the country are attached, often through the use of power of attorney, to the latter two lists in particular. Theodore Rathbone's Liverpool committee gathered in 34.9% of the company's subscribers and 28.4% of the funds from their home city alone, while Lancashire as a whole contributed 42.1% of subscriptions, some 9% more than all of Scotland. The Edinburgh & Glasgow, which was entirely promoted and engineered by Scots (though admittedly with English endorsement of their plans), and was of little obvious advantage to any English railway company, reflects the extent to which Lancashire capital in particular had become blind, seeking the best possible return on investment wherever the project might be located. Rathbone and Edmund Buckley of Manchester were taken on to the board as a result of the large subscriptions gathered in Lancashire, not in an attempt to encourage Lancastrians to come forward.[233] For the first time also a significant amount of London money appeared on a Scottish subscription contract, most from individuals described as 'gentlemen'.

The occupational analysis of the three contracts reveals much less variation between the companies. In each case traders are of overwhelming importance, supplying more than half of the promised capital, as well as being among the prime movers on the committees of management. Second in each company, but a long way behind, came the non-occupational group, consisting in the main of local landowners and slightly swollen in the Edinburgh & Glasgow by its appeal to the London market. Industrial funds were limited, but to most industrialists (except in isolated areas like the Garnock valley) alternative transport was available; while they might desire the railway, not least to break existing transport monopolies, their profits could be more usefully reinvested in their businesses — and in any case the downturn of the economy would engender caution about any long-term financial commitment to railways. Some industrial money did come from Lancashire, but only in the town of Greenock did Scottish manufacturers show any enthusiasm for their local railway. Bankers in their private capacities were even more cautious, while the banks as institutions could not be expected to gamble in scrip, although they might later make loans to established companies. Other sources of appreciable amounts were the lawyers and the mixed group of other professions, many of whom might be expected to hold long-term for a rentier income. It cannot be known how many lawyers were holding scrip on behalf of clients. Already, however, Edinburgh lawyers provide 41.4% of the total legal contribution, giving a foretaste of their prominence in the promotions of the railway mania.

Scottish railway promotion in the late 1830s appears to have been financed above all by the trading classes of Glasgow and Lancashire, with both areas being selective as to which companies they patronised. About half of the total was locally subscribed, limited enthusiasm for the Edinburgh & Glasgow being balanced by heavy support for the Ayrshire. But, although all three lines ran from Glasgow, the city's contribution was less than the combined effort of Liverpool and Manchester: 42.5% of the subscription money came from England, with almost two-thirds of that originating in Lancashire. The Edinburgh & Glasgow in particular marks the start of large-scale English investment in Scottish railways. By the end of the 1840s all major Scottish companies had large groups of English shareholders, and many of the most important, including the Caledonian, the North British and the Edinburgh & Glasgow, had well over half of their capital held south of the border. There could be assorted motives for such English shareholding. Lancastrians could hope that better transport links in, and eventually to, Scotland would promote their own trading interests. Some money, though probably very little, may have been repatriated by exiled Scots for sentimental reasons. Some apparently English subscriptions may have been signed by Scottish residents who happened to be temporarily at an English address. But the great bulk of English subscribers, and later shareholders, were looking for either capital appreciation or a satisfactory rentier income from dividends. Their interest was purely with the financial success of the company, and their lack of concern with the benefits offered by the railway to its local communities made them often the first to complain if financial results were unsatisfactory. Many a Scottish board had to keep a wary eye on a vocal group of English shareholders, which if contented would often supply the directors with enough proxy votes to overcome any Scottish objectors at company meetings, but which if dissatisfied might be powerful enough to have the board ejected. Disgruntled Englishmen led the campaigns which threw John Learmonth out of the Edinburgh & Glasgow in 1846, and which by the end of the decade had dismissed the boards of companies as important as the Caledonian, the Aberdeen and the Scottish Central.[234]

Table 31 classifies the subscriptions of 1837-38 by size, and may be compared with the similar analysis for 1836 in Table 29. The essentially local Glasgow Paisley & Greenock continues the previous pattern of numerous small subscribers, with 57.2% of individuals taking ten or fewer of the £25 shares; many of the important group of traders and professional men who formed the backbone of any subscription list are now contributing over £1,000, as against about £500 on the Dundee & Arbroath. The significant changes from previous distributions appear on the more ambitious Ayrshire and Edinburgh & Glasgow lines. Small applicants are largely ignored, the average subscription rises to £1,463 and £1,736 respectively (or from 12 to 15 times the average for the Arbroath & Forfar), and in both cases well over 60% of the money comes in sums of over £2,000. The heavier capital requirements of these larger companies are being met not by increasing the number of subscribers proportionately to the increase in share capital, but by extracting larger sums from those who do contribute. That such a pattern was possible again suggests that the two companies were not severely pressed to fill up their contracts even at a time when the peak of railway speculation had passed.

The largest subscriptions — those over £5,000 — reveal a distinction between the two companies. The Glasgow merchants who promoted the Ayrshire appear to have more or less consciously decided that £5,000 was an appropriate subscription for a wealthy supporter. Only one subscription — that of Glasgow gentleman John Stenhouse — exceeded that sum, while of the 37 who offered exactly £5,000, 29 came from Glasgow (26 of them being merchants), four from Lancashire/Cheshire, and two each from Lanarkshire and Edinburgh. Of the 30 large subscribers to the Edinburgh & Glasgow on the other hand, only five were Scottish residents, three from Glasgow and two from Edinburgh. The others included twelve from Lancashire/Cheshire, seven from London, five from elsewhere in England and Wales, and one — the military Dr Hankey — from Ireland. The eight who subscribed £10,000 or more included five Londoners and no one living in Scotland. Even allowing for the possible duplication of names discussed above, the concentration on English sources for large subscriptions is striking. Since the promoters were Scots, and would presumably have welcomed massive local support, this may confirm the reluctance or inability of Glasgow to offer the company the same solid backing it had already given to the Ayrshire.

The subscription contracts confirm the transition which was taking place in Scottish railways. The move from projects of strictly local interest to ones of importance for the British economy as a whole is paralleled by a change from financing by local landowners and small traders to increasing dependence on the capital resources of Glasgow, Lancashire and, eventually, London. There was still nothing like a perfect national market for railway capital, in which funds could be channelled from all parts of the country into the projects which needed them, but it was becoming easier, through the use of local committeemen and brokers, to tap resources in the major English cities. Scottish railway development might now be lagging behind the very rapid developments of the mid-1830s in England, but the leading Scottish companies were at last becoming institutions of national significance.

4
From Town to Town: Construction and Operation

I. Expectation and reality: the accounts of the inter-urban railways

Scottish inter-urban prospectuses, like those of all railways projected in the mid-1830s, were overflowing with optimism. Both the Ayrshire and the Edinburgh & Glasgow offered their subscribers the prospect of dividends approaching 10%, while even the remote Arbroath & Forfar expected to yield $6\frac{1}{2}$%. The dividends listed in Table 32 indicate the extent to which they fell short of their targets in the early 1840s, while the years after the mania were to be much worse for all companies unless they had managed to negotiate a satisfactory fixed-interest lease during the years of speculation. It turned out that the promoters and their engineers had underestimated almost everything; unfortunately, although traffic volumes were often much greater than had been anticipated, the costs of construction and operation had escalated even more. As leading engineer John Errington noted:

> While it has in almost every case been proved, that the Traffic has exceeded all previous calculation, the Expense has been in still greater excess.[1]

The imbalance was made worse by the fact that, as compared with the estimates, traffic volume might well increase proportionately more than traffic revenue. Demand elasticities, and the often fierce competition of the canals, meant that the rates charged often had to be lower than those in the estimates.

If traffic volumes were above expectations, it may seem illogical that company officials tended to blame their difficulties on an inadequacy of trade due to the severe economic depression. It was however reasonable to observe that, whatever the traffic might be, it would surely be greater in more prosperous times; and in any case concentration on deficiencies of revenue might distract shareholders' attention from the chronic overspending on the other side of the account which was the principal cause of trouble, and for which the company might legitimately be held responsible. By the summer of 1842 therefore, the Ayrshire board, which in December 1837 had detected the start of a trade revival, was blaming 'the utter prostration under which the industrial energies of the country have so long laboured', with particular reference to the temporary closure of the Blair ironworks.[2] At the same time the Greenock noted 'the stagnation of trade, so universally felt throughout the district';[3] in its next report, the board concentrated on Paisley, where 26,000 people were reported to be unemployed:

> The commercial disasters which have befallen the town of Paisley, and the inactivity which has from the opening of the line pervaded this district, are evidenced in the diminished receipts of all those public bodies in the district whose revenue is contingent on the conveyance of goods and passengers.

Takings from the traffic to Greenock had in fact risen in the previous year, but had failed to compensate for a greater fall in the receipts between Glasgow and Paisley. The Clyde boats, where river dues had fallen by 15%, and the canal had fared equally badly.[4]

Table 32. Dividends Paid by Inter-urban Railways, 1841-44

Company	\multicolumn{8}{c}{Dividend (at % per year rate)}							
	1841		1842		1843		1844	
	A	B	A	B	A	B	A	B
Arbroath & Forfar[a]	3½	3½	0	0	0	0	2½	2½
Dundee & Arbroath[a]	5	5	0	0	4	4	5	5
Edinburgh & Glasgow	—	—	—	5	5	4½	5	4½
Glasgow Paisley & Greenock	—	—	4	1	2	2	2	2
Glasgow Paisley Kilmarnock & Ayr	2⅕	3	3½	2½	2½	2½	4	4½

[a]: declared a dividend once per year
A: spring meeting
B: autumn meeting

Sources: *Railway Times;* More Nisbett Papers, bundle 3, *Reports of Glasgow Paisley Kilmarnock & Ayr Directors;* M. Slaughter, *Railway Intelligence.*

The real problem was excessive expenditure. Table 33 indicates the extent of overspending on capital account, ranging from 74.4% on the Dundee & Arbroath to a staggering and possibly underestimated 205.1% on the Edinburgh Leith & Newhaven. The figures given for actual expenditure are those in the accounts three years after the lines were opened, by which time most if not all of the bills should have been paid. It is possible that they include some expenditure incurred after opening which should not strictly be charged against the cost of creating the railway; on the other hand, given the dilatoriness of most companies in paying for such items as land if they could possibly delay parting with their money, some relevant expenditure may not have yet been recorded. The exception is the Newhaven figure, which was given at the time of the agreement to merge with the Edinburgh & Northern, and thus comes only just after the opening of the Scotland Street tunnel. In all cases the overexpenditure was on a scale which could not conceivably be met either by any extra allowance made in the originally authorised share capital or by the borrowing powers in the act of authorisation; companies had to return to Parliament to extend, and sometimes to double or treble, their

Table 33. Capital Costs per mile of Inter-urban Railways

Company	Estimate		Actual		% Increase
	Date	£	Date	£	
Arbroath & Forfar[a]	1836	4,115	4.1842	8,964	117.9
Dundee & Arbroath[a]	1836	5,241	1.1842	9,140	74.4
Edinburgh & Glasgow	1838	18,994	1.1845	36,548	89.2
Edinburgh Leith & Granton	1836	20,000	7.1847	61,024	205.1
Glasgow Paisley & Greenock[c]	1837	17,022	1.1844	41,068[b]	141.3
Glasgow Paisley Kilmarnock & Ayr[d]	1837	11,562	7.1843	21,908[b]	89.5

[a]: excluding connecting branch at Arbroath
[b]: mileage and costs include half of joint line
[c]: excluding branches
[d]: main line and Kilmarnock branch

Sources: see Table 25.

capital-raising powers. If the dividends forecast in the estimates were to be realised, net traffic revenue would have to exceed its estimates in the same proportions.

Tables 34 and 35 present for the two largest companies the difference between the increasing optimism of the 1830s and the reality of 1844 (which was all the same a much better year than its immediate predecessors). For the Edinburgh & Glasgow, all aspects of traffic had yielded more than foreseen in the prospectus; passenger traffic was remarkably close to the parliamentary estimate, while the shortfall in goods revenue can be attributed to low rates enforced by canal competition. On the Ayrshire, overall revenue was close to Miller's estimate in the prospectus, if not to the inflated figures offered to Parliament — since the prospectus allowed for all rail traffic between Glasgow and Paisley, while the eventual joint line meant that only half came to the company, the results may reasonably be said to have exceeded expectations. In both prospectuses, goods traffic had been underestimated, but the company projectors had swung to over-optimism by the time of their submission to Parliament. In both cases also, working expenditure had reached a more than acceptable level of just over 30% of gross revenue, as against the 33.3% commonly regarded as a reasonable target and forecast by the Ayrshire, or the pessimistic 42.9% drawn by the Edinburgh & Glasgow from the early experience of the Liverpool & Manchester.[5]

Even so, the rate of return on the share capital of these two relatively successful companies was less than half that projected in the prospectus, and less than a third of later company forecasts. The essential difference between the prospectuses and the later realities were in the growth of non-working expenditure (for which no allowance had originally been made at all), and in the enormous expansion of the share capital. Since by far the greatest part of non-working expenditure was the payment of interest on money borrowed, the underlying cause was the same in both cases: companies had to borrow to the permissible limits and to issue further shares in order to pay for the unforeseen extra costs of construction, and the provision of rolling stock and equipment for operating the lines. Some of the extra capital might be required for purposes not covered by the original plans, such as the line through Princes Street Gardens, but almost all of it was simply to cover the enormous gap between estimated expenditure and the actual costs incurred. The 1839 Select Committee on Railways was concerned that:

> In almost every case into which they have enquired, the Parliamentary estimates of the expense of constructing Railways have fallen far short of the actual expenditure.[6]

Table 34. Edinburgh & Glasgow, Traffic Estimates and Reality

	Prospectus Oct. 1835		Parliamentary estimate 1838		Actual Jan. 1844–Jan. 1845	
	£	%	£	%	£	%
Passengers	75,504	81.8	82,628	65.5	82,314	70.1
Parcels	2,062	2.2	2,574	2.0	2,151	1.8
Goods	13,058[a]	14.2	39,356	31.2	27,857	23.7
Mail & miscellaneous	1,633[b]	1.8	1,633[b]	1.3	4,531	3.9
Total	92,258	100.0	126,191	100.0	117,420	100.0
Working expenditure	39,536[c]	42.9			36,292	30.9
Other expenditure[d]	—				23,921	20.4
Total	39,536	42.9			60,213	51.3
Balance	52,719				56,702	
Share Capital	550,000				1406,250	
Rate of return			9.59			4.03

[a]: claimed to omit heavy goods, agriculture produce, lime, manure, cattle
[b]: mail only
[c]: calculated on basis of actual experience of Liverpool & Manchester
[d]: interest, rates and passenger tax

Sources: More Nisbett Papers, bundle 13, Edinburgh and Glasgow Railway prospectus, Oct. 1835; *RT* May 1839; SRO, BR/RAC(S)/1/73, *Abstract of Accounts of the Edinburgh & Glasgow Railway Company.*

The committee identified four main causes — the inexperience of both promoters and engineers, the previously unprecedented scale of the works, the high costs of land and rising prices of labour and materials, and finally over-optimism and:

> the natural desire on the part of the promoters to keep as low as possible the calculations of the anticipated expenditure, and thus to show their undertaking in a favourable light.

It was also possible that several promotions were:

> mere speculations, by projectors totally unconnected with the district, who were little anxious for the ultimate accomplishment of the undertaking so long as they could derive a temporary profit from the preliminary business which such works necessarily occasion.[7]

This reflected a common suspicion of the machinations of lawyers and engineers in particular, and certainly applied to a number of lines floated in London in the 1830s and to many all over the country in the mania period; it does not however describe any of the Scottish railways under consideration in this chapter.

Table 35. *Glasgow Paisley Kilmarnock & Ayr, Traffic Estimates and Reality*

	Prospectus 1836		Parliamentary estimate 1837		Directors' report 12.1837		Actual 1844–45[a]	
	£	%	£	%	£	%	£	%
Passengers	62,713	78.0	84,052	67.7	73,076	64.2	49,578	62.5
Goods	16,967	21.1	40,052	32.3	40,052	35.2	28,787	36.3
Other traffic & misc.	720	0.9	—		700	0.6	1,014	1.3
Total	80,399	100.0	124,104	100.0	113,828	100.0	79,379	100.0
Working expenditure	26,800	33.3	41,367	33.3	37,943	33.3	24,047	30.3
Other expenditure[b]	—		—		—		14,876	18.7
Total	26,800	33.3	41,367	33.3	37,943	33.3	38,923	49.0
Balance	53,599		82,735		75,885		40,456	
Share capital	550,000		625,000		625,000		917,500	
Rate of return		9.56		13.24		12.14		4.41

a: including half of joint line. Year runs from 1 February
b: interest, rates and passenger tax

Sources: More Nisbett Papers, bundle 3, Glasgow Paisley Kilmarnock & Ayr prospectus, 1836; undated [Dec 1837] circular to shareholders: *Report of Proceedings of First General Meeting of Shareholders*, 20 Dec 1837; SRO, RAC(S)/1/11, Reports of Directors, 17 Aug 1844, 8 Mar 1845.

The following discussion analyses the expenditure of the companies on both capital and revenue accounts, and looks at the nature and results of their traffic operations. The available statistics have to be regarded with a certain amount of caution. Most come from the records of the companies themselves, usually as they were presented either to meetings of shareholders or to Parliament. Neither Parliament itself nor any of the various Board of Trade sections which dealt with railways made any serious attempt to check their accuracy, and it is made clear in the parliamentary records that the figures are printed as they were given by the companies. Nor was it easy for shareholders to make a thorough check, except on the rare occasions when a formal committee of inquiry was appointed. Auditing of accounts was erratic, and was not even compulsory before the Company Clauses Act of 1845,[8] after which there was still little guidance given on the manner in which auditing should be done. There were therefore plentiful opportunities for the figures to be wrong, either through incompetent accounting or by deliberate action: the Greenock company, for instance, had by November 1842 spent £2,600 more than its authorised capital but concealed the fact in the accounts.[9] Concern for the accuracy of accounts, accompanied by accusations of fraud and incompetence, followed the excesses of the mania and fuelled the efforts of the shareholders' inquiries at the end of the decade. In most cases their accusations concentrated on the period from 1845 onwards, and it is possible to assume by omission that they believed the pre-mania accounts to be tolerably accurate. The mania also provoked official inquiry, with a House of Lords Select Committee in 1849, but adequate legislation was not passed until 1868 after another series of financial scandals.[10]

Meanwhile the accounts were drawn up in whatever way the directors and their accountant chose, and direct comparisons cannot always be made precisely between various companies' records, or even sometimes between different years for the same company. In particular the line between the capital and revenue accounts could be blurred. Before the railway opened, all expenditure had to be placed against capital, but even after revenue from traffic was coming in, some companies continued to charge items like directors' remuneration or waggon replacement to the capital account. All companies regarded the parliamentary costs of promoting their own successful schemes as capital charges, but some set off the costs of opposing other projects, and even of their own failures, against revenue. Pushing as much as possible on to the capital account involved the risk that further borrowing or even further dilution of the share capital might have to take place. If however it could be done within the existing capital structure, it meant that a larger balance on revenue account would be available for paying dividends; for company managements aiming at a reassuring stability of dividend levels, this became a grave temptation in times of falling receipts. They might indeed go further — as the Edinburgh & Glasgow did when canal competition forced rates unprofitably low — and subsidise the customary 5% or 6% dividend out of capital.[11] Again, though, these practices were commoner after the mania, when companies were faced by both severe financial problems and increasingly enraged shareholders. In the early 1840s the temptations were perhaps less, and shareholders' expectations were imprecise; it may be reasonable to assume that for the most part the directors regarded their accounts as a reasonably accurate summary of company affairs.

The capital expenditure of four of the inter-urban lines of the 1830s is presented in Tables 36-39. Accounts for the Dundee & Arbroath have not survived, while the profligate career of the Edinburgh Leith & Newhaven cannot be unravelled from the sporadic evidence available. The various accounts were not kept to a common pattern, but the broad divisions of expenditure can easily be distinguished. There are some problems. Legal expenses, for instance, which related mainly to parliamentary business and to land conveyancing, are sometimes given separately, sometimes attached to parliamentary costs, and sometimes split between parliamentary and land. It is not always clear whether ballasting has been included with works or with rails. Small payments to road trustees appear to have been made from revenue (except on the Arbroath & Forfar), while large ones were charged against capital; the Arbroath & Forfar accounts also indicate an inexplicably declining total amount paid on account of roads. Salary payments should have been charged to the current account as soon as the line opened; some companies did so more quickly than others.

The general pattern of expenditure is clear. Early payments naturally concentrated on parliamentary costs and the acquisition of land, but by the time that the line was opened a more stable balance had been achieved. Parliamentary costs had been reduced to under 5% of the total, while land accounted for about one-eighth (or rather more at Greenock). The great bulk of money, from 60% to 70% of the whole, had gone to the essential creation of the line in works, rails and stations. Rolling stock took a steadily increasing proportion, rising to about a tenth of the whole, while small amounts, generally under 2% of costs, went to such other headings as interest, salaries, engineers' fees, and payments to road trustees.

The account totals indicate the cumulative capital spending of the company, all of which might reasonably be assumed to have been necessary for the creation and operation of the railway. On the other hand, a figure often bandied about by shareholders and officials, and indeed by Board of Trade officials, was that for the construction cost of the line, and it was much less clear exactly which sections of capital expenditure should be included in it. A good case could be made for excluding peripheral expenses like interest and road trustees, and in particular subscriptions to other railways; rolling stock might be excluded on the grounds that construction costs did not imply the ability to operate the finished line; and some companies even endeavoured to exclude the cost of stations.[12] It was clearly good public relations for companies to make their construction costs appear as low as possible, and once again parliamentary sources simply accepted the figures with which they were presented. Thus Table 33 gives figures for cost per mile based on total capital expenditure, while in 1846 a select committee produced figures for the five inter-urban lines then open which are in three cases lower than the earlier ones in the table. In some cases the committee's figures, presumably provided by the companies, contradict the evidence in their own statistical appendix, which also must have come from the companies. Thus the cost per mile of the Arbroath &

Table 36. Arbroath & Forfar Capital Expenditure

Date	Parl. & Prelim. £	Law £	Land £	Engineer £	Works & Stations £	Rails £	Rolling Stock £	Salaries & Direction £	Roads £	Other Railways	Interest	Miscellaneous £	Total £
14.4.38	2,060		11,498	918[a]	27,614	18,236		492				1,060	60,979
15.4.39	2,160		13,719	874	49,108	22,380		4,381	92			2,249	94,963
15.4.40	2,203		16,465	2,318	58,601	25,654		11,005	144			5,569	121,959
15.4.41	2,703		16,818	2,318	64,269	25,972		13,853	437			7,190	133,560
15.4.42	2,703		17,053	2,318	66,482	26,141		16,016	424			5,588	136,705
15.4.43	2,703		17,387	2,318	67,157	27,133		16,162	424			5,276	139,749
15.4.44	2,703		17,387	2,505	67,415	26,385		16,479	312			7,597	140,783
15.4.45	2,703		17,387	2,505	68,369	26,413		16,642	349			9,272	143,640
	%		%	%	%	%		%	%			%	
15.4.38	3.4		18.9	1.5[a]	45.3	29.9		0.8				1.7	
15.4.39	2.3		14.4	0.9	51.7	23.6		4.6	0.1			2.4	
15.4.40	1.8		13.5	1.9	48.1	21.0		9.0	0.1			4.6	
15.4.41	2.0		12.6	1.7	48.1	19.4		10.4	0.3			5.4	
15.4.42	2.0		12.5	1.7	48.6	19.1		11.7	0.3			4.1	
15.4.43	1.9		12.4	1.7	48.1	19.4		11.6	0.3			3.8	
15.4.44	1.9		12.4	1.8	47.9	18.7		11.7	0.2			5.4	
15.4.45	1.9		12.1	1.7	47.6	18.4		11.6	0.2			6.5	

a: including salaries of clerk and superintendent (transferred to revenue account when line opened)

Sources: SRO, Arbroath & Forfar Minutes, BR/AFR/1/2, 20 June 1839; AFR/1/3, 15 May 1840, 11 June 1841, 14 June 1842; AFR/1/4, 9 June 1843, 10 June 1844, 9 June 1845.

Table 37. Edinburgh & Glasgow Capital Expenditure

Date	Parl. & Prelim. £	Law £	Land £	Engineer £	Works & Stations £	Rails[b] £	Rolling Stock £	Salaries & Direction £	Roads £	Other Railways £	Interest £	Miscellaneous £	Total £
31.12.38	10,558	20,287	22,835	9,141				3,873				1,875	68,569
31.12.39	19,499	20,287[a]	71,924	14,930				6,511			778	5,185	247,235
31.12.40	20,191	24,703	116,520	19,334	107,721	5,266		10,572			2,768	5,850	636,442
31.12.41	20,191	25,442	152,392	22,094	430,838	50,931	46,024	13,262			16,389	8,697	1218,143
31.12.42	23,533	28,371	163,775	26,409	861,721	132,044	111,376	14,064	11,000		19,576	17,772	1513,396
31. 1.44	29,441	28,371[a]	185,851	29,237	965,476	132,256	121,430	14,064	11,000		19,698	16,128	1611,114
31. 1.45	33,129	28,371[a]	219,006	29,280	1023,638	132,256	132,481	14,064	11,000	5,000[c]	19,698	16,388	1686,226
	%	%	%	%	%	%	%	%	%	%	%	%	%
31.12.38	15.4	29.6	33.3	13.3				5.6				2.7	
31.12.39	7.9	8.2[a]	29.1	6.0	43.6			2.8			0.3	2.1	
31.12.40	3.2	3.9	18.3	3.0	67.7	0.8		1.7			0.4	0.9	
31.12.41	1.7	2.2	12.5	1.8	70.7	4.2	3.8	1.1	0.7		1.3	0.7	
31.12.42	1.6	1.9	10.8	1.7	63.8	8.7	7.4	0.9	0.7		1.3	1.2	
31. 1.44	1.8	1.8[a]	11.5	1.8	63.5	8.2	7.5	0.8	0.7		1.2	1.0	
31. 1.45	2.0	1.7[a]	13.0	1.7	62.0	7.8	7.9	0.8	0.7	0.3[c]	1.2	1.0	

a: law expenses for the year included in parliamentary
b: blocks and chairs included in works
c: deposit on £100,000 subscription to Glasgow Dumfries & Carlisle

Sources: SRO, RAC(S)/1/73, Abstracts of Income and Expenditure produced by the Directors of the Edinburgh and Glasgow Railway.

Table 38. Glasgow Paisley & Greenock Capital Expenditure

Date	Parl. & Prelim. £	Law £	Land £	Engineer £	Works & Stations £	Rails £	Rolling Stock £	Salaries & Direction £	Roads	Other Railways[a] £	Interest £	Miscellaneous £	Total £
15. 7.37	6,320	9,920	1,000	2,905									20,145
30.11.38	6,970	9,968	25,559	4,539	11,165	44				11,109	1,527	1,876	73,106
30.11.39	7,228	10,703	56,210	6,269	107,367	13,369	1,372	800		73,498	5,404	4,321	282,664
30.11.40	8,534	11,534	80,314	7,693	220,404	29,601	8,411	1,390		131,001	13,723	10,005	514,111
30.11.41	11,236	11,759	85,591	9,284	276,276	50,482	36,689	1,443		143,764	17,506	17,285	657,532
30.11.42	11,509	11,759	98,762	9,309	302,893	53,359	54,728	1,620		150,910	19,577	17,934	730,289
31. 1.44	13,472	12,935	112,128	10,072	307,743	53,359	60,054	1,662		161,182	19,577	28,103	780,287
31. 1.45	13,705	13,069	125,767	10,072	312,952	53,359	60,716	1,662		162,000	19,656	24,685	797,643
	%	%	%	%	%	%	%	%	%	%	%	%	%
15. 7.37	31.4	49.2	5.0	14.4									
30.11.38	11.2	16.1	41.2	7.3	18.0	0.1						3.0	
30.11.39	2.7	3.9	20.7	2.3	40.0	10.9	0.5	0.3			0.6	1.6	
30.11.40	2.2	3.0	21.0	2.0	57.5	7.7	2.2	0.4			1.4	2.6	
30.11.41	2.2	2.3	16.7	1.8	53.8	9.8	7.1	0.3			2.7	3.4	
30.11.42	2.0	2.0	17.0	1.6	52.3	9.2	9.4	0.3			3.0	3.1	
31. 1.44	2.2	2.1	18.1	1.6	49.7	8.6	9.7	0.3			3.2	4.5	
31. 1.45	2.2	2.1	19.8	1.6	49.2	8.4	9.6	0.3			3.1	3.9	

a: Glasgow & Paisley joint line.
Percentage figures exclude Glasgow & Paisley Joint Line.

Sources: SRO, Glasgow Paisley & Greenock Minute Books, BR/GPG/1/2, 2 July 1838, 18 Dec 1838, 13 July 1839, 21 Dec 1839, 6 July 1840, 17 Dec 1840, 2 July 1841, 5 Jan 1842; GPG/1/3, 1 July 1842, 6 Jan 1843, 13 Sept 1843, 7 Mar 1844, 4 Sept 1844, 12 Mar 1845.

Table 39. Glasgow Paisley Kilmarnock & Ayr Capital Expenditure

Date	Perl. & Prelim & Law £	Land £	Engineer £	Works & Stations £	Rails £	Rolling Stock £	Salaries & Direction £	Roads £	Other Railways[a] £	Interest £	Miscellaneous £	Total £
31.12.38	17,984	8,302	6,379	20,579	3,304		1,535		13,257		4,596	75,936
31.12.39	17,984	51,762	9,478	127,059	83,660	7,455	3,167		75,796	672	7,128	381,905
31.12.40	20,826	83,825	12,041	245,086	143,022	32,166	4,596	18,000	132,372	3,239	4,422	699,595
31.12.41	23,865	95,789	14,297	321,308	158,733	68,735	4,626	20,000	143,982	8,172	−2,207[c]	857,300
31. 1.43	26,570	105,498	15,028	398,936	183,562	72,833	4,626	20,000	151,509	11,854	−1,821	988,595
31. 1.44	27,593	115,102	16,144	424,529	186,544	83,872	4,626	20,000	161,363	13,078	−1,889	1050,962
31. 1.45	28,251	121,046	16,260	427,703	188,167	91,638	4,626	20,000	172,455[b]	13,107	−1,721	1081,532
	%	%	%	%	%	%	%	%	%	%	%	
31.12.38	28.7	13.2	10.2	32.8	5.3		2.4				7.3	
31.12.39	5.9	16.9	3.1	41.5	27.3	2.4	1.0			0.2	2.3	
31.12.40	3.7	14.8	2.1	43.2	25.2	5.7	0.8	3.2		0.6	0.8	
31.12.41	3.3	13.4	2.0	45.0	22.3	9.6	0.6	2.8		1.1	−0.3[c]	
31. 1.43	3.2	12.6	1.8	47.7	21.9	8.7	0.6	2.4		1.4	−0.2	
31. 1.44	3.1	12.9	1.8	47.7	21.0	9.4	0.5	2.2		1.5	−0.2	
31. 1.45	3.1	13.2	1.8	46.5	20.5	10.0	0.5	2.2	1.1[b]	1.4	−0.2	

a: Glasgow & Paisley joint line
b: includes £10,000 deposit on subscription to Glasgow Dumfries & Carlisle
c: includes profit on sale of surplus materials
Percentage figures exclude Glasgow & Paisley Joint Line.

Sources: SRO, BR/RAC(S)/1/11, Glasgow Paisley Kilmarnock & Ayr Reports and Accounts; More Nisbett Papers, bundle 3, Glasgow Paisley Kilmarnock & Ayr Reports of Directors, 23 Aug 1839, 24 Feb 1840, 19 Aug 1840, 17 Feb 1841, 30 Aug 1841, 16 Feb 1842, 30 Aug 1842, 27 Feb 1843, 28 Aug 1843, 22 Feb 1844, 16 Aug 1844, 20 Feb 1845.

Forfar is given in the text as £9,214 and in the appendix as £9,164, while for the Dundee & Arbroath the figures are respectively £8,570 and £9,245. On the more expensive lines, the cost per mile of the Ayrshire was given as £20,607, of the Edinburgh & Glasgow as £35,024, and of the Greenock as £35,451. On the whole these figures compared favourably with those in England, where among the major lines the Great Western had cost £43,885 per mile, the London & Birmingham £38,406, and the London & Brighton £56,981. In the north of the country, the figure for the York & North Midland was £25,924, and that for the Manchester & Leeds no less than £64,582.[13] Since traffic volumes in Scotland would be less than those in the busier parts of England, Scottish construction costs would have to be relatively low if dividends were to bear comparability with those in the south.

The most obvious point about the accounts, emphasised by the companies' frequent recourse to Parliament for the power to issue more shares, was the extent of the overspending when compared with the original estimates. Separate examination of the main items of expenditure may help to indicate why it happened.

II. Ordeal by Parliament

Most pre-authorisation outlay by promoters was lumped together in company capital accounts as 'parliamentary and preliminary expenditure', since there was no clear dividing line between the two categories. The heading covered all costs incident upon setting up a provisional committee, advertising the project, holding meetings, raising subscriptions, drawing up plans, and negotiating the parliamentary hurdles of standing orders and a committee on the merits of the bill: often included as well were payments made to the engineer and his staff before the authorisation of the company. These costs attracted closer scrutiny and more complaints from promoters and shareholders than their comparatively modest proportion of total costs might seem to justify. Firstly, they came at the beginning of a project, when ready cash might be difficult to raise and before the promoters had any powers to borrow money. Secondly, they might turn out to be a complete loss, since they would have to be paid even if the project was refused authorisation. And thirdly, although strictly preliminary costs could be forecast with reasonable accuracy, the expenses of the parliamentary ordeal were both uncertain and, it seemed to promoters, unnecessarily large.

Parliamentary expenditure proper included not only the fees charged for the consideration of a private bill, and the considerable costs of producing the substantial documentation required by standing orders, but also the expenses of lawyers, agents and witnesses before committees. How many of these were required might depend on the extent of any opposition to the bill, since the committee's decision could be crucially influenced by the number and reputation of the expert witnesses on each side. Thus a controversial company like the Edinburgh & Glasgow could spend £25,968 on parliamentary costs by 1843, while the Arbroath & Forfar, which had settled with landowning opponents before

Table 40. Parliamentary and Legal Costs of Inter-urban Railways

Company	Date	Mileage	Parliamentary & preliminary costs		Legal costs		Total	
			£	% of total	£	% of total	£	p.m.	% of total capital expenditure
Arbroath & Forfar	4.1842	15¼	2,703	2.0	—	—	2,703	177	2.0
Edinburgh & Glasgow	1.1845	46	33,129	2.0	28,371	1.7	61,500	1,337	3.7
Glasgow Paisley & Greenock	1.1844	19[a]	13,472	2.2	12,935	2.1	26,407	1,390[a]	4.3
Glasgow Paisley Kilmarnock & Ayr	7.1843	47[a]	n.a.	n.a.	n.a.	n.a.	26,683	568[a]	3.1

a: including half mileage of Glasgow & Paisley joint line: separate parliamentary and legal costs were not recorded in the joint line's accounts

Sources: see Tables 26 to 39.

going to Parliament, had spent only £2,703 (in both cases, however, the sum amounted to 2.3% of the authorised share capital). In 1844 the North British had to spend £17,825 on parliamentary costs, even although the vehemence of the opposition was much less than had been anticipated. In the following year, the expenses of the highly controversial Caledonian amounted to £49,952, although once again this amounted to only 2.4% of authorised share capital.[14] The sum included not only the costs of promoting the Caledonian's own bills, but also expenses incurred in opposing the projects of rivals — an aspect which became increasingly important during the mania years. Shortly after authorisation the Ayrshire gave a breakdown of its parliamentary costs. Of a total of £14,093, 59% had been incurred in preparing the bill and for official fees, $20\frac{1}{2}$% for witnesses, 17% for lawyers' charges, and 3% in opposing the plans of others.[15]

The relationship between the state and the railways in the nineteenth century has been well analysed elsewhere, notably by Parris,[16] and it is not intended to re-examine the subject here. Relations with Parliament or with the government were also, of course, much the same for any British railway company; there was nothing distinctively Scottish about the inherent suspicion with which companies regarded any legislative attempt to restrict their powers or intervene in their affairs, although there could be issues, such as the method of levying passenger tax, on which the Scots were much more indignant than companies in the south. In the 1830s both sides inclined to a laissez-faire relationship. Almost the only relevant general legislation of the 1830s was the extension of passenger tax to the railways in 1832, and the power of enforcing the carriage of mail given to the postmaster-general in 1838.[17] The campaign led by MP James Morrison for much closer state control, particularly of potential abuses of a monopoly position, was unsuccessful, and no general regulatory act for railways was passed during the decade.[18] Even in the early 1840s, the powers given to the Board of Trade amounted to little more than a loose supervision on behalf of public safety and the public interest. In particular there was great reluctance to impose conditions on existing companies which had not been foreseen at the time of their authorisation, and which might seriously affect their profitability. Peel advised the Commons

> to be very careful how it interfered with the profits or management of companies which had been called into existence by the authority, and had invested their money on the faith of Parliament.[19]

Hence the nationalisation option in Gladstone's 1844 act was not to apply to companies authorised before the act, a provision which virtually ensured that the clause would remain ineffective. Even the famous parliamentary trains were not to be made enforceable on an existing company until it next applied to Parliament for a bill. To avoid future problems with such unforeseen legislation, all new companies were thereafter obliged to insert a clause into their bills binding them to follow all future general railway legislation.[20]

The opportunity for Parliament to place some restraints on railway companies came during consideration of individual bills of authorisation. The framework within which consideration took place was the standing orders of the two Houses for dealing with private bills. These orders had developed more or less at random, dealing pragmatically with problems as they arose, and by the 1830s they had achieved a state of confusion in which, even without the growing tide of railway legislation, reform was becoming essential. The railway bills brought another aspect to the fore:

> The expansion of railways for the first time brought more clearly than ever before into the consciousness of Parliament the conception that in private legislation there was an aspect of public, as well as one of private, interest, to which no government could be indifferent.[21]

Pre-mania railways were seldom in competition with each other, and, by their own account at least, were unafraid of any competition which might be offered by road transport or, except for heavy goods, by canals. Both the promoters and Parliament anticipated that the railways would be in or near to monopolistic positions, and that the theoretical restraints of free market competition would not therefore safeguard the public interest. Hence one of the duties of the committees on the bills, which until 1844 were composed of locally interested MPs,[22] was to apply some gentle restraint, often in the form of generous maximum permissible rates, on the freedom of action of the companies. Another was to ensure as far as possible that the line chosen was a good one for the area, since it was not anticipated before the mania that competing lines would ever be authorised.

The elaborate nature of standing orders procedure was thus in part necessary because Parliament had to make the right decisions at the time of authorisation. Early railways had to satisfy conditions which had been imposed piecemeal on canals, and which had become both complicated and highly detailed. The standing orders of the House of Commons laid down a timetable for notice to be given of application for a bill, and of its contents, including a list of parishes affected and changes to be imposed or altered; for the deposit both in the locality and in the Private Bill Office of a map, plan, section, book of reference, estimate of expense, list of subscribers, and list of owners and occupiers of land distinguishing assenters to and dissenters from the plan; for due intervals between the various stages of a bill to allow for full consideration; and, from 1830, for the prohibition of a second reading unless half of the estimated capital required had been subscribed under a binding subscription contract.[23] It was easy for promoters to come to grief somewhere among the details, perhaps by minor errors in drawing up plans, or by an effort to make the strength of financial support look greater than it was; increasingly, as projects became larger and began to encroach on each other's territory, their agents scrupulously examined the plans of rivals in the search for detailed errors with which they might be shot down before a committee. On the other hand, MPs were still not getting all the information which they required. Local members sitting on railway bill committees were accused, often with justification, of letting prejudice overcome impartiality, while independent members, faced in the House with the committee's report, lacked the information on which to make an adequate judgment.

The promotional activity of 1836-37 and the work of a select committee in the former year started the reorganisation of the regulations. Detailed instructions were given to the committees about the information which they were to extract from promoters and present to the House. In 1838 the Lords acted on the composition of committees, referring railway bills to groups of five disinterested members. Similar reform in the Commons was delayed by dispute over the rights of local MPs. In 1840 a compromise added a number of neutral members to each committee, but preserved the rights of the local men. Not until 1844 were groups of competing bills referred to committees of five independent MPs; by the mania, unbiased judgment had become more important than local knowledge.[24]

Other changes were made in order to tighten up the system. From 1837, in an attempt to restrict the promotion of hopeless projects, the subscription contract was required to cover 75% of the estimated costs before the bill could be reported, and a tenth of this subscription had to be physically deposited in the Bank of England or a Scottish bank, or put into government stock. This deposit was reduced to 5% in 1842, making life a little easier for promoters of marginal schemes as the mania approached. In order to allow more time for full scrutiny, the deadline for the deposit of plans was moved in 1840 from 30 November to 31 March in the year before that in which the bill was to be presented. This meant that engineers now had to make their surveys under winter conditions, since they could not act before the harvest was gathered in, and at a time when they were about to give evidence to committees about the previous year's projects; protests from the railway interest led to the restoration of the November deadline in 1842. By then plans had to reach the sheriff-clerks of affected counties and the Railway Department of the Board of Trade; by the end of the year they had also to be sent to town clerks in royal burghs, parish schoolmasters elsewhere, and the Private Bill Office.[25]

Although these changes made Parliament somewhat better informed and certainly increased the impartiality of committees on bills, they did nothing to reduce the costs to the companies; indeed, the deposit requirements and the increasing amount of opposition to be expected from rival projectors threatened to make life progressively more difficult and more expensive. In 1844, for instance, with the standing orders in their final form before the flood of mania projects forced a radical reconsideration, the North British had to fend off a challenge from a group of interests behind which lurked the shadow of the Caledonian. In succession a series of company representatives, many of whom had had to travel from Edinburgh for the occasion, testified to the committee that the company's intentions had been published for three consecutive weeks in the *Edinburgh Gazette,* the *London Gazette* and the local press; that the engineer had prepared his plan and section of the line in full accordance with parliamentary criteria; that application for statements of approval, disapproval or neutrality had been made in writing to all owners, lessees and occupiers of affected land; that a subscription contract covering three-quarters of the estimated cost had been drawn up since the end of the preceding parliamentary session; that three-quarters of the subscribers had paid a 10% deposit on their subscriptions; and that copies of all the appropriate plans and documents had been deposited with all the correct officials by the due dates. Nineteen further individuals, who had witnessed signatures to the original documents, also had to appear in person to testify to the authenticity of the signatures.[26] It was not surprising if promoters began to wonder why such factual evidence could not be covered by written affidavits and presented to the committee by one company agent.

After the hazards of standing orders, promoters still had to satisfy the committee, and subsequently the whole house, that their project was meritorious as well as correctly presented. Here, since the need was to sway opinion rather than merely establish facts, the number of witnesses called could escalate alarmingly. Evidence had to be adduced of the respectability of the promoters, the financial solidarity of their support, the reliability of the estimates, the simplicity of the engineering, the imperative demand of the district for the facilities to be provided, and hence the inevitability of the railway's attracting a vast traffic and making an impressive but not immoral profit. The objections which might be propounded by rival transport interests, disgruntled landowners or hostile local authorities had to be anticipated, and experts kept on hand to answer them if necessary. Sometimes duplication of effort might seem necessary. Six substantial merchants testifying to the value of the line were more impressive than one, and, if it were believed that the opposition had two distinguished engineers ready to disparage the route, it was desirable to have three of your own prepared to testify to the perfection of the planning. Committees had, with the help of any local knowledge or prejudice they might possess, to evaluate this contrary mass of biased evidence. It is not surprising that the results were often accused of being inconsistent and illogical. Not until August 1844, when the newly formed Railway Board started to issue its own reports on proposed railways,[27] did committeemen have access to information which was supposedly both informed and impartial.

Alongside the adjustment of formal procedures, Parliament also established a number of less formal conventions. Most important was the regulation of a new company's borrowing limits. Normal practice had always been for the proposed share capital to be equal to, or more usually slightly above, the engineer's estimate of costs. Few companies left it at this, since even the best engineer could miscalculate or run into unforeseen problems, and powers were also taken to borrow further sums if required. To ensure that unscrupulous projectors could not borrow money without ever building their railway, borrowing powers could not be exercised until half the share capital had been called up and spent.[28] No standard limit on the total amount of permitted borrowing was established by the coal railways of the 1820s and early 1830s. The Garnkirk & Glasgow, overconfident of the abilities of Thomas Grainger, set its share capital precisely at his estimate of £28,497 and failed to allow for any borrowing at all; sadly, the company had to promote an act to raise a further £9,350 of share capital in the following year, and a third act giving borrowing powers of £21,150 in 1830, all before the line was opened.[29] The Ardrossan also failed to take powers to borrow money, although it did set its share capital £1,555 above the engineer's estimate.[30] The Monkland & Kirkintilloch persuaded Parliament not only to set its basic share capital at over £7,000 above the estimate, but also to allow the raising of a further £10,000 in the form of either shares or loans as the company chose.[31] Elsewhere, the ratio of borrowing powers to share capital varied from 54% on the Ballochney to 23% on the Slamannan. As late as 1836 the ratios for the Dundee & Arbroath and the

Edinburgh Leith & Newhaven were 40%, and that for the Arbroath & Forfar was 50%.[32] Thereafter the convention that borrowing powers should be set at one third of share capital was almost universally accepted, and was certainly never exceeded. Even so, Glasgow Paisley Kilmarnock & Ayr agent Patrick Blair appeared embarrassed when cross-examined on the need for any borrowing powers at all:

> I ask you what that 208,000*l*. is to do? — To carry through the Objects of the Act of Parliament.
> But Objects not in the Estimate? — The Objects of the Act of Parliament.
> Not in the Estimate? — They may be in the Estimate.
> How can it be; if everything estimated to be done is brought within 623,000*l*. and you want to use any of that 208,000*l*., do you mean to say that it is in the Estimate? — It is quite possible there may be other Expenses not calculated.
> Is it in the Estimate? — I do not know that there is any thing necessary but what is in the Estimate.
> Then what are you to do with it? — The Estimates themselves may be wrong.
> But then this is One Third to be added to cover the Deficiency that may be on the Estimate? — Yes.[33]

Blair's eventual admission did not augur well for the reliability of his company's plans, but it reflected a truth well known to all railway projectors, and fortunately recognised by Parliament, about the dangers of depending absolutely on the still new science of railway estimating.

Legal costs were often inextricably mixed up with parliamentary expenditure, to the extent that the Edinburgh & Glasgow included the two amounts under a single heading in its accounts in some years, though not in all.[34] Generally, they were only a significant part of costs in the earliest days of a railway, when they relate almost entirely to achieving authorisation and establishing the company, and in particular to the costs of the experienced London lawyers retained to pilot the bill through Parliament. Lawyers, like engineers, were not infrequently accused of deliberately prolonging and complicating parliamentary proceedings in order to increase their own fees, and could easily become scapegoats for failure, as on the Dundee & Perth project:

> Under the peculiar circumstances of the case, and taking into account the inefficiency of the services rendered by these gentlemen, the amount of the claims made by them is far too high.[35]

On the other hand, lawyers were often among the most active and influential railway promoters. The early Scottish system owed much to the Glaswegian brothers Andrew and Dugald Bannatyne:

> The two brothers made themselves remarkable by inaugurating the system of railways in Scotland — the one originating the Edinburgh and Glasgow . . . and the other the Glasgow and Ayrshire . . . Their success in these undertakings led to their employment in parliamentary business to an extent greater than that of any other firm in Scotland.[36]

This encomium may have been an overstatement, but it draws attention to the important promotional activities undertaken by men like the Bannatynes, Learmonth's ally C.F. Davidson, the Dundee firm of Shiell and Small, or the parliamentary agent and anti-passenger tax campaigner Gabriel Lang. Much of the drive and most of the money behind railway promotion may have come from the merchant community, but for the complex and detailed work of planning and organisation a great debt was owed to company lawyers.

In no case did the combined parliamentary and legal costs incurred by a company total over 5% of capital expenditure, even although the sum included some legal expenses incurred in connection with land purchases. The burden which they imposed thus does not appear great; it is reasonable to conclude that

> the financial results of the difficulty of obtaining Acts were not important, and may be dismissed as an annoying interlude in railway financial history.[37]

This did not prevent projectors feeling aggrieved. Parliamentary procedures were prolonged and complicated, and could be unnecessarily expensive. Standing orders were liable to be changed at short notice, and the committee stage of a bill, where much might depend on the prejudices of the committee members, could appear to be something of a lottery. The illogicality of parliamentary procedure could also, of course, serve as a convenient excuse to offer shareholders if something went wrong with a company's progress. On the whole, promoters regarded the parliamentary ordeal as something which had to be endured, with unnecessary complications which added a significant extra burden to their overstretched resources.

III. The acquisition of land

Among the advantages conferred upon companies by parliamentary authorisation was the crucial power of acquiring the necessary land. For many companies, land purchases caused more difficulty and unpremeditated expense than any other aspect of construction. Negotiations had to be completed before construction could begin, and payment would almost invariably have to be made, in spite of all attempts by directors to delay for as long as possible, well before the line could be open and yielding any revenue. Thus land costs made up the greater part of capital expenditure in the early days of a project, and shareholders, under pressure to pay the calls on their shares, often paid them a disproportionate amount of attention.

The price of land was a matter for negotiation between the railway projectors and individual landowners. Railwaymen accepted that they should pay not only for the value of the land, but also something more for loss of amenity and for the inconvenience of intersection, while tenants would also require compensation, including payment for improvements spoiled and crops destroyed. They frequently claimed, however, that landowners were not only placing very high estimates on the agricultural value of their land, but also demanding extortionate sums for the less calculable considerations of inconvenience, amenity and compulsion. In June 1845 a House of Lords Select Committee (which had some of the attributes of a landowners' protection society) declared that payments should certainly include a generous allowance for loss of beauty and enjoyment, and that the final price should then be increased by 50% for the compulsion involved:

> Public advantage may require all these private Considerations to be sacrificed: but as it is the only Ground upon which a Man can justly be deprived of his Property and Enjoyments, so, in the case of Railways, though the Public may be considered ultimately the Gainers, the

immediate Motive to their Construction is the Interest of the Speculators, who have no Right to complain of being obliged to purchase, at a somewhat high Rate, the Means of carrying on their Speculation.

It is also to be observed, that the price of the Land purchased, and the Compensation for that which is injured, form together but a small Proportion of the Sum required for the Construction of a Railway, so that no Apprehension need be entertained of discouraging their Formation by calling upon the Speculators to pay largely for the Rights which they acquire over the property of others.[38]

Such opinions, and the undeniable rapacity of many landowners, encouraged special pleading by representatives of the railway interest that unreasonable land costs had caused many of the difficulties of the system. As early as 1841 John Errington claimed that land 'has been a very heavy item in all Railways, and the exactions in some cases have been a severe tax on the public'.[39] Joseph Mitchell later remembered that:

Railways were opposed in every shape as disturbing innovations, and the cost for land and amenity . . . is an instance of the outrageous extortion practised by proprietors during the early stages of railway construction.[40]

Even in 1928, C.E.R. Sherrington explained that:

It should be remembered that the unduly high costs of our early railways can never be eliminated, and they will always remain a permanent charge against the industrial efficiency of this country, and a lasting reproach against a short-sighted set of nineteenth-century gentlemen.[41]

Landowners were not, however, consistently hostile to railways, or determined to extract every last penny from the projectors. Many, from the mineral owners of the Monklands, through Lindsay Carnegie and Panmure in Angus, to Hope Johnstone and his Annandale neighbours who promoted the Caledonian, were active supporters, and provoked a later writer to the dubious generalisation that:

Broadly speaking, in England it was the landowners who opposed the railways, in Scotland it was the landowners who favoured them, while the landless were indifferent or antagonistic.[42]

This is totally inaccurate about, in particular, the Glaswegian commercial interests who were active in promotion, but it is a useful reminder that the support of landed men who saw the railway as a means of developing their own estates, as a service to the community of which they were the social leaders, or simply as a potentially profitable investment, was important to many early lines.

An enthusiastic landed supporter could be very generous. As far back as 1809, one landowner on the route of the proposed Glasgow to Berwick waggonway had suggested that all the required land should be given free, although there is no evidence that he received much support, while projectors of other waggonways also hoped for landed liberality.[43] 'The truly liberal conduct of Lord Panmure' to the Dundee & Arbroath involved a grant of two-thirds of the land needed at a nominal feu duty.[44] Lord Belhaven gave land free to the Wishaw & Coltness, which was of great value to his estate, on condition that it was not 'turned to any other use than the Act of Incorporation'; in 1846 he invoked this condition in an attempt to block a planned takeover by the Edinburgh & Glasgow.[45] The Ayrshire noted with gratitude the gift of £2,000 worth of land by one of their directors, Colonel Blair of Blair, although the colonel subsequently claimed that he had been misunderstood and was merely prepared to sell cheaply.[46]

The Slamannan met both helpful and hard-bargaining landowners. In 1838 they had taken the risk of starting construction at both ends of their line while still negotiating 'with considerable difficulty' for the connecting land.[47] Next year their plans were held up,

> the lands of Arden . . . having been purchased by an ironmaster [John Wilson], who finds it not convenient to work the minerals, until the supplies he has secured nearer his works begin to fail, the anticipated traffic on it has not yet commenced.[48]

At this point the company, with the line as yet unopened, had run out of money, and reminded landowners of the benefit they would obtain from a coal railway:

> It is not unreasonable to expect that the land and mineral proprietors on the line should furnish a very considerable portion of the sum required for its formation, before the work be continued.[49]

On the other hand, three of the principal landowners had already given land free, and the company later assumed, when planning an extension to Bo'ness, that the Duke of Hamilton would not only be similarly generous but would also contribute £15,000 of the estimated £46,000 capital 'in consideration of the reclamation of about 600 acres of his land which will be effected by the Railway Embankment'.[50] Clearly the landed interest was not universally expected to be grasping.

On the other hand, many landowners saw a railway project either as a threat to normal existence which had to be stopped at almost any cost, or as an opportunity to realise an unexpected windfall profit. John Francis, for the railway interest, drew a distinction:

> There is much to palliate the honest opposition of the landowner. Scenes and spots which are replete with associations of great men and great deeds, cannot be pecuniarily paid for. Sites which bear memories more selfish, yet not less real, have no market value. Homes in which boyhood, manhood, and age have been passed, carry recollections which are almost hallowed. Such places cannot be bought and sold . . . It is the trafficker in sympathies, it is the dealer in haunts and homes, at whom the finger of scorn should be pointed. It is the trader in touching recollections only to be soothed by gold that should be denounced. It is the peer who made the historic memories of his mansion a plea for replenishing an impoverished estate; it is the farmer who made the sacred associations of home an excuse for receiving treble its value; it is the country gentleman who made his opposition the lever by which he procured the money from the proprietors' pockets, who should be shamed . . .[51]

One can only guess where Francis would have placed a landowner like Sir John Hall of Dunglass, who informed the North British that 'the injury done to all those objects which a country gentleman so highly cherishes can never be compensated', but then accepted £12,000 for 57 acres.[52] There could be little doubt, however, about the outraged reaction of the Duke of Buccleuch when in 1837 it was proposed to run the Nithsdale line from Carlisle to Ayrshire in front of one of his numerous houses, Drumlanrig Castle. In his report Locke had noted that:

> Neither have we the certainty that one of the lines could be carried; for, as I was informed at the time the survey was made, a most influential nobleman was opposed to it.[53]

When Provost Kemp of Dumfries, for the projectors, had solicited the duke for help in raising money for the survey, he had received the full force of the ducal opinions on railways in general and this one in particular:

> Having no faith in the permanent prosperity of Railroads already in existence, that they will continue to afford a proper return for the money sunk in making them; and knowing that in almost every instance the Estimate cannot be depended upon either as to the extent of the Capital

> that may be required . . . or as to the Profit to be received . . .
>
> I also consider that the injury which would be done to my property, and the annoyance I should experience from having a Railroad taken thro' it in the manner proposed, would far outbalance any advantages that might be derived from it, either by myself or the Public.[54]

Buccleuch refused to allow surveyors on his land in either 1837 or 1839, even although the object in the latter year was to find a route further away from the house.[55] His continued hostility may have persuaded some wavering subscribers to prefer the rival Annandale route, and may even (though it cannot be proved) have had some influence on the engineering enquiry of Smith and Barlow in 1841. The problem was only eventually overcome when the Glasgow Dumfries & Carlisle built an otherwise unnecessary tunnel through the Drumlanrig property.[56]

Before a company was authorised, a hostile or greedy landowner could be in a powerful bargaining position. If an engineer could avoid his land he would presumably do so, but in many cases a landowner's property might straddle the whole of the only reasonable engineering route; in the case of a great proprietor like Buccleuch, it was virtually impossible to design a route from Carlisle to either Glasgow or Edinburgh which did not cross his land. Since members of both Houses of Parliament, themselves almost all landowners, were already ill at ease at authorising the drastic interference with the rights of private property involved in compulsory purchase powers, vigorous opposition from a major landowner, or a group of smaller ones, might be enough to ensure rejection of a bill. Hence projectors and engineers regularly concentrated on finding the most acceptable route, sometimes at the cost of engineering convenience. Back in 1809 Telford had boasted that his Glasgow to Berwick waggonway 'will be carried without materially affecting, and almost without touching any Parks Lawns or Improved grounds'.[57] Thirty years later the agent for an early east-coast scheme reassured a concerned landowner that 'it is our duty so far as we can to accommodate those whose estates may suffer by our operations'.[58] Grainger and Miller were later described as 'very solicitous about the sentimental feeling of the lairds as regards amenities'.[59] Locke's plans offered the Glasgow Paisley & Greenock two possible routes between Glasgow and Paisley, and the company's selection was based on convenience to landowners. Although there were 130 different landowners in five miles, there were 'fewer objections than might have been expected', and every claim but one was settled either by direct agreement or by mutually acceptable arbitration.[60] The Ayrshire company's agreement to a joint line between Glasgow and Paisley must have been influenced by fears of the opposition they would encounter if they attempted to force a separate line over the same owners' property. Other examples of attempts to buy off or evade particular landowners have been given in Chapters 2 and 3, culminating in the willingness of Kinnaird's Dundee & Perth committee to bridge the Tay rather than attempt to force their way past the opposition of Hay of Seggieden.[61]

Landowners did not always hold the whip hand. Many estates were encumbered with debt or in need of improvement, and the owners were now being offered prices for small parcels of their land well above what they would fetch for agricultural use. In many railway land transactions, an eager buyer was faced by an equally eager seller, and the railway would be well aware that its adversary was also anxious to conclude the bargain. A landowner with significant parliamentary influence may well have been able to force his price unreasonably high, but many more were willing to settle for a modest but welcome windfall profit.[62] And if the railway was authorised before the price for the land was agreed, the balance of negotiating power shifted. If the parties could not come to an agreement, the dispute was referred for settlement to a jury sitting under the sheriff of the county. The landed interest certainly believed that such juries favoured the railways; as Sir John Hall was told by his lawyers,

> The general complaint . . . in all such cases is that, the price or damage allowed by a jury is small, as they mix up with the Case views of public as well as private utility.[63]

Stockbroker John Reid confirmed that juries 'generally award a sum far less than that claimed, — frequently less than a quarter'.[64] Some claims however were clearly ridiculous; one, for £44,000 against the Edinburgh & Glasgow by the Glasgow Lunatic Asylum, which 'would almost lead to a supposition that the claim had been made by the inmates rather than by the directors', was reduced by a jury to an award of £873.[65] Not until the Land and Railway Clauses Consolidation Acts of 1845, which laid down generous criteria for determining compensation including price-fixing on the basis of value to the vendor and an allowance for compulsion, did the landed interest feel more content and the prices paid escalate.[66] Lord Cockburn was uncertain whether even this was sufficient action against the railway juggernaut:

> Self-interest combines all railway speculators into one corporation, which, with its bursting purse, defies resistance, and respects no feelings but its own. Even juries, our former shields, have been obliged to be superseded as the guardians of private interests, because it is found impossible to get fair ones; and Parliament . . . is itself to an alarming degree a company of railway owners.[67]

But even although there were by 1847 some eighty MPs with railway connections,[68] the instincts of a landed parliament ensured that railways were unlikely to obtain their land on the cheap.

Certainly landowners, not unreasonably, were prepared to be hard bargainers. As early as 1825 Thomas Shairp of Houston commissioned a report on the proposed West Lothian Railway, which in principle he supported, and was told that it might be difficult to agree on the amount of compensation to be paid for the line being visible from his drawingroom window. As for intersection damages,

> This is partly real and partly a matter of feeling . . . There is however a *pretium affectionis* which every gentleman is supposed to have for his paternal Estate, and it is usual to give Compensation in money, in cases like the present, when the feeling may be supposed to operate.

He was recommended to ask £94 per acre for agricultural land, but £200 per acre where the land adjoined a turnpike and might some day be used for building, with at least a further £1,000 for the damage to his view.[69] It is unclear how many years' purchase these figures were based on, but they may be compared with a valuer's estimate that rural land on the Ayrshire route in 1837 was worth an average of £80 per acre at thirty years' purchase, which was the normal period although fifty years was not unknown. By 1846 land on the Bo'ness extension of the Slamannan was valued at 35/- to 50/- for arable, 60/- for pasture, and 80/- for garden ground, with a recommended forty years' purchase, giving a range of £70 to £160 per acre.[70] In Angus Sir John Ogilvie sold land on the outskirts of Dundee to the Newtyle company for £185 per acre in 1839 (although he had yet to receive payment for it five years later). Land near the centre of towns was naturally much more expensive, and the Edinburgh Leith & Newhaven paid a 'reasonable' £850 per acre for the three acres of its Scotland Street station site.[71]

Some landowners had more advanced ideas of maximising payments by railways. In 1837 the projectors of the line from Glasgow to Paisley were confronted by Sir John Maxwell of Pollok with his plans for building houses on the agricultural land which they had hoped to acquire, and, since Sir John was too important a man to antagonise, the railway had to pay for the increased land value created by the hypothetical houses.[72] The Duke of Buccleuch's similar claim against the Edinburgh Leith & Granton in 1845 was more solidly based, since his proposed elaborate system of docks and warehouses (which a railway along the foreshore would prevent) had been authorised by private act in 1842.[73] A Glasgow landowner, W.S.S. Craufurd, got a private act in the same year to lay out for feuing his entailed estate beside the Forth & Clyde Canal, which had the incidental advantage of enhancing in value land which lay across the best railway approach to the north of the city.[74]

Other options open to landowners included laying down additional conditions attached to any agreement to sell their land. Some, such as the wayleaves discussed in Chapter 2, had by the 1840s become totally unacceptable to railway promoters.[75] One ploy which was to become common during the mania was to insist that the railway purchased unnecessary ground, or even the whole of an estate, on the plea that the existence of the railway would make it valueless or uninhabitable. As early as 1830 the Monkland & Kirkintilloch admitted that it had bought more land than it had intended;

> In all cases this has been done at the request of the Proprietors, who have objected in several instances to retain small pieces of ground cut off from their properties by the Line of Road.[76]

Ten years later the Slamannan purchased five extra acres of land to avoid having to make a bridge, anticipating that it could lease the surplus for factories — an activity which might well not have been legal under the company's act of incorporation.[77] In his early days as secretary at Greenock, the perceptive Mark Huish warned his board against unnecessary purchases:

> You must set your face against taking the whole of a person's property if part is touched. They no doubt make up very plausible stories about the uselessness of the remainder to them. It is far more useless to us...[78]

But to too many companies the purchase of a few extra acres might seem a quick and easy way of buying off potentially inconvenient opposition.

The Glasgow Paisley & Greenock, indeed, had more trouble than most companies with its land purchases, as is suggested by the sizeable amount which it had to spend. Houston station, a wooden shed with no goods facilities, was built only because Lord Douglas made it the price for withdrawing his opposition.[79] Lord Blantyre, whose land was intersected by three miles of railway including both forty-foot embankments and seventy-foot cuttings, took his opposition to the

House of Lords, and persuaded its standing orders committee to reject the bill. Although the company were able to persuade the whole house to reverse this decision on appeal, they recognised that Blantyre had a serious grievance, that he might persuade Parliament to reject the bill on its merits, and that if the land purchase price had to go to arbitration 'a jury would in such circumstances have been disposed to award very considerable compensation'. Even so, the settlement made at the last possible minute, which gave Blantyre £10,000 for amenity damages as well as the value of the land, the amount of compensation due to tenants, and his own costs, was generous by almost any standards.[80] In the following year the directors aired their grievance:

> From the novelty of undertakings of this nature to any extent in Scotland, there exists in some cases what the Board cannot but conceive an exaggerated estimate of the prices which a Railway ought to pay for the land required in its construction based on a mistaken idea of the permanent injury to be done to the property.[81]

Table 41. Land Costs of Inter-urban Railways (£)

Company	Date	Land Costs Total	p.m.	% of total Capital expenditure
Arbroath & Forfar	4.1842	17,053	1,118	12.5
Edinburgh & Glasgow	1.1845	219,006	4,761	13.0
Glasgow Paisley & Greenock[a]	1.1844	112,128	7,234	18.1
Glasgow Paisley Kilmarnock & Ayr[a]	7.1843	108,588	2,511	12.4

[a]: excluding Glasgow & Paisley joint line
Sources: see Tables 36 to 39.

And even in 1841 Locke was still complaining about landed opposition:

> I have never met with fewer facilities than were granted at Bishopton, and scarcely ever known more concessions made, nor greater compensation given, than have been made by this Company to the owner of the land there.[82]

Clearly, both railway projectors and unenthusiastic landowners often felt aggrieved. The Ayrshire, observing that its land had been 'obtained, on the whole, on reasonable terms', was in a minority.[83] The commoner railway view of rapacious landowners was summarised by *Chambers' Edinburgh Journal*:

> Scarcely any instance is known of a reasonable price being asked for the portion required by the railway proprietors ... One spirit pervades all — a determination to extort as much above the value of the property as they can by any means obtain ... It appears to be only in dealing with an individual, that we have distinct notions of justice. In dealing with the abstraction called a railway propriety, men seem to think themselves entitled to cheat, overreach, and exhort without restraint.[84]

The figures in Table 41 indicate that the inter-urban lines had different experiences. The low costs per mile on the Arbroath & Forfar reflect the advantages of a single-track line in a rural area with mainly friendly landowners. In view of the figures for the other three lines, it is not surprising that the Ayrshire was relatively contented, or that the most vociferous complaints about landowners came from Greenock. The differences cannot be explained by the proportion of urban land or the value of rural land required, and the only conclusion is that the Greenock company was faced by more hostile opponents and that it surrendered more easily — it is remarkable, considering the amounts paid, that in only one case did a Greenock negotiation go to the decision of a sheriff's jury.[85] While the other three companies spent about an eighth of their capital on land, the Greenock proportion rose to more than 18%.

Overall, railways certainly paid more for their land than they had anticipated, even if the highest prices were not to appear until encouraged by the follies of the mania and the advantages given to landowners by the legislation of 1845. On the other hand, the powers of compulsory purchase granted to railways were an outrage to many property owners, particularly when the disturbance was to be created in previously remote and peaceful estates with long family traditions. To other owners, burdened with debt in an economic recession, the opportunities of substantial windfall profits were attractive. In any case the natural instinct was to extract as much from the railway as was possible, and very often the promoters, eager to conclude their deals and anxious about the possible effects of formal opposition, were content to conclude a generous bargain.

IV. Works, engineers and contractors

While the earliest capital expenditure of new companies was concentrated on parliamentary costs and land purchases, from the time of authorisation the emphasis moved inevitably to the actual construction of the line. Once again, accounts break down these costs in different ways from company to company, and not infrequently from year to year in the same company. Three main elements can be distinguished in each case, and are presented in Table 42 for the four inter-urban companies principally under discussion in this chapter. They are firstly the fees and expenses paid to the engineer and his assistants; secondly the costs of earthmoving, tunnelling, and other works required for the roadway, to which companies generally added the expense of stations, engine sheds and other necessary buildings; and thirdly the costs of rails, chairs, sleepers and other requirements for the track.

The table shows that these costs accounted for between three-fifths and three-quarters of all capital expenditure. Since the essential object of all expenditure was to create an operational railway, it might even be argued that the higher the proportion of the total which was spent on physical construction, and the less that had to go on more peripheral items, the greater was the success of the promoters. One striking oddity in the figures is the Ayrshire's high cost per mile for rails, balanced by a low cost for works. It is highly unlikely that companies undertaking construction at the same time would pay significantly different amounts for rails, and it is probable (though unclear in the accounts) that the rails figures include items such as ballasting and/or the labour involved in laying the track which elsewhere are placed under works. With this adjustment, the various

Table 42. Works and Engineering Costs of Inter-urban Railways (£)

Company	Date	Engineering		Earthworks, masonry & stations		Rails		Total		% of total capital costs
		Total	p.mile	Total	p.mile	Total	p.mile	Total	p.mile	
Arbroath & Forfar	4.1842	2,318	152	66,482	4,359	26,141	1,714	94,941	6,226	69.5
Edinburgh & Glasgow	1.1845	29,280	637	1,045,553	22,729	132,256	2,875	1,207,089	26,241	71.6
Glasgow Paisley & Greenock[a]	1.1844	10,072	530	307,743	16,197	53,359	2,808	371,174	19,535	60.0
Glasgow Paisley Kilmarnock & Ayr[a]	7.1843	15,940	317	420,382	8,366	186,373	3,709	622,695	12,392	71.2

a: excluding Glasgow and Paisley joint line

Sources: see Tables 36 to 39.

figures for rails are consistent, remembering that the Arbroath & Forfar was a single-track line with passing places. In each case also the engineering charges are between 2.5% and 2.8% of the total for works and rails together.

The important variation comes in the cost per mile of the works. The Arbroath & Forfar, built on a limited scale through easy country for slow traffic, was not surprisingly cheap to construct, but it is not strictly comparable with the three more ambitious projects radiating from Glasgow. Of these the Ayrshire, even if a transfer of say £900 per mile is made from the figure for rails, was by far the cheapest, reflecting the simplicity of the engineering problems it posed; only an unexpectedly deep bog north of Beith caused any worries at all.[86] Miller's other line, the Edinburgh & Glasgow, cost more than twice as much per mile, due to the conscious willingness of the promoters to incur extra expense to achieve a virtually level line, and to some major items like the Almond and Avon viaducts and the Queen Street tunnel. The Glasgow Paisley & Greenock was surprisingly expensive for what was in effect a branch line; some of the blame could be put on the difficult whinstone cuttings at the Bishopton ridge,[87] but the later shareholders' committee of inquiry was more inclined to suspect corruption by some officials and incompetence by virtually everyone involved. Even so, the proportion of total costs going to works was significantly lower at Greenock than elsewhere, due mainly to its particularly high land costs.

Much of the responsibility during construction fell on the engineer, whose duties included not only designing the line and estimating its probable costs and revenue, but also supervising the details of its construction — an obligation which might continue for some time even after the line was opened. The status of railway engineers was undergoing continuous change in the first half of the nineteenth century. Looking backwards, they stood in direct line of descent from the coal viewer of the previous century, who

> was the most skilled builder and user of Newcomen engines . . ., the most advanced engineer and innovator of rail systems and works connected with them, including tunnels and bridges, and it was within the viewers' profession that the men arose who put improved Watt engines on wheels and developed the locomotive.[88]

The coal viewers had created and managed waggonways as an almost incidental part of their colliery management duties. By the 1830s the growing scale of the railways had led to some specialisation of function; the engineer who designed and constructed a railway — essentially a civil engineering project — was not necessarily expected to have the mechanical skills to provide the locomotives as well (although some men like Robert Stephenson remained perfectly capable of doing both). And, while the viewer had continued to manage the waggonway he had created, the railway engineer moved on to other projects. Something of the old viewer's role, however, was retained after a railway opened by the company's resident engineer, who was responsible for maintenance of track and often of rolling stock, and for the provision of small branches. Often this position was combined with a managerial function, so that, for instance, at the end of the mania Alexander Adie was secretary and engineer of the Monkland & Kirkintilloch, while a decade later Joseph Cochran was engineer and manager of the whole

Monklands amalgamation.[89] Resident engineers, however, although generally worthy men, did not have names which were known to the public at large or which inspired confidence in investors, and companies not uncommonly kept the names of the more distinguished men who had designed their lines on their reports as chief engineers or consulting engineers, long after they had ceased to do any significant work.

The consulting engineer, acidly described by Brunel as a man who was prepared to sell his name but nothing more,[90] stood at the top of the pyramid of status. In essence his function was to examine the plans of others, to suggest modifications if he so wished, and to give a stamp of approval to the project which would reassure hesitant investors. In Scotland he appeared with the need to attract capital from England, and was normally a distinguished engineer whose name would breed confidence in the south: the work of, for instance, George Stephenson for the Ayrshire or numerous notable Englishmen on the route between Glasgow and Edinburgh, has already been noted.[91] At least in these cases it was absolutely clear that the imported Englishmen were merely examining and pronouncing upon the plans of others: the great days of the consulting engineer came with the mania, when projectors sought simply for a famous name to display on the prospectus, in the hope of persuading investors that he would actually attend to every detail of the line. During the mania Robert Stephenson was associated with 34 successful schemes and Joseph Locke with 31, to say nothing of those which failed in spite of them. In Scotland the partnership of Grainger and Miller was named in 37 mania prospectuses.[92] Clearly by then these men could have only limited contact with many of the projects for which they were responsible, while most of the actual work was done by assistants or by the companies' resident engineers.

For the lines authorised in the twenty years from the Monkland & Kirkintilloch to the North British, there was no doubt that the engineer named on the prospectus was the man who actually built the railway, even if on occasion he might have to put up with an English visitor looking at his ideas. His duties included designing and costing the line, ascertaining existing traffic levels and calculating the likely amount on the new railway, giving evidence before parliamentary committees (including evidence against any hostile projects), supervising construction, adjudicating on the adequacy of contractors' work, and not infrequently issuing reassuring reports for the benefit of shareholders.[93] At almost any point a mistake in his calculations might mean the difference between profit and loss for the shareholders; his importance also made him an obvious scapegoat if a company ran into difficulties. Accusations against engineers might range from culpably underestimating costs in order to make an unattractive project look viable, to encouraging unnecessarily expensive works to increase their own fees.

The names of Thomas Grainger and John Miller have appeared with some frequency in previous chapters, and Table 43, which lists the engineers who successfully constructed railways in Scotland, underlines the overwhelming importance of their partnership. Grainger and Miller built over three-fifths of the early Scottish network, including both the pioneering coal lines in the Monklands

Table 43. Engineers of Scottish Railways to 1843

Engineer	Company	Date	Mileage[a] built	% of Total
Blackadder, William	Newtyle & Coupar Angus	1835	4.82	
	Newtyle & Glammis	1835	6.61	
			11.43	4.2
Grainger, Thomas	Monkland & Kirkintilloch	1824	10.78	
	Ballochney	1826	5.00	
	Garnkirk & Glasgow	1826	8.18	
	Wishaw & Coltness	1829	11.03†	
	Polloc & Govan	1830	1.19	
	Paisley & Renfrew	1835	3.00	
	Edinburgh Leith & Newhaven	1836	5.08	
	Arbroath & Forfar	1836	15.25	
			59.51†	21.7
Jardine, James	Edinburgh & Dalkeith	1826	13.06†	
	Ardrossan	1827	9.50	
			24.56†	8.9
Jessop, William	Kilmarnock & Troon	1808	12.00	4.4
Landale, Charles	Dunfermline & Charlestown (reconstruction of Elgin waggonway)	1821	6.00	
	Dundee & Newtyle	1826	10.50	
			16.50	6.0
Locke, Joseph & Errington, John	Glasgow Paisley & Greenock	1837	19.56[b]	7.1
Macneill, John	Slamannan	1835	12.49	
	Wilsontown Morningside & Coltness	1841	8.05	
			20.54	7.5
Miller, John	Dundee & Arbroath	1836	16.75	
	Glasgow Paisley Kilmarnock & Ayr	1837	47.00†[b]	
	Edinburgh & Glasgow	1838	46.00	
			109.75†	40.0
Robson, Neil	Drumpellar	1843	0.86	0.3
	Scotland total		274.71	

[a]: † signs indicate small additional branches of unknown length
[b]: including half of Glasgow & Paisley joint line

Sources: HLRO, HC Deposited Plans of the various railways. For mileages, sources as for Tables 1, 12 and 25.

From Town to Town: Construction and Operation 193

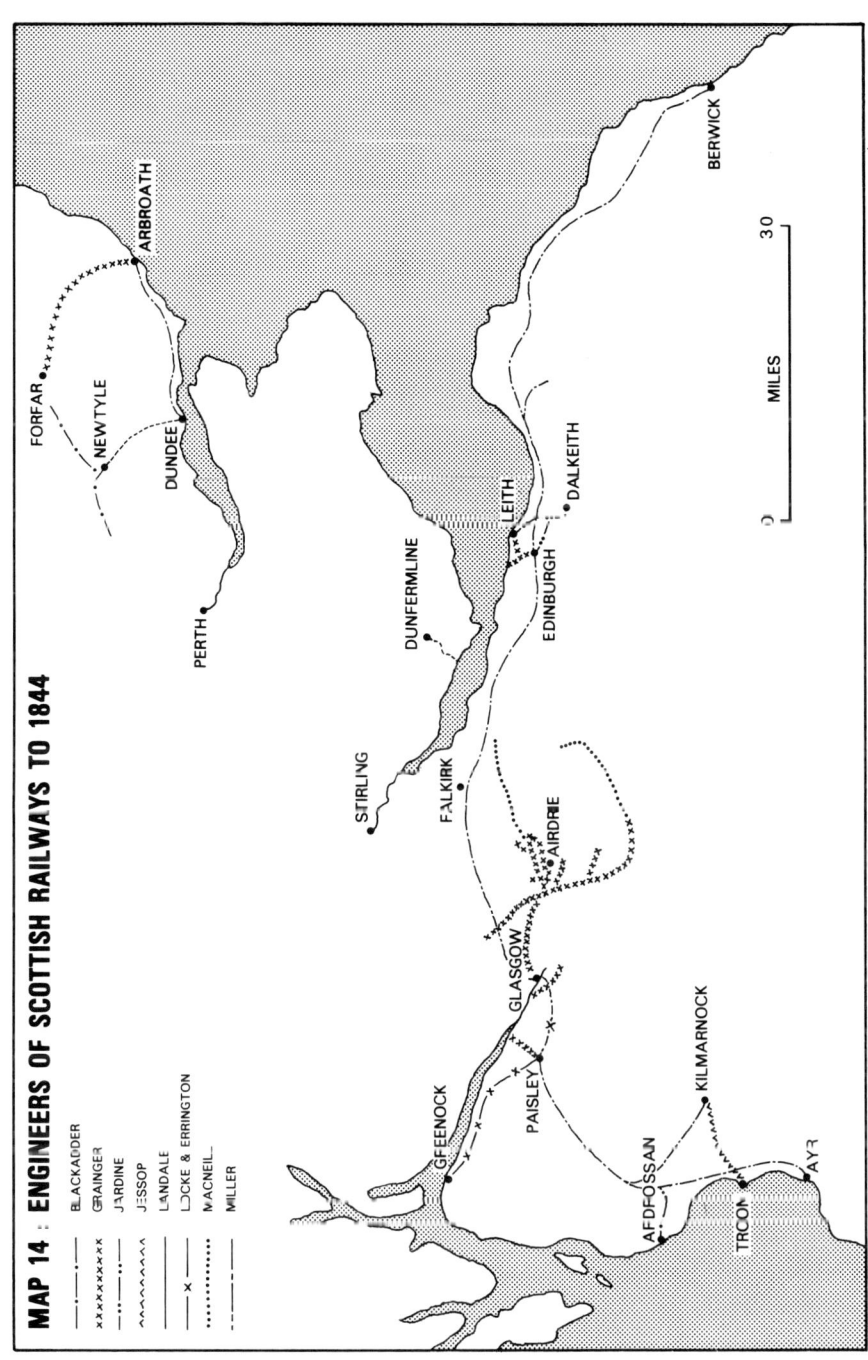

and the major inter-urban routes before the mania which linked Berwick, Edinburgh, Glasgow and Ayr. In 1843 their lines generated 81.1% of the total gross revenue of the Scottish railway system.[94] The one significant project which eluded them, the Greenock line, had in fact been originally awarded to Grainger, before Locke and Errington were brought in to attract English investment. Except in the earliest Monklands days, when Miller was still very much an assistant to the older Grainger, the partners did not work together on a project (their last co-operative venture being the abortive Edinburgh to Glasgow project of 1830-31). The partnership remained intact, with the partners working on separate but never on competing projects, until it was dissolved in 1845.[95] Its monument was the railway network of lowland Scotland, apart from the lines which became known as the Caledonian system.

Thomas Grainger, born in 1794 and educated at Edinburgh University, was in business as an independent civil engineer and surveyor by 1816. His role in the early coal railways was that of propagandist and projector as well as engineer; in particular, he was one of the leaders of the attempt to link Edinburgh and Glasgow.[96] Strangely, as noted before, he appears to have had little concept of the development of a national railway system, if we can judge by his selection of a 4'6" gauge as suitable for coal traffic in the Monklands and a 5'6" one for the mixed traffic of Angus (perhaps made more attractive by the relative cheapness of Angus land and the generosity of Angus landowners). He and Miller proposed to use the latter gauge for their lines from Glasgow to the Clyde coast, until the advent of Locke at Greenock persuaded them to adopt the $4'8\frac{1}{2}"$ Stephenson gauge which had become standard over most of England.[97] Not until the Edinburgh & Glasgow did either partner voluntarily start from the Stephenson gauge, and even there, according to John Leadbetter, the decision was taken under strong pressure from the supervising English engineers.[98] Miller still stated to a select committee in 1846 that a broad gauge of 5'6" or 6' was safer and should be made general, but that he was having to accept majority opinion and alter the Dundee & Arbroath.[99]

Grainger's star went into relative decline after the mid-1830s; although still only in his forties, it appears as if he never fully adapted to the increasing scale of railway operations. Of the inter-urban lines of the 1830s, he built the straightforward Arbroath & Forfar, and then became entangled in the mire of the Edinburgh Leith & Newhaven. By 1841 he was at work on the Fife line which, as the Edinburgh & Northern, became his last major success;[100] even there, he had to survive an attempt by John Gladstone to copy the precedent of Greenock and replace him by Locke. Gladstone's complaint was that Grainger's 'habits are expensive, if not extravagant', and in particular that his Fife estimate of £750,000 for $36\frac{1}{2}$ miles compared very badly with Locke's £800,000 for 72 miles in the much more difficult country between Lancaster and Carlisle. Some of Gladstone's fellow-projectors, however, believed that 'there would be danger in incurring a greater expense were an English engineer employed', and Grainger's contract survived.[101] Grainger might have claimed that a high estimate might merely be a more honest one, but Locke already had a reputation for accuracy — as Francis noted, 'to this gentleman has been awarded the praise of keeping his works within his estimates'[102] — while Grainger's previous record did not inspire confidence. The difference between estimate and performance on the Wishaw & Coltness has been examined above,[103] and Tables 12 and 25 summarise some of his other achievements. In the mania Grainger, like any other engineer who was not actually senile or insane, had plenty of work, but his only successful authorisations in Scotland were the Edinburgh & Northern and two small lines in his favourite coalfield — the Edinburgh & Bathgate and the unbuilt Airdrie & Bathgate Junction. He also built a few lines in West Yorkshire, and numbered among his failures some interesting ventures like the Aberdeen Banff & Elgin and the 1847 version of the Edinburgh & Perth.[104] Meanwhile his diversifying interests were reflected by service as an Edinburgh town councillor and two years as president of the Royal Scottish Society of Arts. In 1852 he died as a result of a train collision, ironically on the Leeds Northern line which he had himself engineered.[105]

Meanwhile John Miller had become unquestionably the leading Scottish railway engineer. Eleven years younger than Grainger, he had trained for the law with Kilmarnock writer C.D. Gairdner (himself a director of the Ardrossan and later a member of the provisional committee of the Glasgow Dumfries & Carlisle), before joining Grainger's office in 1823 and becoming a partner two years later. Oddly enough, in the same year he was helping with the survey for Jardine's Edinburgh to Glasgow line — the one occasion on which his activities seem to have conflicted with Grainger's.[106] By 1836 he was successfully creating the Dundee & Arbroath, in spite of difficulties with the contractors: the directors expressed themselves very satisfied with their engineer.[107] He came to national prominence with the Glasgow Paisley Kilmarnock & Ayr, on which the engineering difficulties were admittedly not great; the board reported that in 1840 'the present satisfactory position of the undertaking is, in a great measure, to be attributed to his zeal and ability in the Company's service', while one director 'had no doubts Mr. Miller's eminent abilities would soon rank him equal to the first English engineer'.[108] The Edinburgh & Glasgow, with its great viaducts and its dramatic tunnel into the centre of Glasgow, confirmed his reputation. By 1844 John Gladstone noted him as 'a man of whom I hear a great deal as a very able Engineer'.[109]

Miller was not immune from the usual criticisms. As the leading Scottish engineer, he certainly took on too much work during the mania, and for some of the more obscure projects to which his name was attached he can only have been the most remote of consultants; even on a project as important as the North British, too much essential work was left to inadequate subordinates, and he must take much of the blame for its sometimes shoddy construction and for the disastrous consequences.[110] He could be arrogant: the Duke of Buccleuch was warned that:

> Mr Miller . . . is said to be a very peremptory person, & there is no getting him to alter his plans or adopt the suggestion of another.[111]

Buccleuch however later considered him to be one of the very few men who had anything sensible to say about the various plans for a railway between Hawick and Carlisle.[112] His estimates inevitably understated the true costs of a project, although they were usually more accurate than Grainger's (see Table 32), and no engineer could reasonably be blamed if, for instance, his directors agreed to pay excessive prices for land. But in 1841, after the opening of the line from Glasgow to Ayr, he had to admit that he was 'at a loss to account for the expenditure of £567,000 in executing the 33 miles of the line between Paisley and Ayr' (to say nothing of substantial bills as yet unpaid), against his original estimate of £485,000.[113] The Ayrshire company retained his services throughout the 1840s, although his peripatetic status may be inferred from the demand by a shareholders' committee of inquiry in 1849 'that the services of a permanent Consulting Engineer be dispensed with'. The board denied that he was simply a consultant: he was employed

> to superintend the execution of the works now in course of construction ... Mr Miller's salary is very moderate; and your Directors are satisfied that it would be a great loss to the Company were they now to be deprived of his assistance.

In spite of this defence, a year later a new company chairman was accusing him, with hindsight, of having charged 'exorbitant' fees.[114] On the North British too, he was faced by shareholders' accusations of extravagance in the engineering department.[115] It is, however, not unreasonable to see in these charges a search for a scapegoat by directors and shareholders faced at the end of the decade with unhealthy balance sheets and non-existent dividends.

Miller's successes during the mania included the North British, its offshoot the Edinburgh & Hawick, the Glasgow Dumfries & Carlisle, the Dundee & Perth, and the Stirling & Dunfermline; he was thus responsible for all the main through lines of southern Scotland with the single exception of the Caledonian. His long list of projects which failed to come to fruition included some adventurous propositions — lines to connect the Irish boats at Portpatrick with Carlisle (the British & Irish Union) and with the Ayrshire railway (the Glasgow & Belfast Union), a pioneering probe into the western Highlands (the Scottish Western, to Oban), and even a flirtation with the atmospheric system between Edinburgh and Leith.[116] In 1850, when his Glasgow Dumfries & Carlisle opened, he retired from railway engineering at the age of 45.[117] It may be that the efforts of the previous six years had exhausted him, but it is likely that his decision was hastened by the often unjustified abuse of disgruntled shareholders. He had made enough money to settle as a country gentleman, purchasing estates which by the time of his death in 1883 amounted to 13,000 acres in Peeblesshire and 2750 acres in Kincardineshire, with a gross annual value of £6,640; from 1868 to 1874 he sat as Liberal MP for Edinburgh.[118] It was a career of considerable achievement, and, as Simmons comments, Miller 'ought to be more widely recognised as one of the great British railway engineers'.[119]

The other engineers who constructed early Scottish railways require, for the most part, only passing attention. The railway work of James Jardine and Charles Landale was limited to the 1820s and has been noticed above;[120] William Blackadder was a Strathmore man whose work was restricted to the two very basic lines in his own locality. John Macneill, on the other hand, became the leading mania railway engineer in Ireland and was to be knighted for his achievements, but his earlier creations of the Slamannan and the Wilsontown line hardly entitle him to a major place in Scottish history; he did also apparently take over the later stages of some lines designed by Grainger, being reported as in charge of construction on the Wishaw & Coltness in 1838 and at Newhaven in 1840.[121] Neil Robson's interest in railways developed as part of a distinguished career in mining engineering, and from his position as engineer to the Monkland Canal for which he designed the ancillary Drumpellar line. His mania lines included half a dozen small companies around Glasgow, notably the Glasgow Barrhead & Neilston Direct and its extension which eventually gave the Caledonian a route to Ardrossan. His main interests, however, remained with minerals, and he became managing director of the coal and iron firm of Merry & Cunninghame, after marrying Agnes Merry. Enthusiasm for railways also had to be tempered by a rival commitment when he became deputy governor of the Forth & Clyde Canal.[122]

Grainger and Miller were the only Scottish engineers whose reputation stood high enough before the mania to attract more than local capital into their projects; even so, several of the occasions on which their work was checked by English engineers to reassure English investors (as with George Stephenson on the Ayrshire, or his son on the North British) have already been noted.[123] The idea was even applied to a minor project like the proposed link from Coupar Angus to Dunkeld in 1838: one of the promoters, suggesting Blackadder as a suitable engineer, added that 'before executing what he proposes you could have some of the eminent Engineers from the English railways to revise it'.[124]

The logical next step was taken by the Greenock committee of management. As they told their shareholders in January 1837,

> the attention of the Committee was early directed to ensure the services of an English engineer of eminence, whose decision in the selection of the line, by being unconnected with the district in any way, would be entitled to public confidence as impartial and uninfluenced by local or individual interests.[125]

This might simply have referred to an English inspection of someone else's plans, and the committee did at first employ Grainger and another Glasgow engineer William McQuistan to make surveys. But through the influence of Patrick Maxwell Stewart, MP for Lancaster, director of the Grand Junction and brother of the recently deceased MP for Renfrewshire, they secured the thirty-year-old Joseph Locke as their engineer.[126] Locke, who had already established a considerable reputation on the Grand Junction and the London & Southampton, at first declared that he was too busy (with, among other things, survey work on the possible routes of a railway from Lancaster to Glasgow), and that the job should go to Grainger 'whose great experience and eminence as an engineer are well known', but in the event he and his partner, the self-effacing and 'invaluable' John Errington, took the contract.[127]

The main contribution of Locke and Errington to early Scottish railways came during the mania, when they engineered the Caledonian and its extensions as far as Aberdeen, as well as being consulting engineers for two of the most ambitious failures — the Scottish Grand Junction, which would have opened up the western Highlands, and the Perth & Inverness.[128] The Greenock line was in some ways not their most successful project. Locke's reputation for accurate estimating is not supported by the figures, though Errington claimed to offer 'proof of the work being carried out when under the engineer's control, to within 5 per cent. of the original estimates'.[129] In January 1840, when the works were well under way, Errington had reported to the directors that

> he had gone carefully over the estimates, and had no hesitation in saying, that all the new works and branches, alterations and extensions, could be executed without additional capital to that authorised by the [original] act.[130]

At the end of the year the company admitted that costs were some 18% above estimate, placing the blame on high arbitration awards and the failure of two contractors.[131] Errington's optimism may account for one contemporary in 1844 placing most of the blame on the junior partner, though he agreed that Locke, as chief engineer, must also accept responsibility:

> Mr Errington's reputation has suffered severely by an under estimate to *a large amount* of the cost of the Greenock & Glasgow Railway.[132]

The original estimate for the line (which had in fact been made by Grainger) had been for £393,000: by September 1843 the board had committed £666,666 of share and loan capital and found it necessary to obtain powers for a further £200,000. The company's committee of inquiry in 1850 was at a loss to explain how

> within five years from the date of the act of incorporation, the Directors and their advisers contrived to squander more than double the estimated cost of the lines, without securing any corresponding solid advantages to which they can point, and that upon a line which Mr. Locke had declared to present no engineering difficulty, and to be one on which no extra cost would be required, and that Mr. Grainger's estimated prices for the work would be sufficient.[133]

The committee was admittedly made up of hostile shareholders, but the directors' reply produced little in the way of satisfactory explanations, and was reduced to impugning the motives for the inquiry.[134]

The engineers could not be blamed if their directors ran up unreasonable parliamentary or legal costs (and sometimes these could mount without the directors being able to do much about it), nor were they responsible if the Greenock board bought off its landowners more expensively than other lines. The excess of land costs over the estimate was £51,000, and a further £103,000 was incurred for extensions which were not in the original plan.[135] The company had, however, spent no less than £325,000 on the works, which presumably were under the engineers' control. The committee of inquiry found it all most unsatisfactory:

> Of the contracts got some are without specifications and schedules of prices, and all without the relevant tenders and plans. And while the payments to contractors greatly exceed, in almost every instance, the tender sum in the contract, it is remarkable that in only one or two cases there should exist an account for the work paid. It is, therefore, impossible that the committee can give any satisfactory report on this expenditure... The work executed is, in many instances, greatly at variance with that detailed in the specifications. Allowance is given to some of the contractors to perform their work in such a manner as to have reduced the cost of labour, while it must at the same time have greatly increased the price of land to the Company. Whether or not proper compensation was allowed by the contractors on these accounts, the committee have no means of judging.[136]

Nor did the critics accept the board's reasons for rejecting a tender for one contract

20% lower than the one accepted, even though the lower one came from a contractor who had already performed satisfactorily elsewhere on the line; in the event the preferred candidate finished six months late and 17% over his estimate.[137] Locke's experiences at Greenock presumably encouraged his later practice of detailed estimates and strict supervision of the works: according to Rolt, he became

> the first man to set a new standard of accuracy in estimating by supplying the contractor with carefully detailed specifications and by exercising a closer control over his subsequent performances.[138]

After his problems with the numerous small contractors on the Greenock line, Locke also turned to dealing with the few large contracting firms which could themselves construct a complete railway, notably the powerful combination of Mackenzie, Brassey and Stephenson which built the Caledonian.[139] Dealing with small subcontractors then became the direct responsibility of the main contractor rather than the engineer.

Locke's work at Greenock practically finished with the opening of the line, although he became for a short time a member of the company board.[140] Errington remained more actively part of the company, accepting the position of manager/engineer; Gourvish notes that 'the Greenock company ... tended to equate operations with the engineering department', so that Errington was clearly in charge 'with the assistance of Captain Huish'.[141] One part of his activities received a glowing tribute from the knowledgeable Sir Frederick Smith:

> The code of regulations, prepared by Mr. Errington, about to be established for working this line ... appears to me to be drawn up in so able a manner that ... it might be taken as the general form for the guidance of companies in this part of the country, where I find very discordant systems in operation.[142]

In July 1841 he refused an offer of a salary increase to £400, 'as he was already perfectly satisfied with both his salary and his duties'. Two years later, when the company was in difficulties, the salary was cut at his own request to £270;[143] admittedly the work was not full-time, as he continued to survey with Locke on the Lancaster & Carlisle and Caledonian routes, and in 1841 was himself conducting a detailed examination of the ground between Stirling and Perth.[144] Unlike most eminent engineers of the period, John Errington appears to have been noted for both modesty and pleasantness of personality. His obituary in the *Proceedings of the Institution of Civil Engineers*, after praising 'his kindness of heart, his great tact, and his affectionate disposition', recommended that 'his example should be carefully studied by the younger Members'. He was not in the class of Locke or Miller as an engineer (even the obituarist could go no further than 'sound talents'), which may explain his apparent satisfaction with the role of junior and often overshadowed partner. But he came into his own in persuading parliamentary committees and dealing with hostile landowners. Joseph Mitchell, who disliked Locke and Errington after being superseded by them on the Scottish Central, attacked him for

> a wonderful degree of diplomatic plausibilities and kind and friendly expressions, for which Mr Errington had a special talent (indeed his principal talent) ...[145]

Whereas Grainger and Miller were a partnership of two individuals who largely went their own ways, Locke and Errington worked together on the same contracts and had complementary talents which made them into a formidable team.

Engineers were inevitably the targets for the wrath of discontented shareholders, and sometimes the scapegoats for buck-passing directors. They were employed for abilities which were supposed to include pricing as well as constructing the railway, and they might well be the only people connected with the project who had either technical expertise or previous railway experience. Directors and shareholders often knew from personal experience what were reasonable prices for land or legal services, and what was a tolerable rate of interest on borrowings (even if in all companies they sometimes had to pay more). But they seldom had experience of large-scale construction, and had to trust to their engineer, who was himself engaged in a relatively new branch of his profession. In practice, engineers underestimated across the board, sometimes from excessive optimism, sometimes from failing to see the extent of potential problems, sometimes perhaps under pressure from projectors who wished to put encouraging figures in the prospectus. Often they underestimated traffic as well, but the eventual balance seldom produced the rosy dividends of the first forecasts. As early as 1837, *Blackwood's Magazine* was reproachful:

> What can have become of the slates and pencils of the engineers? They have seen railways in action these twenty years; it is scarcely possible that every detail of their expenses should not be familiar to them.[146]

The carping continued through to the North British shareholder and future director who

> considered it to be certain that the Company had been cursed with incompetent officials. To a certain extent, he considered their engineer [Miller] to be at the head of them.[147]

The faith placed by companies in their engineers was great, and the later disillusioned criticism was correspondingly bitter. Given the novelty of the projects they were undertaking, the judgment of history may be that of Pollins: 'what is astonishing is that in the early railway age the various engineers were not more inaccurate than they were'.[148]

The engineer was ultimately responsible for the performance of the contractors, and the quality of too many contractors varied from unreliable to reprehensible. Thirty years later, a writer in the *Proceedings of the Institution of Civil Engineers* had unhappy memories:

> The early contractors were, for the most part, men of strong natural abilities, insight into the cost and method of executing work amounting to instinct, low tastes, violent habits, and grasping tenacity of purpose. A contract being once made, it seemed to be regarded as natural that the contractor should set his wits to work to make the most of it. This was to be done, on the one hand, by grinding his labourers . . . and on the other hand by 'scamping' his work.[149]

Scottish contractors in the decade before the mania had not caught up with the increasing scale of railway projects. Large-scale contracting firms did not come to the country until the mania, introduced chiefly by Joseph Locke. Before the Caledonian, however, lines were built by letting out short stretches or specialised projects such as tunnels to a series of individual contractors. The Arbroath & Forfar, for instance, had twelve contractors for a fifteen-mile line.[150] Such men often had limited experience and even more limited financial resources, so that difficulties could push them to the edge of insolvency. And often meticulous scrutiny of their work by the engineer was required to produce a railway of adequate standard.

Complaints against contractors run through the records of almost every company. In 1828 the Ballochney complained that the contractors on their branches had no chance of completion by the agreed date, but were slowing down their rate of work; they insisted that more men should be employed or they would finish the job themselves at the contractors' expense.[151] Similarly in 1839 the Slamannan did take over the work when 'both contractors have failed to fulfil their engagements as to time'.[152] The opening of the Edinburgh & Dalkeith was delayed by 'difficulties which he [the contractor] had not foreseen and which he had not sufficient capital to enable him readily to surmount'.[153] In July 1837 the contractors for rather more than half of the Dundee & Arbroath line went bankrupt; their men, having received no wages, walked out. The contracts were relet to six new individuals, one of whom provoked another strike four months later. As the directors said, it all meant 'much additional and unexpected trouble'.[154] By 1840 Miller was complaining on the Ayrshire of 'the hitherto slow progress of some of the contractors'; one, at Dalry, had no chance of being ready on time

> *if carried on by himself.* If carried on by the Company, *there will be no doubt* . . . I have pointed out to Mr. Milne a mode by which he may have his contract completed in the month of May, and have directed him to adopt it.

The tardiness of the contractor faced by the Beith bog was such that even this sort of close instruction would not serve, and the company finished the contract itself.[155]

As these examples suggest, complaints against contractors during construction were usually about delay. Complaints about expense, as already noted on the Glasgow Paisley & Greenock,[156] tended to come later, often at the time of the series of committees of inquiry mounted by disgruntled shareholders in most companies at the end of the 1840s. Complaints about the quality of the work were less common, even such an unhappy company as the Edinburgh Leith & Granton eventually receiving from the Board of Trade's inspector the tribute that 'the execution of the works reflects great credit upon both the engineer and the contractor'.[157] Pre-mania railways were on the whole reasonably well built, although the Wilsontown Morningside & Coltness, designed as a cheap single track to carry ironstone, and terminating 'in a large field' near Whitburn, was reported at its opening to be in a 'rough irregular state' and to have at least one very badly built bridge. Responsibility was placed on the long illness of the engineer, who had been unable to exercise proper supervision over the contractor; a speed limit of 15 miles per hour was imposed by the Board of Trade, which also implied that if the line had had any serious interest in passenger traffic it would not have been allowed to open.[158] On some more important lines, like the Ayrshire or the Angus railways, the problems were in any case not great; for more complex projects, like the Bishopton ridge on the Greenock line or the great viaducts on the Edinburgh & Glasgow, the engineer had the time to ensure that the work was done properly. Where construction verged on the cheap and shoddy, as on some of the Monklands lines, it was deliberately so, since nothing more elaborate was required for coal waggons conveyed at slow speeds, often by horsepower. The pattern changed with the mania, when there was so much work that some of it had to be given to incompetent contractors, when the principal engineers were too busy to spend enough time on any one project, when financial difficulties after the crash were an encouragement to sometimes unwise cost-cutting, and when directors' pressure to open the line for traffic as soon as possible often failed to allow the earthworks to stabilise. The most notorious case came four months after the opening of the North British, when floods swept away a considerable section of line and destroyed or damaged eleven bridges. Two separate reports to the government heaped substantial portions of blame on Miller, his assistant engineers, the contractors and the directors.[159]

Engineers were often accused, usually well after the event, of doing little to keep expenditure down to a minimum. It was argued that spectacular and expensive works might both enhance an engineer's prestige and increase his fees, and that long drawn out parliamentary contests, which would require him to spend much time in London, were also to his advantage. In fact the engineers' fees, which included payment to their assistants and expenses, do not appear excessive, amounting to less than 2% of total capital expenditure. They were related to the cost of a project rather than merely its length, and ranged from Grainger's £152 per mile for a simple single-track line between Arbroath and Forfar, to £637 per mile for Miller's much larger-scale operations on the Edinburgh & Glasgow (see Table 41). It may be noted that in the 1850s, when Thomas Bouch specialised in providing impecunious projectors in small towns and rural areas with basic lines whose cost had been cut to the bone, he still charged at least £100 per mile for engineering fees, while claiming that the average in Scotland was about £500.[160] In 1842 Miller's rates to the North British were four guineas per day when in Scotland, five guineas when in London, 'the usual fee for assistants', and expenses; if the bill failed, he would receive half the rate for his Scottish work and for the assistants, while if subscriptions were not sufficient to allow a submission to Parliament he would receive expenses only. By July 1844, with the early mania demand for his services well established, his fees were five guineas per day wherever he was, with assistants varying from one to two guineas, and expenses on top.[161] These rates were not unreasonable; if the engineers could be accused of unnecessary escalation of costs, it was because they arguably built their lines too well. Locke came to Scotland from the heavy traffic and planned high speeds of the route between London and Lancashire, and tended to choose a costly direct route in preference to a cheaper deviation which added to overall mileage. Miller, perhaps because his work was so often checked by the Stephensons, seems to have inherited something of George Stephenson's concern for eliminating all possible gradients. One contemporary journal remarked that

> The slowness to believe in the capabilities of the locomotive engine, exhibited by the engineers of Great Britain, is surprising ... Engineers set out in their railway career with the impression that the locomotive was ill-calculated to climb up hill with its load, and that therefore, to work

with advantage, it must work on lines altogether level, or nearly so; hence mountains required to be levelled, valleys filled up, tunnels pierced through rocks, and viaducts raised in the air — gigantic works at a gigantic cost... Here, again, was want of faith in the power of the locomotive engine. The locomotive engine can climb the mountain-side as well as career along the plain.[162]

Hence perhaps the enormous sums spent to achieve a maximum gradient of 1 in 880 on the Edinburgh & Glasgow,[163] and in too many places earthworks stronger than the traffic required and gradients flattened which would have been little or no trouble to contemporary locomotives.

The other major construction cost was the materials of the track itself — the rails, the sleepers or blocks on which they rested, and the chairs and pins necessary to attach one to the other. There was less likelihood of substantial cost variations between lines here, since the major ingredient was the price of iron, and as noted the figures in Table 42 seem consistent with each other. The price of bar iron stayed reasonably steady from late 1836 to the end of 1839 (apart from a dip in the summer of 1837), and then fell steadily until the middle of 1843 (see Figure 3). The figures in Tables 36 to 39 indicate that both the Greenock and the Ayrshire lines purchased the bulk of their rails in 1839, though both also benefited from the falling prices of 1840. The Edinburgh & Glasgow paid for the bulk of their rails in 1842, but had presumably contracted for them earlier at a higher price. The Dundee & Arbroath managed, thanks to Miller's advice, to buy at the bottom of the market in 1837; having paid on earlier contracts up to £14-14/- per ton for rails and £8-19-6d for chairs, they obtained the bulk of their requirements for average prices of about £7-15/- and £7 respectively. On the other hand, they had trouble when their contracted supplier of blocks provided both inferior quality and inadequate quantity, but an arbiter's judgment allowed the company to obtain them elsewhere and charge the original supplier with any increase in expense.[164] In the previous year the Arbroath & Forfar had increased its proposed share capital from £60,000 to £70,000, largely because of 'the present high price of iron', and had paid £15 per ton for rails and £9 for chairs; the total cost was also increased by a decision to use 48lb rather than 42lb rails.[165] In 1839 the Slamannan directors congratulated themselves on having contracted for rails at £10-10/- and £12 per ton, since the current price was at least £2 higher.[166]

The other conceivable way of saving on the cost of rails was to make a given amount of iron go further — in other words, to use lighter track. In fact the weight of rails used was effectively dictated by the amount of traffic anticipated and the speed at which it was to travel. Lines built in the 1820s generally settled for 28lb fishbellied rails, and the horse-drawn Edinburgh & Dalkeith continued to use them until it was taken over by the North British. By the end of the 1830s, however, the Garnkirk company was relaying part of its line with 50lb rails, and the rest with 42lb, while the Ardrossan was preparing to move to 56lb; the Slamannan, at about the same time, had started with 50lb, and the Dundee & Arbroath with 56lb. Meanwhile the Edinburgh & Glasgow and the Ayrshire, as befitted their status as fast main lines, had 75lb rails from the start.[167] In practice there were no gains to be made by scamping on the quality of the rails.

Figure 3. Price of Bar Iron, 1836-44

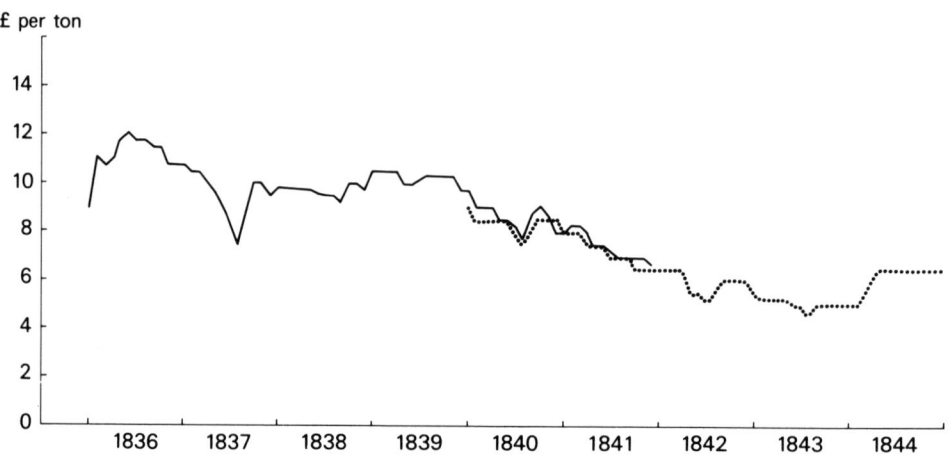

Sources:
——————— English prices from *Mining Journal, Railway and Commercial Gazette*, 23 April 1842.
.... Scottish prices from J. Barclay, *Statistics of the Scotch Iron Trade*, 21.
Rails were estimated to cost from 5/- to 7/6d per ton more than bars.

All things considered, the engineers and even most of the contractors served the pre-mania railways well, producing lines eminently suitable for the traffic anticipated. Although the cost was almost always well above the estimate, in most cases this could be blamed partly on the actions of the directors. Usually, too, some of the fault lay with the contents of the estimates themselves, which often covered simply the basic costs of laying down a line between two termini, omitting such incidental matters as the act of authorisation, legal costs, rolling stock or stations.[168] In 1840 Francis Whishaw protested against the practice:

> Parliamentary estimates for railways have naturally caused much inconvenience and disappointment to a large class of shareholders. No engineer, we are inclined to think, who gave the subject proper consideration, would omit in his estimates the cost of stations, locomotive engines, and carriages; for of what use is the railway without the means of working it? An estimate for this purpose should include every thing necessary to put the railway into complete working condition: the promised returns, without this necessary precaution, are but as glittering bubbles floating for awhile in the atmosphere of attraction, but presently bursting on the ground of deception. Many are the cases in which this essential part of a parliamentary estimate has been omitted, but we shall here notice no individual cases . . .[169]

Although Whishaw claimed to be avoiding individual condemnation, the passage comes during his discussion of the Dundee & Arbroath, and in the following year that company produced an analysis which is compared in Table 44 with Miller's original estimate. For the items covered by Miller the costs had been only 7.4% above expectations (although a figure of £15,131 for payments on account to contractors whose final bills had yet to be settled suggested that these costs would rise further),[170] but a substantial list of other essential items added £36,281 to the expenditure and pushed the total up to 43.8% over estimate. At the same time the capital expenditure of the Arbroath & Forfar had risen to £133,560, or 95.1% above Grainger's incomplete parliamentary estimate (see Table 45). It is unclear how far estimates such as these were deliberately restricted to basic construction in order to keep the figures low, while hoping to persuade investors that they represented the final costs of the railway; engineers could not have believed that the usual allowance of about 10% for contingencies would cover all the missing items, and it seems unlikely that directors would be so naive. Yet, since prospective dividends were calculated in prospectuses on the basis of these estimates, company directors and officials may stand accused of deliberately attempting to mislead their subscribers.

Table 44. *Dundee & Arbroath Capital Expenditure: Estimate and Reality*

	Miller's estimate, 9.2.36 £	Actual expenditure, 1.5.41 £
In original estimate		
Land & cottages	7,545	8,879
Works	32,427	41,827
Rails	52,034	49,546
Engineering & miscellaneous	7,839	7,021
	99,844	107,272
Not in original estimate		
Act of parliament		2,449
Stations		8,869
Workshops & equipment		1,841
Rolling stock		15,437
Embankments		4,174
Legal expenses		2,932
Feu duties		554
Miscellaneous		114
		36,281
Total expenditure		143,553

Sources: HLRO Deposited Plan, HC 1836, Dundee & Arbroath Railway; *RT*, 3 July 1841.

Table 45. *Arbroath & Forfar Capital Expenditure: Estimate and Reality*

	Grainger's estimate 6.8.1835, based on Blackadder survey 1826	Grainger's estimate to parliament, 30.3.1836	Actual expenditure, 15.4.1841
	£	£	£
Land	6,510	6,988	16,818
Works	25,235	20,873	52,994
Rails	17,876	25,711	25,972
Depots and stations		1,000	11,275
Arbroath harbour line	2,660	5,709	—
Miscellaneous	4,962	4,041	7,190
Parliamentary			2,703
Rolling stock			13,853
Engineering			2,318
Road trustees			437
Total	57,243	68,460	133,560

Sources: SRO, Arbroath & Forfar Minutes BR/AFR/1/1, 19 July 1835, AFR/1/2, 11 June 1841; HLRO Deposited Plans, HC 1836, Arbroath & Forfar Railway.

V. Other costs: rolling stock and road trustees

By the time that a company was well established, the purchase of rolling stock accounted for about 7% to 12% of capital expenditure. In choosing their locomotives, directors leant heavily upon the advice of their engineer, so that in Scotland the views of John Miller, who like Locke preferred six-wheel engines with inside frames and outside bearings, were bound to affect the outcome. Both the Ayrshire and the Edinburgh & Glasgow, however, covered both possibilities in the debate between six and four wheels; of the first twenty locomotives on each line, eight and ten respectively were four-wheelers of the type developed by Edward Bury on the London & Birmingham, at a cost of about £1,200 each.[171] The Edinburgh & Glasgow also experimented in 1842 with Davidson's electromagnetic locomotive, reporting that 'the practicability of the scheme is almost placed beyond a doubt', but apparently not following up their interest.[172] By 1844, however, Ayrshire opinion favoured a single model for both engines and carriages for reasons of economy.[173]

The larger companies looked to established English builders for at least some of their stock; the Edinburgh & Glasgow six-wheelers were provided by Hawthorns of Newcastle to a Stephenson design, while Locke's Lancastrian links meant that the best engines at Greenock came from Sharp & Roberts of Manchester.[174] In Angus, support went to Dundee firms. The Dundee & Newtyle's original eccentric bogie locomotives had cost £825 each from J. & C. Carmichael; the example of bogies was not followed anywhere else, since they were considered to be unsafe at high speed.[175] The Dundee & Arbroath paid £1,012 each for six Miller-approved locomotives from the Wallace foundry of Kinmond, Hutton and Steel. Similar engines for the Arbroath & Forfar came from a third firm, the Dundee Foundry Company of James Stirling, whose nephew Patrick was to become one of the country's foremost locomotive designers.[176] By 1840 the Newtyle line was also using three Stirling engines, which had cost £730 each, and had splashed out £1,137 for another from Robert Stephenson.[177]

On the whole it appears that the railways got reasonable value from their locomotives. The persistent troubles on the Dundee & Newtyle probably owed more to cheeseparing on maintenance than to inherent defects in the design, since the company had difficulties with both Carmichael and Stirling machines. The breakdown two days after opening on the Arbroath & Forfar, which meant that the service was horse-drawn for the following ten days, was not a precursor of further trouble.[178] The evidence for general satisfaction is essentially negative — the absence of complaints in the reports made by engineers to shareholders' meetings and, more significantly, the failure of the late 1840s inquiries to include locomotive purchasing in their usually lengthy lists of things which might have been done better.

Passenger rolling stock, particularly before the 1844 act imposed a degree of quality control on the cheapest accommodation, varied widely. At one end came the luxurious first-class carriages of the Ayrshire, where £400 was spent on providing for a mere eighteen passengers;[179] at the other perhaps were the open second-class carriages of the Garnkirk (which had no third class), where £50 provided a space of 13½ feet by six feet for 24 passengers, with doors on one side only 'because of the prevailing north wind'. These, according to Whishaw, were 'by far the most unsightly and uncomfortable we have met with on any passenger-railway in the kingdom'. Surprisingly he claimed that when the company had experimented with 'handsome and well-appointed carriages', traffic had fallen whether or not the fares were raised, and the old carriages had to be brought back; if true, this presumably owed more to the deepening recession than to a distaste for increased comfort.[180] By 1842, two years later, the company stated that their second-class accommodation had cushioned seats, and was covered on top, on the weather side, and at the ends, and commented rather smugly that

> Ten years experience confirmed us in the propriety and good policy of giving comfortable and safe accommodation to every class, without aiming at splendour, which is not wanted by the mass of our passengers, and which our rates of fare (under one penny a mile) will not afford.[181]

Some early carriages owed much to the road transport they aimed to replace. Dundee & Newtyle mixed carriages, built at a cost of £60 to £70 each, 'resemble an ordinary stage-coach, with the addition of an entirely open compartment both before and behind', and were rumoured to be in fact stage-coach bodies mounted on trucks; 'outside passengers secured a commanding view of the countryside, but had to duck their heads at every bridge'.[182] The original Ayrshire first-class carriages were also built on stage-coach principles.[183] At Dalkeith, where the horse-drawn stage-coach tradition was even more appropriate, the carriages were built by the company itself at a notably high cost of £170 each, and also had open seats at front and rear:

> There is a curved splash-guard in front of each outside seat; but we found that they did not

prevent the mud from being cast over in abundance, to the annoyance of ourselves and fellow-passengers.[184]

Third-class accommodation before the 1844 act was commonly in open trucks without seats, known as stand-ups, or by an elegant bowdlerisation as Stanhopes.[185] The degree of comfort might vary with the length of the journey; the Edinburgh & Glasgow declared just before opening that it intended to put stand-ups, taking up to sixty passengers each, on its fast trains, but seated third-class accommodation with forty passengers per carriage on the slow ones — presumably passengers might be expected to survive the discomforts of standing in the open for $2\frac{1}{2}$ hours but not for an hour longer. The Ayrshire also had a mixed bag of third-class carriages, open or covered, with or without seats, but it is not clear on what basis they were distributed between trains.[186]

Waggons required simply to be as cheap as was compatible with being sturdy enough for their job. According to Whishaw in 1840, companies had paid from £15 to £20 each.[187] In 1844 the Dundee & Newtyle received tenders of £17-10/- for wooden waggons or £18-10/- for iron; a decision was delayed under pressure from the company's bankers, presumably because of the railway's general financial crisis rather than because the price was too high.[188] The Wishaw & Coltness, which had paid only £14 per waggon, estimated their annual cost of upkeep at £2-16-7d each, or over 20% of the original cost.[189] The Edinburgh & Dalkeith distinguished between coal waggons at about £16 and general goods waggons at £20, but no evidence has been found to suggest that this distinction was general.[190] On the whole the waggons appear to have done adequately the essentially primitive job for which they were designed; the arguments about damage on many of the Lanarkshire coal lines centred on alleged mistreatment by traders rather than on the basic construction of the vehicle.

Companies on the whole do not appear to have been extravagant in their provision of rolling stock. Since traffic volumes, at least by 1843-44, were generally higher than had been expected, original purchases had not exceeded requirements, and the steady additions to the capital account caused by new purchases reflect the continuing need for expansion of the stock. Although differences of opinion about the best locomotive technology or type of carriage were possible, company officials were in general careful in their purchasing policy and were rarely accused of unnecessary outlay. Admittedly rolling stock purchases were one of the most obvious differences between many original estimates and final accounts, simply because they had been left out of the estimates altogether; but they were not an area to which companies might look for significant economies.

Other aspects of capital expenditure accounted for only small proportions of the total. But the appearance in some accounts of quite substantial payments to road trustees reflects another expense for which allowance was not made during planning. It was clear from the early days of railways that road traffic would be unable to compete for either heavy goods traffic or long-distance passengers once a railway was authorised along the route, and receipts from tolls on British turnpikes fell by 25% between 1837 and 1848. The fall in Scotland, with its less developed railway network, was less dramatic at 6.7% over a similar period, although for the thirteen trusts most affected by railway competition the decline amounted to 16.1%.[191]

Falling receipts posed a serious problem for turnpike trustees, most of whom were landowners along the route of their particular road. The trusts were habitually encumbered by sizeable debts and in Scotland, unlike England, part or all of the debt could be secured not only on the receipts of the trust but also on personal guarantees given by the trustees.[192] Since railway competition meant that the debts could never be paid off from receipts, it was imperative for the trustees either to forestall competing railways or to reach an agreement which would remove the burden of personally guaranteed debt. In the 1820s and early 1830s efforts were concentrated on trying to prevent railway authorisation; the assault on the original Edinburgh to Glasgow proposals noted above was typical.[193] But increasingly such opposition was advanced less in the hope of stopping the railway than with the intention of being bought off. In 1826 the projectors of the Edinburgh & Dalkeith agreed to pay the trustees of the Dalkeith road one penny per ton of goods moved by rail to a maximum of £800 per year and £12,000 overall. The agreement was left out of the act of authorisation as it was believed that the House of Lords would not accept it; nor was it ever ratified by a company meeting after authorisation. By 1834 no payments had yet been made, although the money had been set aside each year and the leading promoters of the company had reluctantly agreed that the agreement should be honoured.[194]

By 1837 the idea of compensation to road trustees was generally accepted, although no clear criteria had been established as to how it should be calculated. A document prepared in 1845 for the Duke of Buccleuch gave some case histories. In 1837, for instance, the Greenock company had undertaken in one case to relieve the road trustees of their personal liability for part of the debt, in another to pay off certain specifically named creditors, and in a third to make up the average income of the trust to its pre-railway level for the next twenty years. The Ayrshire company had agreed a straight cash payment of £18,000 to one trust; elsewhere, where two groups of trustees had become personally liable for sums of £6,765 and £5,300, they were to pay £800 and £1,200 respectively. The Edinburgh & Glasgow arranged to pay the Bathgate road trustees £17,500; in three other cases the trustees were to be relieved of a proportion of their personal liability.[195] In fact by the end of 1844 the company had paid out £18,581 to various road trustees; at some point the payment to the Bathgate trustees had been reduced to £11,000. As the author of the duke's document remarked,

> It does not definitely appear that these different cases were adjusted upon any fixed principle. The respective sets of Trustees appear to have made the best Agreements with the Railway Companies which they could.[196]

These agreements were incorporated in the authorising acts of the various companies. By July 1844, as Learmonth informed the North British shareholders, it was clearly the custom in Scotland, though not in England, to relieve road trustees of some of their personal liability, normally by transferring that liability to the railway company.[197] But events a few months earlier in the House of

Commons committee on the North British bill had introduced doubts. The committee ruled that arrangements to take over parts of the debts of the Haddingtonshire and Berwickshire roads could not be included in the bill; according to the company's London delegation:

> The effect of this resolution will probably be that in future no compensation will be allowed to Road Trustees in similar cases, but as the transactions in the present case were entered into in good faith, and on the basis of the rule which had been enforced in all Railway Acts in Scotland since 1837, there can be no question that they must be adhered to by the company altho' not introduced into the Bill.[198]

The next shareholders' meeting confirmed the agreements to take over half the debt of the Berwickshire turnpikes, and about one-fifth of that of the less affected Haddingtonshire trust — liabilities of respectively £9,375 and £4,300.[199] In the following year an attempt was made to insert a clause in the Railway Clauses (Scotland) Act to confirm the right of trustees to be compensated in respect of personal liability, but not for debts secured on the roads themselves and the tolls taken. The idea backfired, the Commons deciding that no claim for contingent losses by either trustees or turnpike creditors could be sustained.[200]

As far as company finances were concerned, an agreement to take over liability for a debt only mattered if the creditors insisted on repayment, and made no immediate impact on the accounts. To companies trying to keep their capital expenditure figures within bounds, this was more attractive than immediate lump sum payments; the large cash payments made by the Edinburgh & Glasgow and the Ayrshire were to trusts which would be very severely harmed by the railway and whose potential opposition in Parliament was felt to be a possible danger to the bill. Only the Arbroath & Forfar bothered to record small payments to trustees on its capital account.[201] The Edinburgh & Glasgow, while noting the £11,000 payment to the Bathgate trust, did not list separately the numerous small payments, usually of less than £100, in compensation to slightly affected trusts or for repairs to roads damaged by railway constructional activity, and even a payment of £2,000 to the Cumbernauld trustees in 1839 is not specifically recorded.[202] It remains unclear under which heading in the accounts these sums were placed; after the line opened, it is probable that small sums were debited to the revenue account.

Road trustees could also increase railway costs by objecting to specific parts of company proposals. Two examples from the Lanarkshire coalfield indicate what could happen. In 1837 the Stirling and Carlisle Road Trustees obtained an interdict against the Ballochney plan to build a double line of rails across the road. 'Although the amount of debt on the road renders it almost impracticable for the Trustees to undertake any extra expenditure', they offered to pay a third of the cost of a bridge — not an ungenerous offer, but one which lèft the company with an unexpected increase in its own costs.[203] A similar wrangle between the Wishaw & Coltness and the Bothwell Road Trustees led to over a year of acrimonious correspondence. The railway in this case had extended a passing place across the turnpike and on occasion left waggons parked on it. The trustees not surprisingly objected to the extreme inconvenience and danger which the Railway had occasioned to the public, by laying an offset, not communicating at both ends with the main line, but having the west end which protruded into the road open for no other ostensible purpose than to make their road the depot for the Company's stray waggons.

The eventual answer had to be a bridge, and extra expenditure of about £1,000 by the railway company.[204] Another dispute over level crossings caused four years of furious argument between the Arbroath & Forfar and the parallel turnpike.[205] And although the Glasgow Paisley & Greenock received authorisation for two crossings of the main turnpike, they had to provide each crossing with two or three gatekeepers

in consequence of the public not being accustomed to the interruption, and to prevent, as far as possible, the clamour against the inconvenience, and also to obviate the additional risk.[206]

Such impositions were in general more of an annoyance to companies than a serious drain on their finances.

VI. The search for further capital

The disparity between authorised share capital and actual costs of construction meant that companies had to raise more money, either by issuing more shares or by borrowing, and almost invariably by both. In a time of economic depression this might not be easy, and the difficulties encountered at the end of the 1830s by coal railways like the Ballochney, the Slamannan and the Wishaw & Coltness have already been discussed.[207] There was no real difficulty about permission to raise more share capital. The parliamentary procedures might take a considerable time, and committees might be scathing about the inaccurate calculations which had made the application necessary, but authorisation would not be refused if it meant that a partially constructed line could not be completed.

Issuing extra shares could be more of a problem in the depressed conditions of the early 1840s, when the prospects for the railways no longer looked as good as the promises in the prospectuses. New shares were first offered to existing shareholders, who might well feel that they had already taken on a sufficiently large long-term commitment, and who not infrequently failed to keep up the call payments on their original holdings. In 1839 the Arbroath & Forfar found that the holders of about £8,000 of shares were either unable or unwilling to make their payments, forfeiting the shares instead. Although promoter Patrick Chalmers MP agreed to take £2,000 of them, the company still had £6,000 worth on its hands, while little prospect was held out of disposing of them on the depressed open market. Instead the directors agreed to give their personal security to back £6,300 of borrowing.[208] A similar crisis on the Glasgow Paisley & Greenock was solved firstly by running a £35,000 overdraft spread over five different banks and then, since half the share capital had been called up, by the issue of debentures.[209]

Hence companies often found that an extra inducement had to be offered to potential shareholders. This could be to issue the new shares at a price below face value, although even that could do little for such an unattractive investment as the

Dundee & Newtyle — the offer of a £25 share for £20 was of no interest if there was no chance of a dividend and no prospect of capital appreciation. More common was the issue of preference shares with a guaranteed minimum dividend to be paid before ordinary shareholders saw any return at all on their investment. Although this was clearly hard on the original investors, the companies could argue that they were given the first option to buy the new shares, and that the desperate need for funds had to be the overriding consideration. The extent to which a company in difficulties might have to go was indicated by John Gladstone's advice to the secretary of the Slamannan:

> The creation of new shares should be considerable, perhaps one for each two, and to ensure their being taken by the present holders, that the price fixed should *be low*, the money payable by Instalments as distinct from each other as may be practicable, and the Shares *preferent* to the extent of a Dividend of 5 per Cent on their cost.[210]

The Arbroath & Forfar issued 5% preference shares in 1840, to an extent which severely imbalanced the capital structure of the company, and was followed in 1844 by the Edinburgh Leith & Granton (at 5%)[211] and the Glasgow Paisley & Greenock (at 6%). The first two issues were not a great success, in the first case because of shortage of money in the locality and in the other because of doubts whether the line would earn enough even to cover the preferential dividends. At Greenock, however, 17,560 of the 20,000 preference shares were taken up by the old shareholders.[212] By 1842 more than half of the Arbroath & Forfar capital carried a 5% guarantee, and there was no ordinary dividend, although frustrated shareholders claimed that 4% could have been paid had all the stock been ordinary; an attempt to reject the directors' report was however defeated.[213]

Even if all shares were disposed of, there was a limit to the speed with which the money could be called up. By convention, and in some cases by promises made by directors to shareholders, calls were not normally made at less than two-monthly intervals, or for more than 10% of the total at any one time. Certainly, as economic conditions improved and railway enthusiasm moved towards the mania, companies found that it became easier both to dispose of shares and to have the calls paid when due — only one company, the unfortunate Granton line, had a significant proportion of pre-1845 calls still unpaid in 1847. The holders of one-tenth of its stock had apparently decided to forget about their payments, while the next worse record for defaulters was 1% on the Dundee & Arbroath.[214]

Borrowing money was a more attractive option than preference shares, though on the whole directors might have preferred to issue more ordinary capital. But shareholders disliked the watering of share capital, while borrowing, although obviously imposing interest charges for the duration of the loan, at least did not involve the permanent charge against surplus revenue imposed by preference shares. Nor was the rate of interest required to obtain loans any greater than that on preference shares. Most companies in the early 1840s managed to issue their debentures at 4-5% interest, and by 1844 the Wishaw & Coltness committee of inquiry claimed that the majority (although not their own) were paying only $3\frac{1}{2}$%.[215] The records of the Ayrshire confirm that in early 1843 the company was renewing matured 5% loans at 4%, and in 1844 was paying $3\frac{1}{2}$% for five or six years' loans or, in a few cases, $3\frac{3}{4}\%$ for seven or nine years.²¹⁶ There could be a problem in finding willing lenders, and company records often indicate massive juggling with short-term loans, but again the difficulties eased as companies became established.

Legally, borrowing could not take place until half of the share capital had been called up, nor could it go beyond the limits laid down by Parliament, which by the 1840s were established at one-third of the authorised share capital. Tables 46-49 indicate that companies abided by the first rule, but that parliamentary limits were, according to the accounts, broken in one year on the Ayrshire and almost all the time on the Edinburgh & Glasgow and the Arbroath & Forfar. Since similar misdemeanours are known to have been covered up in the Greenock accounts,²¹⁷ it seems that boards and officials were not too concerned about the letter of the law, and that most of them did not bother to hide the fact. Share capital was never allowed to exceed authorisation, although the Edinburgh & Glasgow came perilously close to the limit before the passage of their 1845 act.

The tables also indicate the relationship between the balance on current account and the companies' capital structure. Even if the lines had been constructed within the original share capital, none of them would have paid as well as the promoters hoped. On the trading results of 1844-45, which was in every case the best of the years under consideration, the Ayrshire could have paid 6.7%, the Edinburgh & Glasgow 6.3%, the Arbroath & Forfar 4.5% and the Greenock 3.1%. If construction costs had been kept within the expanded share capitals, the results (as shown in the second last column) would in general have been about two-thirds as good. The final column relates the current account balance to the actual capital raised. The figures are of course entirely hypothetical, in that in no case would it have been possible for a company to spread its profit evenly over all the capital raised. Much of the borrowing was by debentures on which fixed interest of between $3\frac{1}{2}\%$ and 6% had to be paid, while most of the rest was by bank overdrafts where once again interest payments would have to be kept up. On top of that there might be preference shares. Once these commitments had been met, there would be little left for the ordinary shareholders, and the tables add weight to the often-voiced suspicion that some dividends were being subsidised out of capital.

The reasons for overborrowing varied. It could have been that the powers to raise share capital had been fully utilised; in none of the cases under consideration did this happen, although in two an act to extend capital raising powers was passed in the same year as the overborrowing took place. It was more probable that it had to make up for shares which were unsaleable or forfeited, or which the companies felt they dare not put onto a depressed market. On occasion also capital expenditure outran the rate at which calls could reasonably be made on the shares, and short-term borrowing covered the gap. The bulk of borrowing was done on debentures, though these were never allowed to exceed the permissible limits. The Edinburgh & Glasgow accounts distinguish debentures from other loans from the end of 1842; from then until 1845 the debenture total was maintained at £375,000.²¹⁸ Those of the Ayrshire, on the other hand, indicate that from 1840 onwards there was a continual process of repayment of old debentures and issue of

Table 46. *Arbroath & Forfar, Income on Capital Account to 1845*

Date	Authorised capital Share £	Loan £	Paid up share capital £	% of total	% of auth. shares	Borrowing £	% of total	% of auth. loans	Miscellaneous £	% of total	Total £	Balance on current account £	Rate of return on paid up share capital %	Rate of return on total capital %
10.6.39	70,000	35,000	61,865	64.8	88.4	32,784	34.3	93.7	867	0.9	95,517	—	—	—
15.4.40	120,000	40,000	64,043	52.5	53.4	54,879	45.0	137.2	3,068	2.5	121,989	3,068	4.8	2.5
15.4.41	120,000	40,000	85,252	63.8	71.0	44,460	33.3	111.2	3,848	2.9	133,560	2,704	3.2	2.0
15.4.42	120,000	40,000	92,129	66.6	76.8	44,105	31.9	110.3	2,125	1.5	138,359	2,108	2.3	1.5
15.4.43	120,000	40,000	94,154	67.5	78.5	42,691	30.6	106.7	2,905	2.1	139,479	2,321	2.5	1.7
15.4.44	120,000	40,000	95,394	68.0	79.5	40,826	29.1	102.1	4,023	2.9	140,283	2,531	2.7	1.8
15.4.45	120,000	40,000	103,516	72.1	86.3	36,088	25.1	90.2	4,035	2.8	143,640	3,146	3.0	2.2

Sources: as for Tables 25 and 35.

Table 47. *Edinburgh & Glasgow, Income on Capital Account to 1845*

Date	Authorised capital Share £	Loan £	Paid up share capital £	% of total	% of auth. shares	Borrowing £	% of total	% of auth. loans	Miscellaneous £	% of total	Total £	Balance on current account £	Rate of return on paid up share capital %	Rate of return on total capital %
31.12.38	900,000	300,000	142,517	98.7	15.8	—	—	—	1,947	1.3	144,464	—	—	—
31.12.39	900,000	300,000	297,274	98.2	33.0	—	—	—	5,547	1.8	302,821	—	—	—
31.12.40	900,000	300,000	560,727	77.2	62.3	156,670	21.6	52.2	8,883	1.2	726,280	—	—	—
31.12.41	900,000	300,000	833,499	67.6	92.6	386,888	31.4	129.0	12,847	1.0	1233,235	—	—	—
31.12.42	1125,000	375,000	976,428	64.5	86.8	518,399	34.2	138.2	19,537	1.3	1514,364	45,819	4.7	3.0
31. 1.44	1125,000	375,000	1122,803	69.7	99.8	473,407	29.4	126.2	15,371	1.0	1611,580	52,705	4.7	3.3
31. 1.45	1406,250	468,750	1155,029	68.5	82.1	516,088	30.6	110.1	15,440	0.9	1686,557	56,702	4.9	3.4

Sources: as for Tables 25 and 36.

Table 48. Glasgow Paisley & Greenock, Income on Capital Account to 1845

Date	Authorised capital		Paid up share capital			Borrowing			Miscellaneous		Total	Balance on current account	Rate of return on paid up share capital	Rate of return on total capital
	Share £	Loan £	£	% of total	% of auth. shares	£	% of total	% of auth. loans	£	% of total	£	£	%	%
30.11.38	400,000	133,333	88,744	99.3	22.2	—	—	—	634	0.7	89,378	—	—	—
30.11.39	400,000	133,333	215,325	76.3	53.8	65,700	23.3	49.3	1,070	0.4	282,094	—	—	—
30.11.40	400,000	133,333	340,728	72.3	85.2	128,543	27.3	96.4	1,956	0.4	471,185	−465	0.0	0.0
30.11.41	500,000	166,666	415,729	74.3	83.1	142,012	25.4	85.2	1,956	0.3	559,656	7,900	1.9	1.4
30.11.42	500,000	166,666	457,645	73.5	91.5	163,312	26.2	98.0	1,956	0.3	622,872	8,892	1.9	1.4
31. 1.44	650,000	216,666	522,906	76.2	80.4	161,432	23.5	74.5	1,956	0.3	686,253	8,765	1.7	1.3
31. 1.45	650,000	216,666	551,434	71.6	84.8	216,666	28.1	100.0	1,956	0.3	770,016	12,571	2.3	1.6

Sources: as for tables 25 and 37.

Table 49. Glasgow Paisley Kilmarnock & Ayr, Income on Capital Account to 1845

Date	Authorised capital		Paid up share capital			Borrowing			Miscellaneous		Total	Balance on current account	Rate of return on paid up share capital	Rate of return on total capital
	Share £	Loan £	£	% of total	% of auth. shares	£	% of total	% of auth. loans	£	% of total	£	£	%	%
31.12.38	625,000	208,300	115,871	98.7	18.5	—	—	—	1,469	1.3	117,340	—	—	—
31.12.39	625,000	208,300	318,754	83.5	51.0	60,970	16.0	29.3	2,181	0.6	381,905	—	—	—
31.12.40	625,000	208,300	432,495	61.8	69.2	264,488	37.8	127.0	2,612	0.4	699,595	7,589	1.8	1.1
31.12.41	625,000	208,300	581,920	73.3	93.1	208,301	26.3	100.0	3,169	0.4	793,390	20,951	3.6	2.6
31. 1.43	937,500	312,400	710,543	72.4	75.8	266,378	27.1	85.3	4,727	0.5	981,649	22,380	3.1	2.3
31. 1.44	937,500	312,400	752,731	74.0	80.3	259,958	25.6	83.2	4,727	0.5	1017,416	28,548	3.8	2.8
31. 1.45	937,500	312,400	756,906	71.1	80.7	302,646	28.4	96.9	4,727	0.4	1064,279	41,164	5.4	3.9

Sources: as for Tables 25 and 38.

new ones, with little borrowing by any other method taking place.[219] For short-term adjustments of the borrowing total, debentures, which carried a fixed rate of interest and ran for a fixed term of a few years, were not particularly suitable, unless there were large numbers continually expiring. More convenient was either borrowing from the banks or finding some other type of fluid short-term mechanism.

The banks were an obvious source of loans; although they were very cautious about investing either long-term or in unproven projects, they might well be happy to lend to established companies. Such loans, if secured on the assets of the railway and often also on personal guarantees by the directors, should be almost risk-free, carried a reasonably attractive rate of interest, and helped the circulation of the banks' own notes. Banks were also keen to attract the increasingly important general business of the railway companies, and even a company as small as the Arbroath & Forfar found that it could shop around to find the best service.[220] Railways commonly chose to divide their business among several banks, which could with any luck be played off against each other when it came to negotiating assistance. In 1840 the Greenock company had accounts with nine different banks; although five were in credit and four in debit, there was a total net deficit of £27,829.[221]

Most banks would have agreed in principle with the Commercial when it stated in 1844 that loans to railways were to be short-term, limited to successful main line companies, and with no promise of renewal at maturity.[222] In practice, policies varied. The ultra-conservative Bank of Scotland viewed all early railways with suspicion, although it did lend £8,000 to the Polloc & Govan,[223] presumably because William Dixon was regarded as safe. The British Linen Bank virtually avoided railway involvement, while the Union Bank concentrated its lending on the Ayrshire and the Edinburgh & Glasgow.[224] The Royal Bank of Scotland, of which Learmonth was a director from 1830 to 1858, was much more liberal; recipients of its loans included the Ayrshire, the Greenock, the Edinburgh & Glasgow, the Arbroath & Forfar and every railway (apart from the Drumpellar) in Lanarkshire.[225] Some were under conditions which other banks would have rejected; the Slamannan was lent £20,000 before its line was opened, while the £50,000 negotiated in 1843 by the Edinburgh & Glasgow was to last for ten years.[226] On occasion the bank got caught, as in 1840 when the Edinburgh & Dalkeith could not pay the interest on its loan, let alone repay the principle. The more extreme difficulties of the Dundee Union Bank with the Dundee & Newtyle company have already been noted.[227]

Borrowing from the bank did not necessitate formal loan negotiations; if the bank agreed, it could be done by simply running up an overdraft. This method, which had the additional advantage of drawing less attention to overborrowing, was standard practice with the Lanarkshire coal railways,[228] and continued on into the mania when the Royal Bank lent money to at least five companies which never achieved authorisation, let alone calling up half of their share capital.[229] An Edinburgh & Glasgow minute made it quite clear that the company proposed consciously to use the banks to exceed its authorised powers:

> It being now apparent that a sum of money will be required beyond the amount of Capital authorised by the Company's Act to be raised & beyond the amount authorised to be borrowed, & as an Act of Parliament cannot be got to sanction the raising of such additional Sum until next session, it will be necessary to arrange temporary loans from some of the Company's bankers.

A loan of £25,000 had already been agreed with the National Bank, while a further £50,000 was being sought from the Union. Unusually, and perhaps emphasising the questionable nature of the plan, the minute was signed by all the directors.[230]

An alternative method of raising extra money was the loan note — in effect a simple IOU, with no date of maturity and no legal protection for the lender.[231] The Edinburgh & Glasgow had £1,500 outstanding in loan notes at the end of 1841, and £48,500 six months later. A further unspecified amount was noted in 1844, presumably accounting for the £27,900 of borrowing which came from neither debentures nor bank loans. Loan notes probably also account for at least £17,000 of 'loans in anticipation of debentures' recorded in the Greenock minutes.[232] In 1844 both a select committee and the Government's law officers stated their belief that the notes were illegal, but that the transaction often took place in innocence on both sides. Companies, said the committee, should certainly not be allowed such a method of avoiding parliamentary authorisation to raise extra capital, but in the meantime the lenders required protection. Legislation declared the notes illegal, but existing ones were given legal protection and even allowed to be renewed for a further five years.[233]

To the extent that companies were indulging in financial practices of dubious legality which they might wish to keep hidden from both shareholders and the Board of Trade, their accounts have to be regarded with some suspicion. On the other hand they were, apart from the Greenock, generally open about both the use of loan notes and occasional total borrowing beyond their authorisation. Many lines often appeared to be on the verge of financial disaster, but once construction had started none went bankrupt. Only two pre-mania companies failed to complete their main lines — the Ardrossan (whose route fell to the Ayrshire) and the Drumpellar. Elsewhere desperate efforts in the face of an apathetic stock market and increasingly depressed or hostile shareholders raised enough capital to cover the escalated costs of construction. By the upturn of 1843-44 the survival of the companies in the face of adversity was in itself an encouragement to thoughts of further railway expansion.

VII. The costs of operation: expenditure on current account

The current or revenue accounts of the four inter-urban lines which form the focus of this discussion are summarised in Tables 50-53; the figures for expenditure and income will be further broken down in Tables 54-57 and 60-65 respectively. The summary tables indicate that the trend of profitability, as measured crudely by the amount of surplus on the year's trading, was generally

Table 50. Arbroath & Forfar Revenue Account 1839–45, Summary

Date[a]	Income			Expenditure			Balance	Working Exp. as % of Traffic Income
	Traffic	Other	Total	Working	Other	Total		
	£	£	£	£	£	£	£	
1839–40	9,523	217	9,740	4,878	1,794	6,672	3,068	51.2
1840–41	9,191	191	9,382	4,073	2,605	6,678	2,704	44.3
1841–42	9,117	210	9,327	4,413	2,806	7,219	2,108	48.4
1842–43	8,541	253	8,794	3,823	2,650	6,473	2,321	44.8
1843–44	8,021	290	8,311	3,315	2,465	5,780	2,531	41.3
1844–45	8,784	192	8,976	3,548	2,282	5,830	3,146	40.4

a: Periods 3 Jan 1839 – 15 April 1840, then years from 16 April

Sources: as for Table 36.

Table 51. Edinburgh & Glasgow Revenue Account 1842–45, Summary

Date[a]	Income			Expenditure			Balance	Working Exp. as % of Traffic Income
	Traffic	Other	Total	Working	Other	Total		
	£	£	£	£	£	£	£	
1842	93,512	1,090	94,602	25,068	23,715	48,783	45,819	26.8
1843–44	119,940	350	120,290	39,991	27,594	67,585	52,705	33.3
1844–45	116,390	525	116,915	36,292	23,921	60,213	56,702	31.2

a: Periods 21 Feb – 31 Dec 1842; 1 Jan 1843 – 31 Jan 1844; 1 Feb 1844 – 31 Jan 1845

Sources: as for Table 37.

Table 52. Glasgow Paisley & Greenock Revenue Account 1840–45, Summary

Date[a]	Income			Expenditure			Balance	Working Exp. as % of Traffic Income
	Traffic[b]	Other	Total	Working	Other	Total		
	£	£	£	£	£	£	£	
1840	3,717	0	3,717	1,708	2,474	4,182	–465	46.0
1840–41	34,815	98	34,913	17,429	9,584	27,013	7,900	50.1
1841–42	42,075	257	42,332	23,135	10,305	33,440	8,892	55.0
1842–44	48,811	15	48,826	26,363	13,698	40,061	8,765	54.0
1844–45	45,728	157	48,885	20,312	13,002	33,314	12,571	44.4

a: Periods 14 July – 30 Nov 1840; 1 Dec 1840 – 30 Nov 1841; 1 Dec 1841 – 30 Nov 1842; 1 Dec 1842 – 31 Jan 1844; 1 Feb 1844 – 31 Jan 1845
b: including all income from joint line

Source: as for Table 38.

Table 53. *Glasgow Paisley Kilmarnock & Ayr Revenue Account 1839–45, Summary*

Date[a]	Income			Expenditure			Balance	Working Exp. as % of Traffic Income
	Traffic[a]	Other	Total	Working	Other	Total		
	£	£	£	£	£	£	£	
1839–40	22,609	152	22,761	11,672	3,500	15,172	7,589	51.6
1841	52,723	1,359	54,082	22,000	11,131	33,131	20,951	41.7
1842–43	56,644	70	56,714	22,481	11,853	34,334	22,380	39.7
1843–44	65,806	0	65,806	22,193	15,065	37,258	28,548	33.7
1844–45	79,457	0	79,457	24,047	14,246	38,293	41,164	30.3

a: Periods 5 Aug 1839 – 31 Dec 1840; 1 Jan – 31 Dec 1841; 1 Jan 1842 – 31 Jan 1843; then years from 1 Feb
b: including all income from joint line

Sources: as for Table 39.

upwards, although the Arbroath & Forfar had a difficult time around 1842, and in some other cases a drop in income had to be compensated for by a greater fall in expenditure. Income meant almost entirely takings from traffic, with other revenue from, for instance, rents received on odd parcels of surplus railway land making a very small contribution. On the expenditure side of the accounts, however, working costs formed less than two-thirds, and sometimes little more than a half, of the total; other expenses, particularly interest payments, pushed up costs to a level unconsidered in the original estimates.

Current expenditure is analysed in more detail in Tables 54-57. Once again the categories should be regarded as at best approximations since there was no standard pattern of organising the accounts, and companies did not always achieve consistency from one year to the next. In particular, the figures for wages and salaries are not comparable from one company to another. They normally included any honorarium paid to the directors, the salaries of the senior officials, and the wages of the company policemen — a group whose duties varied from line to line, including not only the security functions implied in the name but some or all of the roles of pointsmen, porters, ticket-collectors, signalmen, and other lesser station officials. In some cases payments to other employees are also included, while in others these are subsumed under the departments in which the employees worked.[234] Other items which would refer to more than one department, such as tools, oil and grease, are sometimes listed separately, and have had to be assigned arbitrarily to a specific column; the sums involved are in this case not large. In the Ayrshire and Greenock accounts, expenditure incurred on the joint line is presented as a lump sum, unapportioned to the various departments.

Comments about the balance of working expenditure between the sub-headings can thus only be in terms of broad generalities. The two companies where all wages

Table 54. Arbroath & Forfar Current Expenditure, 1839–45

Date[a]	Working Expenditure												Other Current Expenditure						Total
	Maintenance of way		Coaching		Goods[b]		Locomotive[c]		Wages and Salaries[d]		Miscellaneous		Total		Passenger tax		Interest		Miscellaneous		
	£	%	£	%	£	%	£	%	£	%	£	%	£	%	£	%	£	%	£	%	
1839	344	25.5			93	6.9	625	46.3	288	21.3			1,350	100.0	f						1,350
1839-40	377	7.1	106	2.0	122	2.3	1,427	26.8	1,215	22.8	380	7.1	3,528	66.3	f		1,601	30.1	193	3.6	5,322
1840-41	538	8.1	207	3.1	230	3.4	1,734	26.0	1,127	16.9	332	5.0	4,073	61.0	200	3.0	2,199	32.9	206	3.1	6,678
1841-42	713	9.9	145	2.0	259	3.6	1,451	20.1	1,498	20.8	337	4.7	4,413	61.1	496	6.9	2,100	29.1	210	2.9	7,219
1842-43	554	8.6	116	1.8	227	3.5	1,232	19.0	1,400	21.6	298	4.6	3,823	59.1	240	3.7	2,109	32.6	301	4.7	6,473
1843-44[e]	509	8.8	76	1.3	195	3.4	1,160	20.1	1,083	18.7	289	5.0	3,315	57.4	147	2.5	1,947	33.7	371	6.4	5,780
1844-45	521	8.9	94	1.6	252	4.3	1,334	22.9	1,059	18.2	286	4.9	3,548	60.9	156	2.7	1,732	29.7	394	6.8	5,830

a: Periods as in Table 50
b: includes workshop utensils
c: includes oil and grease
d: includes police and horses
e: accounts reorganised
f: passenger tax compounded and included in miscellaneous working expenditure
Sources: as for Table 36.

Table 55. Edinburgh & Glasgow Current Expenditure, 1842–45

Date[a]	Working Expenditure													Other Current Expenditure						Total
	Maintenance of way		Coaching		Goods		Locomotive[b]		Wages and Salaries[c]		Miscellaneous		Total		Passenger tax		Interest		Miscellaneous		
	£	%	£	%	£	%	£	%	£	%	£	%	£	%	£	%	£	%	£	%	
1842	4,415	9.1	5,042	10.3	3,926	8.0	7,800	16.0	1,366	2.8	2,510	5.1	25,068	51.4	5,148	10.6	18,567	38.1			48,783
1843–44	7,385	10.9	6,133	9.1	9,516	14.1	11,841	17.5	1,948	2.9	3,167	4.7	39,991	59.2	4,139	6.1	23,302	34.5	153	0.2	67,585
1844–45	7,058	11.7	5,315	8.8	4,942	8.2	14,525	24.1	1,572	2.6	2,879	4.8	36,292	60.3	4,075	6.8	19,382	32.2	464	0.8	60,213

a: Periods as in Table 51.
b: includes stationary engine on incline
c: includes engineer's and secretary's departments

Sources: as for Table 37.

Table 56. Glasgow Paisley & Greenock Current Expenditure, 1840–45

Date[a]	Working Expenditure															Other Current Expenditure						Total
	Maintenance of way[b]		Coaching		Goods		Locomotive		Wages and Salaries[c]		Joint line		Miscellaneous		Total		Passenger tax		Interest		Miscellaneous		
	£	%	£	%	£	%	£	%	£	%	£	%	£	%	£	%	£	%	£	%	£	%	
1840–1	300	1.1	1,715	6.3	1,159	4.3	7,188	26.6	964	3.5	2,552	9.4	2,341	8.7	17,429	64.5	2,929	10.8	6,655	24.6			27,013
1841–2	2,703	8.1	3,564	10.7	3,030	9.1	7,227	21.6	1,204	3.5	2,984	8.9	2,423	7.2	23,135	69.2	2,736	8.2	7,569	22.6			33,440
1842–4	3,976	9.9	3,441	8.6	4,827	12.1	7,374	18.4	763	1.9	3,191	8.0	2,711	6.8	26,363	65.8	2,230	5.6	11,098	27.7	350	0.9	40,041
1844–5	2,816	8.5	2,768	8.3	4,870	14.6	5,453	16.4	451	1.4	2,660	8.0	2,578	7.7	20,312	61.0	1,807	5.4	11,195	33.6			33,314

a: Periods as in Table 52
b: includes engineer's department
c: i.e. police. Other salaries and wages included in appropriate departments

Sources: as for Table 38.

Table 57. Glasgow Paisley Kilmarnock & Ayr Current Expenditure, 1841–45

Date[a]	Working Expenditure								Other Current Expenditure			Total
	Maintenance of way	Coaching	Goods	Loco-motive[b]	Wages and Salaries	Joint line	Miscell-aneous	Total	Passenger tax	Interest	Miscell-aneous	
	£ %	£ %	£ %	£ %	£ %	£ %	£ %	£ %	£ %	£ %	£ %	
1841	3,440 10.4	541 1.6	473 1.4	6,962 21.0	6,616 20.0	1,895 5.7	2,073 6.3	22,000 66.4	3,300 10.0	7,831 23.6		33,131
1842–3	3,528 10.3	976 2.8	1,116 3.3	7,347 21.4	7,434 21.7	3,134 9.1	2,079 6.1	22,481 65.5	2,669 7.8	8,908 25.9	276 0.8	34,334
1843–4	3,292 8.8	1,220 3.3	1,208 3.2	6,986 18.8	7,513 20.2	2,541 6.8	1,772 4.8	22,193 59.6	1,700 4.6	13,158 35.3	207 0.6	37,258
1844–5	4,127 10.6	1,026 2.6	1,509 3.9	7,362 18.9	8,068 20.7	1,760 4.5	1,954 5.0	24,047 61.8	2,138 5.5	12,419 31.9	319 0.8	38,923

a: periods as in Table 53
b: includes oil and grease

Sources: as for Table 39.

and salaries were grouped together — the Ayrshire and the Arbroath & Forfar — both found that these payments made up about a third of total working expenditure, while in all cases the locomotive department accounted for a further 30-35%. Maintenance of way, which in newly constructed railways might not be expected to be a great expense, took between one-seventh and one-fifth of the total, with the higher figures including the wages of platelayers and foremen. The coaching (i.e. passenger) and goods departments each accounted for about 3-7% excluding wages, or 14-20% including them, with the goods department almost invariably being the more expensive to operate. The miscellaneous category, in which the main items were printing, advertising, insurance and office expenses, generally made up less than a tenth of the total, although it was for no obvious reason consistently higher at Greenock.

The tables also underline the burden of interest payments; between a quarter and a third of total current expenditure went to service the borrowing which had been undertaken in order to complete construction of the line. It had of course been assumed, at least for prospectus and parliamentary purposes, that the railways would be built within the estimates, or at least within the slightly larger authorised share capitals. Promoters therefore made no allowance for interest payments when forecasting the rate of return which investors might expect. On top of this, passenger duty (included by some companies under working expenditure), rates, feu duties and other miscellaneous items could add up to 5-10% of total costs.[235] Overall, working expenditure amounted only to around 60% of the total; it was the unpremeditated 40% which was the main difference between the promised prosperity of the prospectuses and the unexciting dividends actually realised.

The projectors of the early coal railways had almost casually assumed that working expenditure would amount to about 10% of gross revenue. By the 1830s experience had led to greater realism. Even the most successful of the coal railways spent over 35% of its gross revenue, while in the depression years of 1839 to 1843 the Garnkirk figure rose to 51.3%.[236] Elsewhere the Dundee & Newtyle's steady 80% or more has already been noted, while in 1839 the Edinburgh & Dalkeith reached 72% and the Paisley & Renfrew 86%.[237] In 1835 John Miller allowed 42.85% in the Edinburgh & Glasgow prospectus, reflecting the actual experience of the Liverpool & Manchester; the next year he allowed 33.33% on the Ayrshire.[238] Locke, also claiming Liverpool & Manchester experience as his authority, told a parliamentary inquiry that the allowance should be 50%, although a line predominantly dependent on passenger traffic might reduce it to 33% or 40%.[239] By the time the Scottish inter-urban lines were opened, a figure of about one-third appears to have been accepted as a reasonable target. Tables 50-53 indicate that the Edinburgh & Glasgow achieved it from the start, the Ayrshire improved steadily and satisfactorily to reach 30.3% by 1844-45, the Arbroath & Forfar was always over 40% (though here also the trend was improving), and the Greenock more often than not saw more than half its revenue go in working expenses.

The percentage depended of course not only on the amount of the expenditure but also on the size of the gross revenue; at both Arbroath and Greenock revenue was lower than might have been expected from traffic volumes, since almost all passenger traffic was carried third-class at very low fares enforced either by the poverty of the customers or by the nature of the competition from other transport. Company officials and shareholders were well aware of this, and encouraged efforts to increase the flow of traffic — although some attempts, like the Ayrshire's ventures into shipping, were not entirely profitable. It was however often easier to concentrate on the expenditure side of the equation and to look for ways of cutting costs. It was already clear that the assumption that working expenditure would necessarily bear a fixed relationship to gross revenue was a convenient fiction for the purposes of the prospectus, and that a fall in revenue was unlikely to lead to a corresponding fall in costs. Some costs were more or less decided once the number of trains to be run was agreed; regardless of how much traffic they carried, they would require a given number of men to operate the service, and would cause a more-or-less given amount of damage to the track. Although a fall in demand might be met by taking some carriages or waggons, or even whole trains, out of service, this would mean that expensive capital equipment was being used at less than optimum intensity. For the most part, attempts at cost-cutting had to concentrate on improving the efficiency of the various operating departments of the company. And there was always the danger that a financially pressed company might place an extra burden on the revenue account by trying to economise on capital; cheap rails, waggons and locomotives were likely to lead to enhanced costs for maintenance and repairs.

One practice which certainly evened out the costs of track maintenance, although it might or might not reduce them, was to contract out the job at a fixed annual sum per mile. The Ayrshire, which paid about £85 per mile between Paisley and Ayr, was well pleased with the results.[240] At Greenock the contract was given to resident engineer Davidson, who was also the managing director's nephew-in-law, for £250 per mile (later reduced to £200). The shareholders' committee of inquiry at the end of the decade, which claimed that the Ayrshire paid £95, the Edinburgh & Glasgow £112, and no railway in Britain more than £140, saw the arrangement as part of the corruption which also enmeshed another director, the company's lawyer, Davidson's partner (and brother-in-law), and an accountant who disappeared having made his contribution to the expenses by embezzling £4,000. The committee also claimed that Davidson had failed to do his job properly, although Neil Robson reported to the board that 'all in all, there is not a better maintained piece of Railway to be seen anywhere in this country'. And in 1842 the Glasgow & Paisley joint line, with admittedly much heavier traffic, had been contracted out at £210 per mile.[241]

One area where it often seemed to economy-minded shareholders that cuts could be made was in the bill for wages and salaries. Although in principle companies might approve of a free market in labour, and of paying the lowest wages which would attract the number of men required, in practice the railway labour market (which has been analysed in detail by Dr P.W. Kingsford)[242] contained complications of its own. On the one hand, railway work was socially respectable and eagerly sought after, even in spite of the hard work, long hours, and virtually military discipline, often under superintendents and even company secretaries who were ex-army men. On the other, companies were anxious to recruit reliable labour, since the consequences of error by a careless, drunken or inadequate driver or signalman could be severe, and even lack of politeness by a porter could bring unwelcome publicity. Two letters from the same issue of the *Railway Times* make the point. One writer had descended fuming from the Edinburgh & Glasgow:

> I can speak to the incivility of the servants of the Company. I have travelled upon almost all of the railways in England and Scotland, but I never met any servants so bad; they are disgraceful. The porters are dressed only as the poor creatures you meet with upon landing on the quays. When I have been with them, they go badly as to punctuality . . .

The other had had two cartloads of luggage carried free on the Greenock line, and conveyed to a house one mile beyond the terminus for 3/-: 'I think this instance of the liberality of the Greenock Railway Company, and the courtesy of their servants, should be made known'.[243] The *Railway Times* believed that 'the officers of the Edinburgh and Glasgow Railway would seem peculiarly to require supervision',[244] but it also reported a series of incidents on the Ayrshire, many of which ended with a passenger taking the company to the courts. Successful claimants included a passenger to Arran who missed his boat when the train was late and received the cost of his overnight accommodation; a group of first-class passengers to Paisley who found that no first-class seats were vacant and forced the company to pay for a road carriage; and a regular series of owners of lost luggage — one sheriff remarked of the company's arrangements that 'the wonder is . . . not that loss occurs occasionally, but that it does not occur every day'.[245]

By comparison with most workers, particularly in the country districts along the line in which they recruited most of their unskilled labour, railwaymen were well paid; as John Macneill said of the policemen who were to look after the Slamannan line,

> To render those men careful and attentive to their duty, they should be well paid and houses be built for them along the line of the Railway near their respective stations.[246]

Porters could fare much better than agricultural labourers, while drivers ranked with the most skilled artisans. Tables 58 and 59 indicate that employees were doing well by Scottish standards, although rates paid remained well below those current in England. On the other hand, rigid disciplinary practices, which meant dismissal for a whole range of offences including almost any which impinged upon the general public, meant a high turnover of staff, even although companies preferred if possible to retain an experienced and contented workforce. Companies encouraged the image of loyal service to a railway as a full and worthwhile career for any respectable man, and in return many boards did feel some sense of pride in paying good wages to valued servants.[247]

Hence reducing working expenditure by cutting the wages bill might not be as simple as it appeared to some shareholders. Companies were reluctant to admit that they had been practising overmanning, and that the bill might be reduced by dismissals or by failing to replace men who were in any case leaving. Really hard-pressed boards might have little option. Most of the reduction of the Dundee

Table 58. Weekly Wage Rates for Railwaymen, c. 1840-44

	Scottish Coal Railways	Glasgow Paisley Kilmarnock & Ayr	England
Drivers	18/- to 22/6	21/- to 26/-	28/- to 45/-
Firemen	10/- to 12/-	15/- to 16/-	15/- to 25/-
Guards	15/- to 16/-	18/- to 24/-	24/- to 30/-
Policemen, porters	12/- to 15/-	12/- to 15/-[a]	15/- to 21/-
Platelayers	13/- to 15/-	n.a.	18/- to 21/-
Gatekeepers	10/- to 14/-[a]	12/- to 18/-[a]	12/- to 21/-[a]

[a]: in some cases a house was also included

Sources: F. Whishaw, *The Railways of Great Britain and Ireland*; Shiell & Small, Dundee & Newtyle Sederunt Book; More Nisbett Papers, bundle 3, *Report of the Committee of Inquiry into the Expenditure of the Glasgow Paisley Kilmarnock & Ayr Railway Company*, 22 Feb 1844; PP 1842 [360] XLI, *Second Report of the Officers of the Railway Department*, appx. IV; P.W. Kingsford, *Victorian Railwaymen*, ch. 6; P.S. Bagwell, *The Railwaymen*, 30; M. Robbins, 'The North Midland Railway and its enginemen, 1842-3', *Jnl Transport Hist.*, 4 (1960).

Table 59. Weekly Wage Rates for Non-Railway Occupations, 1843[a]

	All artisans	Masons	Wrights	Weavers	Agricultural labourers	'The Lowest Labourers'
North Lanarkshire[b]	15/5	21/-	14/4	6/10	9/11	6/6
Renfrewshire–Dunbartonshire[b]	15/1	18/-	18/-	6/3	10/5	6/5
North Ayrshire	14/3	19/-	16/3	7/1	10/4	6/2
Midlothian[b]	15/11	18/-	14/-	n.a.	9/9	6/5
Angus coast	12/-	15/-	13/8	8/-	9/6	6/4
Scotland total	13/2	15/8	14/-	6/6	9/1	5/9

[a]: some figures, particularly for masons and wrights, from extremely small samples
[b]: the returns to the Poor Law authorities from which these figures are taken did not include figures from the parishes of Glasgow, Edinburgh and Paisley

Source: I. Levitt & C. Smout, *The State of the Scottish Working-Class in 1843*, 85, 114-16.

& Newtyle's monthly wage bill from £212 in mid-1841 to £150 two years later was accomplished by staff reductions; wages were also cut in the repair shop, although 'they will probably require to be raised again'.[248] Reductions in the rates would have to be considered carefully with respect to wages on rival railways and in other jobs in the area, and even to any possible reaction by the men themselves.

The argument for wage cuts was succinctly put by Ayrshire shareholder James Watson early in 1843, when the prolonged depression had reduced the dividend to $1\frac{1}{4}\%$ for the half-year:

> He was of the opinion that the servants of the Company, from the highest to the lowest, should participate in their want of success as well as in their prosperity. In a bad year every economy should be used, and if it was practicable, something taken off the salaries and wages ... Possibly the directors might find that four men might be made to do the work of five.[249]

At the same time cuts on the Dundee & Newtyle were justified as not meaning a fall in real wages, 'the cheapness of the necessaries of life having caused a general reduction of wages all over the country'.[250] Ayrshire chairman McCall was unimpressed:

> The salaries and wages of those employed on the line were as moderate as they could be made, consistent with efficiency and perfect safety to the public. As to the reduction of the wages, therefore, he thought it was the last thing any one of them should propose, seeing the mischief that had been produced on other lines in consequence of employing inferior servants.[251]

The reference was in particular to events on Hudson's North Midland Railway, where the drivers refused to accept wage cuts and were promptly sacked. Their replacements were men who had for the most part been sacked in disgrace from other companies, one of whom forthwith caused a fatal accident.[252] General Pasley, inspecting the scene for the Railway Department, disliked enforced cuts:

> I consider it essential to the prosperity of passenger railways, as well as of the public safety, that the wages of this important class of men [drivers] should not be screwed down to the lowest scale, at which persons willing to undertake that duty can be found. On the contrary, it is desirable that they should be paid at a much higher scale, and that the wages once agreed upon between them and their employers should not be reduced at all, without the full and free consent of the individuals.

The Department supported their inspector:

> If... the natural desire to curtail expenses, upon lines which are not paying the expected dividend, should induce the Directors or shareholders to carry reductions to such an extent, as to impair the efficiency of the working establishment, and to give the class of engine-drivers and other servants, upon whose sobriety, intelligence, and habitual good conduct the safety of the public depends, just ground for discontent, there is every reason to apprehend a recurrence of accidents, by which the lives of passengers would be sacrificed, public alarm excited, and the value of railway property materially impaired.[253]

In spite of this official rejection of the iron law of wages, and even of allowing free market forces to produce the right number of drivers, the Ayrshire shareholders appointed a committee of inquiry to look for economies. Its report at the beginning of 1844 found that the senior officers and station clerks 'are all on a moderate rate of remuneration' but that cuts of between 3% and 13% might be made in the wages of train staff. The guards were to bear the brunt, since they were well paid by comparison both with other staff and with guards on other railways; the two existing classes at 24/- and 23/- per week were to be redistributed into three groups at 21/-, 20/- and 18/-.[254] A committee member explained:

> Some of the salaries of the superior officers, and wages of the servants of the Company were too low. In the case of the guards, it was a matter of comparison with the enginemen; the latter working 12 hours daily, while the period of labour of the former was much shorter. They conceived therefore the duties of the enginemen so onerous — men in fact on whose care and steadiness depended the safety of the passengers, that they did not interfere with their remuneration.

This was true of existing enginemen, but in fact new drivers and firemen employed in future were to start at rates from 6% to 12.5% below the current ones. A petition from the guards was rejected after Leadbetter claimed that even their new rates were higher than those on the Greenock and the Edinburgh & Glasgow.[255] The difference between wage levels in England and Scotland may be noted: while Ayrshire drivers accepted wages of 24/- to 26/-, their counterparts on the North Midland had struck against a proposed moderate reduction from 42/-.[256] The

Railway Times hoped in 1844 that the reduction in working expenditure at Greenock had not been achieved by undue wage cuts: 'if however there be a Company in which economy is carried to the verge of parsimony . . . it is the Greenock Company'. Two years earlier it had reduced all salaries by 10%: the railway was on the other hand praised for its system of incentives and rewards for good service by employees.[257]

Ayrshire cost-cutting fell upon the wage-earning staff, but elsewhere senior officials could bear the brunt. Wishaw & Coltness resident engineer Robert Dodds offered his resignation in January 1844 when the committee cut his salary to £290, but he was still in office when a shareholders' inquiry reported in July. This proposed a further reduction to £250, a cut for the secretary from £250 to £130, and the abolition of the post of assistant engineer, but no changes in wage rates; in a further criticism of Dodds, the inquiry 'deem it of the utmost importance that the principal Manager or Resident Engineer should be almost constantly on the line'. The board resisted the cuts, and it is unclear if they were enforced.[258] Even the honorarium paid to directors, seldom a very large sum, could be affected by cuts. The Ayrshire meeting of August 1838 voted £600 per year to the board, as well as £500 for the services which they had rendered for nothing in the past; chairman McCall would have been happy to wait until a decent dividend was being paid, and objected vigorously to any retrospective payment. In 1842 the allowance was cut to £400, although as the number of directors was also reduced each individual's share rose from £25 to £33.[259] The Greenock, which introduced a £400 payment to the board in 1838 as 'every other railway made an allowance to their Directors', halved it in 1842.[260] The reduction of the Edinburgh & Glasgow board from 24 to 15 in 1842 also saved a little money, although the main intention was to end the conflict between groups in the two terminal cities.[261]

Cutting down on working expenditure turned out to be less simple than many shareholders hoped. In the earliest days miscalculations might well be made over the number of staff or trains required for maximum profitability, or over the best ways in which they might be organised. After initial adjustments, the number of useful changes which might be made was seldom great; those suggested by the committees of inquiry at the end of the decade, eager as they were to find fault, were often essentially cosmetic and in some cases likely to have damaging effects. The best way to reduce working expenditure as a proportion of income was either to raise rates if demand was sufficiently inelastic, or to increase the total amount of traffic. Experiments in raising rates, particularly for passengers, had not been promising,[262] but most companies found that the improving economic conditions of 1843-44 were making their figures increasingly respectable.

VIII. Current income and the setting of rates

The figures for income from traffic given in Tables 60-65, taken without reference to the escalated costs of construction, are in general a success story, with returns appreciably above anticipated levels. In 1851 John Francis commented on railways that

> If they underestimated their expenses, it is noticeable that they also understated their receipts. Railways have produced results which the wildest prospectus never dared to exhibit. They have found sources of profit which the most vivid imagination never conceived. They have carried millions instead of thousands, and that at a rate so low as to compel traffic where none previously existed. They have created towns, erected manufactories, built churches, peopled villages, filled heaths with houses, given the poor man the luxuries of the rich, placed the wealthy on a level with the poor, enforced a punctuality which was before wanting, have taken the townsman from the smoke of the city, have given the yeoman a glimpse of the town paved with the gold of imagination, have shed a light of life over many a country village, and ... have made the uttermost parts of the land converse with the speed of light.
>
> The boldest prospectus never spoke such language; the most arrant enthusiast never imagined such things. But the first half of the nineteenth century has witnessed them ...[263]

Table 60. *Arbroath & Forfar, Income on Current Account to 1845*

Date[a]	Passengers		Goods		Other Traffic[b]		Other[c]		Total
	£	%	£	%	£	%	£	%	
1839–40	4,269	43.8	5,013	51.5	241	2.5	217	2.2	9,740
1840–41	3,688	39.3	5,226	55.7	276	2.9	191	2.0	9,382
1841–42	3,609	38.7	5,191	55.7	317	3.4	210	2.3	9,327
1842–43	3,139	35.7	4,833	55.0	316	3.6	253	2.9	8,794
1843–44	2,945	35.4	4,858	58.5	267	3.2	290	3.5	8,311
1844–45	3,116	34.7	5,386	60.0	282	3.1	192	2.1	8,976
Total	20,766	38.1	30,507	55.9	1,699	3.1	1,353	2.5	54,530

a: periods as in Table 54
b: includes mail, horses, carriages
c: includes interest, rents

Sources: as in Table 36.

It was a propagandist's exaggeration, but it contained an essence of truth. Francis' emphasis was on the movement of people rather than of goods; in Scotland it was the inter-urban lines of the early 1840s which first accustomed the general public to the idea of passenger rail travel.

The returns, if compared with those for the older-established coal railways during the same period (see Table 21), emphasise both the greater scale of operation and the greater reliance on passenger traffic. Of the older companies, only the Garnkirk & Glasgow and the Edinburgh & Dalkeith averaged over £3,000 per year from passengers in 1842-44, while only the latter company and the tiny and atypical Newtyle & Coupar Angus and Paisley & Renfrew lines took more than half their income from passenger traffic. On the new main lines radiating from Glasgow, passenger revenue ranged from over £25,000 per year on the Greenock line to about £80,000 on the great trunk route to Edinburgh, accounting in each case for between three-fifths and three-quarters of revenue.

Table 61. Dundee & Arbroath, Income on Current Account to 1845

Date[a]	Passengers		Goods		Other Traffic[b]		Other[c]		Total
	£	%	£	%	£	%	£	%	
1841	9,859	78.9	2,603	20.8	n.a.	—	31	0.2	12,493
1842	9,212	75.9	2,620	21.6	n.a.	—	300	2.5	12,132
1843	8,968	73.2	3,290	26.8	n.a.	—	—	—	12,258
1844	9,607	70.1	4,104	29.9	n.a.	—	—	—	13,711
Total	37,646	74.4	12,617	24.9	n.a.	—	331	0.7	50,594

a: Periods are calendar years
b: no separate figures given
c: includes interest, rents

Sources: PP 1842 [360] XLI, *Second Report of Railway Department*, appx. X; PP 1843 [440] XLVII, *Third Report of Railway Department*, appx. II; PP 1844 [551] XLI, *Report of the Officers of the Railway Department for 1843*; PP 1846 [698] XXXIX, *Report of the Railway Department for 1844–45*, appx. II.

Table 62. Edinburgh & Glasgow, Income on Current Account to 1845

Date[a]	Passengers		Goods[d]		Other Traffic[b]		Other[c]		Total
	£	%	£	%	£	%	£	%	
1842	74,010	78.2	17,880	18.9	1,621	1.7	1,090	1.2	94,602
1843–44	82,559	68.6	34,098	28.3	3,285	2.7	350	0.3	120,290
1844–45	82,314	70.4	30,008	25.7	4,006	3.4	525	0.4	116,915
Total	238,883	72.0	81,986	24.7	8,912	2.7	1,965	0.6	331,807

a: Periods as in Table 55
b: includes mail, horses, carriages
c: includes interest, rents
d: includes parcels

Sources: as in Table 37.

Table 63. Glasgow Paisley & Greenock, Income on Current Account to 1845, excluding joint line

Date[a]	Passengers[d]		Goods		Other Traffic[b]		Other[c]		Total
	£	%	£	%	£	%	£	%	
1841	23,633	87.2	2,925	10.8	431	1.6	98	0.4	27,087
1841–42	27,153	74.8	8,542	23.5	355	1.0	257	0.7	36,307
1842–44	28,370	68.1	12,595	30.2	706	1.7	15	0.0	41,686
1844–45	24,401	63.2	13,188	34.1	875	2.3	157	0.4	38,621
Total	103,557	72.1	37,250	25.9	2,367	1.6	527	0.4	143,701

a: Periods as in Table 56
b: includes mail, horses and carriages
c: includes interest, rents
d: includes parcels

Sources: as in Table 38.

Table 64. *Glasgow Paisley Kilmarnock & Ayr, Income on Current Account to 1845 excluding joint line*

Date[a]	Passengers £	%	Goods[d] £	%	Other Traffic[b] £	%	Other[c] £	%	Total
1839–40	18,942	95.1	809	4.1	11	0.1	152	0.8	19,913
1841	34,360	74.1	10,081	21.7	508	1.1	1,359	2.9	46,389
1842–43	34,465	67.9	15,340	30.2	875	1.7	70	0.1	50,750
1843–44	36,663	62.4	21,084	35.9	955	1.6	—	—	58,802
1844–45	42,601	59.0	28,578	39.6	1,014	1.4	—	—	72,193
Total	167,031	67.3	75,892	30.6	3,363	1.4	1,581	0.6	248,047

a: Periods as in Table 57
b: includes mail, horses, carriages
c: includes interest, rents
d: includes parcels
Sources: as in Table 39.

Table 65. *Glasgow & Paisley Joint Line, Income on Current Account to 1845*

Date[a]	Passengers £	%	Goods[d] £	%	Other Traffic[b] £	%	Other[c] £	%	Total
1840	7,372	99.2	—	—	—	—	63	0.8	7,435
1840–41	14,594	93.9	724	4.7	140	0.9	88	0.6	15,546
1841–43	10,437	83.2	1,367	10.9	125	1.0	—	—	12,541
1843–44	13,016	91.6	1,002	7.1	95	0.7	94	0.7	14,206
1844–45	13,953	96.4	418	2.9	—	—	103	0.7	14,475
Total	59,372	92.4	3,511	5.5	360	0.6	348	0.5	64,257

a: Periods 14 July–30 Nov 1840; 1 Dec 1840–30 Nov 1841; 1 Dec 1841–31 Jan 1843; 1 Feb 1843–31 Jan 1844; 1 Feb 1844–31 Jan 1845
b: includes mail, horses, carriages
c: includes interest, rents
d: includes parcels
Sources: as in Tables 52 and 53.

The Dundee & Arbroath, thanks partly to coastal competition but mainly to its totally inadequate facilities for handling goods at Dundee, was the most passenger-dependent of all lines, apart from the special case of the joint line between Glasgow and Paisley, where an 1843 agreement reserved passenger traffic for the railway and heavy goods for the canal.[264] Only the Arbroath & Forfar, on a route where previously there had not even been a road coach, was a small-scale company making over half its income from the movement of goods: as John Miller explained, 'a manufacturing population go about much more than an agricultural population: an agricultural population is not very locomotive'.[265]

For sheer volume of goods traffic, the old Lanarkshire mineral lines still held their supremacy. The Monkland & Kirkintilloch carried over a million tons a year, and the Ballochney and Wishaw & Coltness a third as much each; by 1844 Ayrshire goods traffic had reached only 96,000 tons, while the Edinburgh & Glasgow (which did not bother to record tonnage) had a goods income about 30% greater from a similar length of line.[266] In spite of the very short journeys made by most of its traffic, the goods income of the Monkland & Kirkintilloch was greater than that of the Greenock, or of all the lines in Angus put together, and was within sight of that of the Ayrshire. Goods services on most of the inter-urban lines, with the exceptions of the Arbroath & Forfar and the Glasgow & Paisley joint line, brought in about a quarter of gross revenue; in some cases the original expectations from goods traffic had been even lower, with the promoters of the Ayrshire and of the Edinburgh & Glasgow anticipating that it would provide respectively 21.1% and 14.2% of revenue.[267] The need for improved goods services did figure strongly in company submissions to Parliament. As one example among many may be noted the elaborate evidence given by Alexander Guthrie, overseer to the Duke of Portland, to the committee on the Ayrshire bill about the trade of Kilmarnock. The town, which he claimed to be the fastest growing in Scotland after Dundee, generated enough trade to supply about forty regular carts on the Glasgow road (moving over thirty tons per day), as well as eight to Paisley, four to Irvine, three to Ayr, and others to every significant town in Ayrshire. Even so, trade was restricted by time and expense — the Paisley carts took ten hours and charged about £1 per ton, while those to Ayr and Irvine took four and three hours and charged 9/- and 8/6d respectively. On the other hand, the Kilmarnock & Troon railway had allowed the export of about 150,000 tons of coal per year through Troon harbour.[268] Similar evidence of a large potential goods trade being held back from full development by the inadequacies of existing transport was an inevitable part of all campaigns for new railways.

After authorisation, however, goods traffic generally received much less attention than passengers, both in company minutes and in the railway press. It may be that companies felt that the volume of goods traffic would be less affected than the number of passengers by changes which were within their control. For passenger traffic, demand appeared to respond directly and significantly to changes in price; Table 66 indicates that most companies eventually came to similar conclusions at least about the fare for their cheapest accommodation. The exceptions include at one end the Edinburgh Leith & Newhaven, whose high fares were no more of a deterrent than the inconvenient station at Scotland Street and which had to open its tunnel to have any hope of much traffic, and at the other the Monkland & Kirkintilloch/Ballochney service, where passengers were carried in considerable discomfort as an unimportant accessory to coal traffic. It was of course essential above all things to undercut rival means of transport, which explains the low fares forced upon the Greenock line by the river boats, while the average fare levels of the Glasgow & Paisley joint line reflect the fact that an agreement had been reached with the canal.

In the beginning, with little experience of the effects of various fares on passenger numbers or net profits, rates had to be set by trial and error. The Edinburgh & Glasgow opened with fares for the journey of 8/-, 6/- and 4/-, chosen for little better reason than to 'set coach competition at defiance'.[269] For goods traffic, where the speed advantage of a railway was less valuable than for passengers, undercutting the rivals was the principal consideration, and the choice of rates often seems to have been little more scientific than the methods proposed in the days of horse-drawn waggonways. Factors to be taken into account included the bulk of the commodity, the ease of handling, the risk of damage, and possibly the length of the journey, but above all the nature of road or water competition.

Table 66. Fares and Rates on Scottish Railways, 1844–45 (d. p.mile)

Company	Passengers			Coal p.ton	Lime p.ton	Grain p.ton	Manufactured textiles p.ton
	1st	2nd	3rd				
Arbroath & Forfar	1.9	1.4	0.8	3.3	2.5	3.3	4.6
Ardrossan	2.0	1.5	—	1.5	1.5	2.0	3.0
Ballochney	0.8	0.5	0.4	a	a	b	b
Dundee & Arbroath	1.8	1.4	1.1	n.a.	n.a.	n.a.	n.a.
Dundee & Newtyle	1.8	1.5	1.2	3.1	3.1	2.7	3.8
Edinburgh & Dalkeith	—	0.8	—	1.3^d	1.8^c	3.5^d	3.5^c
Edinburgh & Dalkeith, Leith branch	—	0.8	—	1.8	2.5	2.5	2.5
Edinburgh & Glasgow	2.1	1.6	1.0 } 0.7^k	0.9–1.0	1.5	0.6	1.5
Edinburgh Leith & Granton	—	2.0	—	—	—	—	—
Glasgow Garnkirk & Coatbridge	0.9	0.6	—	e	a	f	f
Glasgow Paisley & Greenock	1.4	0.8	0.6 } 0.3^k	n.a.	n.a.	n.a.	n.a.
Glasgow & Paisley joint line	1.7	1.3	0.9	—	—	—	—
Glasgow Paisley Kilmarnock & Ayr	1.8	1.3	0.9	1.3	1.3	2.0	2.5
Monkland & Kirkintilloch	0.8	0.5	0.4	g	g	h	h
Newtyle & Coupar Angus	1.6	1.2	—	2.5	2.5	2.5	3.0
Slamannan	1.4	1.3	0.9	i	j	1.5	3.0
Wilsontown Morningside & Coltness	1.5	1.0	—	1.0	1.0	3.0	4.0
Wishaw & Coltness	1.1	0.9	—	a	a	b	b

a: 3.75 for 1st mile, 1.75 for 2nd, 1.25 for 3rd–5th, 0.75 for 6th–7th, 0.5 after
b: 2.75 for 1st mile, 2.25 after
c: including inclines
d: 3d extra per journey for incline
e: 3.25 for 1st mile, 1.25 after
f: 4.25 for 1st mile, 2.0 after
g: 1.5 for 1st–4th miles, 0.13 after
h: 1.5 for 1st–3rd miles, 1.0 for 4th, 0.13 after
i: 3.0 for 1st mile, 1.5 after
j: 3.0 for 1st–2nd miles, 1.5 after
k: fourth class

Sources: PP 1845 [614] XXXIX, *Return of Railway Company Charges*: PP 1846 [698] XXXIX, *Report of the Railway Department for 1844–45*, appx. II.

Table 66 indicates how little consensus developed on rates. Some figures are surprising, notably the very low Edinburgh & Glasgow rate for grain, which surely cannot be explained simply in terms of canal competition. The complicated scales applied on almost all the Lanarkshire coal lines reflect a determination to make a profit on the large amount of traffic which travelled a very short distance but still had to be handled at both ends. In some cases it may even have been designed to discourage traffic; the Garnkirk may have regarded moving grain or manufactures for less than a mile as more trouble than it was worth, while the area through which it ran was unlikely to create such traffic in any case. The massive traffic and high dividends of the Monkland & Kirkintilloch reflect the success of a low-rate policy which kept the Garnkirk threat under control, while the high rates on the Dundee & Newtyle were the mark of a company in such financial trouble that it could not afford to risk the time lag between cutting rates and seeing traffic volumes expand.[270]

On the whole, however, companies do not appear to have believed that goods traffic, unlike passenger traffic, was particularly price sensitive. They did of course emphasise the help that the simple existence of the railway would give to industry; the Greenock company, for instance, noted that in its first year both sawmills and cotton mills had been built along the line in spite of the depression.[271] Clearly the cost advantage of rail over road transport could make all the difference to projectors of new enterprises, particularly in mining or heavy industry. But companies doubted whether adjustments to the rates charged on an already existing railway could make much more than a marginal difference to entrepreneurial calculations when compared with the much greater economic forces of an intense depression followed by an upswing into optimism. All other things being equal, they believed, a small cut in rates was unlikely to bring new firms into existence and an increase would not prevent old customers from continuing to send goods by rail. Goods rate fixing in the early days, then, depended on more or less arbitrary decisions by company officials as to what was reasonable and likely to be profitable (sometimes confused by the conflicting interests of some directors in mining, manufacturing or trading). Since companies could not discriminate between customers, the best chance of large entrepreneurs forcing a change to their own advantage was to oppose a company bill in Parliament in the hope of being bought off.[272]

IX. The passenger boom

The most immediate effect of new railways, to contemporary eyes, was the enormous increase in passenger numbers. By the time the inter-urban lines opened, promoters claimed to expect that a new transport facility would generate its own demand. The Edinburgh & Glasgow prospectus compared the single coach which ran between Glasgow and Paisley in 1811 with the hourly services on both road and canal in 1835, and claimed that the opening of the Stockton & Darlington had multiplied passenger numbers along its route twentyfold.[273] The

Ayrshire directors in 1837 were 'not aware of any Railway yet opened on which the traffic in passengers has not been trebled or quadrupled', and in 1841 reported that numbers between Johnstone and the coast were already 89% above estimate.[274] The minister of Dundonald confirmed the trend: since the railway opened

> the whole system of travelling has been completely revolutionized, as if by magic; and, when one hears of the paucity of travellers a few years ago, which was insufficient to support a rustic one-horse conveyance betwixt Irvine and Ayr thrice a-week, and compares it with the numbers who are flocking to and fro at all hours of the day, he is almost led to wonder what moving spirit can have come over the people, and what they can have found to do.[275]

Thomas Grainger generalised to a select committee:

> Whenever a cheap, safe, expeditious, and comfortable means of conveying passengers has been established, either upon land or water, the increase in travelling has in almost every instance far exceeded what could reasonably have been anticipated.[276]

All Grainger's adjectives were important, even if all were not available to each individual passenger, since cheapness and comfort were essentially alternatives. Railways were undeniably expeditious, and increasingly they were regarded as safe. Certainly accidents received wide and unfavourable publicity, and among the Board of Trade's most interventionist activities were the detailed reports made by their inspectors after each occurrence; railway propagandists however made much of the more frequent and sometimes equally horrific accidents which overtook stage coaches. They received an important piece of encouragement in June 1842 when Queen Victoria ventured on her first rail journey.[277] The railway was appreciably cheaper than its rivals, and in the first class at least more comfortable than the stage-coach, if not necessarily than the canal or river boat. The inter-urban lines found that, even in depression, they could tap a vast latent desire for travel at an affordable price.

A breakdown of passenger traffic statistics is given in Table 67. The trends are unsurprising, with most companies recording some falling away in both numbers and receipts as the recession hit bottom in 1842 or 1843 (though the Greenock, which had troubles of its own, saw a dramatic fall in numbers in 1844). The other obvious feature is the importance of third-class traffic. On the Edinburgh & Glasgow and the Ayrshire, well over half of all passengers travelled third, while in Angus the proportion rose to more than three-quarters; on the Greenock, where fares were so low that the second class should also be included in the calculation, almost 90% of passengers travelled for less than a penny per mile. The figures contrast starkly with those for the great English lines, particularly in the south of the country, where third-class passengers were on the whole regarded as an unprofitable nuisance and given very little accommodation. In 1842 the two Arbroath lines were reported to carry between them twice as many third-class passengers as the combined might of the London & Birmingham, the Grand Junction and the Great Western.[278] Two years later the Select Committee on Railways recorded that third-class passengers amounted to 12% of the total on the Grand Junction, 9% on the London & Birmingham, and a mere 8% on the Great Western. Only on some of the northern English lines were proportions comparable to those in Scotland — 51% of passengers on the York & North Midland, and 69% on

Table 67. Passenger Traffic on Inter-urban Railways, 1841–44

Company	Date	Numbers							Receipts							Fares[a] (d.p.mile)		
		1st No.	1st %total	2nd No.	2nd %total	3rd No.	3rd %total	Total No.	1st £	1st %total	2nd £	2nd %total	3rd £	3rd %total	Total £	1st	2nd	3rd
Arbroath & Forfar	1841	2,404	2.4	8,813	9.0	87,219	88.6	98,436	188	5.1	516	14.0	2,979	80.9	3,684	1.86	1.42	0.98
	1842	2,177	2.5	8,419	9.8	74,901	87.6	85,517	165	5.1	495	15.2	2,588	79.7	3,249	1.91	1.46	1.03
	1843	2,343	2.7	9,843	11.2	75,739	86.1	87,925	157	5.4	524	17.9	2,251	76.7	2,933	1.77	1.32	0.87
	1844	2,002	2.2	10,040	11.0	79,396	86.8	91,448	146	4.7	567	18.4	2,374	76.9	3,086	1.93	1.45	0.93
	Total	8,926	2.5	37,115	10.2	317,255	87.3	363,326	656	5.1	2,102	16.2	10,192	78.7	12,952			
Dundee & Arbroath	1841	11,164	4.1	49,727	18.4	209,928	77.5	270,819	n.a.		n.a.		n.a.		9,861	2.16[b] / 1.79	1.79[b] / 1.43	1.07
	1842	9,551	3.8	41,709	16.6	200,249	79.6	251,505	n.a.		n.a.		n.a.		9,212	"	"	"
	1843	10,384	4.4	47,182	19.9	179,768	75.7	237,334	907	10.1	2,331	26.0	5,729	63.9	8,967	"	"	1.43[b] / 1.07
	1844	10,598	4.1	61,573	23.9	185,246	72.0	257,417	902	9.4	2,894	30.1	5,882	61.2	9,607	"	"	"
	Total	41,697	4.1	200,191	19.7	775,191	76.2	1017,075	—		—		—		37,647			
Edinburgh & Glasgow	1842	86,027	15.8	135,982	24.9	323,514	59.3	545,523	n.a.		n.a.		n.a.		73,556	2.09	1.56	0.85
	1843	90,997	15.0	155,690	25.7	360,056	59.3	606,743	26,284	34.6	24,456	32.2	25,348	33.3	76,051	"	"	"
	1844	92,555	12.4	160,593	21.5	493,634	66.1	746,782	n.a.		n.a.		n.a.		90,933	"	"	"
	Total	269,579	14.2	452,265	23.8	1177,204	62.0	1899,048	—		—		—		240,540			
Glasgow Paisley & Greenock	1842	85,489	14.6	279,304	47.7	220,378[c]	37.7	585,171	n.a.		n.a.		n.a.		29,202	n.a.	n.a.	n.a.
	1843	54,272	8.9	166,450	27.3	388,856	63.8	609,578	5,838	23.5	9,106	36.6	9,906	39.9	24,848	1.62	0.82	0.35
	1844	48,617	10.9	185,927	41.6	141,162[d] / 70,795	31.6[d] / 15.9	446,501	5,373	21.6	11,060	44.4	6,685[d] / 1,770	26.9[d] / 7.1	24,887	1.38	0.84	0.53[d] / 0.27
	Total	188,378	11.5	631,681	38.5	721,191	43.9	1641,250	—		—		—		78,937			
Glasgow Paisley Kilmarnock & Ayr	1841	77,387	13.2	233,366	39.8	275,331	47.0	586,084	10,249	25.9	18,197	46.0	11,125	28.1	39,571	1.88	1.25	0.88
	1842	60,617	8.6	249,916	35.6	390,796	55.7	701,349	n.a.		n.a.		n.a.		39,307	"	"	"
	1843	63,653	7.7	275,767	33.5	483,318	58.7	822,738	n.a.		n.a.		n.a.		41,707	"	"	"
	1844	73,339	9.6	275,430	36.2	412,988	54.2	761,757	n.a.		n.a.		n.a.		47,132	"	"	"
	Total	274,996	9.6	1034,479	36.0	1562,433	54.4	2871,928	—		—		—		167,717			

a: Average given where different fare levels are recorded during the year
b: Higher rates charged on mail trains
c: Third class provided during the second half of the year only
d: Lower figures refer to passengers carried on goods trains

Sources: PP 1842 [360] XLI, *Second Report of Railway Department*, appx. X; PP 1843 [440] XLVII, *Third Report of Railway Department*, appx. II; PP 1844 [551] XLI, *Report of the Officers of the Railway Department for 1843*; PP 1846 [698] XXXIX, *Report of the Railway Department for 1844–45*, appx. II.

the Manchester & Leeds, travelled third class.[279] The average United Kingdom railway carried 18.4% of its passengers first class at 2.73d per mile, 46% second class at 1.75d, and 35.6% third class at 1.15d;[280] Table 67 emphasises just how different were both the class balance and the fare structures of even the major Scottish lines.[281] The Committee's verdict that 'in this Country what is called the High-fare System ordinarily prevails' was a judgment based on events in the south, as had been Robert Stephenson's view in 1839:

> The first-class passengers, such as merchants and professional men, and so on, probably will take the first-class carriages; to them money is not of so much importance as time, consequently I would keep those fares high; but to a tradesman, to whom money and time are both important, a different scale ought to be adopted; and I think there is a class of people who have not yet had the advantage from the railway which they ought, that is the labouring classes.[282]

This reflected a system in which the fastest trains might consist of first-class carriages only, while the third class might be restricted to a solitary and inconveniently timed train each day. In Scotland, however, virtually all trains on the main lines catered from the beginning for all three classes.

The difference was one of necessity. As the *Railway Times* observed:

> Three causes, thin population, comparative poverty, and more economical habits in the rich, combine to prevent passenger railways in Scotland paying so well as the English lines.[283]

Scotland, like the industrial areas of northern England, was on average poorer than the south, and had in particular fewer of the sort of men who might be expected to travel first class. Hence passenger services had to concentrate on quantity rather than quality; as Mark Huish noted:

> In England the principal revenue of the railways is derived from the better class of passengers; in Scotland it is derived almost entirely from the poorer class of passengers.[284]

Lindsay Carnegie categorised the classes for the benefit of a select committee; in the first class 'the gentlemen of the country and mercantile gentlemen going backwards and forwards', in the second 'those who are compelled to go for purpose of business in winter or bad weather', and in the third 'labourers, artizans, fishwomen, and the lower class of society, the poorest'.[285] At Greenock Huish identified a yet poorer group of potential customers: 'a very peculiar class of travellers, a vast number of Irish reapers and persons of that description', for whom the company proposed to introduce a fare of 1/- per journey or 0.53d per mile. This service had to await a change in the passenger duty regulations before it could hope to pay, and then, thanks to the poverty of the customers and the competition of the river boats, was introduced at 0.35d per mile. Later even this was cut, when in 1844 the company introduced a fourth class, attached to the goods trains and paying 0.27d per mile or 6d per journey from Greenock to Glasgow.[286] The Dundee & Newtyle also had specially low fares, in its case for sheep-shearers.[287]

These rates, and the more comprehensive lists returned to Parliament in 1845-46 and given in Table 67, confirm that not only did Scottish railways cater for third-class passengers in large numbers, but they also offered extremely cheap fares. Only some of the Angus lines and the partially open and horse-drawn Edinburgh Leith & Granton did not have third-class fares which met the penny per mile stipulation for parliamentary trains under the 1844 act; in almost all cases, rather than introduce a parliamentary class, companies brought their third-class fares down the marginal amount required or altered the nature of the service to conform with the act. Only the Edinburgh & Glasgow had a more complex structure, with a third class at 1.04d per mile, a parliamentary class at 1d, and a fourth class at 0.65d (which presumably failed to meet some of the other parliamentary requirements).[288] Elsewhere, the limited passenger traffic of the coal railways was conducted at astonishingly low fares. The Monkland & Kirkintilloch and the Ballochney both offered the full panoply of three classes, but even first class cost less than a penny per mile. It was also impossible to exceed a penny per mile on the two-class Garnkirk or the single-class Edinburgh & Dalkeith (which had earlier also had a first class, still at just under a penny per mile, but had found that 'the numbers are so few it is scarcely worth referring to').[289] Such rates may be compared with those in England, where in 1842 the lowest fare on the London & Birmingham was 1.5d per mile, on the Grand Junction 1.3d, on the Eastern Counties 1.46d, on the London & Brighton (which had no third class) 2.0d, and even on the Manchester & Leeds, with its large third-class traffic, 1.37d.[290]

Scottish railways thus fulfilled the hopes expressed by numerous official bodies, that third-class fares and the provision of services would facilitate travel by the poor. As the 1840 Select Committee remarked,

> To convey the labourer cheaply and rapidly to that spot where his labour might be most highly remunerated, was frequently stated to be one great benefit which would be derived from opening these new channels of intercourse, while it was added that the health and enjoyment of the mechanics, artizans, and poor inhabitants of the large towns would be promoted, by the facility which they would be enabled to remove themselves or their families into healthier districts and less crowded habitations.

Such cheap provision had been made more necessary because the advent of a railway often ended the old traffic 'by waggons, vans, and carts, which afforded a cheap though dilatory mode of travelling to the labourer and his family'.[291] Sometimes indeed the fare levels startled visiting Englishmen; Whishaw, visiting the Arbroath & Forfar, claimed that 'with the low fares adopted on this line, it is more economical for the poor man to ride than to walk'.[292] Scottish companies were not of course simply being altruistic. Their potential passengers were for the most part highly price-conscious, and a small alteration in fare might substantially affect passenger numbers and returns. According to Thomas Grainger,

> on the railways in Scotland, and not a few in England . . . a low rate of charge is quite essential to the very existence of a passenger-traffic.

The Garnkirk & Glasgow had been 'among the first in Scotland to show to what extent a passenger-traffic could be created by carrying at a low rate', but rising prices on the line had led to a disproportionate fall in passenger numbers (see Table 68).[293] Conversely, claimed Whishaw,

> It is notorious, particularly in Scotland, that wherever railway-fares have been lowered, the traffic has invariably increased to an amazing extent.[294]

At Greenock, Huish expected remarkable results from the fare structure enforced by river competition:

> The peculiarities of our trade will force us to carry at a lower fare than is charged on any railway

From Town to Town: Construction and Operation 241

in the kingdom; at the same time we have every reason to expect that we shall have a larger traffic than any railway; that is, a larger number of passengers, multiplied by the number of miles travelled.[295]

Even the Edinburgh & Glasgow, which could expect a higher proportion of first and second-class passengers than most lines, found that there were gains when canal competition briefly forced the cheapest fares below $\frac{1}{2}$d per mile. John Learmonth told the shareholders:

A great many passengers were now taken along the whole line as low as 1s.6d, a measure which had been found not only a benefit to the public, but a great advantage to the Company. They derive a large sum of profit from such passengers, who were greatly on the increase, and he was not sure, even if the competition were given up, that they ought not to discontinue the low fares [*sic* — either Learmonth or the reporter appears to have inserted an extra negative]. He was one of those who thought that the most profit was to be made by low rates.[296]

Although some company expectations, such as those expressed by Huish, were excessively optimistic, all Scottish companies were to depend on the mass provision of cheap passenger travel for an important part of their profits.

Table 68. Garnkirk & Glasgow, Passenger Traffic, 1836-39

	Second-class fare for $8\frac{1}{4}$ miles d.	No. of passengers
April 1836	6	12,733
April 1837	8	9,594
April 1838	8	9,798
April 1839	10	6,989

Source: PP 1839(517)X, *Select Committee on Railways, Second Report*, evidence, Q.3455 (T. Grainger).

One problem with the provision of cheap services for the benefit of the poor was that they might be used by travellers who could well afford to pay for better accommodation. Railway companies felt that awareness of correct social status and behaviour ought to lead the affluent and the adequately well-off into the first and second-class carriages automatically, but they were well aware that many passengers were tempted to put economy above propriety. *Chambers' Journal* pointed out that on occasion it was possible to prefer the cheap facilities for other reasons:

The third-class, designed for the poorest travellers, are, in fine weather, much used by the rich also. To this two considerations conduce — the superiority of these carriages for a look-out and for the enjoyment of the open air in fine weather, and the common regard for economy . . .[297]

The Edinburgh & Glasgow was even accused of making the second class deliberately uncomfortable by leaving the sides open to the weather, in order to force people into the first class, while a *Railway Times* correspondent claimed that

There would be no necessity for first-class carriages if it depended on Glasgow men. The money-takers know you are a stranger if you ask for a first-class ticket.[298]

The Greenock line, with its very cheap fares, suffered particularly badly, and was reported to be taking only $\frac{5}{8}$d per passenger mile.[299] One of their expected sources of first and second-class fares was the numerous Glaswegians who took a family house on the Clyde coast for part or all of the summer, and commuted to their business. The company secretary appeared before a select committee:

> What description of persons travel by your railway?
> — A great proportion of the passengers are composed of people of some means, who go to live at watering places below Greenock in the summer.
> Do they go by the sixpenny train? — Perpetually.[300]

Chambers' was irate:

> The shabby rich, by the disposition they show to make use of these trains, are the sole cause of their being made less comfortable than they otherwise need to be. We have been astounded to hear that men worth scores of thousands have not scrupled to use the third-class carriages on the Greenock railway: some have even purchased camp-stools on which to seat themselves in these carriages. It should be held up to universal contempt, as a practice not only mean in itself, but inhumane, as it tends to deprive the poor of comforts that otherwise would flow to them.

Three months later:

> Certain magistrates of Glasgow have been observed to content themselves with a 'stand' in the cheapest part of the train from their own city to Edinburgh. The motives here may not be unmixedly good; and we have already expressed our regret that the frequenting of third-class carriages by wealthy people should have had the effect of lowering the character of the accommodation for the poor, directors being naturally anxious to drive as many of the rich as possible to the higher-priced vehicles.[301]

The comments indicated the only solution apparent to the companies — to make the third-class carriages sufficiently unattractive to deter the rich from entering, which, if they were not already deterred by the prospect of standing in the open for more than forty miles in possibly uncongenial company, might not be easy.[302] No Scot went as far as the chairman of the Northern & Eastern Railway in England, who suggested inserting chimney-sweeps in their working clothes into the carriages in order to drive the respectable out,[303] but often concern was expressed lest the accommodation was too comfortable. *Fraser's Magazine* believed that the debate demonstrated the fallacy of providing any cheap accommodation at all:

> It is all very well for authors to describe in glowing terms the miseries and insults to which the third-class passengers on railways are exposed. The reality is quite the reverse. Otherwise how should we hear at railway meetings the reiterated and piteous complaints of directors that the rich will persist in going into these vehicles; merchants, bankers, dignitaries of the church, members of parliament, ... ask for third-class tickets ... The practical effect of running third-class carriages is easy to be understood. It makes the railway shareholders pay the difference between a second and third-class ticket to every traveller who on such terms will consent to go into the third-class carriage. There is no conceivable reason why the railway shareholder should pay this money to the traveller any more than any other individual in the community.[304]

Few if any Scottish lines would have shared the apparent belief that abolition of the third class would transfer all its passengers to the second. All the same, it was possible to profess pride in the standards of the third class. The Arbroath & Forfar claimed that 'the comfort of the third-class passengers has been an object of particular attention with the Directors',[305] and they had provided some seats, but the line in any case depended more than any other on third-class traffic, and its limited number of other passengers would mostly be known local men on whom social convention might be expected to have an influence.

More representative were the views of Ayrshire chairman James McCall. The company's third-class carriages were

> without seats simply because, were they otherwise, those who took the second class, would crowd into them, and without the higher charges of the first and second class, the Company could not afford to keep the third class on line. It was unpleasant to hear it said that the Directors were injuring the feelings of the working people.[306]

In 1844 the Greenock secretary noted that the introduction of a sixpenny fare had transferred numerous passengers from the first and second classes, and that restricting the cheap rate to the slow trains had not helped: 'Passengers who can well afford the difference of fare prefer standing an hour, on the Greenock line, to paying 6d'.[307] The only difference between the second and third class was the absence of seats in the latter, and the company were considering removing the roofs from the third class in order to drive passengers back to the second.[308] As far as he was concerned, the proposed rules for parliamentary trains 'would have a most prejudicial effect on the receipts of railway companies'.[309] Reaction to this attitude was forcibly expressed in a parliamentary motion by Greenock's Radical MP Robert Wallace, who asked the Railway Commissioners

> to ascertain whether passengers of all ages and both sexes, because they may not be wealthy, should be forced during journeys to stand huddled together, in any number, and to any extent it may please the executive staff at railway stations to crowd them, to their great personal discomfort, to the injury of their clothes, and the preventing of their taking, to or from market, their small wares, or farm produce, which passengers are not sheltered from the weather, and thereby are liable to be melted with heat in the summer, and frozen with cold in the winter, as the seasons alternate.[310]

When the 1844 act introduced parliamentary trains, the provisions on fares of less than a penny per mile may have been aimed at the main English lines, but those on quality of accommodation were as applicable or more so in Scotland.

The universal Scottish provision of cheap passenger travel met a very large potential demand. The Ayrshire's estimates had been fairly optimistic, but when they opened their first piece of line from Ayr to Irvine, numbers in the first $6\frac{1}{2}$ months were twice what had been forecast for a year.[311] Fares could of course be too low for a company's own good; the Greenock had originally calculated on receiving 1/7d per passenger, but at the start of 1843 was taking only 9d, which could ruin the accounting of a largely passenger-dependent line.[312] But most lines could expect only a limited number of first and second-class travellers, and had to cater for the poor in order to make a reasonable return on their passenger business. Since companies made no effort to allocate passenger working expenses between the various classes, and indeed could not have done so with any accuracy even if they had wanted to, the exact amount of profit due to the poorer travellers cannot be calculated. Scottish companies however had no doubt that it was a traffic which should be encouraged.

From the beginning also, the railways of the late 1830s realised the potential demand for travel for pleasure. The boat services on the Clyde had already shown the way, as the ex-provost of Paisley indicated to the parliamentary committee on the Glasgow Paisley Kilmarnock & Ayr bill:

> Does it fall within your knowledge whether great Numbers of Persons go from Glasgow to

> those Watering Places on the Firth of Clyde during the Summer Months? — There are an immense number.
>
> All Ranks and Classes of Persons? — I should say, generally speaking, the Exception is their not going.[313]

Throughout the period the seaside was the great attraction. As Greenock chairman R.D. Ker announced at the opening of the joint line, Glaswegians

> may be brought within the bracing influence of the sea breeze all along the Western coast, a recreation of which they stand much in need.[314]

On the Arbroath & Forfar, 'the highlanders come down to bathe in the summer'; in Dundee 'we have a watering place called the Ferry, where they go to bathe . . . It is an immense benefit to such a population as that of Dundee'.[315] In 1841 the Saturday afternoon trains from Glasgow to Greenock, and the return trips early on Monday morning, 'are at present regularly crowded with sea-bathing folk . . . seldom carrying less than 500 people'. Higher up the social scale came the Glaswegian businessmen who rented a seaside house for their families for the whole summer, some of whom caused such distress to the Greenock company by travelling third class in spite of their ability to pay more. Similar behaviour occurred on the east coast, where middle-class citizens of Dundee and Perth took houses along the Angus coast. Some of these people, the companies hoped, would become permanent householders on the line, and therefore permanent commuters.[316] As Leadbetter said at the start of work on the Ayrshire:

> Many gentlemen would erect along the coast marine villas, induced by the near connection with the city of Glasgow to select those delightful situations which were to be found in the county of Ayr.[317]

On the Arbroath & Forfar this process was seen even before the line opened:

> This undertaking has given an impetus to building in this quarter most unprecedented, Sir John Ogilvy has already feud at a high rate nearly a dozen of stances for houses, some of which are in active progress; and applications are making every day for additional lots. Some of the portions of ground through which the railway runs, have sold at almost incredible prices, and property along the whole line has risen prodigiously.[318]

One newspaper examined the prospects during the construction of the Dundee & Arbroath:

> The proceedings along the entire line promise, ere many years pass over, to convert the Railway into a street, through a continuous line of busy villages, and literally to convert a desert waste into a home for thousands, of whom many, very many, will experience at least comparative happiness.[319]

Blackwood's envisaged a future which had to await the motor car for its fulfilment:

> How much would all this be improved if . . . the great body, at least, of the better order of the population, could make the country a permanent residence, sleep every night out of town, in cottages scattered at all distances round it, in every picturesque and pleasant spot within twenty or thirty miles of the great cities.[320]

In the early 1840s, however, the economic depression meant that the prospects for large-scale housebuilding along the railways were not perhaps as good as the companies might have hoped.

Even although third-class fares were by any standards low, companies showed keen interest in offering yet cheaper rates for excursion traffic. As early as 1834 the Garnkirk & Glasgow ran what was probably the first British excursion train, the Dundee & Newtyle promised them from its early days, and by the end of the 1830s they were common on several lines.[321] The Edinburgh & Dalkeith conveyed particularly large numbers on Musselburgh race days; one report of a holiday in April 1842, when the line had carried 6,661 passengers or 700 more than the previous record, suggested that such business was not necessarily profitable:

> Much credit is due to the different parties who furnished a healthful recreation to such members of the working classes; for the extraordinary expenses on such occasions generally outrun the remuneration.[322]

The Dalkeith company, however, had particular problems with the taxman, being charged duty both as a passenger railway and for the horses which it used for traction.

Passenger duty at $\frac{1}{8}$d per passenger mile was bound to discourage the provision of excursion facilities. The change in 1842 to a 5% levy on gross passenger receipts[323] was much more encouraging:

> This modification of the mileage duty will afford great facilities in making arrangements for special trains and pleasure excursions, which your Directors are of opinion should receive every encouragement, as they consider it to be a public duty to afford to the working classes the means of health and recreation at a cheap rate; and they feel satisfied, that in pursuing a liberal policy in this respect, the revenue of the railway will not sustain any loss.[324]

The possibilities were still restricted, partly by the number of people with some money to spare in a time of economic hardship but even more by the sabbatarian principles which prevented the running of excursions on the one day, apart from public holidays, on which the mass of the population would be free to travel. Even so, excursions became a regular feature of railway planning; as *Chambers' Journal* observed, 'shorter country trips to the neighbourhood of large towns are *always* to be had at a cost far below that under the old system'.[325] As early as 1840 arrangements had to be made for 14,700 passengers in two days for Paisley races on the Glasgow & Paisley joint line, as well as 8,800 on the Paisley & Renfrew.[326] Queen Victoria's visit to Edinburgh in 1842 pushed the weekly receipts on the Edinburgh & Glasgow from £1,618 to £6,500, while a month later the Slamannan — an unlikely excursion railway — took 1,500 passengers in three trains on a trip to Linlithgow and for safety reasons brought them all back in one enormous train with 110 carriages and five locomotives.[327] On 14 July 1843 the Ayrshire took thirty carriages full of Glasgow teetotallers to the Burns country — 'on some of the carriages passengers were clustering outside like bees' — while another 300 Glaswegians went by the Edinburgh & Glasgow and Edinburgh & Dalkeith lines to Dalkeith Palace where, although no one had warned the Duke of Buccleuch's agents that they were coming, they were allowed in and were reported to have behaved very well.[328] By October 1844 a circular letter to companies from the Railway Department expressed concern about the scale of excursion traffic:

> Much danger to the passengers is incurred on these occasions, from the unmanageable size of the trains, travelling at a high rate of speed, and without guards in proportion to the number of carriages and passengers... Danger is to be apprehended, unless the size of the trains be considerably diminished, or their rate of speed lessened...
>
> The primary object of each company is to carry passengers generally, according to the published time tables;... in no case should the trains so published be postponed or delayed, or

246 *The Origins of the Scottish Railway System*

otherwise interfered with, by casual trains, however beneficial to a particular section of the public, or profitable to the Railway Company.[329]

Such strictures were however more appropriate to the enormous excursion traffic of lines like the London & Brighton than to the more limited Scottish traffic.

X. Rendering unto Caesar: the passenger duty

At the beginning of the 1840s, it appeared that the extension of third-class services would be made impossible by the operation of the passenger tax. The principle of such a tax was not really in dispute; taxation had been applied to stage coaches since 1776, and, although railway proprietors could claim that their heavy expenditure on land and works made their case different, they could not complain too bitterly when in 1832 comparatively modest taxation was extended to their own passenger services.[330] Nor could they object to being taxed on the number of passengers actually carried, when coaches were taxed on passenger places, whether filled or not: one estimate suggested that railways paid $\frac{1}{8}$d per passenger mile, stage coaches $\frac{1}{4}$d and post-horses $\frac{3}{4}$d.[331] The problem was the method of assessing the duty. A rate of $\frac{1}{2}$d per mile for every four passengers carried, rather than a percentage levy on gross or net passenger receipts, was no great deterrent to the high-fare main lines of England. But in Scotland, where the third class almost invariably and the second class quite frequently were carried for a penny per mile or less, such taxation might make it unattractive for mineral lines to provide any passenger service at all and impossible for passenger-based lines to make any profit.

For most of the 1830s, the practice in Scotland was not so harsh as the theory. The Inland Revenue allowed railways in poorer areas to compound for the duty by paying an agreed annual lump sum, reviewed every second or third year, and the composition paid was normally much less than the duty theoretically due. All the Scottish companies, with the unimportant exception of the Dunfermline & Charlestown, compounded for their tax, and the 1839 Select Committee, which could see no general pattern in the ways by which the amount of composition was determined, produced statistics to underline the extent to which composition helped the Scots (see Table 69).[332] Apart from the Monkland & Kirkintilloch, which would have done better to pay the ordinary duty, and the Ballochney, which for some reason was prepared to pay for passengers it did not carry, the savings were generally substantial. Overall from the inception of the tax in 1832 until April 1839, Scottish companies paid £1,883 instead of a theoretical £10,563,[333] and on this basis the inter-urban lines would have little to fear from the levy when they opened.

The crucial change came in October 1838, when the Inland Revenue decided to end composition: with very few exceptions, the concession was withdrawn as the agreements became due for renewal. The duty paid by Scottish companies trebled from 1838 to 1839 (see Table 70), partly due to increased traffic but mainly because of the tax change, and they complained vigorously to Lord Seymour's

Select Committees in 1839 and 1840. In 1839 the Arbroath & Forfar petitioned the Committee for a tax based on a percentage of gross passenger receipts. In support Grainger claimed:

> It appears to me that a more unjust and unfair mode of levying a tax can hardly be conceived, and is wholly inapplicable to the case of railway conveyance in general, but more particularly to the railways in Scotland, and not a few in England, where a low rate of charge is quite essential to the very existence of a passenger-traffic.[334]

Table 69. Passenger Duty Payments, 1837–39

Company	Period	Passenger miles	Amount of composition	Amount due if not compounded	Composition as % of amount due
		m	£	£ s d	%
Arbroath & Forfar	1.38 1.39	264,992	10	138 0 4	7.5
Ardrossan	10.37–10.38	132,904	20	69 4 5	28.9
Ballochney	10.37–10.38	0	5	0 0 0	—
Dundee & Arbroath	7.38–5.39	415,624	29	216 9 5	12.5
Dundee & Newtyle	4.37–4.38	493,752	60	257 3 3	23.3
Edinburgh & Dalkeith	10.37–10.38	1784,360	250	910 2 1	27.5
Edinburgh & Dalkeith, Leith branch	5.38–5.39	157,056	40	81 16 0	48.9
Garnkirk & Glasgow	10.37–10.38	1148,696	150	598 5 7	25.1
Kilmarnock & Troon	10.37–10.38	63,912	30	33 5 9	90.1
Monkland & Kirkintilloch	10.37–10.38	5,928	5	3 1 9	162.3
Newtyle & Coupar Angus	3.38–3.39	120,616	20	62 16 5	31.8
Paisley & Renfrew	4.37–4.38	409,296	80	213 3 6	37.5
Wishaw & Coltness	10.37–10.38	23,040	5	12 0 0	41.7

Source: PP 1839 (222,517) X, *Select Committee on Railways, Second Report*, appx. 26.

The company hoped to halve their third-class fares on Saturdays, but in that case almost a third of the revenue would go to the government. A percentage levy would give both the company and the government an interest in increasing the amount of traffic, whereas under the existing system it could pay a company to carry fewer passengers. In the following year Lindsay Carnegie, who claimed that railways 'are worse rated than the stage-coach proprietors', was totally pessimistic:

> What effect will the imposition of the tax have upon your railway in the carriage of the lower class of passengers? — I think it would stop our railroad altogether; I do not know how we could go on with it.[335]

George Duncan of the Dundee & Arbroath, which had just paid £414 in tax for seven months against a previous annual composition of £50, admitted that the full rate was 'far beyond anything we had any idea of', and foresaw the total abolition of the third class even although 'the lower classes are those to whom we must principally look as the principal travellers upon our railway'.[336] He and Lindsay Carnegie both rejected the possibility of evading a percentage levy by lowering fares and raising goods rates, since in that case road carriers would again be

Table 70. Passenger Duty paid by Scottish Railway Companies, 1833–45 (£)

Company	1833	1834	1835	1836	1837	1838	1839	1840	1841	1842	1843	1844	1845
Arbroath & Forfar	—	—	—	—	—	—	10*	200	398	300	147	146	33
Ardrossan	—	1*	1*	15*	15*	20*	56	64	165	115	81	89	132
Ballochney	—	—	—	5*	5*	5*	14	69	90	30	12	8	21
Dundee & Newtyle	—	10*	10*	60*	60*	150*	243	323	290	222	131	132	142
Dundee & Arbroath	—	—	—	—	—	—	50*	879	1,023	797	447	478	506
Dunfermline & Charlestown	—	25	29	25	31	30	26	28	29	24	14	13	12
Edinburgh & Dalkeith	20*	20*	200*	200*	250*	250*	786	614	563	511	280	298	276
Edinburgh & Dalkeith, Leith branch	—	—	3*	5*	13*	30*	76	79	68	66	45	43	38
Edinburgh & Glasgow	—	—	—	—	—	—	—	—	—	4,941	3,798	4,043	4,796
Edinburgh Leith & Granton	—	—	—	—	—	—	—	—	—	—	70	71	68
Garnkirk & Glasgow	10*	10*	10*	10*	150*	150*	469	445	603	343	218	319	56
Glasgow Paisley & Greenock	—	—	—	—	—	—	—	404	2,928	3,420	1,256	1,048	592
Glasgow & Paisley Joint Line	—	—	—	—	—	—	—	—	976	1,277	569	755	444
Glasgow Paisley Kilmarnock & Ayr	—	—	—	—	—	—	151	1,778	3,860	2,765	1,774	2,070	2,375
Kilmarnock & Troon	—	10*	10*	15*	15*	30*	43	43	36	31	19	21	17
Monkland & Kirkintilloch	—	—	—	5*	5*	5*	7	43	90	43	6	8	6
Newtyle & Coupar Angus	—	—	—	—	—	40*	56	65	57	43	25	24	32
Newtyle & Glammis	—	—	—	—	—	—	69	65	56	35	17	18	20
North British	—	—	—	—	—	—	—	—	—	—	—	—	57
Paisley & Renfrew	—	—	—	—	5*	45*	213	208	144	67	34	37	39
Slamannan	—	—	—	—	—	—	—	94	257	77	24	24	38
Wishaw & Coltness	—	5*	5*	5*	5*	5*	10	22	22	17	50	104	42
Total	30	76	267	345	553	760	2,279	5,421	11,657	15,125	9,017	9,751	9,743

*Compounded. Other figures for 1839 are partially compounded. Years run from 5 January. In a few cases the figures are, for no clear reason, incompatible with those in the source used for Table 69.

Sources: PP 1843 (151)XXX, *Returns of Passenger Duty Paid to 5th January 1843*; PP 1846 (687)XIV, *Select Committee on Railway Acts Enactments*, 'Second Report', appendix 14, returns of passenger duty 1841–46.

competitive — for the Dundee & Arbroath at least, without a goods depot in central Dundee, this fear was clearly realistic.[337] For the Greenock, Mark Huish summed up:

> The Scotch railways ... as a body, feel this tax more than any other, inasmuch as they are compelled, from the comparative thinness of the population, and the poverty of the districts, to carry at a much lower fare than the great English railways.

Thanks to river competition, his company already charged less for first class than the London & Birmingham did for third; hopes of introducing a third-class fare of 1/- for $22\frac{1}{2}$ miles would be made impossible by a tax which took about a quarter of the gross receipts.[338]

The Garnkirk & Glasgow appealed for a continuation of their composition at the old rate of £150, as otherwise the necessary increase in fares could end their passenger service since 'the class of passengers using the railway (principally mechanics and labourers) could not afford it'. The fare increases of 1838-39, which led to a 28.7% fall in passenger numbers (see Table 68), could plausibly be blamed on the trebling of the company's duty liability. Grainger considered that the duty, coupled with the absence of taxation on canals, was substantially responsible for the fact that £100 of Garnkirk stock stood at £76 in the market, while a similar holding in the Monkland Canal was worth £1,200. The company's calculations did not impress Thomas Pender, the Tax Office's comptroller in Scotland, who described them as 'founded purely on hypothetical grounds, and in my opinion not to be depended upon'; his main concern was for the protection of stage coaches, and this led him to suggest that the tax, even if compounded, should be at least £350 per year.[339]

The Scottish companies also had to face the fact that, apart from some possible support in parts of northern England, they had to campaign on their own for a change in the tax. The view of officialdom, which in general seemed more concerned about the railway threat to road services, was summarised by Henry Wickham, chairman of the Board of Stamps and Taxes:

> The duty of one eighth of a penny per mile is so low a rate, and bears so small a proportion to the fares charged by the railway companies, that it cannot possibly have the effect imputed to it of preventing a great reduction of such fares in favour of poor passengers.[340]

Allowing that there was to be a passenger tax at all, Wickham's view would have been accepted by the high-fare main lines in England, which would be far better served by the existing system than by a 5% levy on gross passenger receipts. The Scots had first to demonstrate that the duty did discourage the provision of cheap services, and then to play on the expressed official desire to encourage such facilities to obtain a change in the method of levying the tax.

By 1840 the Scottish companies had all had the experience of paying uncompounded tax. Table 70 indicates that the national total payment had trebled from 1838 to 1839, and was to rise by a further 138% in the following year, and that comparatively little of the increase was due to the opening of new lines. At the beginning of the year representatives of the seven largest Scottish companies met in Glasgow to form an anti-tax association: ten weeks later, eleven companies signed their petition to the Treasury against the duty. The *Railway Times* appreciated their efforts: 'If the English companies have been culpably negligent in this matter, the Scottish have been characteristically active and persevering'.[341] Six companies sent witnesses to testify before Seymour's 1840 Select Committee, and its report suggested concern for the effects on the poor, if not for the finances of the railways:

> In proportion as Railway communication is extended throughout the country, the unequal pressure of this Tax will be more severely felt, inasmuch as it will be found to limit the accommodation which Railways might otherwise beneficially afford to the labouring classes . . . Your Committee believe that Parliament would deeply regret to find that the Tax imposed upon Railway Passengers had a tendency to deprive the labouring classes of these promised advantages, and especially when it is seen that in those parts of the country where the pressure of this tax is most severely felt, the revenue derived from it is insignificant in amount.[342]

From the point of view of the Exchequer, particularly at a time when many other sources of government income were under consideration for reduction or abolition, the last point was important. Almost three-quarters of the revenue from the tax came from the five leading English companies — the London & Birmingham, Grand Junction, Liverpool & Manchester, Great Western and London & Southampton — while in 1839 Scottish payments accounted for only 3.4% of the British total.[343] The Committee proposed a graduated percentage duty, levying $7\frac{1}{2}$% on fares of over 2.45d per mile, 5% between 1.4d and 2.45d, and $2\frac{1}{2}$% on lower rates, and emphasised the need for urgent action before some companies had to abandon cheap services. Such a conclusion delighted the Scots, but the opposition of the major English companies, coupled with parliamentary inertia, prevented any immediate legislation.[344]

In 1842 much publicity was given to an open letter from Gabriel Lang, the Glasgow lawyer who had become a leading anti-tax campaigner, to Chancellor of the Exchequer Goulburn. His arguments were familiar, but well documented with statistics. Composition had meant that up to April 1839 Scottish companies had received a remission of over 80% of the tax due:

> During this period, the short lines of passenger Railways in the north of England, and in Scotland, enjoyed a transient gleam of prosperity, and their shares were in general at a premium. Since 1839 the full amount of Mileage-duty has been exacted, and the passenger Railways of Scotland are, *without a single exception, at a discount*.[345]

Had the full duty been charged in the mid-thirties, it would have prevented the construction of the Scottish passenger lines 'which are now proving of such advantage to every person in that country, except the shareholders!'. The effect of the tax, as a percentage of gross passenger receipts, varied from 4.7-5.8% on English main lines, and from 10-17% in Scotland, where low fares and limited traffic also meant higher working expenses.[346] A comparison of three presumably carefully selected lines made his point about the effect on share values (see Table 71).

Lang added some evidence about the effects of the tax on third-class passengers. In November 1841 the Glasgow & Paisley joint line had been forced by canal competition to cut third-class fares from 6d to 4d for seven miles, resulting in more than 50% more passengers and a small **increase** in gross receipts but, thanks to the tax, a 7.4% fall in net passenger revenue. Conversely the Edinburgh & Dalkeith had found that a fare increase from 6d to 8d, provoked by the end of composition, had removed over a quarter of the passengers and a small amount of gross receipts, but had increased net receipts by 3.7%. The existing system, Lang concluded, had a built-in incentive to carry fewer passengers and 'gives the Railway companies an interest opposed, not only to the revenue, but to the convenience of the great mass of the people'.[347] Such arguments landed on increasingly fertile official ground; concern about the provision of travelling facilities for the poor was growing rapidly as the railways drove out of business many of the carts, waggons and

Table 71. Effect of Passenger Duty on Selected Lines, 1841-42

Company	Income £	Working expenses %	Passenger duty £	Net revenue £	Effect on shares
London & Birmingham	100	30	5	75	94% premium
Glasgow Paisley & Greenock	100	38	13	49	12% discount
Dundee & Newtyle	100	66	10	24	'unsaleable at any price'

Source: G.H. Lang, *Letter to the Right Hon. Henry Goulburn, M.P. . . .*, 9.

'caravans' upon which the poor had previously relied. By the end of 1842 the government had accepted the force of the arguments for change, and introduced a 5% levy on gross passenger revenue, in an act which also reduced the tax burden on stage-coaches. Fifteen railway companies, of which only one was in southern England, presented Lang with a piece of plate in recognition of his work.[348]

Welcome though the tax change was to the Scottish companies, there remained an opinion that any tax at all on the third class discriminated unfairly against both the poor passenger and the poor railway. Lang, this time in an open letter to Gladstone at the Board of Trade, again summarised the case:

> One of the chief advantages to be expected from the establishment of railways, was the cheap and rapid conveyance of the labouring classes to the places where their labour was most required, and would be most highly remunerated . . . Railway conveyance is a luxury to the rich, and affords them the means of increased enjoyment, but to the poor it is necessary, in order to procure the means of living. It is therefore unfortunate that the railways which have charged the lowest fares and carried the greatest number of the labouring classes have been the most unsuccessful . . . Before the establishment of railways, the poorer classes did not travel in taxed conveyances, but in untaxed canal boats, waggons and carts so that the duty on railway passengers was the first tax ever imposed on the movement of the poor.[349]

By 1844 Lang and his associates were knocking at an open door. The Board of Trade, concerned about the mobility of the poor for both social and economic reasons, were increasingly determined to make the major English lines offer some cheap facilities. Part of the compensation for the imposition of the parliamentary train was the remission of duty on its passengers.[350] The Scots, who for the most part already carried their third-class passengers at fares of less than a penny per mile in trains which travelled at twelve miles per hour or more and stopped at all stations, were substantial if incidental beneficiaries of the act, although some had to provide their third-class carriages with seats and roofs before qualifying for tax exemption. The credit for the change was claimed by a number of Scottish interests, including even the directors of the Greenock company when in 1850 they were called to account by their shareholders,[351] but in fact the Board of Trade appears to have been essentially concerned with English traffic.

The 1844 act removed the passenger tax from the forefront of controversy, creating a levy which the companies disliked no more than they disliked taxation in general. Table 72 indicates the effect on Scottish companies. The percentage figures for 1841, although high, are lower than those given in evidence to the select committee of 1839 and 1840, perhaps reflecting the effect of fare increases; they fall in 1842, when the 5% levy applied for the latter part of the year, and then settled around (but, according to the companies' statistics, not often exactly at) the 5% figure.

XI. Rendering unto God: the sabbatarian question

By 1844 the Scottish companies had achieved a reasonably satisfactory arrangement about what they should render unto Caesar; they had still to sort out their relationship with God, or at least with Scottish sabbatarianism.[352] Any proposal to run trains on Sundays came into direct conflict with the strong early nineteenth-century revival of a strict sabbatarianism, based on a rigid interpretation of the fourth commandment. One minister claimed that sabbath observance was enforceable in Scottish law, both by an act of 1661 and by the Westminster Confession, which had had the force of law in Scotland since 1690 and in which the sabbath was declared to be

> when men ... do not only observe an holy rest all the day from their own works, words, and thoughts about their own wordly employments and recreations, but are also taken up the whole time in the public and private exercises of ... worship, and in the duties of necessity and mercy.[353]

Sabbatarians were not prepared to allow that regular train services could conceivably be either necessary or merciful; indeed, Scottish stage-coach and canal services had been required by custom, and in some views by law, to stop on Sundays. In England and Wales, with the exception of a few minor lines in remote areas, Sunday services were accepted as normal; in Scotland, they would be a substantial breach with tradition.[354]

The accepted leader of the opponents of Sunday train services was Sir Andrew Agnew of Lochnaw, who sat as a 'moderate reform' Tory MP for Wigtownshire from 1830 to 1837. He became conspicuous in Parliament solely for his absolute advocacy of sabbath observance, and his bills to prohibit all Sunday labour except works of necessity and mercy soon became an annual feature of the parliamentary timetable. On successive appearances Agnew's bill was defeated on second reading by 73-79, 125-161 and 43-75; finally in 1837 it was carried by 110 to 66, but while it was in committee William IV died, Agnew lost his seat at the consequent election, and no one else took up the bill. In 1839 he became the first chairman of the Scottish Society for Promoting the Due Observance of the Lord's Day, and in his first speech in office declared war on 'the threatened invasion of Sabbath-breaking customs from England by the railways'.[355]

Agnew's first success was an 1835 agreement with the Post Office that the Wigtownshire mail coach would not run on Sundays — the only mail service in Britain to be thus restricted[356] — and one of the first campaigns of his Lord's Day Society was an attempt to end all movement or delivery of mail on Sundays. The mail, however, was not an easy target. Once the postmaster-general had determined to use the railways to move the mail, he had, by an act of 1838, absolute power to decide what mail trains should run.[357] He was unlikely to stop Sunday mail services simply because the type of transport used had changed; it was more probable that he would be influenced by the commercial community than by sabbatarian pressure groups; and, being based in London, he was less likely to be affected by the views of even a sizeable section of the Scottish people. In effect the enforcement of Sunday mail services had been decided before any of the major Scottish lines were opened. Mail trains ran on all through lines, regardless of the views of company directors, and only occasionally, on a peripheral route such as the Greenock line, was it possible for local interests to have the service suspended. A great sabbatarian meeting congratulated the Greenock board, reiterated that Sunday mails broke the commandments, and agreed that

> It being notorious that all such Sabbath desecration has been greatly aggravated by the opening of Railways in England, every legitimate means should be used to prevent the introduction of the Railway Mails on the Lord's Day in Scotland.

The postmaster-general replied that 'with every desire to meet the wishes of the Memorialists, his Lordship could not interfere with the course of the public services'.[358] The directors of the more important Ayrshire line failed to persuade the Post Office to stop the Sunday Glasgow to Belfast mail, but since the company refused to carry passengers on the trains the Post Office agreed to pay the whole cost of the service. The *Railway Times* was shocked by such a concession to 'the ultra-godly agitators of Ayrshire'.[359] But the mails had to be carried wherever and whenever the postmaster-general chose, whatever either Agnew's society or sabbatarian railway directors might think.

There was, however, no point of government policy involved in the Sunday carriage of goods or passengers, and companies were free to make up their own minds. Goods traffic was in fact not a problem. No company, whatever the individual principles of its directors, thought that the expedition of goods services was important enough to defy Scottish convention, and before 1850 no goods train ever moved on a Sunday. The coal railways in general also refrained from Sunday passenger services; for most of them, the limited amount of custom which might be expected would in any case have been unlikely to cover the costs of providing a service and making employees work on a day when no minerals were being moved. The one exception, a service on the Edinburgh & Dalkeith, was soon withdrawn after some Sunday rowdyism in Dalkeith had swung local opinion against it.[360] The Garnkirk & Glasgow wished to enforce Sunday closure in its bye-laws, but

Table 72. Passenger Duty and Gross Passenger Revenue, 1841–44

Company	1841 Revenue £	1841 Tax £	1841 Tax % rev.	1842 Revenue £	1842 Tax £	1842 Tax % rev.	1843 Revenue £	1843 Tax £	1843 Tax % rev.	1844 Revenue £	1844 Tax £	1844 Tax % rev.
Arbroath & Forfar	3,684	398	10.8	3,249	300	9.2	2,933	147	5.0	3,086	146	4.7
Ardrossan	2,182	165	7.6	1,683	115	6.8	1,726	81	4.7	1,808	89	4.9
Ballochney	661	90	13.6	531	30	5.6	197	12	6.1	78	8	10.3
Dundee & Arbroath	9,859	1,023	10.4	9,212	797	8.7	8,967	447	5.0	9,607	478	5.0
Dundee & Newtyle	2,937	290	9.9	2,757	222	8.1	2,630	131	5.0	2,651	132	5.0
Dunfermline & Charlestown	n.a.	29	—	341	24	7.0	285	14	4.9	248	13	5.2
Edinburgh & Dalkeith	n.a.	563	—	5,811	511	8.8	5,637	280	5.0	5,511	298	5.4
Edinburgh & Dalkeith, Leith branch	n.a.	68	—	878	66	7.5	905	45	5.0	846	43	5.1
Edinburgh & Glasgow	—	—	—	73,556	4,941	6.7	76,051	3,798	5.0	80,933	4,043	5.0
Edinburgh Leith & Granton	—	—	—	—	—	—	1,211	70	5.8	1,410	71	5.0
Garnkirk & Glasgow	4,523	603	13.3	3,658	343	9.4	4,549	218	4.8	6,492	319	4.9
Glasgow Paisley & Greenock[a]	n.a.	3,416	—	29,202	4,059	13.9	24,848	1,541	6.2	24,887	1,426	5.7
Glasgow Paisley Kilmarnock & Ayr[a]	39,571	4,348	11.0	37,307	4,004	10.7	41,697	2,059	4.9	47,132	2,448	5.2
Kilmarnock & Troon	n.a.	36	—	n.a.	31	—	n.a.	19	—	n.a.	21	—
Monkland & Kirkintilloch	n.a.	90	—	n.a.	43	—	n.a.	6	—	n.a.	8	—
Newtyle & Coupar Angus	n.a.	57	—	n.a.	43	—	494	25	5.1	492	24	4.9
Newtyle & Glammis	n.a.	56	—	n.a.	35	—	n.a.	17	—	n.a.	18	—
Paisley & Renfrew	n.a.	144	—	n.a.	67	—	n.a.	34	—	n.a.	37	—
Slamannan	n.a.	257	—	n.a.	77	—	644	24	3.7	632	24	3.8
Wishaw & Coltness	n.a.	22	—	n.a.	17	—	1,162	50	4.3	2,343	104	4.4
Total		11,655			15,725			9,018			9,750	

a: including share of Glasgow & Paisley joint line

Sources: as for Tables 36 to 39, 61.

was advised by the Board of Trade that this might be open to legal challenge — however, a simple notice issued by the company might be expected to achieve the desired result.[361]

At first the advent of inter-urban lines seemed unlikely to break the pattern. Sabbatarian feeling was strong in the boardrooms of both the Greenock and the Ayrshire. Greenock shareholders supported Sunday closure on grounds of both religious observance and necessary rest for the workforce, and in 1847 some 3,600 residents of the town were to sign a memorial stating that the closure had caused them no inconvenience.[362] The Ayrshire directors did not even consult their shareholders, but pro-trains sentiment was not strong enough to force debate at a shareholders' meeting.[363] Thus the decision of the Edinburgh & Glasgow board to institute two Sunday trains in each direction from the opening of the line in 1842 was indeed 'a great and most startling *innovation*'.[364] For most of 1841 the board postponed a final decision on the question, while they were bombarded with memorials and petitions from kirk sessions and sabbatarian groups. When they finally approved the service by ten votes to eight, John Leadbetter resigned his chairmanship in protest. His successor was John Learmonth, even although he too had voted against Sunday trains.[365]

Given that the majority of the board did not have sabbatarian views, their decision to run Sunday trains was taken on purely commercial grounds. One of the arguments most often advanced against Sunday services was the effect on the workforce, both by overworking employees and by forcing them to work in spite of any religious reservations they might have. It was claimed that Sunday work would discourage respectable men from taking up railway employment, and, quite reasonably, that the railwayman's normal working week of twelve to fourteen hours on each of six days was as long as was compatible with health and efficiency. Agnew even asked rhetorically 'whether the numerous railway accidents in England are attributable to the denial of Sabbath rest to the people employed'.[366] But since the mails had to be conveyed in any case, and some company servants had therefore to carry out Sunday work, the majority of the board could see no reason why the mail trains should not also take passengers.

Reactions were predictable. The anti-sabbatarian *Railway Times* optimistically claimed that 'the anti-Sunday travelling agitation may be looked on as at an end. It is high time that it were'.[367] The Edinburgh Police Commissioners inquired of their English colleagues what results might be expected, and were reassured from Manchester that there had been no sabbath desecration, no increase in drunkenness, no rioting and no complaints from residents: 'the more facilities there are offered to the labouring classes for innocent recreation, the fewer cases of crime are brought before the police'. On the other side, some Glasgow clergy threatened users of the trains with excommunication, and the very first passengers from Glasgow were welcomed at Haymarket station by a hellfire sermon on the platform.[368]

More considered sabbatarian reaction was of two kinds. One was a partial boycott of the railway. The supporters of the newly formed Committee for Opposing Sabbath Traffic on the Edinburgh & Glasgow Railway agreed not only to avoid the sacrilegious Sunday trains, but also to give preference to rival methods of weekday travel. The committee spent £112 on promoting two new road coaches between the two cities, but these were routed via Hamilton, nowhere near the railway, and depended on intermediate rather than inter-city traffic. Within a year the committee, having spent £461 against an income of £212, was appealing desperately for funds, while Agnew's debt-laden Lord's Day Society was unable to help.[369]

The other sabbatarian tactic was to attempt to persuade shareholders to reverse the decision. Agnew, displaying according to his biographer 'no small degree of moral courage', bought a few shares in the Edinburgh & Glasgow in order to qualify himself to speak and vote at company meetings, and was soon exhorting his supporters to do the same. The logic was simple:

> Railways have hitherto desecrated the Sabbath, not by the necessity of the law, but by a vicious administration on the part of the proprietors; and what they can do, they can undo.[370]

For the next five years the greater part of the company's statutory half-yearly meetings was taken up with prolonged theological disputation, to the point that where the *Railway Times* suggested with some justification that shareholders' supervision of the normal running of the company was being made impossible.[371] Sabbatarian leaders were however well aware that theological points alone were unlikely to persuade many waverers at an essentially commercial meeting. Humanitarian arguments, such as the need not to overwork staff, might make some converts, particularly when an accident caused by an exhausted driver or signalman could have severe financial consequences for the company. But it might also be possible to persuade shareholders that in Scotland, unlike England, public opposition to Sunday trains was such that they could not be made to pay. During the first few months after opening the average Sunday train carried 1,025 passengers who paid £74. It is not possible from surviving records to separate out the cost of the service and, given the state of railway accounting, it may not have been possible at the time, but it may be noted that an attempt by some shareholders to get a breakdown of Sunday train finances was vigorously opposed by the directors and rejected by a company meeting.[372] Certainly no attempt was made in the years before the mania to extend the service beyond the mail trains. Even although one of the arguments put forward for Sunday services was the need to provide the poor with access to the fresh air of the countryside, Sunday excursion trains would have been seen as inexcusable laxity even by Scots who were not strictly sabbatarian; a twice-daily train, including third-class carriages, was felt to be sufficient provision.

The Edinburgh & Glasgow board could feel confident that, although they might have to sit through interminable debates at shareholders' meetings, the final vote would support their policy. The crucial point was the massive dependence of the company on English capital. Over 60% of the shares were held south of the border, with over 40% in Lancashire; and in 1842 a meeting of the Lancashire shareholders voted overwhelmingly in favour of the trains.[373] English shareholders, not surprisingly, seldom travelled north to company meetings, and those who did tried to concentrate discussion on commercial matters. Most of the absentees gave their proxy votes to the directors to use as they pleased. At most meetings over 90% of the votes cast were proxies, and often over half were under the direct control of the board. As long as the directors retained the confidence of English shareholders in their general management of the company, they were highly unlikely to be defeated on the Sunday question.

Thus in the period before the mania the Edinburgh & Glasgow, the only company with a majority English shareholding, was the only one to offer a Sunday service, with the intriguing exception of the Ardrossan, which after the Disruption of 1843 offered trains

> for the accommodation of the adhering population in going to, and returning from, the public ministrations of clergymen belonging to the Free Protesting Church.[374]

Elsewhere sabbatarian principles, regard for Scottish tradition, or simply doubts about potential profitability held sway. After the mania the new main lines, mostly with a majority of English share capital, gradually introduced more Sunday services; it was ironic that the 1846 campaign by Lancashire shareholders in the Edinburgh & Glasgow to oust Learmonth's board for mismanagement was made successful by an alliance of convenience with the sabbatarians, the price of which was the ending of Sunday services.[375] For twenty years from 1846, the company which had pioneered the breach in the Scottish sabbath was one of the very few main lines not to offer Sunday passenger services.

XII. Relations with other transport

Railways had to take into account the extent and nature of the competition offered by other transport on land and water. Road competition was the easiest to overcome, since the railways had clear advantages in speed, cost and, for passengers, comfort; as the turnpike trustees found, a parallel railway made immediate and devastating inroads into their revenue. One commentator in 1841 explained:

> Travelling by railway is somewhat cheaper than other methods of travelling: it is nearly three times as expeditious; and from the greater space that can be appropriated to each passenger, certainly less fatiguing. In cheapness, in speed, and in comfort, it is preferable, at least, to any other public conveyance ... Where there is a railroad, other conveyances are comparatively deserted; so much so, that we often hear of the 'monopoly' of the railroads. This 'monopoly' ... is given by nothing but superiority: railroads have no legal privileges; there is no proscription of turnpike roads; no penalty for travelling by coach or with post-horses. There are still roads, coaches, horses and harness, and capital to set them in motion, but the voluntary choice of the public has led to the adoption of the new rather than the old methods of conveyance.[376]

Before an 1837 select committee, 'many post and coach masters ... stated the ruin which is daily approaching them, and their utter inability to compete'[377] — a situation which might be made worse if turnpike trustees endeavoured to recoup lost revenue by raising tolls. But, according to a report made to turnpike users in Midlothian,

> It is well known to every person accustomed to travel the turnpike roads, that the toll-duties exacted at the gates are heavy in no ordinary degree; but coach proprietors in particular, find them a burden not easily borne.

Specifically the tolls at Cramond 'cannot possibly be increased for even at present they are loudly and justly complained of'.[378] Turnpike trustees had to concentrate on trying to stop railway authorisation or, more profitably, on negotiating compensation. Some in their private capacities accepted the railway; at the foundation of the Edinburgh & Dalkeith,

> Col Wauchope . . . objected to having *both* the public Road and the Railway going through the Niddrie estate and as he was thus to make an election between the two he determined in favour of the Railway.[379]

As far as road transport operators were concerned, there was little they could do, especially after the failure of steam road vehicles, and none of the main lines radiating from Glasgow was ever seriously concerned about road competition.

For some smaller lines there was, or at least was claimed to be, more of a problem. The horse-drawn Edinburgh & Dalkeith always felt threatened by road carriers, while the Edinburgh Leith & Newhaven lost potential traffic to the more conveniently routed horse buses. In Angus road carriers caused the railways some concern. While the Dundee & Arbroath lacked a goods depot in Dundee it claimed to be unable to compete with the carriers, while, according to Lindsay Carnegie in 1840, the Arbroath & Forfar, which had originally set its rates at half those charged by the carriers, had been forced to lower them further when the carriers retaliated with cuts of their own. Even if the railway forced the carriers out of business, he suggested, any rise in rates would bring them back;[380] since by 1844 his company had by a considerable margin the highest rates in Scotland, this fear turned out to be unfounded. Most carriers found it more profitable to carry goods from the hinterland to railway stations, while stage coach proprietors could not realistically hope to compete with the trains on either speed or price. As already discussed, the other major road interest, the turnpike trustees, soon turned from attempting to block the authorisation of railways to extracting the maximum possible compensation, sometimes with great success.[381]

Canal competition was a much more serious matter, and before the mania at least canal companies had some confidence in their ability to see off the new competitor. In retrospect the balance of advantage had moved to the railways by the time of the lesser mania of the mid-1830s. Not only did the scale of railway enterprise increase, but proposals for new canals disappeared. The last canal building authorised in Scotland was for the half-mile Forth & Cart in 1836 and the Campsie branch of the Forth & Clyde in 1837, and only the former was built. The last great scheme was the Marquess of Breadalbane's plan in the early 1840s to link Loch Tay, Loch Lomond and the Clyde, which was succeeded during the mania by his more promising but ultimately unsuccessful Scottish Grand Junction Railway.[382]

At this time, in 1835, appeared a report of some interest. The two distinguished engineers who produced it, John Macneill and Neil Robson, were both to become notable railway builders, but their report to the projectors of a branch canal from the Forth & Clyde to Stirling argued strongly for the advantages of canals over railways. They made their priorities clear:

> The first consideration — to which all others are subordinate — is the return to Shareholders

or parties advancing their money for the execution of the improved communication;

The *second* consideration, is the comparative benefit of Railway and Canal communication to the landed proprietors, tenants, and residenters of the country through which the proposed undertaking is to proceed;

And *thirdly*, the general advantages to be derived by the country at large, from the proposed measure, are to be considered.[383]

A railway, it was accepted, would have a clear advantage in speed, taking 40 minutes for the proposed $11\frac{1}{2}$-mile route against 70 minutes by the improved canal boats being pioneered by William Houston between Glasgow and Johnstone. But all other advantages lay with a canal. For goods traffic,

it is acknowledged, that on the most improved railway there can be no competition, so far as regards the conveyance of heavy articles, with a Canal route . . . The most ardent admirers of improved Railway communication, at once give up the palm of merit as regards the carriage or transit of heavy articles, to Canal conveyance.

Even for light goods and passengers, they argued, cheaper fares and greater comfort would outweigh reduced speed. Fares on the Liverpool & Manchester (still the universally quoted standard railway), at an average of over 2d per mile, compared badly with the Paisley canal fares of $\frac{3}{4}$d per mile in the first class and $\frac{1}{2}$d per mile in the second. The continuing uncertainties of locomotive traction are indicated by the fact that they could claim that

The accommodation afforded by a railway, is almost exclusively confined to the towns situated at its extremities. The difficulty and the loss of power and time occasioned by any stopping of the locomotive engines, and then recovering their speed, prevents any stoppage except for taking in coal and water.

As a final point, the cost per mile of constructing the canal was estimated to be only 20% that of the Liverpool & Manchester. At an estimated cost of £100,000, an annual gross revenue of £11,125, and a mere $11\frac{1}{4}$% allowed for working expenditure, they anticipated a return to shareholders of about 10%.[384]

It was the last blast of the canal-builders' trumpet, and future construction in Scotland was to go into railways. But the existing canals in central Scotland still anticipated holding off the railway attack on their traffic, and the early railway projectors suspected that they might be right. The Edinburgh & Glasgow prospectus of October 1835 calculated its traffic estimates by comparison with existing road services only; it was assumed that heavy goods would stay with the canal, and that much of the light goods and passenger traffic might do so as well:

It is probable that a small portion of it might flow into this new channel, as the rapidity of the conveyance would . . . partly make up for the greater cheapness of water transport . . . There is *enough of trade for both*.[385]

By the time John Leadbetter appeared before the House of Lords committee on the bill in 1838 he was more optimistic, but still had no expectation of driving the canal out of business.[386] The 1840 Select Committee considered that

As far as regards the heavy merchandize, it appears probable that the canals will always secure the public against any unreasonable demands on the part of the Railway Companies.[387]

Appearing before it, Charles King of the Garnkirk believed that the Monkland Canal would retain passenger traffic as, although slow, it charged a penny less than the railway and offered respectable travellers better accommodation 'as they are separated from the lower class'.[388]

The Glasgow Paisley & Johnstone Canal was under severe threat from the joint line of the Ayrshire and Greenock companies, whose authorisation it vigorously opposed;[389] it had built up a volume of traffic which was worth contending for, and which was steadily increasing (see Table 73). The passenger service — perhaps the most successful on any British canal — had been assiduously cultivated, with fast boats running from Glasgow to Paisley in 55 minutes (which, according to the canal manager, could be reduced further), and covered accommodation for all passengers. The inconvenient location of the basin at Port Eglinton was partially offset by a company omnibus into the centre of Glasgow, which at a fare of 2d was used by only one passenger in six, and made an annual loss of £130. The fare reductions of early 1835, forced by the appearance of a new rival road coach company, had kept passenger numbers climbing, even if at the cost of a fall in revenue for a year.[390]

The opening of the railways led to fierce price competition. Between Glasgow and Paisley railway tactics were made more difficult because they had to be agreed by the two companies involved. On the other hand, failures could be blamed on the other partner. As chairman James McCall told Ayrshire shareholders in August 1841:

> The Company ... had been for some time past put to very considerable expense, arising partly from their preparations for the goods traffic, and partly from running, at the request of the Joint Committee, special trains between Glasgow and Paisley. These special trains had been put on the line as a means of counteracting the successful competition of the canal, arising from the low fares charged by that mode of conveyance, but the plan, which had been chiefly promoted by the Greenock Railway Company, had not succeeded, and had caused a great additional expense, without any corresponding advantage.[391]

The canal proprietors were charging fares almost 50% below the railway, and 'have since carried almost as many passengers as they ever did'.[392] After some dispute between the two railways, third-class fares were cut to match the canal; Huish, for the Greenock, estimated that each passenger brought in 0.57d per mile but created expenses of 0.61d, and would have preferred to challenge the canal's goods traffic by a 20-30% rates cut in that section.[393] Competition increased, to the delight of travellers but the distress of both canal and railway management, until in late 1842 both sides were carrying passengers between Glasgow and Paisley for 2d. Peace was made in 1843; the canal agreed effectively to end its passenger and parcels services from the start of July in return for an annual payment of £1,367, and the relieved railways doubled the third-class fare to 4d.[394] Goods traffic, apart from parcels, was not formally covered by the agreement but appears to have been left to the canal; before July 1843 it accounted for less than a tenth of the joint line's revenue, but thereafter it disappears from the accounts altogether.[395]

On the short distance between Glasgow and Paisley, the extra time taken by canal passengers was not of great significance. Between Edinburgh and Glasgow the time advantage was more important for all but the poorest. Even so, the two canals — the Forth & Clyde and the Union — were prepared to fight. By 1843 canal cabin fares of 3/- and steerage at 2/- between the cities had induced the railway to start a fourth class at 2/6d; in May 1844 the canal steerage fares went down to 1/4d by day or 1/2d by night, with further reductions in July.

Competition for goods meant that by February 1843 most items were making the inter-city journey for 2/4d per ton by rail or 2/- by canal.[396] The canal's price advantage meant that only about a quarter of the railway's revenue came from goods, and much of this was for light goods and parcels where speed might be important. Among the incidental victims of the competition were the Edinburgh & Dalkeith and the coalowners of Midlothian, who found themselves undercut in the Edinburgh market.[397] Edinburgh & Glasgow shareholders complained vigorously about the 'insane competition between it and the canal companies', while 'the revenue of the Union Canal Company was literally annihilated' and the railway dividend was

> only maintained by running down their plant, and their Dividend also must have been paid out of Capital or materially reduced.[398]

Table 73. *Traffic on the Glasgow Paisley & Johnstone Canal, 1830–36*

Year	passengers		Fares (d) Glasgow-Paisley		No. of boats p.week each way	Goods	
	No.	% increase	Cabin	Steerage		Tons	% increase
1830–31	79,455		9	6	24	48,191	
1831–32	148,516	86.9	9	6	42	51,193	6.2
1832–33	240,062	61.6	9	6	54	53,194	3.9
1833–34	307,275	28.0	9	6	72	57,853	8.8
1834–35	373,290	21.4	6	4	72	60,510	4.6
1835–36	423,186	13.4	6	4	n.a.	67,305	11.2

Note: years run from 1 October.

Sources: More Nisbett papers, bundle 13, *Prospectus of a Canal intended to connect the Town of Stirling with the Cities of Edinburgh & Glasgow, and the Towns of Ayr, Greenock and Port Glasgow;* PP HL 1837 (146) XVIII, *House of Lords Committee on the Glasgow, Paisley, Kilmarnock & Ayr Railway Bill,* evidence, pp.22–3 (Robert Farquharson); C. Hadfield, *British Canals: an Illustrated History,* 168; *NSA* VII (Renfrewshire), 279 (Paisley).

The canal efforts were sufficiently effective for them to secure such favourable terms for a proposed merger in 1846 that the railway shareholders rebelled and threw out Learmonth's board. The new directors concluded an agreement which ended four years of warfare by giving all high-value and light goods to the railway, while the canal was to have a price advantage on bulk goods of 1/8d per ton between the cities. The effect was that the rate for many goods, which had been 15/- per ton when the canal had a virtual monopoly before 1842 and had fallen as low as 2/-, settled at 6/8d by canal or 8/4d by rail.[399]

To some extent, John Leadbetter's forecast had been fulfilled — there was enough traffic for all, or at least neither side had been able to drive the other out. For the canals passenger traffic was a losing battle, particularly as the provision of cheap trains increased and carriages became more comfortable, and the Paisley canal may well have been wise to get out while it could still extract a reasonable annual sum for so doing. But, as the Monkland Canal and its rival railways had discovered earlier, neither side could establish a monopoly in goods. For conveyors of heavy bulk goods who had easy access to canal facilities, they were undeniably cheaper, and their relative slowness might not matter. Before the mania the canals also had the important access to the Clyde which was denied to any rail user east of Glasgow, and that to the Forth where the only rail connection was with the Leith branch of the Edinburgh & Dalkeith. After the mania the canals were gradually crushed by the increasing power of the expanded railway system. It was a slow process, but by 1870 all the canals in central Scotland had sold out to their conquerors, and had become the property of railway companies.[400]

The Greenock and Ayrshire companies had to concern themselves with other water transport as well as with the canal. For both, the Clyde boats offered both competition for passengers to the resorts on the Ayrshire and Renfrewshire coasts, and possible co-operation in conveying traffic to Arran, Bute and the far side of the Firth, or even to western England, Ireland, or points beyond. The threat was greater to the Greenock line, since the curve of the coast meant that the railway was much the shorter route to Ayr or Ardrossan; either might profit from co-operation, depending on whether and where boats could be persuaded to exchange traffic with the railway rather than undertaking the tedious river journey to Glasgow.

Before the arrival of the railway, the Greenock boats had already proved their efficiency by driving all the previous twenty-four stage-coaches off the road, and were carrying some 750,000 passengers per year between Greenock and either Glasgow or Port Glasgow. In the railway's first year, it took 427,548 passengers at a minimum fare of 1/6d for $22\frac{1}{2}$ miles, against 1/- on the boat.[401] By 1844, when the alteration of the passenger tax had allowed the introduction of a third class at 1/- (met by a boat fare of 8d), the company secretary claimed that 611,000 passengers had used the train against 312,000 on the boat in the previous year. Since the train took one hour against $2\frac{1}{2}$ on the river, the overall preference appeared to be for speed rather than economy.[402] The railway board noted that

> This railway, from its first projection, has always been regarded more as a speedy means of transit for the inhabitants of Glasgow residing at the watering-places in the summer, than in any other light; and by the railway route they gain at least $2\frac{1}{2}$ hours every visit.[403]

When the boats originally cut their fares to meet the railway threat, the company believed that this had to be a short-term measure; by 1843 they accepted that the competition of 'the most perfectly organised [boats] in Europe' was presenting them with an enduring problem.[404] Although the Greenock line retained a large passenger traffic, it had to be done by charging extremely low fares and accepting a level of return which would not have satisfied better-placed companies.

The hope of the Greenock railway, expressed with some confidence before opening, was that Firth of Clyde boats would agree to operate in co-operation with the company:

> By starting steamers from Greenock on the arrival of each train to the numerous ports to which they now run, a great saving of time will be effected ... In this union of purpose, the Board is led to believe the many if not all, the steam companies ... will cordially join.[405]

The established steamer companies, however, saw no reason to surrender any part of their traffic, and even when they called in at Greenock they consciously avoided making connections with the trains. Nor was the location of Greenock station, away from the pier, an encouragement to passengers. By 1841 the railway had decided to charter vessels and in effect to guarantee their owners against loss. To encourage passengers, the waterborne part of their trip was heavily subsidised – at a time when a first-class passenger paid 2/6d from Glasgow to Greenock, he could go on to Helensburgh for no extra charge or to Rothesay for 3d: the company, however, had agreed to pay the boat-owners 1/– for every first-class passenger. In 1842 some directors and shareholders, acting as individuals, went further and took over an existing steamboat company to create the 'Railway Steampacket Company', thus guaranteeing the railway some connecting boat services.[406]

The results were unsatisfactory. Company chairman Ker described the move as a defensive expedient, to be ended if agreement could be reached with independent boat operators.[407] They, on the other hand, suspected the railway of empire-building. Gabriel Lang, in this case a knowledgeable and comparatively unbiased witness, observed that the Clyde steamers were

> already diminished to less than half their usual number by Railway competition, and it is certain will be further reduced next summer, when Third class carriages are introduced on the Greenock Railway.[408]

An appeal to the Railway Department to stop the company 'indirectly becoming steam-boat proprietors ... and acquiring a monopoly of the whole passenger traffic on the Clyde' was rejected as nothing illegal was being done, and the Department approved of the fact that 'a certain class of passengers were conveyed ... at lower fares than had been charged before the opening of the railway'.[409] Although a few small operators now agreed to co-operate, most did not. While a price war between boats on the Glasgow to Largs run had cut the fare to 3d, no company would take passengers from Greenock to Largs for the same price.[410] The large Castle Steamboat Company offered only a connection with its oldest boat, a near-derelict capable of a mere six miles per hour. Early in 1844 the Helensburgh boats broke off their previous agreement.[411] Meanwhile shareholders were restive, accusing the directors of trying to increase the number of passengers without considering whether they were profitable. The railway company was in an almost impossible position. Fares were too low for satisfactory profitability, yet if they were raised, too many passengers might be driven back to the boats. Further cuts in the hope of attracting a proportionately greater increase of trade might again be met by counter-cuts on the boats. For traffic beyond Greenock, it was doubtful whether any conceivable fare level could make up for the change from train to boat and for the limited choice of boats available, particularly when the independent steamer companies were investing in faster vessels. In a last effort, the Railway Steampacket Company introduced faster boats of its own in 1844, but its operation closed down in 1847 as the railway company moved into a merger with the Caledonian.[412]

The Ayrshire company hoped for much from boat connections. It was a reasonable expectation that boats from Lancashire or Ireland might be happy to call at Ardrossan if a good rail connection to Glasgow was available, rather than facing a long trip up the overcrowded Clyde. Even before opening, the board reported that a company had been formed to run steamers between Troon and Liverpool, another was under consideration for the Ardrossan-Belfast run, and a third proposed to connect one or other harbour to the Western Isles. 'There is little doubt,' they concluded, 'that Ardrossan will be one of the most important Packet Stations in the west of Scotland.'[413] In fact the service provided by the steamer *Fire King* when the railway opened was from Ardrossan to Liverpool and, like that at Greenock, was organised by railway directors and shareholders acting individually.[414] Within three months, however, the boat had been sold to the private firm of J. & G. Burns; recriminations about the extent to which directors' mismanagement was involved rumbled on for another three years.[415] Although Burns continued the service, and another boat ran to Belfast with financial help from the railway, the directors had again to account for disappointing results:

> It is at all times difficult to induce the public to adopt new lines of communication; and the circumstances of there being only two departures in the week by the Fire King, and three departures in the week by the Ayrshire Lassie [to Belfast], and of its being difficult for travellers at a distance to remember or ascertain when these took place, operated as a disadvantage in the competition with the Glasgow and Clyde boats.[416]

The unsatisfactory service was allowed to fold.

The company still believe that an efficient service which could capture a large part of the traffic between Glasgow and Lancashire was practicable. In 1843, with the encouragement of a memorial 'numerously signed by Shareholders', they issued a circular:

> It is obvious, that neither the Board of Directors nor the company as such, possess the power of providing this Steamer; but the Proprietors individually may lend their assistance, and the Board highly approve of the attempt being made to procure Subscribers to a Steam-Vessel of proper size and dimensions, and also of the plan of allowing 1s. of each Dividend to form a Guarantee Fund, without which, it is apprehended, it may be difficult, if not impossible, to effect the formation of a Steam-Boat Company.

The idea had come from the Preston & Wyre Railway, which hoped to be the other beneficiary of a service from Ardrossan to Fleetwood, and whose shareholders had already agreed to a guaranteed fund.[417] It was expected that the boat journey would take twelve hours as against twenty-four from Glasgow to Liverpool, as well as avoiding 'the enormous port charges at Liverpool'.[418] The two groups eventually bought a boat each, with more than 80% of Ayrshire shareholders subscribing; the dissentients included chairman James McCall, who thought that the lesson of the *Fire King* should have made more impact, and that the Ayrshire should concentrate on achieving the Nithsdale link to Carlisle and the English railways.[419] By the summer of 1844 two boats were running to Fleetwood, one to Belfast, and one calling at the coastal ports to Stranraer; the Post Office, however, had refused to transfer the Portpatrick mail service to Ardrossan. In the following year a second Belfast boat, allowing a daily service, and a new one to Londonderry were added.[420] All but the Fleetwood boats extracted generous concessions from the railway company before they agreed to co-operate. Meanwhile a Select Committee concluded that the railway companies were exceeding their powers, and that their ownership of steamboats or coaches 'is obviously improper'.[421]

Relations between the two Clyde coast railway companies and the steamship lines generally worked in favour of the boats. Between Glasgow and Greenock, in a situation of direct competition, the lower overheads and greater versatility of the boats enabled them to force fares down to a level which any other railway would have regarded as intolerable. Beyond Greenock and Ardrossan it was more complicated. The railway idea, that guaranteed boat connections would create substantial extra traffic and might indeed work to the benefit of both trains and boats, was fine in principle. But the Greenock company found that independent operators were unwilling to surrender the Glasgow to Greenock part of their run unless the terms offered by the railway ensured tham a substantial profit. Hence although the railway charged minimal sums (or even nothing at all) to have them taken on to Helensburgh, Dunoon or Rothesay, up to 45% of the total fare might have to be given to boat proprietors.[422] In the summer, when the vast seasonal demand allowed small independent operators to run from Glasgow to Rothesay for a fare of 6d, no conceivable combination of rail and boat could compete. In 1844 it was stated that Greenock shareholders had foregone eighteen months' dividend to pay for the Railway Steampacket Company, financing not only its capital costs but also working losses which were, for instance, costing £40 per week on the Rothesay route.[423]

The guarantee extracted by steamer operators from the Ayrshire had to compensate for the risks of opening up new routes, and had to be high enough to offset any profits which might reasonably be expected from running on a different route. Boats running along the coast southwards from Ayr extracted commissions of 25–33%, while in 1843 the weekly cost of running the railway's own boat to Fleetwood was estimated at £163–10–0.[424] By 1846 the company claimed, though in unspecific terms, that the boats were doing well.[425] While some initial subsidising of Irish boats might well have been worthwhile to divert as much as possible of a permanent sea traffic on to the railway, the Fleetwood service was little more than a troublesome distraction. It created a crucial division of interest in the company at a time when full attention should have been concentrated on promoting the rail link by Nithsdale to England.

5
The Battle for the Border

I. English trade and border topography

THE railways constructed before the mania fell, almost without exception, into two groups. There was first the long tradition, stretching back to the waggonways, of short lines promoted by coal and ironmasters, with some help from potential consumers, for the movement of minerals. In the later 1830s these had been joined by inter-urban links, promoted primarily by merchants in the central belt and by landowners in Angus, and marking a move from the railway as an adjunct to other entrepreneurial interests of the promoters to a primary area of investment and profit in its own right. But still no company had a main line of more than fifty miles, and only two (the Edinburgh & Glasgow at £900,000 and the Ayrshire at £625,000) had an initial share capital of over £500,000. Only the former, indeed, thanks to the importance of the terminal towns and their related harbours, could be compared in importance to the great lines built at the same time in England. Alongside these two types, however, there existed plans for more ambitious schemes, at least in terms of the distance to be covered. Long-distance projects had been pioneered in Scotland by Stevenson and by Telford's proposal to link Glasgow and Berwick by waggonway. The later hopes of extending the system to Aberdeen and Inverness might also be regarded as extending the conceptual scale of the railway, even if very little in practice was done to pursue them. But the long-distance projects on which most attention was focused during the decade between the lesser and greater manias were those to connect the railway networks of Scotland and England.

Not surprisingly, interest in a trans-border railway tended to increase with proximity to the area. By 1838 London, for instance, had or had authorised rail links as far afield as Exeter, Lancaster and Newcastle: trade with Scotland was a relatively limited part of the capital's total concerns, and it might well seem adequate to communicate with eastern Scotland by sea, or with Glasgow by rail and 'the splendid steamships which run to Liverpool in sixteen or twenty hours from Greenock'.[1] For the many who believed that imminent improvements in steamships would soon make them as fast as the railways, there was almost no incentive to change. But in the other great centre of English railway investment,

Lancashire, trade to the north was more significant, while in Scotland itself the crucial economic links were with the south. It was inconvenient, not to say costly, that all trade between Glasgow and Newcastle, or Edinburgh and Lancashire, had to be transshipped either between sea and rail or between sea and canal. Even between Glasgow and Lancashire, where many Glasgow merchants (and for a time the Ayrshire Railway Company itself) had interests in the steamships,[2] a railway would not merely allow greater speed and, for passengers, greater comfort, but it would introduce a vital element of competition and perhaps force a reduction in the rates charged by the shippers. Thus even in the depths of recession around 1840 planning went ahead on both sides of the border; Grinling hardly exaggerates when he says that in 1839 'railway projection elsewhere was now entirely at a standstill' — only twenty minor projects for railway construction were authorised in the next four years in the whole United Kingdom — and he is right to point to the contrasting enthusiasm of border planners.[3] Although the recession might inhibit any immediate attempts to raise funds, the proven capacities of the railway as exemplified by the 1830s trunk lines kept the projectors active against a time of economic recovery.

The main problems of the border were geographical. The topography of the area permitted four serious routes for a trunk railway (see Map 15). From Newcastle an east coast line could be built by Berwick to Edinburgh. From Carlisle on the west a line might go up Nithsdale to join the Ayrshire line at either Kilmarnock or Ayr; another might with some difficulty use Eskdale or Liddesdale to reach Hawick, and then cross the Pentlands to Edinburgh; or a third might go by Annandale and Clydesdale to Glasgow, in this case having the extra advantage of an easily added Edinburgh branch from somewhere near Lanark. In any of the three the hills south of Carlisle formed an additional difficulty. The barrier of the Cheviots effectively prevented an intermediate crossing of the border by a trunk line, though this did not prevent the appearance of a series of plans for winding up Northumberland valleys and tunnelling for miles under the mountains. These were supported and in some cases sponsored by the Newcastle & Carlisle, which would gain much through traffic if a line went northwards from Hexham or Gillsland, but would lose out altogether if railways were built from both Newcastle and Carlisle. But the hills meant that a single central trunk line was impracticable — the only line ever to cross the central Cheviots, the Border Counties of 1862,[4] could be seen as part of North British manoeuvres against the North-Eastern or just possibly as a desirable local service, but never as a fast or profitable route for through traffic. It was impossible to provide a single line which would be equally satisfactory to Glasgow and Edinburgh, Lancashire and Tyneside.

On the other hand it was generally believed that there was only enough traffic for one line. Joseph Locke put it bluntly to the board of the Grand Junction in 1836: 'Two great lines from Scotland to England cannot pay'.[5] Two years later Stockton & Darlington pioneer Edward Pease used almost the same phrase in the House of Commons: 'the very nature of the country pointed out that not more than one line could by possibility pay'.[6] It was believed that, since the border country and the Southern Uplands of Scotland promised to generate little traffic, all the traffic between the industrial areas north and south of the border would have to be brought on to a single line in order to achieve profitability. Intermediate traffic was indeed unpromising. The east-coast route had some hopes in Berwickshire and the agricultural Lothians, but on the other hand offered an exceedingly circuitous route between Glasgow and Lancashire, the two main sources of long-distance traffic. Similarly, the Nithsdale route in the west would pick up traffic in Ayrshire and Dumfries, but was unattractive to Edinburgh and eastern England. A Hawick line would serve the towns of the central Borders, and convey coal from the Duke of Buccleuch's Canonbie colliery and possibly from the field at Plashetts, just over the border, but offered a devious route to Glasgow as well as difficult engineering problems. Overall the Annandale route had the advantage of serving all destinations but Tyneside, but it traversed an even more deserted area than the others and was faced with the barrier of Beattock summit.

A connection from Edinburgh and/or Glasgow to England had figured in the plans of enthusiasts from the earliest days of the locomotive, with little detailed attention being given to problems of routes and traffic. Thomas Gray's 'General Iron Railway' of 1820 was to extend to the two Scottish cities, though points further north were disregarded while England and Ireland were to be given comprehensive systems.[7] By 1825 Charles Maclaren, never a man to restrain his enthusiasm, was proposing a complete Scottish network, with two connections across the border:

> One of the great roads from England must obviously enter by the east coast, and the other by Carlisle. We may suppose the former continued through the middle of East Lothian to Edinburgh; thence through Linlithgowshire to Stirling, after sending off a branch to Glasgow; from Stirling along the Allan and the Earn to Perth; and thence by the coast to Dundee, Montrose and Aberdeen. From Perth, however, a branch scarcely less important than the main trunk might be carried through the fertile valley of Strathmore in the line marked out by Mr. Stevenson. The course of the Tweed will offer a separate line into Midlothian, and perhaps into Lanarkshire. The most advantageous line for the western main trunk would probably be from Carlisle to Dumfries; thence along the valley of the Nith into Ayrshire . . . and by Ayr, Irvine and Paisley, to Glasgow. A branch line would, of course, run along the coast to Wigtown and Stranraer. It is not improbable, that by rails of a cheaper kind . . . the benefits of this improvement may even be extended to the Highlands . . .[8]

It is notable how closely Maclaren predicted the main structure of the Scottish system, apart from the Caledonian, as it was created by 1850, although the Stranraer line had to wait until 1861, and the Tweed valley, thanks to the defeat of the Caledonian Extension project during the mania, never became a major artery for through traffic.

Apart from these border crossings as parts of grand paper projects, the 1820s witnessed some more specific plans. In 1823 Sir John Sinclair solicited the views of Robert Stevenson on an Edinburgh-London link, and was told that 'an iron railway would not only be much more practicable, but more commodious and useful for general intercourse than a canal', although the engineer did not commit himself on the financial prospects.[9] In 1824 William Bell's railway plan to link London and Edinburgh via his native Newcastle was dismissed by the *Tyne Mercury* as 'of all the irrational schemes that have yet been broached, surely the most absurd'.[10] This description applied better to another Edinburgh-London project of the same year, which was apparently designed by drawing a straight line between the two cities on a contourless map, and claiming that a combination of locomotive and stationary engines would provide an efficient service.[11] None of the schemes got beyond the vaguest planning stage.

These proposals came out of the speculative atmosphere of the mid-1820s; by the next similar period, a decade later, long-distance railways were under construction in England, and it could not be held that an Anglo-Scottish line was technologically impossible, even if it might legitimately be doubted whether it would pay. Table 74 summarises the schemes projected for the border, the most promising of which became the combatants in a series of parliamentary battles before and during the mania. Three of the fifteen (nos. 2, 4 and 5) represent the three most straightforward routes, by the east coast, Annandale and Nithsdale, where distinguished engineers with powerful support were in action from the mid-1830s. It is apparently surprising that no proposal was made for the Carlisle-Hawick-Edinburgh route until it was taken up by the North British in 1844; even then Parliament was in the event only asked to approve the Edinburgh to Hawick section. In fact rivalry between the North British and the Caledonian, coupled with the wavering sympathies of the Duke of Buccleuch (who owned much of the route), prevented a line between Hawick and Carlisle until 1862.[12]

Most of the remaining schemes represent attempts by Newcastle interests, often with support from the border towns around Hawick and Galashiels, to find a way across the Cheviots. For Newcastle there was a real fear that any of the Carlisle lines would result in the main trade arteries of the country keeping to the west: in 1839 the mayor of the city observed that Blackmore's line offered the only hope of Newcastle remaining on the direct route between London and Glasgow, and the town council agreed to share the costs of the survey with the Newcastle & Carlisle company.[13] Blackmore produced perhaps the most hopeful of the central lines, with connections to both Glasgow and Edinburgh, an individual route across the Cheviots which avoided any need for the lengthy tunnels of the Carter Bar proposals, and, according to the engineer, copious supplies of coal, ironstone, lead and freestone waiting to be developed.[14] The Roxburgh Commissioners of Supply and a meeting of Midlothian landowners both formally supported the line, although the former apparently had doubts about Blackmore himself, since they wanted the government to appoint an engineer.[15] The other reasonably serious central contender was the line via Wooler and Kelso surveyed by George Remington, in 1836 for John Rennie and in 1838 on his own account.[16] But these lines depend on assumptions about the importance of Newcastle, and interests in the great commercial areas of London, Lancashire and Glasgow did not deem Newcastle of sufficient importance to give preference to a devious and expensive line with enormous engineering difficulties. The next survey commissioned by the Newcastle & Carlisle from Nicholas Wood and George Johnson in 1843 abandoned the city of Newcastle in favour of a line to Edinburgh via Hawick from Gillsland, twenty miles east of Carlisle.[17] This line, promoted at the beginning of the mania as the Central Union, still claimed to be an all-purpose Anglo-Scottish

Table 74. Border Railway Projects, 1833–45

Railway[a]	Date of Plan or Prospectus	Route	Engineer	Promoters/ Supporters
1 'Newcastle, Galashiels, Edinburgh & Glasgow'	1833	Newcastle–Carter Bar–Melrose–Galashiels–Edinburgh. Branch: Melrose–Peebles–Lanark–Glasgow		Glasgow merchants et al
2a 'Nithsdale'	1835	Carlisle–Dumfries–Cumnock–Kilmarnock or Ayr–Glasgow	Joseph Locke, later Grainger and Miller John Miller	Grand Junction Ry, later Glasgow Paisley Kilmarnock & Ayr Ry.
b Glasgow Dumfries & Carlisle*	1844	Carlisle–Dumfries–Cumnock–Kilmarnock (Glasgow Paisley Kilmarnock & Ayr)		
3 Great Anglo-Caledonian	11.1835	Leeds–Newcastle–Gillsland–Langholm–Moffat–Glasgow		Support from Newcastle & Carlisle Ry.
4a Edinburgh Haddington & Dunbar	3.1836	Dunbar–Haddington–Edinburgh. Branches to Edinburgh markets, Portobello, Dunbar harbour	Grainger & Miller + George Stephenson (consulting)	Edinburgh committee, chairman John Learmonth, secretary Charles Davidson. Support from George Hudson, Edinburgh & Glasgow Ry.
b. Great North British, or Edinburgh Dunbar Berwick & Newcastle	9.1838	Newcastle–Alnmouth–Berwick–Dunbar–Edinburgh		
c. North British	4.1842	Dunbar–Edinburgh	John Miller	
d. North British*	8.1842	Berwick–Dunbar–Edinburgh	John Miller	
5a 'Annandale'	1836	Carlisle–Annan–Beattock–Lanark–Glasgow. Branch: Lanark–Edinburgh	Joseph Locke	Committee under J. J. Hope Johnstone, later also Grand Junction Ry et al.
b Caledonian*	2.1844	Carlisle–Annandale–Beattock–Carstairs–Glasgow. Branches: Carstairs–Edinburgh, Coatbridge–Castlecary	Joseph Locke	

The Battle for the Border

6a	Tyne Edinburgh & Glasgow or Tyne & Edinburgh	6.1836	Hexham–Redesdale–Carter Bar–Melrose–Galashiels–Edinburgh. Branch: Galashiels–Glasgow		Stephen Reed, Newcastle
b	'Hexham & Edinburgh'	12.1838 and 1843	Hexham–Kielder–Note o' th'Gate–Melrose–Galashiels–Edinburgh. Poss. branch: Abbotsford–Biggar–Glasgow		Newcastle & Carlisle Ry and Newcastle Town Council
7	Grand Eastern Union	7.1836	Newcastle–Morpeth–Berwick–Dunbar	Stephen Reed	
8	'Newcastle Kelso & Edinburgh'	7.1836	Newcastle–Morpeth–Wooler Kelso–Lauder–Edinburgh	John Blackmore, based on Reed	
9	Newcastle-upon-Tyne Edinburgh & Glasgow	9.1838	Newcastle–Morpeth–Wooler–Kelso–Galashiels–Edinburgh	Matthias Dunn, Robert Hawthorne & John Dobson, of Newcastle	
		12.1836	Newcastle–Redesdale–Carter Bar–Jedburgh–Peebles–Edinburgh. Branches to Glasgow, Hawick, Selkirk Kelso	George Remington for John Rennie George Remington	
10	Central Union, or United Central	11.1843	Gillsland–Liddesdale–Hawick–Galashiels–Edinburgh	Joshua Richardson	
11	'Edinburgh Hawick & Carlisle' later Edinburgh & Hawick	12.1844	Carlisle–Langholm–Hawick–Galashiels–Edinburgh	Nicholas Wood & George Johnson John Miller	Newcastle & Carlisle Ry North British Ry. Hawick–Carlisle section withdrawn before submission to Parliament
12	Newcastle-upon-Tyne Hawick Edinburgh & Glasgow Jn, or Newcastle Edinburgh & Direct Glasgow Jn.	10.1845	Newcastle–Belsay–Redesdale–Carter Bar–Hawick–(Edin & Hawick, Cal Extension). Branches to Hexham, Bellingham	John & Benjamin Green + John Miller (consulting)	Landowners on route

13	London & Edinburgh Direct and Darlington & Hawick Jn.	10.1845	Darlington–Blanchland–Hexham–Hawick, by undefined route	
14	Direct Newcastle Edinburgh & Glasgow	11.1845	Newcastle–Jedburgh–Melrose–Edinburgh. Branches Melrose–Glasgow, and to Bellingham, Kelso, Hawick	W. B. Pritchard + Sir John Macneill & Jas. Thomson (consulting)	Provisional committee of 277 miscellaneous names. Secretary: Henry Lake

a: Company names in inverted commas were not officially attached to projects at the time, but are for ease of identification. 'Nithsdale' and 'Annandale' were in almost universal usage: other lines were often referred to by their engineers' names – Remington's, Blackmore's etc.

*: indicates companies authorised 1844–46

Sources: HLRO, HC deposited plans of various railways; Aberdeen University, OD.MISC.25, Notes by Prof. A. C. O'Dell, *Railway Times*; W. W. Tomlinson, *The North-Eastern Railway*; J. Locke, *Report on the Proposed Line of Railway from Carlisle to Glasgow and Edinburgh by Annandale*; G. Graham, *The Caledonian Railway*; N. Wood & G. Johnson, *Report on a Central Line of Railway into Scotland...*; SRO, LIB(S)/6/6/224 p.81, *Prospectus of Great North British Railway Company*; SRO, GD104/348, J. Blackmore, *Report to the Directors of the Newcastle-on-Tyne and Carlisle Railway*; J. S. MacLean, *The Newcastle and Carlisle Railway 1825–1862*.

The Battle for the Border

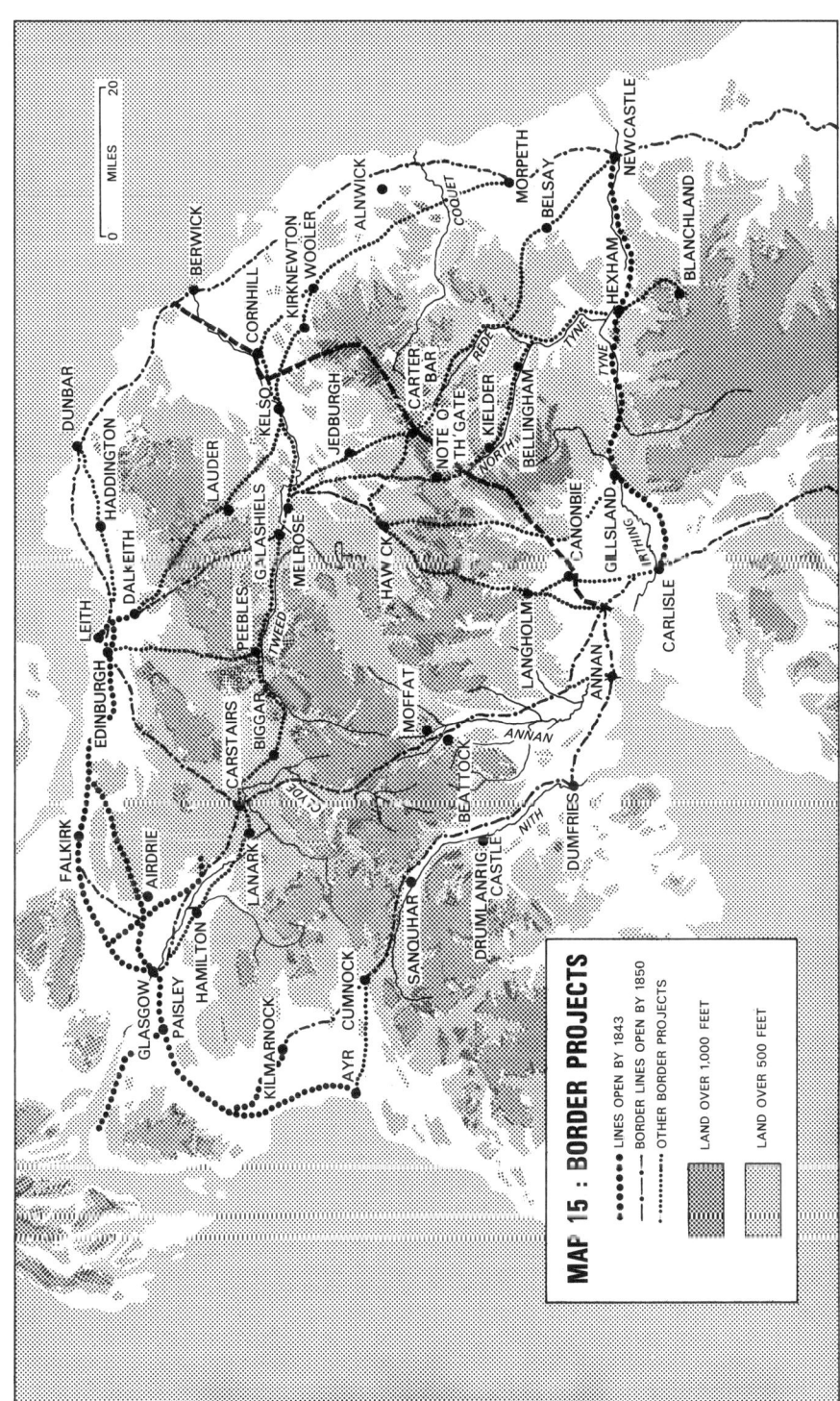

link, but looks in practice like an effort to collect the crumbs of the Carlisle-Edinburgh trade on the assumption that direct lines would be built from Carlisle to Glasgow and from Newcastle to Edinburgh by the coast. The estimated cost of the line was £1.9 million, although the proposed share capital was only £1.8 million. Conveniently assuming that all lines would cost £20,000 per mile whatever their route, the projectors estimated the sum required for the Caledonian at £2.1 million, and for the east coast line from Newcastle to Edinburgh at £2.53 million.[18] The mania produced two more Newcastle-based central line projects (12 and 14 in Table 74), as well as the eccentric London & Edinburgh Direct, clambering by an uncertain route over the hills from Darlington.[19] And as late as 1852 another Direct North Railway was floated, to run from Doncaster to meet the North British branch at Galashiels.[20]

II. Joseph Locke and the Smith-Barlow Commission

Of all the engineers connected with the trans-border projects, only Grainger and Miller were Scots. It is probably true that they were the only Scottish engineers with the ability and the reputation to be entrusted with extensive lines of considerable engineering difficulty; though men like James Jardine and George Buchanan had a good local reputation in Scotland, they had not the experience for projects of this magnitude. Nor had they names which would instil confidence in English investors, and indeed for this purpose Grainger and Miller had once again to submit to having George Stephenson look over their shoulders. On the other hand, most of the English engineers were Tyneside men whose reputation was also essentially local, and most indeed were associated with the unsuccessful attempts to find a central line from Newcastle. The exceptions — the figures of national repute — were Stephenson and Joseph Locke, two of the three men (George Hudson being the third) whom Grinling rather unfairly considers to have been the only real enthusiasts for a trans-border line in the late 1830s.[21] Stephenson, from his connection with the Edinburgh & Glasgow projects, already had knowledge of Scottish railways: as the engineer of a series of lines in Northumbria he was not surprisingly in favour of a route from Newcastle to Edinburgh. His perennial concern with gradients convinced him that it must go by the coast.

Locke, on the other hand, was a newcomer to Scotland, his first work on the border predating his connection with the Glasgow Paisley & Greenock. In the 1820s he had been one of Stephenson's assistants on the Liverpool & Manchester and the London & Birmingham, but the two men had come into conflict on the Grand Junction, where Locke had eventually replaced his mentor as chief engineer. In the 1830s he consolidated a reputation for honest work and estimates which were more accurate than most on the London & South-Western, the Manchester & Sheffield, and the various piecemeal extensions of the Grand Junction northward to Lancaster.[22] It was these last contracts which confirmed the interest of both Locke and the Grand Junction company in developing a west-coast route to Glasgow.

Late in 1835 Locke, on behalf of the Grand Junction, surveyed the country between Carlisle and Glasgow, published his report in the following February. His first attempt followed Annandale and the line of Telford's mail road over Beattock summit, but a ten-mile incline on a gradient of 1 in 75 appeared an absolute deterrent. Locke turned to Nithsdale, and approved the longer line with the lower summit, also making much of the extra traffic which might be expected from Ayrshire. Not only was he convinced that there could be only one Anglo-Scottish line, but he now felt that 'if only one is to be made, there can be no question as to the route it ought to take'.[23] To many people Locke's report seemed to settle the issue; George Graham, later the Caledonian's resident engineer for 51 years, gave his verdict half a century later:

> The above report would now be thought little of, but considering the small amount of knowledge possessed of railway affairs, it became an important and oft-quoted document.[24]

The report also led to a flurry of counter-activity by Annandale landowners, led by county MP John James Hope Johnstone and his factor Charles Stewart, who were primarily concerned with the effect on their own locality. As Stewart observed,

> The railway up Annandale would be of incalculable importance to its prosperity . . . It would, especially by bringing coal and lime, increase in a great degree, perhaps double, the value of its cultivated productions in no distant time . . . If it was understood that the Grand Junction Railway Company were really to bring forward Mr Locke's plan for immediate subscription of shares, then I should think it would be necessary to put some caveat to stop them, if possible.[25]

An Annandale committee was formed to sponsor a further survey of their area. Fortunately for them, Locke was not a man to insist that his first opinions must be right, and Stewart was able to persuade him to undertake another examination of Annandale.[26] Fortunately also, the Grand Junction did not act immediately on Locke's report (after all, the way in which a railway might be taken from Lancaster to Carlisle was far from clear), and by 1837 the onset of depression meant that the company was in no hurry to start operations in Scotland. As its chairman John Moss told Hope Johnstone, in reply to a request for support for the new survey:

> The Committee are unanimous in recommending that no steps be taken until an alteration takes place in the Money Market — great inconvenience is now experienced in Lancashire by the Calls made to carry on the numerous Railroads now in execution . . . Of the importance of a line of Railway to Scotland no one can doubt, but I question very much whether any new line would go down with the public just now.[27]

The Annandale committee had a breathing space.

Locke's second report, published in November 1837, accepted that an Annandale railway might be practicable, since the new survey suggested that the gradient on Beattock bank could be reduced to 1 in 93. The engineering difficulties would still be severe, but the route's shorter distance to Glasgow was an obvious advantage, as was the potential for a convenient branch to Edinburgh to an engineer who believed that there could not profitably be both a railway from Carlisle and one along the east coast.[28] Improving technology was increasingly casting doubt on the degree of difficulty involved in severe gradients, and on the conventional wisdom which said that it was worthwhile to add an extra mile of distance to avoid a twenty-foot increase in altitude. George Stephenson, perhaps unwilling to credit the extent of change since the *Rocket*, continued to insist on the best possible gradients above all other considerations: it was a further sign of Locke's emancipation from his tutor that he came to prefer the routes by Beattock and Shap to their longer but easier alternatives. In his evidence on Beattock he could even refer back to Stephenson's most famous creation:

> The inclination is similar to those on the Liverpool and Manchester Railway, which are worked by assistant locomotive engines. Of the facilities of drawing up considerable weights at moderate speeds there is no doubt; in short the ascent involves nothing but more power and more time. In the descent, however, there is more danger, and this is a question of importance. Perfect machinery and perfect watchfulness on the part of the attendants leave no room for apprehension. A train of passengers, on an inclined plane of 1 in 93, may be kept under perfect control by ordinary means, and therefore we should not be disposed to go so far out of our way to avoid it, as if the plane were steeper, and were to be worked by a fixed engine. If I were asked, however, how far I would be prepared to go round to avoid such a plane, I should, before giving my answer, desire to know . . . whether the additional cost of the longer line would compensate for the practical disadvantage to be encountered on the other . . . A plane like this ought not to be adopted without sufficient reason. You cannot expect it to be so economically worked, nor so certain in its operation as a line of equal length that is free from such a plane.[29]

These comments were hardly a wholehearted endorsement of Annandale, and Locke's realistic concern with braking power was an uncomfortable reminder to enthusiasts who thought that the only problem with summits was ascending them. As he also pointed out, the survey was limited to engineering problems, and contained no estimates whatever of costs of construction or potential revenue: by implication, the easier engineering and greater local traffic of Nithsdale might still be preferable. Nor had the Annandale committee, in spite of a claim in April 1837 that they already had almost enough subscriptions to cover the cost of the survey, been able to pay Locke to examine the ground between Biggar and Glasgow, although no difficulties were expected there.[30] But at least, as far as Hope Johnstone's group were concerned, Locke had said that Annandale was practicable, and the debate could be reopened.

At the end of 1837 it seems probable that Locke was genuinely undecided between the two routes; although the debate might now be better informed on the engineering prospects, it was no nearer to an agreed solution than it had been before his 1835 report. In November 1839 John Leadbetter, for the Nithsdale interest, was still quoting the first anti-Annandale report and claiming that in the second report 'Mr Locke was writing against his own convictions'.[31] But in the meantime further surveys by Locke's assistants MacCallum and Dundas had reduced the Beattock gradient to 1 in 103, and early in 1840 the engineer produced a third report positively recommending Annandale.[32] Locke and the Grand Junction had now severed all possible connection with Nithsdale; the counterblast in favour of that route came instead from the Ayrshire company's engineer John Miller.

Uncertainty over the route of the border railway may account in part for the considerable caution with which the Scots approached their decisions. The Edinburgh projectors of an east-coast line decided in March 1836 to promote their railway only as far as Dunbar, although an English link was envisaged for the future: it was left for a group of Newcastle engineers to produce, a few months later, their plan for a Grand Eastern Union Railway to fill in the gap.[33] In the west, where there was no attraction in lines going only part of the way, projectors looked to the government for help either with finance or with examination of the routes. It was not unknown for the government to finance transport improvements (though in fact no Scottish railway benefited), and, as noted above, promoters from the Borders in 1821 to the Highlands in the 1840s hoped for help.[34] In 1836 Andrew Bannatyne, Ayrshire company lawyer and Nithsdale projector, put the case:

> These parties who have hitherto taken an interest in the project have never contemplated the possibility of carrying it into execution without the aid of Government. Even although we could concentrate the whole land traffic of Scotland on one line, it would, I fear, scarcely yield a low rate of interest on the cost of the undertaking; far less would it yield a return such as is required to induce capitalists to embark in a scheme of the sort. But it has always seemed to them likely that Government might be induced by a powerful representation by a large body of the North of England and Scotch landed proprietors, and by the mercantile body in the west of Scotland, to devote a portion of the public money to the improvement of the means of communication between Scotland and England. The object seems one of national importance, similar to what has already received the aid of government, as for instance in the cases of the Carlisle and Glasgow Road, the Holyhead Road, the packet stations on St. George's Channel, &c.[35]

Two years later the Ayrshire directors were still qualifying their support for an extension to Dumfries by insisting that construction 'must be encouraged by the Government, otherwise it will not be undertaken by private enterprise'.[36] On the other hand, the results of earlier government help had not been to the taste of all commentators; the Newcastle & Carlisle, according to *Fraser's Magazine*, had been

> paid by government loans; which will never be repaid, nor were ever intended to be repaid, as far as the principal is concerned, and only for a few years of interest: for, in such cases the nation is a regular milch cow.[37]

No financial help was forthcoming for Anglo-Scottish schemes, the government perhaps judging correctly that commercial interests would in fact be willing to undertake the risks.

The other possible method of government intervention was in the selection of the route of the railway. Although a House of Commons Select Committee had declared in 1836 that

> Upon the whole . . . it appears decidedly best to leave Railways in Great Britain, like all other undertakings, to be decided upon according to the judgement and interest of those who are willing to embark their capital in them, subject only to the scrutiny and control of Parliament,[38]

it was not clear how far in fact the state would or could apply strictly laissez-faire criteria in its dealings with the companies. James Morrison MP and others were already pressing for a degree of planning and control, and possibly even for a state system of railways.[39] The need for an authorising act to give, if nothing else, compulsory powers for land purchase and limited liability status, meant that Parliament was inevitably involved in any railway project, and the process of increasing if reluctant intervention to regulate safety, maximum rates and potential abuses of monopoly powers was already under way, leading towards the powers latent in the 1844 act.[40] Certainly private projectors could not be expected to look beyond the interests of their own projects; was there not a case for the state at least to determine the main lines of railway which would best serve the national interest? *Blackwood's Magazine* proposed that all major routes should be planned by a committee of engineers not involved with the direction of any railway companies:

> They will continue to be driven through lines of country totally unfitting for them, if they are not put under guidance, and the result will be a constant succession of bankrupt companies, with all the misery that accrues from individual failure, and all the disturbance that belongs to ruinous public speculations.[41]

On the border, Locke's indecision in 1837 led him to suggest a commission of inquiry to determine the best route:

> I might even go so far as to suggest that Government should institute this inquiry: and when the best line is discovered, it is probable that the enterprising spirit of the two countries, aided by such means as the Government might think fit to furnish, would soon carry the project into execution.[42]

In the following year another interested individual, J. Hodgson Hinde, MP for Newcastle, chairman of the Newcastle & North Shields Railway, and later a director of the Caledonian, moved in the Commons for such a commission to find the best route in purely engineering terms — the commercial prospects were not a matter on which a government-sponsored body should be expected to pronounce.[43] His proposition was opposed on grounds of public cost and on the combination of laissez-faire and chauvinism implicit in the claim that

> It was much better to leave these great works to private capital, and not to mix up the government in them as was the case in France.[44]

After a tied vote, the speaker decided against Hinde. He explained in a letter to the press that it was more necessary to find a line which would pay than simply the best engineering line, that the government should only intervene if asked directly by the companies, that the 1836-37 Select Committee on Railroads ('one of the best Committees that ever sat') had opposed state intervention, and that the precedent might lead to other similar requests 'supported by canvass and jobbing'. The *Railway Times*, suggesting that canvass and jobbing were not unknown under existing parliamentary procedures, was not impressed.[45]

By 1839 continuing uncertainty both in Parliament and among projectors as to the best long-distance routes allowed two engineers, Colonel Sir Frederick Smith, the first Inspector-General of Railways for the Board of Trade, and Professor Barlow of the Royal Military Academy at Woolwich, to be appointed as a Commission on Railway Communication between London, Dublin, Edinburgh and Glasgow. Their report on the border schemes in March 1841 came to apparently positive conclusions while in fact leaving enough qualifications and loopholes to permit continuing debate. With some difficulty, which they attributed to the financial stringencies of most promoters, they gathered information on sixteen proposals or variations for lines north of Lancaster and Newcastle (see Table 75):

> The information furnished to us has, with few exceptions, been very incomplete, and it has only been by repeatedly urging the projecters or promoters of some of the competing lines, that we have been enabled to obtain from them even the meagre documents with which we have been supplied.[46]

Table 75. Border Routes considered by the Smith-Barlow Commission, 1841

	Route	Engineer	Miles	Equivalent mileage allowing for gradients	Tunnels (yds)
1	Lancaster–Cumberland coast–Carlisle	G. Stephenson	n.a.	n.a.	n.a.
2a	Lancaster–Kent valley–Carlisle	—	n.a.	n.a.	4400
b	Lancaster–Lune valley–Carlisle	J. Locke	n.a.	n.a.	2420
c	Lancaster–Kendal–Orton–Carlisle[a]	J. Locke	70.21	n.a.	2244
3a	Newcastle–Howick–Berwick–Edinburgh	G. Stephenson	117.78	123.41	3522
b	Newcastle–Morpeth–Berwick–Edinburgh[b]	(G. Stephenson)	116.53	122.66	2022
4	Newcastle–Morpeth–Wooler–Kelso–Edinburgh	G. Remington	104.13[c]	110.23[c]	n.a.
5	Newcastle–Hexham–Kielder–Melrose–Edinburgh	J. Blackmore	115.95	130.24	5390
6a	Carlisle–Lockerbie–Beattock–Glasgow	J. Locke	100.93	116.98	—
b	Carlisle–Annan–Beattock–Glasgow[d]	—	105.69	121.85	—
7a	Carlisle–Nithsdale–Ayr–Glasgow	J. Miller	127.73	133.11	n.a.
b	Carlisle–Nithsdale–Kilmarnock–Dalry–Glasgow	J. Miller	125.11	130.05	n.a.
c	Carlisle–Nithsdale–Kilmarnock–Beith–Glasgow	—	120.75	124.68	n.a.
8a	Symington–West Linton–Edinburgh[e]	McCallum & Dundas	31.38	38.85	—
b	Thankerton–Auchengray–Edinburgh[e]	McCallum & Dundas	n.a.	n.a.	2706
c	Thankerton–Midcalder–Ratho–Edinburgh[f]	J. Miller	37.51	44.30	–

a: A compromise between 2a and 2b, serving Kendal but avoiding Shap Fell
b: An improvement, accepted by Stephenson, on his original line in order to serve Morpeth
c: To Dalkeith only
d: An Annandale variation, serving Annan and allowing a shorter branch to Dumfries, but adding almost five miles to the main line
e: Alternative lines, on opposite sides of the Pentlands, for the Edinburgh branch of the Annandale route
f: An attempt by the Edinburgh and Glasgow, which it would join at Ratho, to salvage some traffic if the Annandale line was built

Source: PP 1841 (132) XXV, *Commisssion on Railway Communication between London, Dublin, Edinburgh and Glasgow, Fourth Report,* 6, 52–4.

In the event they failed to obtain detailed estimates for three potentially serious contenders — Remington's Wooler route, the Cumberland coast link from Lancaster to Carlisle proposed by George Stephenson (who believed Shap Fell to be impassable), and, most surprisingly, the Nithsdale line. It cannot be known to what extent this affected their final judgement, but it does indicate surprising lack of preparation by Miller and the Ayrshire company.

The estimates which were produced were often deemed unsatisfactory. The commissioners could see no reason why tunnelling should cost £50 per yard at Shap but only £30 at Note o'th' Gate on Blackmore's line; by assuming that Blackmore's figure was based on an inadequate survey and was too low, they by implication confirmed Locke's reputation for reliable estimates. Similarly, while the various figures for land costs (£40 per acre north-east of Symington, £70 in the Lune Valley, £150 in Annandale and around Kielder, £212 on the Scottish section of the east coast route) owed something to the quality of the land required, they too often owed more to the optimism or inexperience of promoters.[47]

The report effectively put paid to the Remington and Blackmore projects; though Blackmore was still campaigning for his line in 1843, his Newcastle & Carlisle backers switched their support to the Gillsland route which became the Central Union, and which one opponent dismissed as a 'nonsensical project'. Although Blackmore's line, by using part of the Newcastle & Carlisle, needed less new construction than some others, everything else was against it. Its gradients more than offset any reduction of mileage, it served few people en route (an estimated 22,525 against 36,008 on the east coast), and the summit tunnel was substantial enough to mean a long delay before opening. Remington's line, though comparatively short, required works 'of a most formidable character', and would cost so much to construct that it could not conceivably pay.[48]

The commissioners' conclusions on the three main contenders were clear enough, but room was still left for dispute. The key assumption was on the old argument between one line or two:

> We are led to believe, from all the information that we have been enabled to collect, that the amount of traffic which in the present state of the commercial and other relations of England and Scotland may be expected, is not such as would be likely to afford an adequate return for the construction of two distinct lines of Railway, the one from Darlington to Edinburgh, and the other from Lancaster to Glasgow.[49]

In an earlier report on the routes between Lancaster and Carlisle in October 1840, Smith and Barlow had shown that they were not averse to severe gradients when they preferred the Shap route to the long detour by the Cumberland coast.[50] Given this and their adherence to the single-line theory, their final verdict for Annandale, with its connections to both Glasgow and Edinburgh, was inevitable. But while Hope Johnstone and the Grand Junction rejoiced, the verdict was sufficiently qualified to encourage the others. If they had considered two lines practicable, reported the commissioners, they would certainly have supported the east coast route, and in the west would have decided between Annandale and Nithsdale purely as routes to Glasgow; and in this case the 'decidedly superior . . . mechanical properties' of Nithsdale might well have offset the greater length and correspondingly greater cost to passengers. And even the decision for Annandale depended on the construction of the railway through the inhospitable country between Lancaster and Carlisle: if this was not undertaken while projectors appeared for the east-coast line, the latter should have preference.[51]

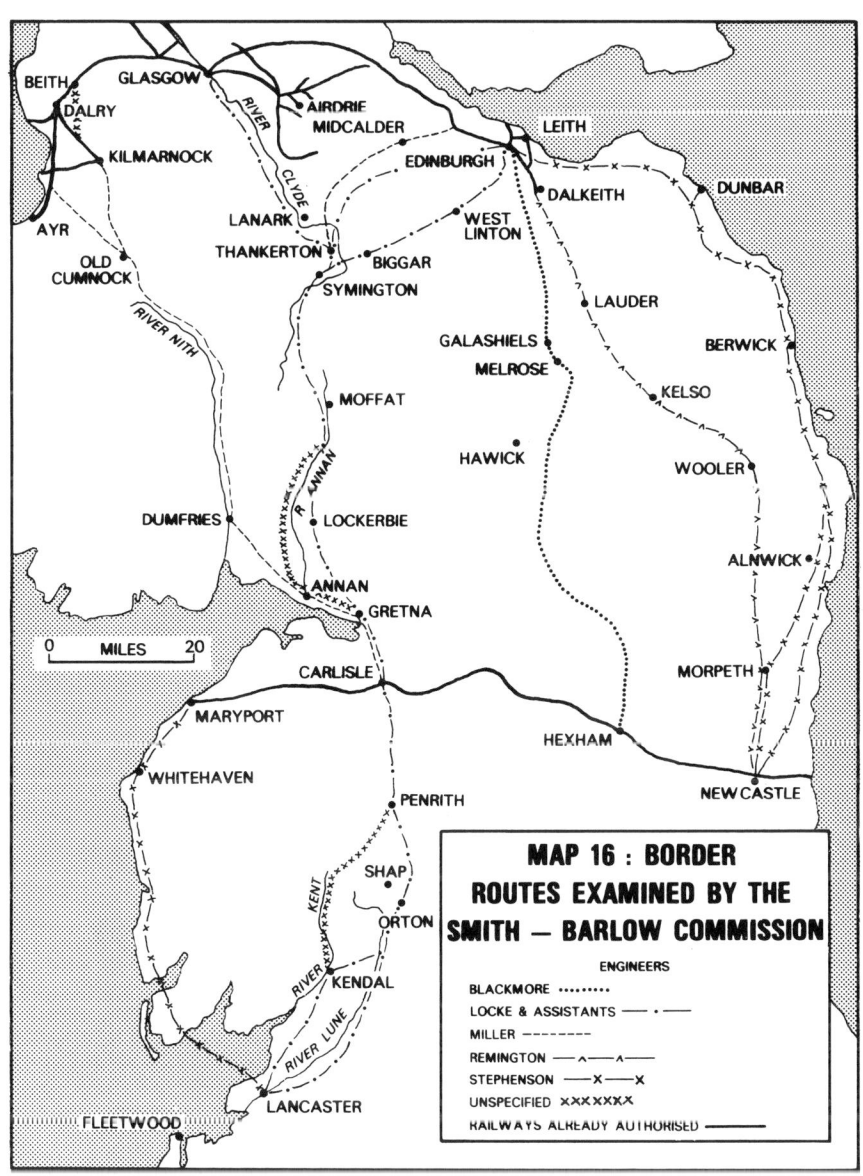

Table 76. Border Railway Projects – Distances between Principal Towns

Project	Glasgow –Carlisle m.	Glasgow –Newcastle m.	Edinburgh –Carlisle m.	Edinburgh –Newcastle m.	New Line to be built m.
Nithsdale (7b)	125.1	189.6	171.1	237.1	102.5
Annandale (6a + 8a)	100.7	165.2	97.6	162.1	124.3
Blackmore's (5)	153.3	162.0	135.4	116.0	95.4
East Coast (3b)	227.0	162.5	181.0	116.5	116.5
Nithsdale and East Coast (7b + 3b)	125.1	162.5	171.1	116.5	209.0

Source: Calculated from Table 75.

The effect of the Smith-Barlow report was to give some degree of encouragement to everyone but Blackmore, Remington and the Newcastle & Carlisle. Although the Annandale project was clearly in the lead, there was some encouragement for the east coast, particularly as the recession was inhibiting action in the Cumberland hills. The east-coast projectors, and even more those in Nithsdale, now had to persuade parliament and the public that there would be enough traffic for two lines — that, for instance, the commissioners' conclusions had been pessimistic because they had been reached in a time of depression, or that they had not sufficiently taken into account the increase in traffic volumes which the creation of a railway had been shown to produce. The alliance of the east coast and Nithsdale routes, aimed at stopping the Annandale project, dates from the Smith-Barlow report, even if it was not formally consummated until the agreement for mutual cooperation signed in March 1844 by chairmen John Learmonth of the North British and John Leadbetter of the Glasgow Dumfries & Carlisle.[52]

III. Supporters and committeemen

It was to be three years from the publication of the Smith-Barlow report to the authorisation of the first trans-border railway. The delay was in part due to the caution of investors who were prepared to wait and see whether the progress made by recently opened lines on both sides of the border would suggest that there was profit to be made by extending the system. It was also of course due to a shortage of money, a decline in the demand for transport, and a general if temporary lowering of expectations caused by the intensity of the economic depression. In Scotland the two key growth industries, coal and iron, had expanded so rapidly that by 1840 they were suffering from over-production, and required a period of consolidation while demand caught up with supply; meanwhile the mid-1830s boom in shipbuilding orders had been satisfied, while the decline in construction work as earlier railway lines were completed added to the unemployment of resources. A further blow came from the American tariff increases of 1842, affecting one of the main trading areas of the Glasgow merchants who were so important as railway projectors.⁵³ At such a time little money was available for any scheme, particularly not for those involving a substantial time lag between investment and return.

All that projectors could do in such circumstances was to perfect their plans, consolidate their support, negotiate with potential opposition, and keep their ideas before the public. By 1844 the groups of projectors associated with the east coast, Annandale and Nithsdale had been organised into respectively the North British, Caledonian and Glasgow Dumfries & Carlisle companies, while the Newcastle & Carlisle's final protégé, the Central Union, faintly kept alive the idea of a central route.

To the general public, the most obvious result of these labours was the production of a prospectus, which not only presented the inevitable promises of future prosperity, but also gave some estimate of likely costs and traffic volumes, and, in the names of the provisional committee, an indication of the strength of support. In many companies, particularly as the mania approached, much time was spent assembling a provisional committee of appropriately impressive credentials. The names had above all to indicate substantial and influential local support, certainly from leading gentry and if possible also from the aristocracy, since this was a criterion of much importance to parliamentary committees. But it was also desirable to have some men of known commercial experience and wealth, one or two of whom might come from Lancashire or London to give encouragement to English investors. The implications of support given by the presence of one or two directors of established railway companies would also be reassuring. The ideal committee might thus contain a few aristocrats, the major landowners on the route, the local MPs, the provosts of the towns principally affected, the chairmen of connecting railways, an imposing array of merchants and industrialists, and a leavening of lawyers, bankers and high-ranking military officers. If this could not be achieved, the list must still be as imposing as possible. Substitutes could be found — for an unwilling provost, a bailie or dean of guild might serve; for a railway chairman, one of his directors. The list might be extended by peers and MPs with no obvious local connection, by lesser landowners, and if it appeared that an important town was in danger of being unrepresented, by the most influential citizen who could be persuaded to countenance the project. By the time of the mania, the quest for impressive lists had led to elephantiasis of committees, with one trans-border speculation, the London & Edinburgh Direct, presenting a list of 277 names to the public.⁵⁴ Some voices were raised against the practice, including that of John Gladstone when considering names for the Edinburgh & Northern:

> I think the object not only to have persons of known respectability on the Council, but such as from situation interest and connection, would be working men and desirous to forward the undertaking.

This would mean the exclusion of such men as

> Mr Stewart of Baladrum, Mr Bannerman M.P. for Aberdeen (who I have understood reduced to poverty) and Chalmers of Anstruther [sic] who unless you know them to be connected with Fifeshire could not attend the Committees or be of any essential service.⁵⁵

Though the company's list was eventually limited to men with a direct interest in Fife or Edinburgh, it still contained 44 names, and was clearly too big to function as an efficient executive.⁵⁶ What was happening in almost all promotions was summarised by a contemporary commentator:

> In nineteen cases out of twenty, members of Provisional Committees are in truth nothing more than a party of gentlemen, who certify to the public their approval of a certain scheme, but who, in reality, have no voice or influence whatever in the matter: this is wholly engrossed by the Committees of Management, who are the responsible parties for every act done in the concern.⁵⁷

The committee of management, of perhaps a dozen members, was mainly composed of the more commercially experienced, if less socially distinguished, members of the provisional committee. The names of the full provisional committee, however, rather than just those of the committee of management, were used to attract investors, the less informed of whom may not have realised the difference until the committee of management resurfaced more or less unchanged after authorisation as the board of directors.

There had not always been such emphasis on extending committees. The Ayrshire Railway had been quite content to issue its original prospectus with a provisional committee of twelve, all ranked as 'esquire' (though two earls appeared on the first board of directors).⁵⁸ Its protégé, the Glasgow Dumfries & Carlisle, was at first almost as modest, entering the field in March 1844 with a list of only 22 names, some two-thirds of whom were Glasgow merchants, while eight were already and three more were shortly to become directors of the Ayrshire. One innovation was the creation of the posts of governor and deputy governor, perhaps to suggest the sort of financial security associated with the banks, and perhaps to provide positions of appropriate status for the committee's two aristocrats. One name missing was that of Ayrshire chairman James McCall, who had chaired the first committee meeting on 9 March, but then resigned in order to assume a tenous appearance of neutrality so that 'his services might be more useful in any negotiations between the Competing Western Lines'. By the time the prospectus was issued to the public, John Leadbetter, formerly of the Edinburgh & Glasgow, was in the chair. The list also included the provosts of Paisley, Dumfries, Kilmarnock and Ayr (no final decision having been taken as to whether Kilmarnock or Ayr was to be on the through line to England), but sadly the Lord Provost of Glasgow was committed to the Caledonian.⁵⁹

Within five weeks, however, inflation of the committee had set in, a list of 41 names was issued, and, in an effort to widen their geographical appeal, it was agreed to rescind the minimum qualification for a committeeman of forty £25 shares for 'Gentlemen unconnected with Glasgow'. Most of the newcomers were landowners in Ayrshire or Dumfriesshire; they included the Marquis of Queensberry and William Ewart MP, both of whom had been on the Caledonian committee but had defected to the rival line which would serve respectively their estates and their constituents. The committee's resolution to find some support in Yorkshire, perhaps to counterbalance the Caledonian's solid Lancastrian backing, brought the prize of George Hudson (the only English resident on the committee), while the anti-Caledonian common cause also persuaded John Learmonth to serve.⁶⁰

The role of the Ayrshire company in the promotion was well known from the start, but it was still comforting to be able to state in the prospectus that the Ayrshire directors intended to contribute £200,000 to the new line. As yet, no other existing company was formally involved, although by the end of 1844 the Edinburgh & Glasgow had agreed to subscribe £100,000. The promoters were unreasonably optimistic, however, about support from the southern companies which were in fact already committed to the Caledonian:

> There is also every reason to hope that the Glasgow Dumfries, and Carlisle Railway will be not only cordially welcomed, but actively encouraged by the Proprietors of the Lancashire and Cumberland lines of railway, to which it will prove so important a feeder.

According to the Nithsdale promoters, the Smith-Barlow assumption of a single trans-border line had already become irrelevant:

> Edinburgh and the east of Scotland have refused to accept anything but a direct line along the east coast from Newcastle through the agricultural counties of Northumberland, Berwick and East Lothian, and terminating at the North Bridge, Edinburgh.[61]

Clearly neither Glasgow nor Lancashire would accept such a line as their own link. But on the east coast the North British was to some extent detached from the arguments of the west. Its prospectus was the first of the trans-border bids to appear, in August 1843, and eleven months later the company was authorised a year ahead of any of its rivals. The prospective committee contained 57 names, all but three resident on or very close to the route. In social rank the list rose no higher than a solitary baronet, but concentrated on the landed gentry of East Lothian and Berwickshire and on the lawyers, bankers, merchants and accountants of Edinburgh. Eight members, including Learmonth and his deputy Eagle Henderson were directors of the Edinburgh & Glasgow; few of the others ever had any connection with any other railway. The senior magistrates of Edinburgh, Leith and Berwick all appeared, while the absence of the provost of Dunbar was offset by the presence of two of his bailies. No committeeman lived south of Berwick, and only one, Edinburgh & Glasgow director John Sligo, in Glasgow;[62] there was apparently no need to place committeemen in the great financial centres. The North British, with no rival for the east-coast route and coming into the market ahead of the west-coast challengers, did not feel the need to pack the committee with impressive but non-working outsiders. Parliament would have to consider the scheme on its merits, rather than by comparison with others.

North British confidence was boosted by the support of a man whose name did not originally appear on the provisional committee, although he became a director after the act was passed. George Hudson had been keenly interested in the project ever since the tentative Edinburgh to Dunbar line of 1836 had been replaced by the Berwick project of 1838. Indeed, in 1838 the company's lawyers claimed that the new plan had no connection with the old Dunbar project, but had originated in England, presumably with Hudson.[63] By the early 1840s Hudson had promised Learmonth his support for both the North British and the Newcastle & Berwick which would connect it to the rest of his empire to the south. Hudson's services to the North British were substantial — according to one historian of the company, 'without his support the North British Railway would have foundered' — and included the raising of enough money to enable the parliamentary deposit to be paid.[64] After authorisation his fellow directors voted 'the English Director' £50 in appreciation of his efforts, with the carefully minuted qualification that

> the allowance set apart for the English Director is only now granted in consequence of the great trouble Mr. Hudson has taken in forwarding the present measure, but shall not be taken as a precedent with regard to any subsequent Director.[65]

Hudson's stay on the board was brief, and by 1846 the company was fighting off his unwelcome attempts at a takeover,[66] but his earlier support and the consequent backing of major English interests had been invaluable. Similarly the connection with the Edinburgh & Glasgow, cemented by substantial overlap on the boards and by the Princes Street Gardens link, meant support from the largest existing Scottish company. Although the Edinburgh & Glasgow's main aim was to stop the Caledonian, which by linking to both Glasgow and Edinburgh would cut it out of almost all trans-border traffic, it must also have considered that, if the North British were to become the only Anglo-Scottish route, all rail traffic between Glasgow and England would pass over its own line. Although in 1838 it was reported to be willing to subscribe to a Nithsdale line,[67] and although Learmonth later joined that line's committee, a single east-coast line would have suited it even better than one to the west.

The Central Union, always the weakest competitor in the race, drew most of its 38 committeemen from Newcastle and its hinterland. Of the fourteen Scots, all but the Earl of Stair (the committee's only peer) came from the border area around Hawick and Galashiels, and even there local government was represented only by two town clerks and a bailie. The mayor of Carlisle, however, gave the line the backing which he denied to the Caledonian, while one of his predecessors in the civic chair briefly found it compatible to join both companies' committees: surprisingly, it was the Caledonian from which he then chose to resign, perhaps because he was also a former director of the Newcastle & Carlisle. The list also included the Central Union's engineers, Nicholas Wood and George Johnson.[68] Overall, however, it was not a committee which would force opponents to revise the opinion, expressed by a Caledonian projector, that the Central Union was 'a nonsensical project'.[69]

The Caledonian prospectus, issued like those of the Nithsdale line and the Central Union in March 1844, includes the first of what may be called Scottish mania committees, calculated to impress MPs and distant investors as well as local interests. The first Directors' Minute Book of the company opens with the record of 'a Preliminary Meeting of Gentlemen favourable to the construction of the Scottish Line of Railway recommended by the Government Commissioners' in March 1841, held significantly in London, and attended by nine men of whom five were MPs. Apart from agreeing to produce a prospectus, appointing Locke and Errington as engineers, and establishing a sub-committee 'to wait on Government to ascertain its views in respect to the undertaking and also what support may be expected from them', they drew up a list of 124 individuals who might be approached to serve on the provisional committee. Their list started with 29 noblemen, led by the Dukes of Buccleuch, Hamilton, Montrose, Richmond and Sutherland, and 29 MPs, of whom only seven represented constituencies on the route. The Edinburgh MPs were notably missing, presumably because their sympathies were known to lie elsewhere, and in fact only eight men from the city, none enjoying either elective office or high social status, were to be approached.[70] The Caledonian made considerable efforts to persuade Edinburgh that it was seen as a destination of comparable importance to Glasgow, rather than as the terminus of a branch line. For a time the company office was sited in the capital, and the ceremony to mark the start of construction was held at Carstairs, where three 'first sods' were cut, one facing towards each terminus.[71] Even so, Edinburgh's support was at best modest. Of the whole list, only 64 names — 51.6% of the total — were clearly connected with the route; most of the rest were either Scottish aristocrats or MPs with constituencies ranging as far afield as Inverness, Dover and Antrim.[72]

The provisional committee as it appeared in the prospectus contained 70 names.[73] None of the dukes had agreed to serve. Montrose, who inclined to the Nithsdale, had consulted Buccleuch, who was totally opposed to the Nithsdale but willing to assent to the Caledonian without giving it active support: 'I should be very sorry to have any shares in it, as I do not think it is likely to pay'.[74] The Caledonian committee was still impressive. Five aristocrats were headed in rank by the Marquis of Queensbury (shortly to defect to the Nithsdale), and included the chairman, Lord Belhaven. There were twelve MPs, including the seven with constituencies on the line, while three were Scots with English constituencies and two represented the Manchester connection. There were five provosts or lord provosts, and an ex-mayor of Carlisle; in Glasgow, a city more identified in the popular mind with the Nithsdale, the Caledonian captured both MPs, Lord Provost James Lumsden, two bailies, the dean of guild and the trades convener. Other railway companies were represented in strength. From Lancashire came John Moss, chairman of the Grand Junction which agreed to subscribe £200,000; he was supported by the deputy chairman of the Lancaster & Carlisle and the chairmen of two smaller Lancashire companies, each of which subscribed £50,000. Mark Sprot, chairman of the Garnkirk & Glasgow, was joined in a sort of block booking by the complete boards of both the Monkland & Kirkintilloch and the Wishaw & Coltness; the Caledonian proposed to run over sections of all three lines on its way into Glasgow. The remaining third of the committee was made up of landowners along the line and of leading commercial men from Glasgow with important names like Tennant, Dixon, Houldsworth, Dennistoun and Campbell of Blythswood. From these last groups came the men who were to be the active promoters and later the directors — the amount of work anticipated from most committeemen was indicated when the quorum for meetings was set at three. Apart from a weakness in Edinburgh, where the committee contained only a retired colonel and an accountant, the Caledonian list was well calculated to impress.[75] An accusation made at an Ayrshire meeting that this 'formidable array of names' were not genuine subscribers but had merely agreed to contribute to the cost of the 1842 survey was denied with vigour, but in terms which suggested that the accusation had originally been true.[76] Certainly at the end of 1844 the secretary found it necessary to point out that under the recent Companies Act no one could be on the committee without holding a share in the company, and to suggest that all committeemen should acquire a minimum holding of twenty shares.[77]

IV. Western delays and the authorisation of the North British

At first sight the three-year delay between the Smith-Barlow report and the publication of the western lines' prospectuses is surprising, particularly as there might well have been an advantage in being first in the field, and as the commissioners had stated that if action were not taken quickly, preference might be given to the east coast. But in 1841-43 there were reasons, even apart from the worst trade depression the country had yet experienced, for companies to be cautious. The Nithsdale projectors had failed to find, or even to seek with any great enthusiasm, support from English companies (Hudson's adherence only coming after the prospectus was first issued, and then being of minimal practical significance). It was therefore essentially dependent on Glasgow commercial interests who were facing financial stringencies, and on the recently opened Ayrshire railway which was itself dependent on the same Glaswegians and on traffic receipts whose long-term value was as yet unclear. Since the Caledonian was also in the field, the Glasgow Dumfries & Carlisle could not even hope to mobilise all the capital which Glasgow might be willing to invest in railways.

The Nithsdale promoters also had to decide whether their line was to set out southwards from Kilmarnock or Ayr, a point of serious internal dispute. It was not until April 1844, after the prospectus was issued, that John Miller reported in favour of Kilmarnock, with a junction line thence to Monkton and thus on to Ayr; it was then agreed that the Ayrshire company should itself build from Kilmarnock to Cumnock, with the Glasgow Dumfries & Carlisle running thence to Carlisle.[78] It was also true that some of the Ayrshire interest were dubious about any extension to Carlisle, since their company had invested heavily in Ardrossan harbour and in developing the steamship service to Liverpool and Fleetwood, whose prospects could only be damaged by a rail link. The views of some Caledonian supporters perhaps deliberately encouraged such doubts:

> Considering the ease and perfection of steam navigation from the Clyde to Liverpool on the one hand, and from Leith, Dundee &c., to London on the other, ... there can be little conveyance of goods, and ... it is mainly from passengers a revenue for a railroad in Scotland is to be derived.[79]

The argument was put forward to support the case for the Caledonian route as being the shortest, since passengers would attach considerable importance to reducing the time of the journey. Even so, it might carry some conviction amongst the already doubt-ridden Ayrshire projectors. For all these reasons it took the Ayrshire from 1838, when they promised support to a local initiative on behalf of a Nithsdale line, until 1844 to launch their protégé company. In 1843, for instance, the Ayrshire board could only tell the shareholders that 'proceedings ... have been anxiously observed' and that 'every consideration of sound policy and expediency' indicated the Nithsdale line. The company was also at this time becoming distracted by its own high expenditure levels; at the following half-yearly meeting it appointed a committee of inquiry into its own affairs.[80]

The delay from the Smith-Barlow report until the appearance of the Caledonian prospectus is more surprising, since the company had the greatest interest in taking immediate action. One anonymous supporter reacted to the report by calling for the immediate implementation of the Commissioners' 'strong and decided opinion, in favour of the Line recommended by Mr. Locke'.[81] And the Annandale landowners and their allies were not idle: meetings were held and support canvassed along the route, while continual negotiations took place with the Grand Junction and other components of the route south of Carlisle. One problem certainly was that the severe depression had by 1842 (in which year Locke presented yet another report on the route) prevented any progress on the Lancaster & Carlisle, and the Lancashire interest might well be reluctant to place money north of Carlisle until the intervening line was settled and the company to construct it was authorised.[82] Both the Caledonian and the Glasgow Dumfries & Carlisle were dependent on a satisfactory route being found through the Cumbrian hills, and many observers, including *Chambers' Edinburgh Journal*, which strongly advocated the east-coast route, believed this to be impossible:

> The project of putting forward a line of railroad from Lancaster through Westmoreland and Cumberland has been a good deal agitated, and various have been the surveys made for it, but there is every reason for believing that no capitalists could be got to risk money in such an undertaking, the physical difficulties of the countryside being almost insurmountable . . . There is only *one* line which can possibly answer reasonable expectations, and that is the plain and obvious one — *from Darlington to Newcastle, Newcastle to Berwick, and Berwick to Edinburgh*.[83]

Caledonian supporters were well aware of the need to move at least as rapidly as the North British. As an 1841 pamphlet said,

> It is only necessary that the inhabitants of Glasgow, Paisley, Greenock and the other Manufacturing and Mercantile Towns, should come forward without delay, and strenuously co-operate with the Land Owners along the Clydesdale Line — the Committee at Carlisle — the Great English Railway Companies and the other influential parties interested, in taking immediate steps for securing its completion.[84]

It was optimistic to assume that Paisley and Greenock should prefer the Caledonian to its Nithsdale rival. But in practice many potential supporters in Glasgow and Lanarkshire may have held back, even when the depression began to lift, from the Caledonian because of the difficulties of Beattock, or from any western railway because of the problems in the Cumberland hills. In 1839 Lord Provost Henry Dunlop of Glasgow held the Caledonian embrace at bay, insisting that any new survey would have to be impartial between the two routes to Carlisle.[85] In the same year the Annandale projectors claimed that the owners of 95% of the required land in Dumfriesshire had subscribed to their survey,[86] but a few years later it seemed that other counties on the line were less enthusiastic. The encouragement given by the Smith-Barlow report and Locke's resurvey of 1842 quickly died away. By March 1843 Mark Huish of the Grand Junction, who complained particularly of lack of support from Carlisle, saw the campaign as being in abeyance: 'However, there is no use greeting. We made a bold push'.[87] Charles Stewart grumbled to John Marr, the town clerk of Lanark, that 'you Lanarkshire people have shown a strange coolness and indifference in not supporting us in what is of great importance to you'.[88] Even in Glasgow, although the Town Council agreed in spite of Bailie Leadbetter's vigorous opposition to subscribe £200 to the 1842 resurvey,[89] further support came forward only slowly, and the distinguished array of merchants on the provisional committee were for the most part not recruited until 1844. The fact that over half the Caledonian's subscription capital eventually came from England confirms that the Scots were not providing the degree of support anticipated, and it may have been necessary to delay at least until Lancashire interests had worked out how to reach Carlisle.

There is also an impression of some sort of hiatus in the management of the Caledonian project between 1841 and 1844. Neither Hope Johnstone nor Stewart was present at the 'preliminary meeting' of Caledonian projectors in March 1841, nor was either of them appointed to any of the sub-committees then established, although these did include men not at the meeting. That meeting was chaired by William Lockhart, MP for Lanarkshire:[90] when the project was revitalised early in 1844, the first chairman of the newly established company was Lord Belhaven, with joint deputy chairmanships going to Dumfriesshire landowner Lieutenant-Colonel William Graham of Mossknow, and to the Grand Junction's John Moss.[91] Hope Johnstone, although apparently always a power in the project, was not formally in command until he became chairman after the act of authorisation of 1845. Similarly, although numerous meetings to raise support were held in 1841 along the route and as far afield as Liverpool, press reports of further efforts in 1842 and 1843 are hard to discover. The only meeting of importance in those two years noted by Butt and Ward was held in Birmingham under the chairmanship of George Carr Glyn of the London & Birmingham, and was mainly a consideration of Locke's latest report on events to the north.[92] That John Learmonth could claim at the end of 1842 that the western rival 'might almost be said to be abandoned' may be put down to his his perpetual over-optimism, but by March 1843 the *Railway Times* was using very similar phraseology. By September it regarded the Annandale line as moribund, while the North British appeared certain of authorisation.[93]

The Caledonian projectors failed to seize the advantage given to them by Smith and Barlow. Immediate action, with the production of a prospectus giving detailed estimates, might have secured the benefits offered by the commissioners' support and by their presumption in favour of a single border railway. But depression and the lack of certain financial support made both the Scottish projectors, and more crucially the Grand Junction, sufficiently timid to delay, and the impetus was lost. To use a military metaphor which became popular in mania circles, they failed to occupy the ground. Their final sign of confusion came in 1844, when only two committeemen appeared for the meeting which was to instruct their legal agent to oppose the North British bill, and could thus only send the recommendation of an inquorate committee. Four months later, having presented their petition against the North British scheme to Parliament, they failed to appear to give supporting evidence.[94] By the time the Caledonian committee had re-established itself as an efficient campaigning body, the North British bill was almost through to authorisation, and the Glasgow Dumfries & Carlisle could plausibly argue that, since it appeared that there were to be two border railways, they were at least as entitled as the Caledonian to build the western one.

The east-coast projectors were much better organised. Their efforts in the late 1830s had admittedly done little to disturb their opponents. Blackmore reported to the Newcastle & Carlisle that a coast line would not be able to compete with shipping, and that Stephenson was in desperation planning branches up all the river valleys to attract some traffic from the hinterland.[95] One concerned landowner was assured by his lawyers that 'although the Survey of the Rail Road proceeds, it

Table 77. *C. F. Davidson's Comparisons of Routes to link London, Edinburgh and Glasgow, 1838*

Route	Summit height	Mileage built	Mileage to build	Distances London– Edinburgh	London –Glasgow	Cost
	ft.	m.	m.	m.	m.	£'000
East coast	370	332	114	399	446	1824
Richardson's (Carter Bar)	1370[a]	285	149	387	413	2384
Locke's (Shap and Annandale, incl. Edinburgh branch)	1000	236	190	398	397	3040
Stephenson's western (Cumberland coast and Nithsdale)	570	342	150	492	411	2400

a: or 780 feet in tunnel

Source: SRO, GD206/1/63A/5, Hall of Dunglass Muniments, C. F. Davidson, draft 'Report on Proposed Railway between Newcastle & Edinburgh & Glasgow', 31 July 1838.

is very probable that it will never be executed, — for it will certainly never pay the Shareholders'.[96] And the Great North British project of 1838 had indeed gone into abeyance in face of the increasing depression. The projectors had not, however, given up, and the enquiries of Smith and Barlow found them better prepared than any of their rivals. George Stephenson's report on the east coast from Newcastle to Edinburgh was by far the most detailed engineering submission made to the commissioners,[97] while the indefatigable Charles Davidson, later to be the North British secretary, bombarded them with statistics for estimated traffic by rail and sea, with 'Remarks on the Herring and Ling Fisheries', and with 'Statistical Observations on the Trade and Manufactures of the Principal Towns on the East Coast of Scotland and on the Facilities for Commercial Intercourse with England'. All of these the commissioners obligingly printed in the appendices to their report, and some later reappeared in the North British prospectus as figures from the commissioners' inquiry, with the implication that Smith and Barlow had accepted their accuracy.[98] Table 77 indicates one of Davidson's earlier conclusions, by which his line had an advantage on cost, on summit levels, and on mileage required to be built to link the Scottish and English rail systems, and only suffered on the length of the journey to Glasgow. He also claimed that the west coast, unlike the east, was adequately catered for by existing transport:

> The slow progress of steam navigation on the east coast ... is in a great measure to be ascribed to the want of public confidence in it as a safe and regular means of conveyance for passengers ... Upon the west coast the passage has seldom been interrupted; and now that a Railway has been opened from Glasgow to Ardrossan, the communication with Lancashire has been rendered quite sufficient for the trade, both in passengers and goods.[99]

On the other hand, the depression seems to have caused a certain failure of nerve even among Learmonth and his colleagues. The first meeting of the North British, held in the offices of his other company, the Edinburgh & Glasgow, endorsed a line only as far as Dunbar, for which the prospectus was issued in January 1842. This proposed to build $28\frac{1}{2}$ miles of line for a capital of £500,000, and anticipated a net revenue of £58,222 and a return on capital of about $11\frac{1}{2}\%$, even after allowing a comparatively reasonable 40% for working expenses. Although the bleak border lands had been left for future consideration, 'the great importance of establishing a line of railway to England' was emphasised, not least by another scathing attack on the Caledonian:

> From the magnitude and cost of the works, and the smallness of the population and traffic, the prospects of success held out for it by the Commissioners are so small, as to render it improbable either that the line will be subscribed for, or that the requisite security for its completion within a definite time can be given.

The North British line, on the other hand,

> having been approved not only by some of the Best Engineers of the present day but also by the Government Commissioners, it may be safely assumed, that no material difference of opinion can exist regarding it.[100]

Tait's Edinburgh Magazine reflected Edinburgh commercial opinion when it insisted that the extension to Berwick and a junction with the English system should not be delayed:

> Upon this project it depends whether Edinburgh is to continue retrograde, as it has done for the last quarter of a century, or again spring into new life and vigour.[101]

The projectors went to work with enthusiasm. At their first meeting they agreed to approach four of the principal aristocrats of south-eastern Scotland (Buccleuch, Dalhousie, Lauderdale and Melville), all of whom were to give their assent. The prospectus was sent to all shareholders in the Edinburgh & Glasgow, and to all prominent citizens and public companies in Edinburgh itself, while the city was divided into sections each of which was to be canvassed by two or three committeemen. An agreement that each committeeman was to dispose of one hundred shares apart from his own purchases was soon rescinded. The plans of the old Great North British were purchased for £2262, although no payment was actually made until 1845.[102] But progress was unsatisfactory, and was not helped by a running dispute between the main committee, which favoured a route near the coast of the Firth of Forth, and a group led by the Marquis of Tweeddale and supported by the opinion of Smith and Barlow, which insisted that the line should pass through Haddington.[103] The committee had agreed not to raise money in England until they had guarantees of £50,000 from Scotland. In June 1842 Learmonth, Davidson and another committeeman visited England, seeing

The Battle for the Border 293

among others Hudson, George Stephenson and George Carr Glyn, and returning with little hope of raising English money while the recession lasted; English advice had been that the prospectus should not be launched there until there was a certainty of raising at least £100,000 south of the border. By October the committee had raised only £79,500 of the £300,000 required to go to Parliament, and, as the Caledonian could not possibly be ready for 1843, it was agreed 'to suspend operations for a season'.[104]

By May 1843 the alliance with Hudson was confirmed, and, as the North British committee gratefully minuted, 'Mr. Hudson likewise promised that he and his friends would do everything in their power to assist the parties here in their undertaking' — assistance which included Hudson's private subscription for £50,000, an obligation which he later got the York & North Midland Railway to take off his hands.[105] Thus fortified, and having realised that a line to Dunbar was not an attractive investment, the North British recovered its courage and in its new prospectus in August 1843 proposed a line to meet Hudson at Berwick. This time their project came out in a time of recovery, when, as they told their subscribers, 'the extreme cheapness of materials and labour, and the favourable state of the money-market' made railways a much more attractive investment.[106] And above all they had gained a year's advantage over their western rivals.

Progress to authorisation on 4 July 1844 was remarkably smooth. The ineffectiveness of Caledonian opposition has been noted already, and its committee was eventually reduced to asking the House of Commons committee on competing bills to agree that, if the North British were authorised, that would not be a reason for opposition to the Caledonian in 1845.[107] The Grand Junction's hope that a petition by four English companies for a new government commission to prevent competing border lines 'may possibly stop the Berwick line for one year which is all that we require' remained unfulfilled.[108] The only other opposing railway, the Edinburgh & Dalkeith (which was backed by 'a powerful body in the House of Peers'), was literally bought off when the North British agreed to purchase it for £113,000[109] — a reasonable bargain for the shareholders of a small company whose inconvenient station and leisurely horse-drawn operations made it difficult to compete with road vehicles for either coal or passenger traffic. Landowners along the line were seldom totally hostile, although they might drive hard bargains: Sir John Hall of Dunglass, for instance, after observing that 'the injury done to all those objects which a country gentleman so highly cherishes can never be compensated', accepted £12,000 for 57 acres of land and damages to amenity, against an original company offer of £7,000.[110]

To the company's surprise, the second reading of their bill was totally unopposed.[111] Although the parliamentary costs, later estimated at £17,825, were substantial, and considerably greater than any previously incurred in a single year by any Scottish company, the directors were happy to note that the expense was

> certainly not so great as it would have been had the contest been carried on next Session against the projected Caledonian line.[112]

North British wellwishers may later even have felt that the bill's passage through Parliament was too easy. Nine months from prospectus to authorisation was rapid progress for an important pre-mania railway, and the projectors arguably spent too much time raising funds and not enough ironing out the snags in their plans. Even as late as June 1844 Miller had to appear before the parliamentary committee to obtain approval of a last-minute deviation of the route.[113] The lack of a comprehensive scrutiny before Parliament by hostile interests allowed the company to start construction with inadequate plans: coupled with undue haste by the directors and incompetence by some contractors, this set the company on the impecunious path which yielded only one dividend in its first decade.

V. The triumph of the Caledonian

The authorisation of the North British changed the terms of the discussion in the west. Both the Caledonian and the Glasgow Dumfries & Carlisle now had to argue that there should be two border railways, against the beliefs of those like the Railway Board member G. R. Porter, who still believed in a single Anglo-Scottish line, and in concentrating efforts on linking Newcastle and Berwick, even though

> It is precisely in that quarter that a railway is least needed, by reason of the numerous, powerful, and well appointed steam ships by means of which the traffic between England and Scotland has been so very greatly extended and is now carried on.[114]

If there were to be two lines, Edinburgh's link to the south appeared to have swung the western balance back to the Nithsdale route, with its easier engineering and its substantial intermediate traffic in Ayrshire. Porter believed that 'all which the projectors of the Caledonian Railway propose to accomplish will be done by the completion of the Eastern line' apart from a link to Lancashire: if the western sea services were considered inadequate, this link should certainly be by the Glasgow Dumfries & Carlisle.[115] At the end of 1844 the Nithsdale route received further formal support from a source which had alway been sympathetic, when the Edinburgh & Glasgow company agreed to subscribe £100,000 to the Glasgow Dumfries & Carlisle.[116] The Nithsdale provisional committee confidently informed its subscribers that the construction of the North British 'removes the only ground on which the Royal Commissioners had preferred a Central line'.[117]

Discussions at the end of 1844 consolidated the anti-Caledonian alliance, which had been geographically sensible from the start, and which had been formalised in the spring of the year by Learmonth for the North British and Leadbetter for the Ayrshire.[118] In September the North British determined to upgrade its newly acquired Dalkeith line and extend it to Galashiels, in the process buying out an existing independent project for a Galashiels Railway for £1200. By mid-October the directors had agreed to go to Hawick: although they were clearly aiming at a connection to Carlisle, thus establishing a direct route between Edinburgh and Lancashire and destroying another argument for the Caledonian, funds would not permit them to apply for the section of the route beyond Hawick.[119] Within a week of this decision, the Nithsdale projectors proposed that they would build the missing Hawick-Carlisle link, subject to North British agreement and parliamentary rejection of the Caledonian. By the spring of 1845 a formal alliance of the

North British, the Glasgow Dumfries & Carlisle, the Ayrshire and the Edinburgh & Glasgow had agreed to guarantee to the Board of Trade that the Nithsdale company would build the Hawick-Carlisle link, and to form a joint purse of up to £30,000 for expenses in opposing the Caledonian; the first two companies also agreed jointly to guarantee a 4% dividend to shareholders in the Hawick and Carlisle line.[120]

Not everyone, even apart from the Caledonian promoters, was impressed. According to Lord Dalhousie, in a memorandum drafted a fortnight after he became President of the Board of Trade:

> That the Caledonian is in itself preferable to either of its competitors simply seems to be admitted in the fact that they have felt it necessary to combine, so as to form one scheme, in order to stand a comparison with it . . . A combination of the two would be less expedient than the sanctioning of the Caledonian alone.[121]

Opposition also came from George Hudson, concerned about the effect of the Edinburgh & Hawick line on coal supplies to the Borders from Northumberland, and presumably also about the possibility of the North British diverting much of its English trade to the west of the country and away from his lines. But Learmonth, secure in his company's authorisation, was now prepared to be independent of the Railway King, and the North British board noted that 'Mr Hudson of York does not bind himself to this opposition, and the meeting did not consider the opposition to be important'.[122] The directors went ahead with the promotion of a theoretically independent Edinburgh & Hawick company, for which all of the £400,000 capital was subscribed by the North British in the names of the directors. They had estimated that the line would pay even without the Carlisle extension, yielding a gross revenue of £48,978 and a net £29,387, or an adequate if unspectacular 7.3%. After authorisation the shares were to carry a guarantee of 4%, and to be allocated to existing North British shareholders.[123]

One Caledonian supporter was convinced that the Hawick line was not a serious proposal, and would be abandoned if the Caledonian were rejected: 'it is well known that they [the North British and Glasgow Dumfries & Carlisle] have brought this proposal forward solely for the purpose, as they think, of assisting their opposition to the Caledonian'.[124] It was certainly true that the prime aim of all the allies was the destruction of the Caledonian threat. The Edinburgh & Glasgow might reasonably have been hostile to the Hawick plan, which if successful would prevent it having any share of the Edinburgh-Carlisle traffic, but it was much more afraid that 'the very unproductive character' of the Caledonian main line would force the company to compete for traffic between Edinburgh and Glasgow by offering cheap rates via Carstairs.[125] The alliance certainly promised well for the Ayrshire-Nithsdale route, which by now was not only stressing its superior gradients and its greater local traffic, but also comparing its northward links over the proposed Glasgow Junction favourably with the Caledonian dependence on inadequate Lanarkshire mineral lines. The Ayrshire even offered to cut eight miles off the through route by building a new direct line between Glasgow and Kilmarnock.[126]

The early months of 1845, however, dealt several blows to the alliance. In

January Edinburgh Town Council, with only two dissentients, petitioned in favour of the Caledonian. The first reason offered by Lord Provost Adam Black was that, in the council's opinion, engineering difficulties south of Hawick ensured that no line would reach Carlisle by that route, and that support should be given to the most direct practicable line between Edinburgh and the west of England. Probably more important in fact was the Caledonian promise of supplies of high quality 'parrot' coal for 8/6d per ton against the existing price of 20/-, while the clinching factor was an agreement between the railway and the Edinburgh Water Company (whose directors included both Lord Provost Black and Provost Reoch of Leith) to lay a main supply pipe along the line.[127] As the council petition declared:

> Your memorialists consider this a most important and peculiar feature in favour of the Edinburgh branch of the undertaking, tending to diminish the expense of one of the necessaries of life, and to give facilities for increasing the comfort of the poorer classes, for whose accommodation public baths are about to be erected.[128]

The Caledonian, mindful of the public relations value of the agreement, agreed to charge only £5 per year for rent if the project was carried out by a public trust, while the water company subsequently took the railway's chairman and vice-chairman on to its board.[129] The company now had valuable council support from Glasgow, Edinburgh and Leith, and its situation in Glasgow improved in the same month when the Merchants' House refused to support the Ayrshire company's opposition to the Caledonian bill: 'the more railways which have their terminus at Glasgow the better for the community'.[130]

Two months later came a further setback for the Nithsdale interest, with the publication of the Railway Board's *Report on the Schemes for Extending Railway Communication in Scotland*. The Smith-Barlow report was receding into history, and there was a loss of conviction in the Caledonian's continuing description of the Glasgow Dumfries & Carlisle as

> a scheme got up in opposition to the recommendation of the Government Commissioners, by comparatively a few interested parties, to serve their own ends.[131]

But the Railway Board, as part of a general examination of all the Scottish schemes put forward for parliamentary consideration in 1845, now restated the case for the Caledonian as the main western line across the border.[132] It was not usual for any branch of the government to state a preference for any particular railway route: such intervention had been rejected in the 1830s as unwarranted interference with private enterprise, and the Smith-Barlow enquiry had been regarded as a very exceptional case. The Railway Board, however, during its eleven-month existence, was much more prepared to offer detailed guidance to Parliament, and indeed for most of its life prepared a critical report on each plan before it came under parliamentary scrutiny.[133] The Caledonian was fortunate that both the major 1840s reports on trans-border lines favoured the shortest practicable route north from Carlisle against either easier gradients or greater intermediate traffic. The Board's ministerial overlord, Dalhousie, gave grudging approval: 'I do not think that we can say that the Caledonian line is so hopelessly devoid of all prospect of remuneration as to make it our duty to report against it'.

Dalhousie still believed that another border line was probably unnecessary, but accepted that to forbid any construction when three companies wished to undertake it would be an unjustified restraint upon the rights of private capital.[134] Certainly he did not offer the Caledonian the wholehearted support which some of its agents had claimed and which had worried a North British supporter:

> The agents of the Caledonian declare roundly that they have the declared support of *Dalhousie* & the govt. I wrote to ask him if this was the case. He tells me that *it is a lie* as I was quite sure it must be.[135]

The Railway Board recommended that the rival projects should be restricted to serving local traffic, with the Ayrshire extending only as far as Cumnock and the North British stopping at Hawick. Dalhousie accepted the value of the latter from his own knowledge:

> The country is of course well known to me . . . I feel strongly that this line cannot be palpably objectionable if I can arrive at the conclusion that it ought to be allowed to go on: because it is a line which crosses a part of my property in a very disagreeable manner, and naturally & necessarily my bias was against it.[136]

The Railway Board report was in practice to ensure passage of the Caledonian bill. The company already had an advantage in parliamentary terms over its rivals, since it had close connections with Lancashire interests in general, and with a number of Lancashire and London MPs in particular, while its rivals had no English parliamentary links at all, since Hudson disliked any plan which would divert traffic from the east coast to any route by Carlisle. While Scottish MPs might vote on the basis of some knowledge of the different routes, and on the practical interests of themselves or their constituents, the great majority of English members would have little reason to prefer one scheme to another. Thus they were easily swayed both by the lobbying of pro-Caledonian members, and perhaps even more by the apparently impartial report of the Railway Board. Most southern English members accepted the desirability of a line to Glasgow, but were quite prepared to accept official guidance as to its route.

During 1844 the Caledonian committee had also substantially put its house in order. In the first place, the crucial support of the connecting English companies had been consolidated. Although most of the names on the committee of management, and indeed the name chosen for the company itself, were emphatically Scottish, and although the *Railway Times* claimed that the company was

> dissimilar to . . . any other, having been formed from the first of a party of enterprising noblemen and gentlemen, proprietors in the district through which the line passes,[137]

the English connections were essential. At the March 1844 meeting which approved the prospectus and the estimated cost of £1,800,000, confirmation was given of the £350,000 subscription to be made by the Grand Junction and its associated English Railways, while a further £250,000 was confidently anticipated from individuals already connected with the company. This ensured a sound financial start, even although the London & Birmingham decided to offer only approval and not subscription money.[138] Within a week of the appearance of the prospectus, the London market had taken up a further £400,000 of scrip.[139] Although the Lancaster & Carlisle declared early in 1845 that it had in fact voted no subscription and did not intend to take any shares,[140] such a setback could no longer damage the progress of the Caledonian, which was steadily gathering momentum under the impulse of approval from the Railway Board and from English financial interests, and of the early enthusiasm of the impending mania.

Another problem which the Caledonian committee did not solve until after the prospectus appeared was the approach to Glasgow. The possibilities were to construct a line of their own, to use the existing Lanarkshire coal lines to reach the Garnkirk's station, or to reach the south side of the Clyde over the newly projected Clydesdale Junction, which was a revival of the old plan to link the Polloc & Govan to the coal and iron area around Wishaw (see Map 17). If they were to use another line, the Caledonian directors would clearly require their trains to be given priority over the smaller companies. Already in the spring of 1844 Caledonian representatives had tried unsuccessfully to pressure the Lanarkshire companies into taking shares in the railway, since 'the present opportunity affords the only hope of securing the Railway communication from England through this district'. Though the financially straitened coal companies did not subscribe, they did agree to endorse the scheme in the Caledonian prospectus.[141]

By the summer negotiations were in progress for running powers into Glasgow. Locke, whose advice was generally accepted without question by the committee (they had agreed, for instance, to accept his route for the Edinburgh branch before they had even read his report),[142] favoured either the use of the Clydesdale Junction or an independent line, but in fact negotiations concentrated on reaching Glasgow over the Wishaw & Coltness, the Garnkirk & Glasgow, and a short connecting section of the Monkland & Kirkintilloch. After much hard bargaining a complex agreement was reached in August with the Wishaw & Coltness, by which the Caledonian agreed to pay over 36% of gross receipts from any of its traffic which travelled for less than five miles on the Wishaw line, and 30% of gross receipts from its other traffic, in return for which the line was to be converted to standard gauge and doubled throughout. Similar agreements soon followed with the other two companies.[143] In all cases the smaller companies concentrated on ensuring that their local traffic was not included in the agreements, but it was inevitable that, if the Caledonian offered remotely reasonable terms, they would have to agree. The threat of an independent Caledonian line, which might even send off branches, through their area was conclusive: there was no way small companies could stand up to the pressures exerted by a competitor with the influence and financial power of the Caledonian, even before its authorisation.

In early 1845, particularly after the Railway Board report, Caledonian confidence increased. Meetings of the committee of management were held in London, so that members might be on hand for parliamentary business. They were generally presided over by company vice-chairman Colonel Graham, supported mainly by the two most active MPs, Hope Johnstone and Lockhart, and the enthusiastic Edinburgh accountant Charles Barstow; Belhaven, now a lame-duck chairman since he had determined not to be on the board after authorisation, remained for the most part in Scotland to conduct negotiations with landowners and road trustees.[144] By the end of February the committee were purchasing rails, and had accepted the tender of McKenzie, Brassey and Stephenson to build the line for £1,275,000. By May the long-drawn-out negotiations with the trustees of the Glasgow to Carlisle road had been brought to a satisfactory if expensive conclusion, and the last of a series of recalcitrant landowners (who included committeeman Sir William Jardine) had been bought off or had agreed to go to arbitration. The company's minutes in the last months before authorisation on 16 July 1845 show few signs of apprehension.[145]

The Caledonian indeed could now afford to look beyond the achievement of its own main line. The Railway Board report had identified a group of lines, all engineered by Locke and Errington, which it called the Caledonian system: they included the Clydesdale Junction, the Caledonian & Dunbartonshire Junction (from Glasgow along the north bank of the Clyde to Loch Lomond), and three companies which together formed the route from the Caledonian to Aberdeen (the Scottish Central, the Scottish Midland Junction and the Aberdeen).[146] These, although clearly of great value to Caledonian planning, were independent promotions, and asked nothing of the Caledonian company but general support. In March 1845, however, the Caledonian committee initiated a project which broke all the precedents of Scottish railway planning, and which, although projected as an independent company, was clearly to be directly under parental control. Locke was sent to survey a route from Ayr to the main trunk near Lanark, and thence to Peebles, Galashiels and Kelso, near where it would link with a branch of Hudson's projected Newcastle & Berwick. The new company was to be called, with stunning lack of subtlety, the Caledonian Extension.[147]

With the Extension plan, the Caledonian abandoned the principle of the natural territory of companies, which all promoters were liable to invoke when rivals pushed branches in their direction. One of the most jealously protected rights of any railway company was its monopoly in the immediate vicinity of its route; early companies had punctiliously avoided each other's routes, and expected to be shielded by Parliament if any aggression on their territory was proposed. Even where companies were largely competing for the same traffic, as with the Caledonian and the Glasgow Dumfries & Carlisle, their routes met only at the towns which were the inevitable common terminals. The Caledonian Extension cut straight across the territory claimed by both the Ayrshire, through its Nithsdale protégé, and by the North British offshoot to Hawick. It offered an alternative route between Glasgow and the Ayrshire ports in competition with the Ayrshire company: it would allow Hudson a direct route to Glasgow, and even a devious one to Edinburgh, independent of the North British, and strengthen his hand in negotiations with the independent-minded John Learmonth. It would give the Caledonian effective control of, and the Grand Junction access to, virtually the entire Scottish lowlands. As the Caledonian directors told their shareholders, the Extension would

> in connection with the Branches to Edinburgh and Glasgow of the Caledonian, form direct and continuous communications between Edinburgh and Ayr in the one direction, and between Glasgow, Berwick, and Newcastle in the other, and will contribute very materially to increase the revenue of the Caledonian Company.[148]

Yet because it ran from west to east across the country, it could claim to be catering for traffic which was not helped by the existing or planned north-south border lines. The Extension plan, with its direct threat to the interests of the Ayrshire, the North British, and the through traffic of the Edinburgh & Glasgow, was the idea of a very confident company.

There was little that the opposition could do now. The Edinburgh & Hawick plan could be achieved, and indeed, as long as it could not be extended to Carlisle, the Caledonian had little apparent interest in trying to prevent it. By May 1845, however, the Caledonian had nevertheless come out in opposition, reversing an earlier decision made when they feared that North British pressure might persuade the Board of Trade to oppose their own Edinburgh branch. £2600 was spent on fruitless opposition, the cost being shared between the Caledonian proper and the Extension.[149] But there was little hope for the Glasgow Dumfries & Carlisle. It was virtually certain that Parliament would sanction only one line between Glasgow and Carlisle, and the Caledonian now had the support of two official reports in four years, of powerful English interests, and increasingly of civic leaders in Glasgow and Edinburgh — a fact which it rubbed in by inviting MPs Macaulay of Edinburgh and Dennistoun of Glasgow to join Hope Johnstone and Lockhart in presenting the bill.[150] And in February 1845 the crucial landowner, the Duke of Buccleuch, came out in favour of the Caledonian. Buccleuch's main interest was in obtaining a through route from Edinburgh to Carlisle by Hawick, which would traverse his extensive estates in the Borders and improve the prospects of his two coalfields, by connecting Canonbie to the railway system and by giving Dalkeith access to southern markets. Since the Glasgow Dumfries & Carlisle proposed to construct a line from Carlisle to Hawick, to link up with its North British allies, the duke was prepared to countenance the company in spite of his horror at its interference with his Dumfriesshire home at Drumlanrig; as his legal adviser John Gibson agreed, 'I should care very little for this Railway were it not so much connected with the Hawick & Carlisle line'.[151]

The Nithsdale project's lawyer had, however, made it clear that his company would withdraw from the Hawick-Carlisle route if the Caledonian were passed, since it believed that a monopoly of the traffic between Edinburgh and Carlisle would be necessary for financial success. The Railway Board report convinced the duke that the Caledonian would succeed, and he determined formally to oppose the Glasgow Dumfries & Carlisle, 'thus saving them the risk & expense of what would appear an almost hopeless contest'.[152] Gibson, who believed that this decision might cost Buccleuch £10,000 a year, proposed a jointly owned line from Hawick to Carlisle, which would allow the Caledonian to ensure that it did not take all the through traffic; while the Glasgow Dumfries & Carlisle, he thought, would agree, the Caledonian, 'spoiled with success & patronage, have hitherto been unapproachable'.[153] Equally unsuccessful was a desperate plea from Leadbetter, who was

> convinced the decision of the Board of Trade must have been given in consequence of erroneous information which we have not had any opportunity of seeing or rebutting,

and who believed that, if Buccleuch would only assent or even remain neutral, the Glasgow Dumfries & Carlisle could still win in committee.[154] Buccleuch's lead was

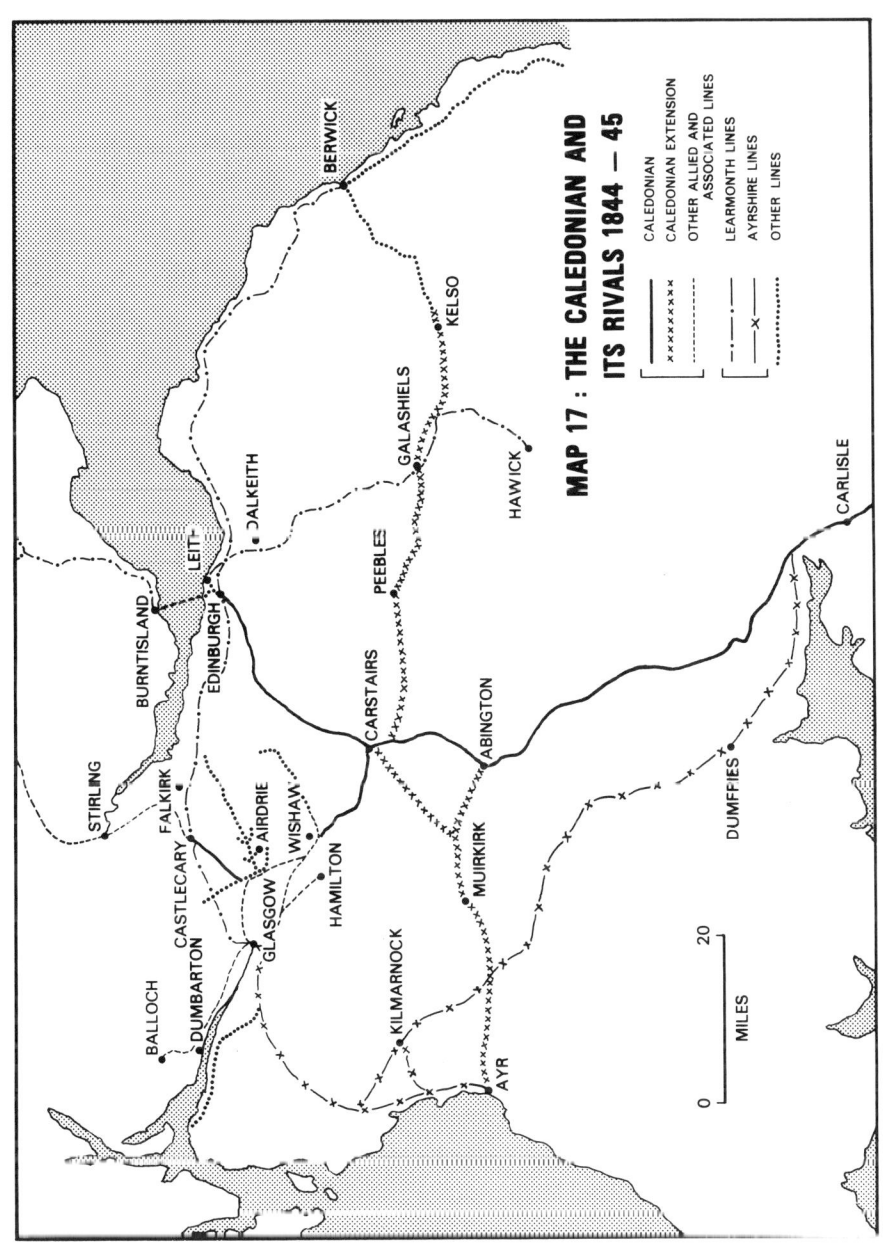

followed by the principal landowner between Carlisle and the border, Sir James Graham of Netherby.[155] While the opposition of two landowners of such importance was enough to kill the chances of the Glasgow Dumfries & Carlisle, it was a further misfortune for the company that both Buccleuch and Graham were cabinet ministers.[156]

The Nithsdale promoters battled on. In March 1845 they suggested that if the Caledonian got their main line to Glasgow, they should drop their Edinburgh branch in favour of a line by Hawick (possibly shared between the two companies), and allow the Glasgow Dumfries & Carlisle to go forward as a feeder. In May they suggested that the Caledonian should abandon its proposed branch to Dumfries, that both sides should end opposition to each other's plans, and that they should build a joint line from Carlisle to the border.[157] It was now clear that such suggestions were made out of weakness, in a despairing effort to salvage something from the session, and the Caledonian inevitably rejected them out of hand. Even John Gibson developed some sympathy for the Nithsdale efforts:

> While their opponents, the Caledonian, have been most powerfully & cheerfully supported, they have had to fight every inch of their way, with hardly any encouragement, & I am surprised they think of going on.[158]

At the end of May, after the Commons committee had favoured the Caledonian, with the Glasgow Dumfries & Carlisle existing as a feeder if at all, the Caledonian wrote to its rivals requesting in effect an unconditional surrender:

> The Directors of the Caledonian Railway, presuming that after what fell from the Committee today, all past feelings of rivalry and hostility between them and your clients will cease; entertain a most anxious desire to fall in with the views indicated, though not fully explained by the Committee . . .[159]

The Nithsdale reply (which was never inserted in the space left for it in the Caledonian minute book) was of such a nature that the Caledonian board resolved to make no further contact. A week later they learned that the preamble of their bill had been approved and that of the Nithsdale rejected. Although strenuous opposition to the Caledonian continued, forcing its parliamentary costs up to a very expensive level, there were no more serious obstacles before authorisation.[160]

The parliamentary session of 1845 was a triumph for the Caledonian. Their only loss was their planned Dumfries branch, and even that was due to doubts about the best route. Against that could be placed not only their own main line to Glasgow and Edinburgh, but the authorisation of the Clydesdale Junction and the various lines reaching to Aberdeen and Dundee. Successful opposition had been mounted against not only the Glasgow Dumfries & Carlisle but also the Glasgow Junction link between the Ayrshire and the Edinburgh & Glasgow, and the latter company's branch to Stirling. No railway had been permitted to cross the empty lands south of Hawick. For the following session the Caledonian looked forward to blocking its possible rivals by sending out branches to Dumfries and Langholm (but positively not to Hawick), and to encouraging the Extension as well as other nominally independent connections to Portpatrick and Dumbarton. The board's satisfaction appeared in their first report:

This satisfactory result was not attained without a long and arduous contest which was excited, conducted, and in a great measure paid for, by some of the principal Railway Companies previously established in Scotland. Of other opposition there was little or none.

The combination of the powerful bodies above alluded to, the pertinacity of their opposition, the expense unsparingly incurred by them, and the sacrifices they were willing to have made in the construction of costly works to supply the defects of *their* System of Railway communication afford evidences not to be mistaken that the struggle was looked upon by them as highly important. All parties, in fact, were aware that it involved the command of the general railway traffic of Scotland.

The last session of Parliament presents an uninterrupted series of successes in favour of what the Board of Trade aptly designated the "Caledonian System" of railway communication in Scotland.[161]

For the Caledonian's rivals the prospects ranged from moderate to gloomy. John Learmonth's dream of an Edinburgh-based empire which would extend to control the main railway system of Scotland was over: at best he could look forward to a prolonged conflict with a rival possessed of at least as much power as his own companies. The Edinburgh & Glasgow now had to fear competition on the intercity route; price-cutting wars were a feature of the years following the mania, though the extra mileage on the Caledonian route by Carstairs meant that passengers at least would stay with the old line. Learmonth's new Edinburgh & Northern company in Fife was unchallenged for most of the county's local traffic, but the inconvenience of the Burntisland ferry made it an unattractive alternative to the 'Caledonian system' route by Stirling and Perth for northern traffic. Much effort was to go into attempts to detach the key link in the route, the Scottish Central, from the Caledonian interest. Unless this could be done, Learmonth's system would effectively be confined to eastern Scotland south of the Tay, apart from its main line into Glasgow.

The North British had at least an unchallenged position on the east coast, and 1845 brought the authorisation of Hudson's Newcastle & Berwick to complete the connection to the south. It also saw approval of the line to Hawick, staking the North British claim to the central Borders.[162] Although there was as yet to be no extension to Carlisle, it had been the company's own decision not to apply for the line; its doubtful prospects of profitability and the heavy expenditure required on the main line to Berwick counselled caution. Although the North British supported the efforts of the Nithsdale promoters to fill in the gap, their willingness to leave the route to another group of projectors whose chances of authorisation in face of the Caledonian were at best doubtful suggests that they did not yet attach the highest priority to the Carlisle line. For the moment they would be content if the issue remained open, which meant that the Caledonian must be prevented from constructing a blocking branch to Canonbie or Langholm. In fact it was to be another fourteen years before the North British line to Carlisle was authorised.[163]

The promoters of the Glasgow Dumfries & Carlisle had suffered total defeat. Conventional wisdom had declared that there was enough traffic for one transborder railway, while enthusiasts allowed that there might be two. The Nithsdale projectors now had to persuade Parliament and the investing public that there could reasonably be three. And, even if the Caledonian's Dumfries branch had for the moment been rejected, any Nithsdale line would require to obtain running powers over the Caledonian main line from Gretna to Carlisle. In Glasgow, no link had been achieved to the Edinburgh & Glasgow, and, while Learmonth's companies would remain anti-Caledonian, they could hardly be expected to give the Nithsdale as much support in future as they had when they hoped that its success would stop the Caledonian altogether. Logically it seemed that the Ayrshire company was to be restricted to serving south-western Scotland, with any English traffic depending on what could be retained by the shipping routes from Ayr and Ardrossan. Fortunately for the company, however, the country was moving into the full enthusiasm of the railway mania, when logic was in short supply. Although most of the projects of 1846 (including the Caledonian Extension) foundered, generally from a failure to raise funds after the mania speculation itself collapsed, the growing belief in railways and the financial support of the Ayrshire and the Glasgow merchants saw the Glasgow Dumfries & Carlisle through to authorisation.[164] By the end of 1846 one trans-border railway, the North British, had opened, and two more were under construction.

6
Summary and Conclusions

I. Scottish railways in 1844

THE previous chapters of this book have charted the development of the Scottish railway system from the first waggonways to the eve of the railway mania, and have examined the achievements, problems and financial results of the various companies. This final chapter recapitulates the main themes of the discussion, along with brief consideration of one or two other aspects which fit more easily into the argument at this point.

The authorisation of the North British in July 1844 is as good a symbol as any of the coming of age of the Scottish railway system. It followed a pause in parliamentary activity; no significant new Scottish company had been authorised for six years. Nor was there much constructional activity — after the opening of the Ayrshire's Kilmarnock branch in April 1843 work was continuing only on a few extensions to the Lanarkshire coal system, of which the most important was the Wilsontown line, and on the apparently perpetual building site between Edinburgh and Granton. Further lines were of course being planned, particularly across the border and northwards to Stirling, Perth and Dundee. In 1844 it was already obvious that the North British and the associated extension of the Edinburgh & Glasgow through Princes Street Gardens were only the first authorisations in a series which would create a comprehensive rail network in lowland Scotland, although it was not yet apparent that promotional enthusiasm would develop into speculative mania.

A few miles east of Edinburgh the North British crossed the line of the original Tranent & Cockenzie waggonway. In the century and a quarter between the two projects, the railway had developed from an experimental horse-and-gravity powered adjunct to coal mines into a central feature of the economy. Already the leading railway companies were among the largest in Scotland, and they affected directly or indirectly most of the population of the central belt. Already transport costs had been significantly reduced in many areas, particularly for collieries not served by canals, and already the pent-up demand for passenger travel had become apparent. 'Railway' had become the cant work for modern — in 1835 cloth manufacturer Henry Ballantyne of Galashiels even marketed a 'railway check'.[1]

Yet, while the main lines of the 1830s had given the English railway system a degree of cohesion and maturity, in Scotland the railway was still largely of potential rather than actual consequence. In 1844 it stood on the threshold of importance: in the next few years it was first to disrupt the financial system of the country, and then to become arguably the leading sector in the continuing expansion of the economy. Men like the Stephensons, as canonised by Samuel Smiles,[2] were to be held up as models of the great Victorian virtues of self-help and self-improvement, while others like George Hudson, and in a small way John Learmonth, were to enjoy a briefer period of spectacular success. The railway had come a long way from the days of the York Buildings Company, but in Scotland especially it had still much further to go.

The creation of the early railways was above all an example of free market capitalism in operation. By the early nineteenth century public opinion, nurtured by an increasing acceptance of the ideas of Adam Smith and the less gloomy of the classical economists, was strongly in favour of leaving economic growth to the benevolence of free market forces; the development of the railways came from a series of separate decisions by promoters in response to their perceptions of market opportunities. Neither government nor Parliament played any important part beyond passing the necessary acts of authorisation, while any concept of central planning of railway development in the national interest, as was shortly to take place in Belgium, was (insofar as it was considered at all) promptly rejected. Promoters' opportunities arose primarily from the growing industrialisation of the Scottish economy. From the start there was the need to convey bulky raw materials, especially coal, to urban, industrial or maritime markets, either in co-operation with water transport or increasingly in competition with it. From the 1830s there also developed a trade in general goods and passengers between the towns of central Scotland, which in turn encouraged the desire for a more convenient connection to English markets than was offered by coastal shipping.

II. The conveyance of coal

For most of the pre-mania period the overwhelming stimulus to railway development was coal, and coalowners and users were the main beneficiaries of the new transport. Demand for the product existed and was increasing, but might well be frustrated by high freight costs. Road carriage for more than a few miles could multiply the price to the consumer by five or ten times, and effectively prohibited the exploitation of resources in areas like Cumnock or Slamannan. In industrial England coal was the main incentive for the construction of canals, but in Scotland only the Monkland Canal was a major coal carrier before the advent of the railways. Water transport was commendably efficient at conveying large quantities of coal at reasonable expense, but it was less suited to gathering its cargoes at the coalfield end. Coal production, unlike iron, could not be consolidated at a few major works, but had to take place where the coal was found, at numerous pits, each with a relatively limited output, scattered haphazardly across the coalfield areas. Since new shafts were regularly sunk and old ones closed as they were worked out, their transport system had to be flexible, with collieries being linked to the main network at low cost. The solution was waggonways, whose overwhelming purpose was the short-distance conveyance of coal, generally in order to extend the catchment area of canals or coastal shipping, but also on occasion running directly to customers who, like the expanding iron industry, happened to be located on the coalfields. The concept of rail transport as subsidiary to water survived into the Union Canal's West Lothian project of 1824-25, and into the Forth & Clyde's role in the promotion of the Monkland & Kirkintilloch and the Ballochney.

Water-linked waggonways reduced coalowners' freight costs to a level with which road transport could not hope to compete, but they did little to promote competition in the provision of transport services. The main beneficiaries were coalowners who were located at an inconvenient distance from water transport, who could now compete on more equal terms with better located rivals; coastal and canal coal carriers, who had much improved access to supplies; and steam shipping in general, since bunkering coal became more readily available at a series of ports on the Firths of Clyde and Forth. Although coal prices did fall in the cities, customers of the Monkland Canal suspected that its high dividends reflected an abuse of a monopoly position – a belief presumably confirmed when the opening of the Monkland & Kirkintilloch enabled the Forth & Clyde Canal to compete in the Glasgow market and within two years halved the price of coal in the city.[3] Even so, the Monkland & Kirkintilloch and the Ballochney were initially simply further horse-worked extensions of the canal system, differing little if at all from earlier waggonways, and it is not surprising that some authors have included them in their accounts of waggonway development.[4]

The conceptual breakthrough, directly challenging the canals, had to await the adoption of the locomotive, and was made by the Garnkirk & Glasgow, which offered a through route from the coalfield to the city without the inconvenience of transshipment. As the system of coalfield lines extended through North Lanarkshire, most coalowners had the choice of supplying Glasgow either by rail or by canal, and Glaswegian consumers gained not only by the continuing increase in supply but by the much reduced prices caused by greater competition. The canals were perfectly capable of price reduction to remain competitive – the promoters of the Edinburgh & Glasgow were initially to accept the view that they could not compete with the canal for the bulk of inter-city heavy goods traffic[5] – and they could also retain their advantage in the bunkering trade since the Lanarkshire railways had as yet no connection to Glasgow harbour. To compete effectively, railways had to concentrate on flexibility, on the reduction of transshipments made possible by extending sidings into collieries and works, and on the greater speed of the locomotive.

Coal traffic was a major consideration for all railways authorised before 1836, and the overwhelming purpose of all those which achieved commercial success. Profitability, indeed, was geared to a large mineral traffic. The less prosperous of the early railways included the Newtyle lines, dependent for coal traffic on the demand of small towns and their rural hinterland; the Paisley and Renfrew, whose principal urban centre could be as well or better supplied with coal by water transport; and even the Slamannan, since mining development in its area was neither as rapid nor as extensive as the projectors had hoped. By contrast the dividends of the major coal-carrying lines, such as the Monkland & Kirkintilloch and the Ballochney, were steadily over 10%, while even less successful Lanarkshire lines, such as the Wishaw & Coltness or the Wilsontown, showed sufficient promise to be able to negotiate profitable leases to larger companies during the mania.[6]

The initial stimulus to railway development in Scotland thus came from the increasingly voracious demand of the economy for coal, but in itself this demand required a rail system of no great geographical extent or technological sophistication. The principal Scottish market, Glasgow, could perfectly well be supplied by sea or by the Monkland Canal; Edinburgh had long imported coal by sea from Tyneside, in spite of the need for road cartage from Leith, and from 1822 also received waterborne supplies via the Union Canal. For these cities, and for the Monklands iron industry, waggonways as a supplement to water transport were all that was required. For the more remote cities of Aberdeen and Dundee, the coastal trade in coal was under no threat from land transport. And although landlocked small towns and rural areas might be crying out for better supplies of coal and lime, the failure of the various schemes for canals or long-distance waggonways suggests that they offered little prospect of profitable trade,[7] and the record of the Dundee & Newtyle also implies that other projectors in similar areas had been wise to hold back. Before 1830, rails were still essentially an auxiliary part of a freight transport system firmly based on water.

Waggonway technology — essentially the provision of a roadway and rails suitable for horse-drawn carts — was simple, not expensive, and already in use in England and on the Continent before it came to Scotland. It was also adequate for the limited requirements of coalowners and consumers, and it is unsurprising that technological change before the mid-1820s was limited to experiments with types of rail and sizes of waggon. Even the advent of the locomotive had little advantage to offer if the sole purpose was to convey minerals over short distances to water; the waggonways of Fife and the Lothians continued to operate perfectly satisfactorily without locomotives until, after the mania, their areas were invaded by steam railways which had not been built primarily for the conveyance of coal. The exception was the Garnkirk, which offered locomotive speed as one of the attractions in its direct confrontation with the Monkland Canal. But the Garnkirk was not, in national economic terms, a strictly necessary railway. Without it, Glasgow would still have been adequately supplied with coal, while the owners along its route could easily have connected themselves to the canal by road or waggonway. The line had, of course, advantages; it reduced prices to consumers by introducing genuine competition among suppliers, and as the rail system extended it enabled an increasing number of coalowners to reach the Glasgow market without the inconvenience of transshipment to the canal. But, had the purpose of the railway remained simply the carriage of coal to towns, harbours and ironworks, there would have been no pressing reason for its technology to advance beyond that of the horse-drawn waggonway.

III. Diversification: general freight and passengers

By the 1830s promoters had realised the possibility, and in some cases the necessity, of diversifying beyond mineral traffic to the conveyance of passengers and general freight. Earlier lines had, incidentally and almost by accident, indicated the potential demand. Even the Kilmarnock & Troon and the Dunfermline & Charlestown, running between ports and substantial towns, had carried other goods and a few passengers alongside their essential coal traffic. The three Newtyle lines had to depend on non-mineral trade for the bulk of their income, even if its volume was insufficient to save them from financial chaos. Even railways with ample coal traffic diversified: by the early 1840s the Edinburgh & Dalkeith took more than half its revenue from passengers, while the Garnkirk's steam trains conveyed over 300,000 passengers per year between Glasgow and the Monklands area.[8] By this time the increasing reliability of locomotives, and the demand for passenger services revealed by the main English lines (and above all by the Liverpool & Manchester), had led to the promotion of Scottish inter-urban lines for which mineral traffic made up a relatively small part of their anticipated traffic. Perhaps the key change came when the Edinburgh & Glasgow preferred a fast flat route north of the coalfield to a hillier one through the middle of it, thus emphasising the new concentration on speedy passenger services.[9]

The new interest in inter-urban traffic did not mean the end of concern with the coalfields. The Ayrshire in particular, in the continued absence of a canal between Johnstone and Ardrossan, would provide the North Ayrshire coalfield with both a link to Glasgow and an export trade from the Firth of Clyde ports. The Edinburgh & Glasgow omitted all mention of coal from its prospectus (perhaps to avoid direct confrontation with canal interests by implying that heavy traffic would continue to travel by water), but in fact it received appreciable coal traffic from its link to the Monkland & Kirkintilloch – particularly as the management of the Monkland lines became increasingly divorced from the canal interest. The Edinburgh Leith & Granton was in theory well placed to profit from the sea-coal trade to the capital, whether or not Granton succeeded in supplanting Leith as its principal port. Even the Greenock company professed an interest in coal, since the joint line to Paisley was expected 'to open up an extensive mineral and agricultural district'.[10] Meanwhile a few coalfield lines, notably the Wilsontown, were authorised during the depression, while private individuals and companies continued to build waggonways, in the form of horse-drawn private railways and sidings which were in most cases linked to the coal railway network. The Monkland Canal's Drumpellar Railway was simply a waggonway with such additional status as might be conferred by parliamentary authorisation. Although the surviving records of the inter-urban companies do not distinguish between coal and other goods traffic, their mineral trade was sufficiently successful for both the

Ayrshire and the Edinburgh & Glasgow to propose numerous further branches and subsidiary companies in the coalfields during the mania.

The new lines might be intended for more diversified traffic, but there was nothing particularly adventurous about them. The conceptual precedents had been established in England, and for the most part they concentrated on established traffic arteries, following routes where road and water carriers had already demonstrated that there was a demand for transport services. There were small exceptions: the projectors of Strathmore were willing to accept the opportunity cost of tying up money in lines which would pay little or no dividend for the sake of 'the indirect advantages which such undertakings, whether lucrative to the promoters or not, never fail to secure',[11] and in the event the shareholders of the Arbroath & Forfar were to be at least a little better rewarded than those of the Newtyle lines. The Edinburgh Leith & Newhaven, designed primarily as an urban passenger railway, was as such an innovation in Scotland (in England the London & Greenwich had been authorised three years earlier), but followed a route well pioneered by horse buses. The other inter-urban lines were designed for traffic which had already been substantially developed by canal or estuarial services. On the whole there was little enthusiasm for expanding into new territory, and certainly very little sign of the belief which became apparent during the mania that almost any railway anywhere in the country would, merely by its very existence, create enough traffic to make a profit.

This failure to open up new traffic routes is reflected in the limited impact of the early railways on the Scottish economy; by 1844 they had merely demonstrated a potential for creating change rather than the actual achievement. According to Mitchell,

> The introduction of railways in Britain did not have a very great immediate effect on the economy, the main effects not being felt until during and after the great mania of the 1840s,

and the railways 'did not to any significant extent simply make possible what had previously been impossible'.[12] If this applied to the main lines of England, it was even more accurate about the more embryonic Scottish system, where most of the 260 or so miles open before the mania was in small lines of essentially local significance. Of these, perhaps only some of the coalfield lines, the Strathmore companies and the Garnock valley section of the Ayrshire were in areas where the inadequacies of existing transport had been serious enough to inhibit economic activity. Elsewhere the new lines were in competition with existing services which, particularly for freight traffic, had coped tolerably with previous demands.

The early railways offered their customers above all a saving of time. Where the only competition was from road carriers, they could be confident of also offering a cheaper service, but canal and river shipping was quite able to conduct and, as the Greenock discovered, perhaps to win, a price-cutting war. Apart from lines specifically constructed for mineral traffic, the railways found that their speed made them attractive for passenger and light goods traffic; indeed, the most successful lines in England 'maintained high dividends [over 6%] in the 1840s by concentrating on first-class passengers and high-tariff freight'.[13] In Scotland, even between Glasgow and Edinburgh, such traffic was not plentiful enough to keep a railway solvent, and attention had to be given both to third-class passengers and to the fight for a share of heavy goods traffic. Although the Edinburgh & Glasgow expected the bulk of the coal trade to remain on the canals, both it and the Glasgow & Paisley joint line conducted intense struggles against canal competition until mutual damage enforced compromise.

For freight traffic, however, Scottish railways suffered from numerous defects. Too often goods yard facilities were neglected or inadequate, notably on the Dundee & Arbroath which could handle very little goods traffic at the Dundee end. Directors and senior management often paid comparatively little attention to the freight department, whose superintendent was seldom one of the most important officials. Above all, there was the disconnected nature of the system, which, thanks largely to the lack of links within the cities, was still much more fragmented than a glance at Map 18 might suggest. In Glasgow, although the Glasgow & Paisley joint line was connected to the unimportant Polloc & Govan at Tradeston, it had no link across the Clyde to either the Garnkirk or the Edinburgh & Glasgow. Similarly in Edinburgh the Edinburgh & Glasgow had no connection to the Dalkeith line, nor until the opening of the Scotland Street tunnel in 1847 to the Edinburgh Leith & Granton. Until the promotion of the little Slamannan Junction in 1844, nothing was done about the one-mile gap at Causewayend between the Edinburgh & Glasgow and the Slamannan, thus ensuring that the latter's Edinburgh traffic would continue to be conveyed by the Union Canal. In both Dundee and Arbroath, physical gaps in the system remained to prevent through rail traffic. While the system remained thus disjointed, sending goods by rail might well involve more transshipments than using water transport. Even the Monklands lines, built specifically for the mineral trade and with direct access to Glasgow over the Garnkirk, had as yet no rail link to either the Clyde or the Forth, so that owners sending coal for export or for bunkering might well choose to send it by canal. Recognition of these weaknesses may explain why the parliamentary estimates of both the Ayrshire and the Edinburgh & Glasgow expected less than a third of revenue to come from goods traffic. Not until the end of the 1840s, when the main-line network offered through routes from Aberdeen in the north to the main English centres in the south, and when mania promotions had proliferated branches throughout the industrial areas, could the railways offer a more reliable, more flexible and cheaper freight service which could be seen as clearly superior for most purposes to water. Not until 1852 did freight receipts exceed passenger revenue on British railways.[14]

The inadequacy of the provision for freight is reflected in the modest welcome which the railways initially received from the iron industry at a time when it was undergoing rapid expansion. With some outstanding exceptions like John Wilson and the Bairds, ironmasters showed only a limited interest in pre-mania railways. Admittedly their capital resources might be committed to further expansion in their own industry, but also many of the works were located on or near the Monkland Canal and continued to give preference to water transport. Further expansion could take place on the same sites and use established carriers. The Garnock valley was opened up by the Ayrshire railway – John Leadbetter even

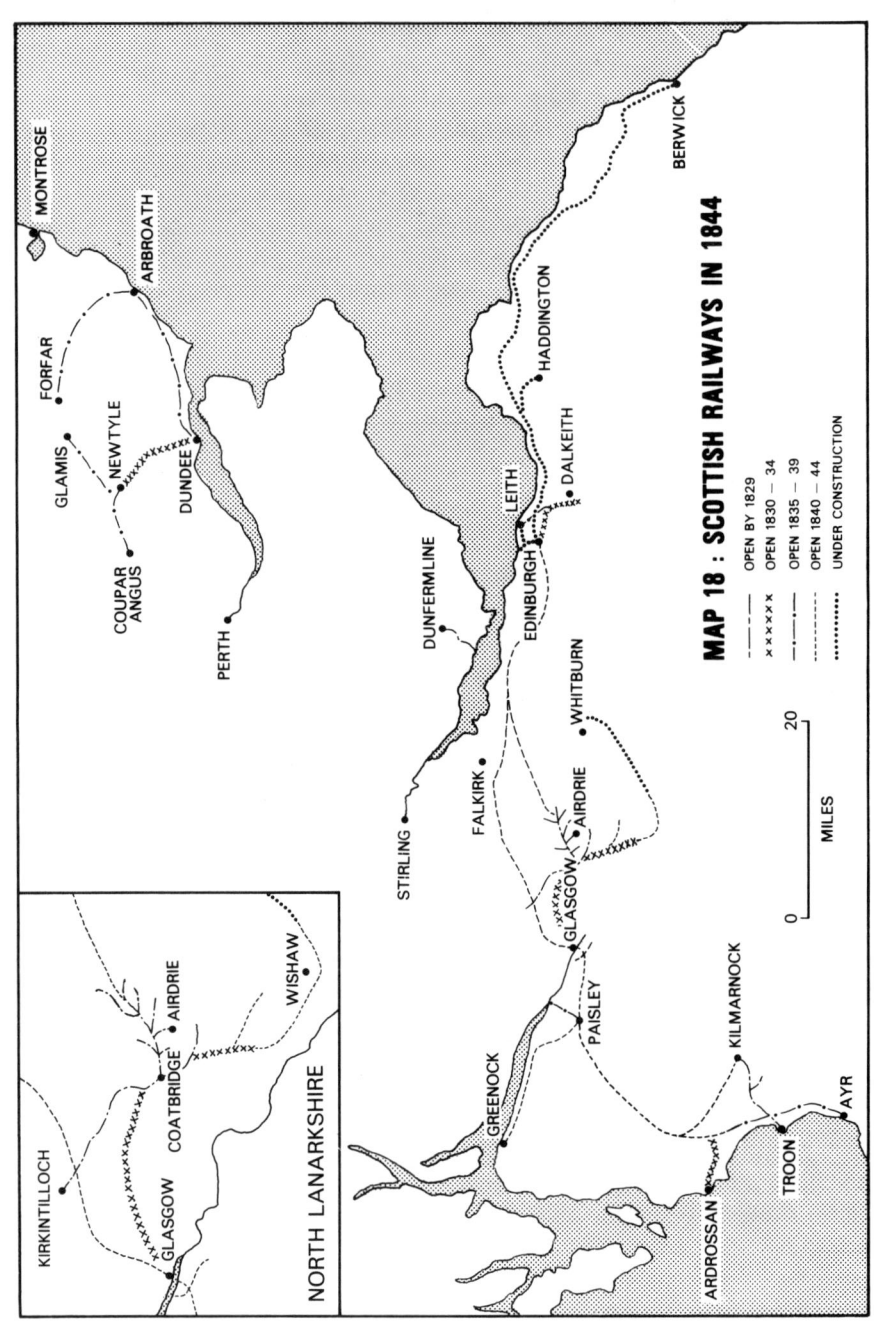

expected it to become another Monklands, particularly once the railway had reached Carlisle[15] – but in their early days the ironworks might have been better served if the planned canal from Ardrossan to Johnstone had been completed. On the whole ironworks used waggonways of short hauls on coal railways to bring fuel to the furnaces, but found that as yet the railway system was not sufficiently comprehensive to give satisfactory access to many markets. Other goods traffic, apart from coal, was also slow to develop. Stone and lime might leave the quarries by rail but were often taken in the traditional manner to water transport for dispersal beyond the local area. Agricultural traffic was generally a small source of revenue for most companies, although landowners and farmers who lived close to a line like the Arbroath & Forfar, or later the Caledonian, gained invaluable access to markets and to supplies of lime and fertilisers.

Another potentially important impact on national freight movement was the demands posed by the construction of the railways themselves, materials for which might come by road, by water or by already existing railway lines. According to Deane and Cole,

> The impact on the U.K. economy of expenditure on railway construction began to be significantly large in the second half of the thirties . . . By the triennium 1838-40 expenditure on construction and railway stock was running at a rate of over £$10\frac{1}{2}$ m. per annum.[16]

The share of Scotland in the British railway system by the end of 1843 was about one-twelfth, whether measured by capital authorised or by the amount expended on construction and rolling stock, which was not obviously out of line with the importance of Scotland in the national economy as a whole. Total Scottish expenditure since 1824 was some £4.7 million, heavily weighted towards the end of the period; £3.9 million of it had been spent by companies authorised since 1836, while more must have gone on extensions, improvements and replacements by older lines.[17] Details of the expenditure are not easy to come by, but it is clear that not all of the money was spent within the Scottish economy. Because the Scottish iron industry lacked both experience and ability in the manufacture of rails, much Scottish permanent way was purchased in England. Similarly, from the Garnkirk's 1831 purchase of two engines from Robert Stephenson onwards, and perhaps sometimes under the influence of English consulting engineers, locomotives were often purchased from Newcastle or Lancashire firms.[18] Of 91 locomotives in use before 1845, 55 were English and 36 Scottish-built. England was also the source for many of the pre-mania carriages and waggons.[19] The main impact of early railway construction on the Scottish economy may have been more through the payment of wages and salaries than through the purchase of materials and equipment.

There is little evidence in the surviving records to suggest that the levels at which goods rates should be set received peculiarly close attention from company management. Decisions appear to have been reached by pragmatic tinkering rather than by scientific attempts to calculate the levels which might give maximum profitabilty. Where competition was not to be feared, rates were presumably set at a level where it was believed (perhaps by instinct) that the equation between the volume of traffic attracted, the revenue per ton-mile and the costs of carriage yielded the maximum return. In practice, especially on the coal railways, the companies might also have to allow for pressures or even threats from important customers or from directors whose mineral interests might conflict with their duty to the railway; they would also have to consider whether a reduction in rates, or a change to a sliding scale which charged short journeys more per ton-mile than long ones, would create sufficient new traffic to compensate for any reduction in income from existing customers. If effective competition by canal or river existed, rates were established primarily by the effort to undercut the rival, with competition sometimes being taken well beyond the bounds of commercial common sense.

The major traffic success of the inter-urban lines was in the conveyance of passengers, which supplied all but the Arbroath & Forfar with over 70% of their pre-mania receipts.[20] The disjointed structure of the system, while inconvenient, was not a serious deterrent to passengers who could transfer themselves between stations with relative ease, while the companies made more serious efforts to establish centrally located passenger stations than to provide adequate freight terminals. For passengers, locomotive speed gave the railways an overwhelming advantage over all rival transport, and, as the Liverpool & Manchester had already demonstrated, it could also bring out a level of demand whose existence had been previously unexpected. By the later 1830s promoters were allowing for this increase in their planning; by 1844-45 passenger revenue on the Edinburgh & Glasgow was almost exactly equal to the forecast in the parliamentary estimate. The Ayrshire had overestimated its passenger income by 41%, but the shortfall was due to low fare levels and an unexpected preference for the third class rather than to a lack of passenger numbers.

The most distinctive features of Scottish passenger services were, in fact, the low fare structures and the emphasis on third-class provision. These came not simply from company benevolence (although the companies clearly wished their customers to think that they did), but from the relative poverty of the Scottish travelling public and the regrettable willingness of the well-off to save money by travelling in a class below that indicated by their social status. The railways were thus forced to build their passenger revenue on quantity rather than quality of travellers. Inevitably, Scottish companies led the eventually successful campaign to alter the government's passenger duty from a fixed sum per passenger to a percentage levy on gross passenger receipts, and then to remove the duty altogether from services which satisfied the conditions laid down for parliamentary trains. Passenger fare levels were often considered with rather more care than goods rates, perhaps because of the belief, for which there was circumstantial evidence, that passenger traffic was more price elastic than freight, which, it was felt, depended more on the general economic state of the country. One of the more important contributions of the Scots to the development of railways was their demonstration that satisfactory profits could be made, even in a time of depression, from the conveyance of large numbers of low-fare passengers. Although fares could go too low, as the Greenock discovered under the pressure of river competition, by 1844 most companies agreed with John Learmonth that the low-fare policy which had been forced on them actually had positive advantages.²¹

IV. Connections and extensions

With varying degrees of success, the separate inter-urban lines demonstrated the possibilities of using railways for passenger and general freight traffic as well as coal, at least along well-established traffic arteries. The next requirements were to connect the various lines into a coherent system, by bridging the gaps between existing lines and by altering some of Grainger and Miller's unorthodox gauges to the Stephensonian 4' 8½", and to extend the network to serve further areas of Scotland and to link with the railways of England.

Within Scotland there was more planning than action. Projects to link Edinburgh to Perth and Dundee did not advance beyond preliminary surveys, while hopes of reaching Aberdeen or Inverness remained embryonic. The Angus lines stayed detached both from each other and from the main lines to the south. After July 1844 the first gaps in the system were bridged by the authorisations of the Slamannan Junction and the line through Princes Street Gardens. The most obvious hiatus remained in Glasgow, where Admiralty opposition during the mania meant that there was no railway bridge across the Clyde and no direct link from the Edinburgh & Glasgow to the Ayrshire and Greenock lines until the opening of the City of Glasgow Union in the 1870s.²² Perhaps even more important was the continuing inadequacy of links to the harbours of Glasgow and Leith. After the failures of Grainger's Blythswood tunnel and Jardine's street tramway in the early 1830s, no further serious effort was made to reach the Broomielaw; railway access to the Clyde was limited to the south bank, and for Lanarkshire coal was not achieved until the opening of the Clydesdale Junction in 1849. In the east the tedious progress of the Granton company meant that the first physical link between Leith and the main railways was not made until the junction of the North British and the Edinburgh & Dalkeith in 1846. A coherent network embracing most of the significant towns, harbours and coalfields of lowland Scotland had to await the completion of projects which survived the collapse of the mania.

Long before the consolidation of the Scottish lines was under way, attention had also turned to trans-border connections to England. There was no essential difference of purpose between the creation of a Scottish network and the first moves towards establishing a British one. Increasingly, economic interests were thinking in terms of more unified national markets, with greater regional specialisation of production being made possible by faster, cheaper and more comprehensive transport. The trans-border lines completed in the late 1840s, which marked 'the greatest step towards economic integration since 1707',²³ were a logical development from the moment that promoters' ambitions rose above the limits of the simple coal waggonway.

Even so, trans-border lines marked a considerable increase in the scale of uncertainties and difficulties facing the promoters. The main competitors, the coastal shipping lines, also had the advantage of steam power; but although they might expect to retain heavy goods traffic on convenient runs, such as the trade in Tyneside coal, their use too often involved problems of transshipment, and they were unable to provide suitable links between east and west coast industrial areas. For the traffic in passengers and light goods which the experience of existing English main lines had suggested would be the mainstay of trans-border trade, the speed and flexibility of the railway offered good chances of success. Less clear was the potential size of the traffic. No one could be certain of the extent of the demand for long-distance Anglo-Scottish services, nor of the amount which could be diverted from the sea routes, while the sparsely populated border country offered little hope of intermediate traffic. It was not surprising that the commonly held view was that, at best, only one trans-border railway could be made to pay. Given that option, there could be no agreed opinion on the best route to follow. Neither the east nor the west coast route could satisfy both Glasgow and Edinburgh, while the only practical central line, by Annandale, had to overcome the engineering difficulties of both Beattock Bank and Shap Fell, and traversed an even more deserted tract of hill country than its rivals. Whichever route was selected, the distance to be covered and the constructional difficulties imposed by the terrain were greater than anything previously faced by Scottish projectors and engineers, and the capital required would be well in excess of that raised by the largest existing Scottish company, the Edinburgh & Glasgow. Even if the physical problems could be surmounted, it was by no means clear where the money was to come from.

Thus, although the Scottish economy probably stood to gain most from a trans-border rail link, Scottish projectors not surprisingly remained cautious. It was no longer possible, as it had been with coal railways, to promote a railway simply because it would benefit the other interests of the projectors, nor could the capital come from a relatively small group of local men and potential users. For inter-urban or trans-border projects with share capitals of £500,000 or £1 million, promoters had to be confident that the prospects and the returns would satisfy a wide range of investors who were looking only for a dividend income or capital appreciation; railways had ceased to be auxiliaries to mining or manufacturing, and had become important business and major investment opportunities in their own right. And in the depression years from 1837 to 1843 it required remarkable confidence to be sure that a trans-border railway would be profitable.

Hence, although the Ayrshire contemplated the Nithsdale route with great theoretical enthusiasm from the start, and certainly from the time of Locke's first report, their continuing caution was shown both by their desire to involve government money and by their long and ultimately futile efforts to establish a satisfactory steamer service to Lancashire. On the other coast, Learmonth's group got cold feet to such an extent that they were reluctant to build beyond Dunbar, while in the centre Hope Johnstone's Annandale squirearchy generally retained their enthusiasm but had no hope of themselves raising the capital for the Caledonian. The essential initiatives came from England, and above all from the Grand Junction company. Once it had encouraged a series of subsidiary companies to push northwards from industrial Lancashire, its logical target was not Carlisle but Glasgow. It was the Grand Junction which sent Locke to survey the western routes, which guided the Caledonian promoters through many of the technicalities of establishing their company, and which encouraged English confidence to such an extent that almost four-fifths of Caledonian subscriptions came from south of the border.[24] George Hudson performed a similar role in encouraging, prodding and helping to finance the North British; indeed, whereas John Moss joined only the provisional committee of the Caledonian, Hudson remained for a time on the North British board after authorisation. The achievement of links between England and Scotland owed a great deal to rivalries between major English companies. Only the Glasgow Dumfries & Carlisle managed to cross the border without substantial English assistance; in its case, sponsorship by the Ayrshire provided experience of railway promotion, while the irrationality of the mania eventually assured it of the finance which in more normal times could not have been expected to appear for a third trans-border line.

V. Costs and estimates

A recurrent theme in this book has been the disparities between estimates and results, both in the costs of creating the railways and in the eventual returns from traffic. Promoters were to a considerable extent learning their craft by trial and error; the experience of earlier forms of transport (roads and canals), and even of lines already in operation in England, offered parallels which were neither sufficiently numerous nor sufficiently exact to form a reliable basis for calculation. Although projectors habitually presented their estimates to Parliament and the public with an air of precision, they were working within wide margins of error over numerous variables, and as the scale of their operations increased, so too did the potential for serious inaccuracy.

The principal costs of creating a railway were the expenditures on the works and on the purchase of land. Land acquisition regularly created more ill-feeling and more unpremeditated expense than any other item, and over the years it became increasingly contentious. For the waggonways, which were generally built on the land of a single owner or of a few individuals all of whom were interested in the line, there was seldom a problem. From the limited evidence available, early waggonways cost about £1000 per mile to construct; this figure probably does not include any allowance for land since, although a proprietor may have had some notional figure for the opportunity cost of using the land for a waggonway rather than for any other purpose, he is unlikely to have included in his calculations any formal sum for the purchase of his own land. Surviving estimates for long-distance waggonway projects show that (apart from Robertson Buchanan's hope of free donations from Dumfriesshire landowners) estimates ranged from £76 per mile for Stevenson's moorland Roxburgh & Selkirk to a remarkable £814 for the double-track Midlothian on the valuable land near Edinburgh (see Table 8, page 40).

Since these waggonways remained unbuilt, the accuracy of the estimates is a matter of conjecture. Similarly, while the estimates for the coal railways have survived, a detailed breakdown of their actual costs has not. The estimates underline the local variations in land prices, ranging from under £200 per mile on the Newtyle lines, to £250-300 in the Monklands, and up to £562 for the coalfield and suburban land of the Garnkirk. The totally urban Polloc & Govan allowed over £3500 per mile, or almost half of its total estimate, for land.[25] Only for the Dundee & Arbroath and the Arbroath & Forfar have both estimates and actual costs survived; on the former Miller's allowance of £450 per mile (even although much of the land was to come virtually free from Lord Panmure) was 17.8% too low, while on the latter Grainger's £458 estimate became, in spite of generally well-disposed landowners, an actual expenditure of £1118.[26] These figures may be cautiously compared with the payments made by the main inter-urban lines (see Table 41, page 187), where even the relatively economical Ayrshire had to expend over £2500 per mile.

It is not surprising that promoters' complaints about the unreasonable prices exacted for land date mainly from the late 1830s and 1840s. Longer lines were more likely to be faced by a number of hostile landowners, who were increasingly realising (perhaps from English examples) the importance of the land to the railway companies, and thus the extent to which the price might be pushed above agricultural levels even before allowance was made for the element of compulsion and for loss of amenity. Unless a landowner was known to favour the project, or to be himself urgently in need of money and therefore anxious to conclude a sale, companies could anticipate paying well above the value which had been suggested in their engineers' estimates. Although arbitration provisions were available to solve disputes, committee members (who, rather than the engineers, were in charge of land negotiations with the help of their lawyers) might prefer simply to pay high prices rather than risk incurring influential opposition to the passage of their bill through Parliament. Even so, the astonishing £7234 per mile paid by the Greenock company with a minimum of litigation (which equalled 18.1% of total expenditure as against the 12-13% of other main lines, even although works costs at Greenock were also above average)[27] suggests a committee which was being reckless with its subscribers' money in order to get the line into operation.

Growing concern about land costs can be seen in, for example, the views of *Chambers' Edinburgh Journal*, which strongly supported railway expansion. In June 1838 it was restrained:

> It is much to be feared that the spirit of enterprise which prompts many of these speculations may give way, or be allowed to wax cool, before the opposition which, on various pretexts, is raised against them. Private rights must not of course be expected too early to yield even in the certainty of great public advantages. Full compensation ought to be given in every case where a private right, even of the most sentimental kind, is invaded.

By October its patience appeared to be running out:

> It appears to be only in dealing with an individual, when we can think how *we* should like to be used so, that we have distinct notions of justice. In dealing with the abstraction called a railway proprietory, men seem to think themselves entitled to cheat, over-reach, and extort without restraint.[28]

The importance of the rights of private property was an integral part of the social framework of nineteenth-century Britain, enshrined in the law and, perhaps more crucially, in the beliefs of the upper and middle classes. Only the canals had previously posed the sort of conflict between these rights and the public interest which was implied by compulsory purchase, and it was now clear that the railways were to be much more widespread and pervasive than the canals. It was a further complication that, although the railways might claim to be institutions of public benefit, they were also private corporations dedicated to making profits for their shareholders, and thus might appear to be claiming that the rights of one type of private property should be allowed to supersede those of another. The problem indicated an underlying contradiction between the preservation of individual freedoms and the advancement of the interests of the community as a whole, which was almost inevitably solved by compromise; if promoters could satisfy Parliament of the value of their scheme, they would get their land but, particularly after the Clauses legislation of 1845, they would pay handsomely for their interference with hostile landowners. Similar conflicts were to dog much later Victorian reform, and similar solutions (though often with local government as the reforming party) were applied in areas such as public sanitation, street improvements and slum clearance.

Parliamentary costs often caused a degree of resentment among promoters and shareholders which appears disproportionate in view of their relatively small role in total expenditure. They were to some extent related to land acquisition, since the prime reason for mineral lines at least to obtain parliamentary authorisation was the need for powers of compulsory purchase: it was not entirely clear whether a passenger service required specific parliamentary approval, but every line from the Monkland & Kirkintilloch onwards included an appropriate clause in its act. The need for parliamentary authorisation did allow companies to obtain limited liability status without having to promote a special bill for the purpose – an advantage which became increasingly important with the growing scale of railway operations and the need to attract blind investment. Resentment of parliamentary procedures came partly from their cost, which in the early days was inflated by unnecessarily meticulous regulations which, for instance, enforced the appearance in London of numerous superfluous witnesses to matters of form, and later by (in the view of railway promoters) an undue sensitivity to the interests of landowners and other opponents. Even more exasperating to promoters was the uncertainty of the parliamentary process. Companies felt that their prospects might depend unfairly on the composition of the committee on their bill; they also resented the danger of a project on which much money and effort had been expended being lost on standing orders without any consideration of its actual merits because of a minor technical error. On the whole, however, the parliamentary hurdle was unlikely to catch a well-prepared scheme in the days before the vicious inter-company competition of the mania.

Other aspects of government intervention caused little serious complaint. The state was in any case inclined, with the possible exception of the measures latent in Gladstone's 1844 act, to leave railway affairs as far as possible to private enterprise. The main powers of intervention, exercised through the Railway Department of the Board of Trade, concerned the safety of the public; although the companies opposed an 1841 bill to strengthen these regulations (which failed with the fall of Melbourne's government),[29] they had no serious complaint about powers of inspection which seldom inconvenienced them, and which might even help to persuade potential customers of the security of rail travel. The efforts of the Smith-Barlow commission and of the Railway Department report of 1845 to influence the route of a trans-border railway, although presenting invaluable propaganda to the Caledonian, were in the event of little significance, since by 1850 all three of the serious contenders had in any case been built. Even the passenger duty was not opposed in principle, since taxation of public passenger transport had long been accepted, but only in respect of the method by which it was calculated. On the whole the railway interest strongly approved of the government's inclination to keep as far as possible out of involvement with the railways.

The greatest part (commonly over 70%) of the capital costs of creating a railway was incurred on the physical construction of the line – earthworks, roadway formation, rails, masonry and, in most cases, stations. These costs also underwent a steady increase over time, both per mile and, as later lines were generally longer, in total. Almost all of the £1000 or so spent on each mile of waggonway must have gone into the works, while the works estimates for the long-distance waggonway projects vary wildly between Stevenson's figure of £1167 per mile for the Roxburgh & Selkirk and his £3423 per mile for the Midlothian (which was to have a double line of rails and much heavier earthworks).[30] The estimates for the early coal railways again suggest that they were not seen as a radical advance on the waggonways. For their main lines, the Monkland & Kirkintilloch allowed £1696 per mile on works, the Garnkirk £2410 and the Dundee & Newtyle £1735.[31] In the event, however, the estimates for total capital expenditure on these three companies fell short of actuality by respectively 75%, 246% and 263% (see Table 12, page 52). There is no evidence to show exactly where this over-expenditure took place, and some was certainly incurred on land costs, on rolling stock (for which only a minimal allowance was made in the estimates), and on items such as stations which had been omitted altogether. Even so, these companies have left no evidence of complaints about land costs, nor did they proliferate branches which might have legitimately pushed up capital expenditure. It appears likely that their costs were forced up mainly by the higher standards of works construction required for the use of locomotives, by the heavier traffic which they anticipated (and except perhaps at Newtyle achieved), and by the standards necessary for parliamentary approval.

By the mid-1830s promoters had adapted their ideas, and for extremely basic works construction the Paisley & Renfrew and the Slamannan allowed respectively £3590 and £4323 per mile.[32] In Angus John Miller once again proved to be a more accurate estimator than his partner; his works allowance of £5042 per mile on the Dundee & Arbroath was only 8.2% too low, while Grainger's £3055 per mile for the apparently straightforward Arbroath & Forfar was an underestimate of 69.5%. The extent to which costs had escalated is also indicated by the fact that the Wilsontown line, although built for relatively light mineral traffic to the minimum standard acceptable for locomotives and considered by the Railway Department's inspector to be unsuitable for regular passenger traffic, cost in total £9408 per mile, or almost four times Baird's allowance for the West Lothian project in an adjacent and very similar area seventeen years earlier. Works, including rails and stations, were inevitably more expensive on the main inter-urban lines, which were designed for more frequent, faster and often heavier trains. By three years after opening, they had cost about £12,100 on the Ayrshire, £16,000 on the Greenock, and no less than £26,300 on the Edinburgh & Glasgow, reflecting Miller's concentration on a fast level line.[33] Although the estimates for these lines are unknown, it is clear that they must have been greatly exceeded. Not only did the three companies exhaust their original share capital and borrowing rights, but they had to return to Parliament for permission to raise further substantial sums to complete their lines. By the end of 1841, and with its main line still unopened, expenditure on works and rails for the Edinburgh & Glasgow had already exceeded the total authorised share capital of the company. Including their shares of the joint line, the Ayrshire (with the Kilmarnock branch unopened) and the Greenock reached a similar situation late in 1842.

Other costs, apart from land and works, could in general have made only a relatively small contribution to the discrepancy between estimates and costs. Admittedly, some of the coal railway estimates, as submitted to Parliament, made no pretence of being complete, since they omitted such essential items as parliamentary costs, engineering fees and even occasionally land.[34] Most of the coal railways omitted any mention of rolling stock, with only the Garnkirk and the Dundee & Newtyle stating a token sum (respectively £400 and £500) for the purchase of waggons.[35] The companies might claim that, at the time they were before Parliament, they intended to operate primarily as toll roads, with traders providing their own waggons, and that they had as yet taken no positive decision to purchase locomotives; in any case, it meant that large unannounced sums had to be found before the lines could operate. The inter-urban lines' estimates must have included an allowance for rolling stock; it is unlikely that it had been seriously exceeded before the mania, since traffic levels had yet to rise far above the parliamentary estimates. Hence the 7% to 10% of total costs incurred here was probably well in line with the original expectations. Other relatively small expenses, on parliamentary and legal costs, engineering fees, salaries, payments to road trustees and interest on capital account, seldom amounted individually to more than 2% of the costs of any inter-urban company, and never collectively to over 10%. Nor is it likely that rails were unexpectedly costly, since the price of iron when most of the calculations were made in 1835-36, was higher than in subsequent years (see Figure 3, page 204). For the inter-urban lines at least, the source of most unpremeditated over-expenditure lay clearly in the major cost areas of land and works.

The responsibility for drawing up the estimates rested with the company engineer, who must therefore at first sight be blamed for their defects. For some aspects, however, he might well be guiltless. Extra costs incurred by decisions to build new branches or to improve the line to meet the needs of unpredictably heavy traffic were entirely reasonable; the Monkland & Kirkintilloch, for example, could hardly have planned at the outset for the traffic of a network covering much of the Lanarkshire coalfield. Too often, however, money raised ostensibly for new lines was in fact applied in part or in whole to the completion of the original design. Nor could the engineer be held responsible if his committee of management agreed to pay excessive prices for land. It may even be unclear where the responsibility lay when essential items were omitted altogether from the estimate, as happened all too often on Grainger's lines. The engineer could hardly expect that the committee would overlook the absence of allowances for items as important as rolling stock or parliamentary expenses, and it may be that in some cases he was specifically asked to estimate only for the areas to which his professional competence was directly relevant. But there can normally be no doubt of his responsibility for underestimation on the cost of the works as laid out in the original plan. Even if some of the expense stemmed from unexpected difficulties on the route or from the incompetence or dilatoriness of contractors, the engineer's duties included the supervision of both surveying and construction. It might be argued that the novelty of railways, and in particular of the scale of work involved, might excuse substantial errors. Yet there is little indication that engineers became more accurate with experience: neither Grainger's work at Arbroath nor Miller's for the Edinburgh & Glasgow suggested that they had learned caution from their earlier experiences, while Locke's approval of the Greenock estimates (admittedly originally drawn up by Grainger) belied his later reputation for accuracy.

It is more difficult to compare expectations and results for companies once they were in operation. The traffic estimates of the coal railways were only expressed in the most general terms, and exact comparisons with the outcome cannot be made, although in the Monklands at least coal volumes were clearly satisfactory. For the main inter-urban lines, the revenue expectations of the Edinburgh & Glasgow estimates had almost been achieved by 1844; those of the Ayrshire still were some distance away,[36] although this probably was more due to low levels of fares and rates than to an inadequate volume of traffic. By 1844 the main lines had only been open for a few years, most of which had been in economic recession, and it was reasonable to anticipate that their revenue would increase as the economy recovered and as the railway system was connected and extended.

The most serious underestimates on current account were in the figures allowed by the coal railways for working expenditure, where estimates of 10% to 15% of gross revenue became realities ranging generally between 35% and 55%, with occasional disastrous returns of over 80% on the Dundee & Newtyle and the Paisley & Renfrew. The inter-urban lines allowed at least a third of gross revenue; while the two largest companies kept under their target, the excessively low fares of the Greenock meant that working expenditure often exceeded 50%. But the item on all revenue accounts which made the difference between a handsome dividend and a disappointing one (or even none at all) was non-working expenditure, and in particular interest payments on money borrowed. Since projectors had always assumed, at least for public consumption, that their lines would be constructed within the allowed share capital, no such item could of course appear in the estimates. The effects of consistent underestimation on capital account, however, were firstly to impose interest payments equal to between 20% and 40% of total current expenditure, and secondly in most cases to enforce further issues of share capital.

The scale of continued miscalculation by engineers and committees leads to the unprovable suspicion that it was to some extent deliberate. Pressures certainly existed to keep estimates low. Apparently low estimates of cost would enable the committee to make their project attractive to potential subscribers by promising high dividends without having to produce improbably high assessments of potential traffic. They might also, if believed by Parliament, ease the process of authorisation by reducing any doubts about a line's ability to pay its way. Engineers might believe that committees would prefer to employ a man who claimed to be able to construct their line economically, and certainly there is no evidence that persistent underestimating made it more difficult for a pre-mania engineer to obtain employment. The escalation of costs must be divided, in uncertain proportions, between three possible causes — legitimate extensions and improvements to the original plan; unforeseen increases due to misfortune, miscalculation or incompetence; and deliberate understatement, usually with the aim of attracting investors. In most companies all three made their appearance.

VI. Investors and promoters

The increasing cost of railway projects meant that more and more attention had to be paid to the sources of investment capital. The largest initial share capital of any of the coal railways was £86,000 for the Slamannan; the largest total authorisation among them before the mania was the £240,000 reached by the Wishaw & Coltness in July 1844. In contrast, the initial share capital authorisation at Greenock was £400,000, for the Ayrshire £625,000, and for the Edinburgh & Glasgow £900,000, and in each case actual capital expenditure had exceeded these allowances by between 70% and 100% before the end of 1844. Hence there had to be an emphasis firstly on attracting subscriptions to start a project, and later on raising further funds the requirement for which was itself often a sign of original miscalculations.

Since waggonway construction was generally undertaken by coal and ironmasters, either as individuals or as firms, to aid their business by extending its market (for producers of coal) or by reducing the cost of supplies (for consumers), investment was undertaken with a view to the profits of the enterprise as a whole and not of the waggonway by itself. For lines which were commonly under two miles long, and rarely exceeded five miles, construction was cheap enough not to require outside financing, while even the longer and more costly Kilmarnock & Troon was well within the resources of the Portlands. Such individual financing continued into the 1830s and 1840s with the Polloc & Govan, the Drumpellar, and the numerous private sidings attached to the railway system.

The early coal railways extended both the service which they provided and the source of their subscription capital to relatively small groups of mineral owners, users and traders, who were still often more interested in the use of the railway than in the level of their dividends. In the Monklands, for instance, the primary aim was to get coal to the Glasgow market more cheaply than could be done at the monopolistic rates on the Monkland Canal. Elsewhere, even in the 1840s customers of the Dundee & Newtyle were prepared in effect to subsidise the company by taking shares on which they knew they would never see a return, in order to preserve from bankruptcy a concern which they believed to be of value to their other interests. Gradually, however, as some of the coal railways demonstrated their capacity for making profits, and as more money had to be raised for extensions and improvements, the pattern of shareholding widened to include investors interested primarily or solely in a rentier income. The role of the railways was widening from the original function of reducing the freight costs of minerals to include a market status in its own right; equally, the cost of constructing them was increasing to the point at which it was unreasonable to expect them to be financed simply by a group of potential users.

In 1836 the *Circular to Bankers* declared that 'no considerable railway can be completed that depends on local money for its outlays'.[37] The truth of the comment depends on the definition of 'considerable'. The Dundee & Arbroath and Arbroath & Forfar raised most and possibly all of their subscriptions locally, as far as can be determined from the numerous recognisable names on their lists, while in 1845 the Glasgow Barrhead & Neilston Direct raised 74% of its £130,950 subscription list in the locality[38] and could certainly have done even better if other funds had not come easily from elsewhere. But the discussion in Chapter 3, section VII, and the tables in the appendix indicate the extent to which the three principal Scottish lines had to look outside their own areas (even although in each case the area included Glasgow) for subscribers. The Ayrshire, which had by far the greatest local support, still took 26.9% of its money and 24.1% of its subscribers from elsewhere. For the Greenock the corresponding figures were 44.3% and 34.2%; for the Edinburgh & Glasgow 69.3% and 66.3%. Even among local shareholders, an increasing number were much more interested in dividend income and capital appreciation than in the personal use of the railway services, while the increasing number located in England were solely interested in the financial performance of the company. Directors had increasingly to keep in mind the need to satisfy distant or uncommitted investors.

As the nature of shareholding changed, so too did the background of railway promoters. The promoter of a waggonway was almost invariably also the owner and principal or sole user. For most of the coal railways, the main promoters were still coalowners, assisted by coal consumers such as Charles Tennant and some of the major ironmasters, and by a group of Glasgow lawyers who often acted as secretaries or law agents of the new companies. The leading promoters, at least in the early days of the companies, were still among both the largest users and the main shareholders. The essential change comes with the inter-urban lines. The

Summary and Conclusions 325

role played by both mineral owners and industrialists declines sharply; on the whole, the earlier lines and the canals had satisfied their essential needs, and any surplus capital they might have was better employed in further investment in their other enterprises. Instead, from Grainger's Edinburgh to Glasgow project of 1830 onwards, the lead in both promotional activity and the provision of subscription capital passes above all to the merchant community, with support from landowners along the routes of the various lines. Merchant involvement might indicate a desire to reduce freight costs in particular trades, or simply to increase the total volume of goods traded, but it often also indicated an interest in railway investment as a generator of financial profit in its own right – again a sign that the railway was now much more than simply a servicing agency for other economic activities.

Table 78 indicates both the increasing amounts of capital which the financial system was being asked to supply, and the cyclical pattern of the requirements. Even when new company promotions almost ceased in the depressed years from 1839 to 1843, the demands of existing companies for more money to compensate for the inadequacies of their original estimates kept the market in railway securities active; more new capital was authorised in 1842 alone than in the whole of the period up to 1834. The authorised total is a little greater than actual requirements, since the West Lothian of 1825 and the Rutherglen of 1831 never

Table 78. Capital authorised per year, 1824-44

Year	Share capital		Loan capital		Total	
	New companies	Old companies	New companies	Old companies		
	£	£	£	£	£	%
1824	3200	—	10000[a]	—	42000	0.53
1825	40700	—	20000	—	60700	0.76
1826	147054	—	20000	—	167054	2.09
1827	95658	9350	—	—	105008	1.32
1828	—	—	—	—	—	—
1829	60000	80575	20000	10000	170575	2.14
1830	10000	10000	5000	41150	66150	0.83
1831	15000	36000	5000	15000	71000	0.89
1832	—	—	—	—	—	—
1833	—	—	—	20000[a]	20000	0.25
1834	—	—	—	8053	8053	0.10
1835	144200	—	41600	10000[a]	195800	2.45
1836	270000	100000	105000	—	475000	5.95
1837	1025000	29000	341633	—	1395333	17.48
1838	900000	89198	300000	—	1289198	16.15
1839	—	217568		56333	273901	3.43
1840	—	326700		80000	406700	5.09
1841	70000	220000	23000	73333	386333	4.84
1842	—	587500	—	189100	776600	9.73
1843	26000	276000	8666	78667	389333	4.88
1844	900000	366550	300000	118183	1684733	21.10
Total	3735612	2348441	1199899	699819	7983471	100.00

a: might be raised as share or loan capital
Sources: Local and Personal Acts

Table 79. Share Capital Called Up by Inter-Urban Railways, 1836-44

Company	1836 £	1837 £	1838 £	1839 £	1840 £	1841 £	1842 £	1843 £	1844 £	Total £
Arbroath & Forfar	12600	36400	21000	—	32000	—	—	—	—	102000
Dundee & Arbroath	15000	30000	55000	—	—	—	—	—	—	100000
Edinburgh & Glasgow	—	—	450000	—	180000	270000	90000	135000	—	1125000
Edinburgh Leith & Granton	15000	10000	10000	10000	20000	25000	—	—	20000	120000
Glasgow Paisley & Greenock	—	—	96000	128000	144000	64000	52000	42000	—	526000
Glasgow Paisley Kilmarnock & Ayr	—	a	a	a	a	a	62500	—	—	687500

a: Glasgow Paisley Kilmarnock & Ayr called up £625000 between 1837 and 1841: figures for individual years not available.

Source: PP 1847-48(731)LXIII, *Return of Calls Made by Each Railway Company*.

got beyond preliminary and parliamentary expenditure; companies which completed their lines, however, almost invariably required every penny to which they were entitled. The figures in Table 79 are a reminder of the time lags between authorisation and construction, and in particular of the problem that subscription money promised in the economic optimism of 1836-37 had to be produced in the depths of the subsequent recession. The figures were gathered by the companies themselves and should (especially in the case of the Granton) be treated with some caution; they do, however, suggest the tardiness of some companies in paying their bills, since both the Greenock and the Edinburgh & Glasgow were still calling up money in 1843, long after their lines had opened. They also underline the tedious progress of the Granton line, which had exhausted its share calls by 1841, did not obtain an act for more capital until 1844, and by the end of that year had still received, in share and loan capital, less than half of its final expenditure.

Inevitably such capital requirements made an impact on the marketing of stock in Scotland. In the 1820s railways were still on the whole distrusted by the investing public, which felt more confidence in the shares of banks, insurance companies and public utilities. Besides, the coal railways, being financed chiefly by the close knit groups who were both their promoters and their main users, had no large turnover of shares, and could usually make any necessary transfers through personal contacts. By the expansion of the mid-1830s the general abundance of capital, a greater acceptance of the potential of the railway, and the low prevailing rates of interest made railway shares more attractive to outside investors. The Edinburgh & Glasgow in particular profited by the slowing down of English investment opportunities by 1837 to such an extent that five-eighths of its subscriptions came from south of the border. The greater geographical and numerical spread of shareholders, and the growing tendency to hold shares short-term and for purely investment reasons, enforced more organised marketing. For the collection of subscriptions, it meant that companies required agents not only in their immediate areas, but as far afield as Liverpool and London. For the transfer of shares, it meant the intensification of the informal co-operation between brokers in Edinburgh and Glasgow which already existed for dealings in banking and other shares; the process had started which was to lead, under the pressure of the mania, to the founding of the Scottish stock exchanges.[39]

VII. Managerial developments

The increasing scale and complexity of railways enforced corresponding changes in management and organisation. Managing a waggonway meant primarily maintaining the track and rolling stock in working order, with little concern about the profitability of the operation taken in isolation, and the task might form a small part of the duties of a works manager or be delegated to a subordinate. The advent of the locomotive, the increasing complexity of operations as more and more branches and sidings brought in traffic from more and more sources, the appearance of passenger services, the growing interest of the Board of Trade in safety and the protection of the public interest, and the need to make a profit for shareholders as well as to provide a service to users, all necessitated attention to increased efficiency. Even so, according to Pollins,

> In many ways the operating activities of the 1830s and 1840s look chaotic, and one has an impression that not much thought had been given to the question of working the lines.[40]

Certainly no very coherent pattern of management emerged before the late 1840s.

The locomotive was perhaps the most potent single force for operational change. Even without appreciable passenger traffic, a locomotive line could not operate with the haphazard running practices which survived on the Edinburgh & Dalkeith; both safety and efficiency demanded accurate time-keeping, careful maintenance, and increasingly a total separation of horse and locomotive traffic so that horses were steadily forced off all main lines.[41] The importance of both civil engineering in constructing the line and of mechanical engineering in building and maintaining the locomotives, coupled with the ignorance of almost all promoters and directors of the technicalities of the subject, may account for the importance attached to these departments in the larger companies, symbolised by the deference awarded to Errington when he joined the staff of the newly opened Greenock line.

Too often there was no clear hierarchy of management in a company, nor might there even be a clear delineation between the duties of the various officers. On the coal lines a managerial structure began to emerge, based to some extent on the practice of canals, road carrying firms, or perhaps any other commercial enterprise of which directors and officials had detailed knowledge. At its most basic this might still centre on an all-powerful factotum like James Moffat of the Ardrossan Railway, but elsewhere there emerged over time the various roles of manager, secretary, accountant or cashier, resident engineer and various departmental superintendents. On the Lanarkshire lines two or more of these posts might well be filled by one man, but the minute books of the Ayrshire and the Edinburgh & Glasgow bear witness to the appointment of a numerous and apparently impressive list of managers and supervisors.[42] All too often, however, the various departments showed signs of operating as independent empires with the minimum of reference to each other. It was not always evident who was the superior officer, and it is sometimes difficult to avoid the conclusion that the real power might depend on forces of personality rather than official status. At Greenock, for instance, Mark Huish as secretary was clearly the most prominent official before the line opened in March 1841, while John Errington as manager/resident engineer held sway (rather than the new secretary Wyndham Harding) after Huish's departure to the Grand Junction in July;[43] the intervening four months may have seen an interesting study in power politics. Not until the late 1840s were company organisations tightened up, and a hierarchy under the overall supervision of a general manager established.[44]

Time also had to pass in order to solve the problems of managerial inexperience. Men with knowledge of finance and general administration might be attracted from almost any branch of commerce, but experience in traffic operation could come only from canal and road carriers and even then might not be strictly comparable; few railways could equal the mutual advantage which came from the London & Birmingham's close association with Pickfords.[45] Many senior officials came from the forces — the navy in particular was being run down in the 1830s — and found that, since railway staff were organised with something approaching military discipline, their habits of command and skills in man-management might well be useful.[46] Some, from the Royal Engineers, might even have useful technical knowledge; the most outstanding acquisition of this kind was to come in 1847 when the Caledonian appointed as secretary Captain Joshua Coddington, who had followed service in the Engineers with a period as an Inspector of Railways, and had indeed just completed an unflattering report on the construction of the rival North British.[47] Often, however, railway companies found that it was necessary to train up their own management which, although men did move between companies, entailed a risk of inbreeding and complacency.

Ultimate power rested in theory with the company shareholders. By 1842 shareholders in some English companies, such as the North Midland or the London & Brighton, had become sufficiently restive to impose major changes on the boards;[48] in Scotland, however, in spite of incidents like the sabbatarian battle on the Edinburgh & Glasgow or the committee of inquiry set up by the Wishaw & Coltness in 1844, shareholders' revolts did not appear until after the mania. Actual power therefore was in the hands of the directors, who too often combined enthusiastic amateurism with a lack of knowledge of how to manage a railway. Most had come to the fore during the promotion of the railway, and were qualified by having business reasons for wanting it to succeed, by owning land along the line, by investing a substantial sum of money, or simply by having a name which might encourage other potential investors. Knowledge or experience of running a railway was not a prerequisite. During the promotion most had been content to leave the technicalities to their chosen engineers and lawyers, but all too often some at least felt competent to intervene in the detailed management of a working railway. Company minute books regularly indicate that poorly attended board meetings, often composed of the same few enthusiasts, deliberated not only about broad company policy but also about day-to-day details of operations that were properly the province of the paid officials. Under such circumstances an official would need to be both self-confident and strong-minded to carry out a consistent policy. Gourvish suggests that

> Management problems and deficiencies were . . . at their worst in the 1830s and early 1840s, when the companies found profits relatively easy to come by.[49]

Although the main Scottish companies were not making the same level of profits as the most successful English ones, their management practices equally showed little sign of change before the post-mania crash led to the introduction of more hard-headed directors, an established hierarchy of officials, and a clearer dividing line between the duties of the two.

VIII. English influences

A recurring theme in early Scottish railway history is the impact of English influences. Formally the border, or at least the border towns of Carlisle and Berwick, marked an apparently clear division between two national railway systems until the great rationalisation of 1923; the minor lines sponsored by the North British in Northumberland and Cumberland in the 1860s and the Caledonian's Solway Junction made no real breach in the principle of geographical separation.[50] North of the border, the main lines were built by companies based in Glasgow or Edinburgh, with predominantly or entirely Scottish directors and officials; to the south were English companies based on Lancashire or London, York or Newcastle. No English company took permanent possession of a Scottish one, although there were numerous rumours, some negotiations, and occasional near-misses like the participation of the London & North-Western and the Lancaster & Carlisle with the Caledonian in a short-lived post-mania lease of the Scottish Central.[51]

A degree of separateness about the Scottish system was recognised by both investors and the railway press. The latter, which until the founding of the *Scottish Railway Gazette* in 1845 was entirely English-based, tended to reflect a natural assumption of English leadership and superiority. The comparatively slow development of the Scottish system was remarked upon, though often with understanding:

> It was not to be expected that Scotland — although comparatively unsuited for railway undertakings both from the nature of the country and its less dense and stirring population — would suffer the example of her wealthier neighbour to be unproductive of a generous rivalry in this most important department of art. And if the more cautious Scots have not suffered themselves to be led away to the extent of their enthusiastic brethren of the south, in giving credence to every plausible scheme of the kind that has been brought before the public, their prudence is to be commended, and their patronage esteemed the more valuable, when, as in the instance before us [the Edinburgh & Glasgow], they have embarked, heart and hand, in carrying out a great national work.[52]

But two years later it was 'much to be regretted that the spirit which prevails in England, has not extended itself to Scotland'. In specific instances the Scots might be praised, for achievements ranging from their leadership of the anti-passenger tax campaign to the form of presentation of the accounts of the Arbroath & Forfar. But a tribute to 'the Scotch capitalists, who have a great reputation for shrewdness' was provoked by a report that they were having the sense to buy into lines like the Great Western and London & Brighton, while leaving the stock of the less prosperous Edinburgh & Glasgow to be 'almost exclusively' held in England.[53] The common assumption, unless some moral was being drawn especially for English benefit, was that Scottish railways tended to be following rather tamely and without conspicuous success behind English practices. All too often the Scots, by the deference which they paid to English expertise, appeared to accept the verdict.

English involvement went right back to the beginning with the activities of the York Buildings Company at Tranent, and much subsequent waggonway development followed Tyneside precedents. The Tyneside coalfield stimulated development in another and very different way, since its seaborne trade to Leith forced Fife and Lothians coalowners into waggonway construction in order to compete in the Edinburgh market. In the early nineteenth century the two English lines traditionally regarded as seminal for railway development — the Stockton & Darlington and the Liverpool & Manchester — were as influential in Scotland as they were south of the border. The Stockton & Darlington was in concept only a fairly lengthy coal-to-water waggonway, but the extent of its traffic and above all Stephenson's introduction of steam power made it the exemplar for the increasing ambition of the coal railways. Its promoters' aim of challenging the power of the Tyneside coal interests by opening up new collieries[54] may also have found an echo in the minds of Scottish coalowners who were not served by existing transport. Many Scots, like the Rev. James Adamson, went to see for themselves; a comparison of Adamson's subsequent pamphlet, or of Grainger and Miller's report to the projectors of the Wishaw & Coltness, with the Highland Society essays of a few years earlier indicates the change which the Stockton & Darlington brought to the concept of the railway.[55]

The coal railways of Lanarkshire would however presumably have been built even without the example of the Stockton & Darlington. The Monkland & Kirkintilloch was after all authorised and even vaguely contemplating the use of steam power before the Stockton line was open, and the Ballochney before it had proved successful. It may be that knowledge of Stephenson's work encouraged promoters to build their lines strong enough for locomotives, even when they intended to start operation with horses, and thus to increase costs to the point where it became necessary to attract outside capital. The success of the Stockton company, and of other English coal lines, might also help to persuade dubious investors that such railways could be profitable, but in Scotland the profitability of the first Monklands lines, and even more the help which they gave to the coal and iron interests of the promoters, might be even more influential with later projectors. The Stockton & Darlington was much quoted in Scottish promotional literature, but if it had not existed, almost the same points could have been made by reference to the Monkland & Kirkintilloch, and even in some instances to the Kilmarnock & Troon.

There could however have been no Scottish substitute for the Liverpool & Manchester. In the late 1820s some lines, in both Scotland and England, had demonstrated that useful income could be made from carrying general goods and even passengers. The Liverpool & Manchester showed from its opening that prosperity could be achieved by concentrating on these trades, and established in particular the extent of the demand for inter-urban passenger transport. Immediately its success became the chief argument of all promoters for extending the system:

> The companies promoted in the 1830s, including the main lines, relied on the operating statistics of the Liverpool & Manchester, even though that railway had been open for only a short time and its conditions were not likely to be closely paralleled elsewhere.[56]

The only conceivably comparable route in Scotland was that between Edinburgh and Glasgow, and even in 1831 the projectors of the Grainger line were implying comparisons:

> The brilliant success which has attended the Liverpool and Manchester Railway, infinitely

exceeding the expectations even of the most sanguine, has proved, beyond doubt, not only the utility of that undertaking in a national point of view, but has shown the immense advantage which the towns in connection with it have derived from the superiority of the conveyance which it has established between them.[57]

Even with this example, and increasingly those of other English lines, to refer to, it took eight years to achieve the authorisation of the Edinburgh & Glasgow over canal and landed opposition.

For pre-mania Scottish projectors the Liverpool & Manchester remained the paradigm of modern railways. Both Miller and Locke justified their estimates by reference to it, and directors comforted their shareholders with allusions to its profitability. Promoters of new lines in the mid-1830s referred to it constantly to demonstrate that prospective investors had nothing to fear. In England the company's success, coupled with the growth of the economy in a period of good harvests, the availability of surplus capital, the continuing improvements of locomotive technology and the increased demand for better transport, stimulated the main line expansion of the mid-1830s. Scotland, a relatively poorer country, shared in the boom to only a limited extent, and the few new main lines might have been created in any case, but it seems probable that the continuing expansion in England was important for maintaining some impetus in Scottish promotional activity.

So far, Scottish respect for English practices could have been achieved without a single Englishman crossing the border. In practice Scottish deference to English experience often extended to the employment of English experts, particularly engineers. The engineers of most of the waggonways are unknown, and were probably employees of the collieries or works with which the lines were connected, but it is likely that some guidance was sought from visiting Tyneside coal viewers. More ambitious projects in the early nineteenth century on occasion called in either an English engineer, such as Jessop at Kilmarnock, or a southern-based Scot, such as Telford and Rennie on the two Berwick projects.[58] And George Johnson, who reconstructed part of the Elgin waggonway, is probably the same Newcastle man who later collaborated with Nicholas Wood on the projected Central Union across the border.

In the 1820s the importation of English engineers declined. The reputation of Robert Stevenson — perhaps the most imaginative of all early Scottish railway engineers, even if his vision never fully embraced the locomotive — and the rising ability of Thomas Grainger could satisfy the projectors of both the vague but ambitious long-distance schemes and the primitive coal railways of Lanarkshire. Since the latter could raise almost all the required capital from parties locally interested, and the former never matured their plans sufficiently to make any serious attempt at promotion, no serious assault had to be contemplated either on other investors in Scotland or on the English capital market. Hence, since indigenous talent could be trusted to create an adequate line, there was no need for an impressive but possibly expensive English name to reassure outside investors.

As already noted, the importation of English engineers resumed with the need for English subscribers to the more expensive lines of the 1830s. In most cases credibility could be achieved by obtaining an expert statement of approval of plans drawn up by a resident Scot. Thus a series of Englishmen — Vignolles, Rastrick, Walker, Leather and above all the Stephensons — made their mark on Scottish railway practice. In these circumstances it is tempting to suspect Scottish engineers of drawing up their plans with half an eye on the views of their prospective English overseers. Certainly the most important of the Scots, John Miller, built lines which accorded very well with Stephensonian ideas, incurring considerable expense in order to build as level a line as possible; the Edinburgh & Glasgow between Haymarket and Cowlairs was the most level main line in Britain, with only one point at which the gradient exceeded 1 in 800.[59] There is however no evidence that Miller was in any way acting against his own inclinations, and it may be that the Ayrshire or the North British invited the Stephensons to report not only because of the respect their views would command but also because they were likely to approve of a Miller design. On the Edinburgh & Glasgow the final version had to be reported on by Rastrick and Locke, George Stephenson being ruled out by the committee's promise to use an engineer who had not previously been on the route.[60]

Essentially Scottish railway engineering was not innovative. Miller built, on the whole, well-engineered lines, but he built them on principles already established in England, particularly by the Stephensons. Other Scots — Jardine, Robson, Blackadder, even Grainger — were constructing little more than high-class waggonways. The most innovative development in pre-mania Scotland was the Caledonian's proposal to make a direct up-and-over assault on the hills between Annandale and Clydesdale, which owed much to the support of the English Grand Junction company and to the vision of its English engineer, Joseph Locke. Indigenous innovation had to await the work in the 1850s and 1860s of Joseph Mitchell in the Highlands or Thomas Bouch on cheap connections to lowland towns.

By the time Locke was at work in Scotland, English involvement in Scottish railways had extended to the large-scale provision of capital. Some money had gone into the coal railways, presumably attracted by the high dividends of the more successful Monklands lines and often belonging to English residents like John Gladstone with strong Scottish connections. George Stephenson had remarked on the shortage of native Scottish capital for railway schemes even in the 1820s,[61] and by the time of the lesser mania of the mid-1830s English capital was essential if more than a few small lines were to be built. Although the two Arbroath lines were constructed from local resources, the fate of the Dundee & Perth and Edinburgh & Dunbar projects demonstrated that outside capital would be required even for quite short lines.

English capital was not ready to flood into Scotland in the speculative years of 1835-36, since the most attractive propositions, in terms both of investors' knowledge about them and of prospective profitability, were the main lines still being constructed in the south. As a result Scotland's participation in the promotional enthusiasm, like that of the remoter regions of England, was severely restricted, and came at the end of the period — perhaps picking up the remaining crumbs of investors' financial resources.⁶² Serious efforts at attracting English capital started with the subscription lists of the Ayrshire, the Greenock and the Edinburgh & Glasgow; in the event the three companies were promised respectively 15.1%, 37.6% and 62.7% of their subscription money from south of the border.⁶³ In 1838 the Edinburgh & Glasgow appointed two Lancastrian directors 'in consequence of the large amount of stock held at present in Lancashire', and when Locke joined the Greenock board one of the stated advantages was that he lived 'in England where so many of the shareholders reside'.⁶⁴ By the time of the trans-border projects, the need for English capital to supplement home resources was imperative, and the combination of incipient mania and prospective help to English trading interests made it easy to obtain. Even the Glasgow, Dumfries & Carlisle, fighting a rearguard battle with no formal English allies, raised 22.5% of its money in England. The North British figure was 42.2% (with 26.7% coming from Hudson's strongholds in Yorkshire and Derbyshire), while the view of the Caledonian as an English company in disguise received support when 77.8% of its subscriptions came from the south. All the other new companies of 1845, apart from the little Glasgow Barrhead & Neilston Direct, depended heavily on English subscriptions, the proportions varying from about 30% on the Scottish Central and Edinburgh & Northern to a massive 78.3% on the Scottish Midland Junction.⁶⁵

Normally the existence of a large body of English shareholders strengthened the control of the directors over the company. As noted in the context of Sunday trains, English shareholders seldom took an active part in company meetings as long as they considered the financial results to be tolerably satisfactory, and most of them gave their proxies to the directors to use at their discretion. On the other hand the existence of powerful groups of shareholders geographically remote from the day-to-day running of the company and with no real interest in the nature of its services as against its financial results could create a focus for discontent if times were bad. Strong localised groups, like the Bristol shareholders in the Newhaven line, the Lancastrians in the Edinburgh & Glasgow or the Londoners in the Caledonian, were reminders to directors that they held office on suffrance. When the mania collapsed, English suspicions that Scottish management was either deliberately exploiting English investors for the benefit of Scottish rail-users, or else was simply and crassly incompetent, were encouraged by sections of the railway press. English groups took the lead in the shareholders' revolts which first threw Learmonth's board out of the Edinburgh & Glasgow, and by 1850 created boardroom upheavals in virtually all main-line Scottish companies.⁶⁶

In 1845 the Scottish railway system remained formally independent of England, in that the companies were Scottish-based and run by Scottish directors and officials, but there were powerful links with and influences from the south. The extent of English shareholding meant that Scottish shares were clearly part of a British national financial market, and were in competition for further funds with rivals in the south; an increasing number of Scots were also investing in English railways. It also ensured that the companies had to be run primarily for financial profit, since distant shareholders would not forgive poor results for the sake of the service provided. The advent of large-scale English involvement emphasised the increasing separation of shareholding from effective control, and of the railways themselves from the industrial and commercial concerns upon whose trade they depended.

Anglo-Scottish co-operation was inevitably strengthened by the advent of serious trans-border schemes. Sea trade with England had already been of interest to the Ayrshire in particular, and was sufficiently developed for both east and west-coast projectors to claim that their rivals could not hope to compete with the efficient steamships on their side of the country. Now the quest was for allies by land. The close links of the Caledonian with the Grand Junction and the Lancaster & Carlisle were clearly essential to its very existence. The rather more uneasy association of the North British with the Hudson empire, which was strained at various times by the North British interest in the Carlisle route, by the Newcastle & Berwick's flirtation with the Caledonian Extension plan and by Hudson's later attempts at takeover, survived because co-operation remained in the interest of both parties. The Ayrshire, which had formed an alliance of no great value with the Preston & Wyre for its venture into shipping services, failed to find English allies either before or after it reached Carlisle (as the Glasgow & South-Western) in 1850, and it suffered in consequence.[67] Scottish main-line companies were becoming conscious that, however independent they might be, it was also important that they were links in the long railway chains stretching out from London.

IX. Forward to the mania

The details of the railway mania which developed from 1844 lie chronologically beyond the bounds of this study. It was a British, and to a great extent an English-centred, phenomenon, especially after rampant speculation had taken the place of rational expansion of the railway system. But most of the projects put forward in 1844 for parliamentary consideration in the following year were for lines along logical routes with apparently good prospects of profitability; most indeed achieved authorisation and came through the traumatic post-mania years to eventual commercial viability. The amount of subscription capital offered to Scottish schemes by both Scottish and English investors suggests that Scottish railways were not inherently unattractive propositions, by comparison either with English railways or with non-railway investments. In so far as there was logic in investors' decisions, the record of existing Scottish companies, for all their over-expenditure and disappointing dividends, must have been regarded as moderately encouraging.

Cyclical recovery after 1842 might in itself have been enough to arouse expansionist ideas. It was accompanied by a move to cheap money, enhanced by the Bank Charter Act of 1844 and by the Bank of England's increasing intervention in the money market.[68] In 1843 it was claimed that some £20–25 million was lying inactive in the City of London, and

It was remarked that there had been a larger continuation of a plentiful supply of money than had occurred in the memory of the oldest capitalists.[69]

Overseas investment was unpopular, following the formal repudiation of their debts by three of the United States and the less formal failure to pay of six more and of Spain, Portugal and several Latin American countries. For the first time in eighty years, 3% Consols rose to par, and in September 1844 the discount rate fell to $2\frac{1}{2}$%.[70] As returns on government stocks declined, industrial and commercial investments appeared more attractive. In Scotland, although the main expansion of the textile industries was over, there were still opportunities in coal, iron, shipbuilding and engineering — the iron industry, for instance, doubled its output between 1840 and 1848 and was planning the investment which permitted a further 37% expansion in the following four years[71] — and money was to go into all these areas. It was arguable, however, that industrial overcapacity had been one cause of the preceding depression, and that further investment needed to be regarded with some caution. Early in 1844 the *Morning Chronicle* summed up the investors' dilemma:

> For two years our capitalists have been anxiously awaiting for a revival of trade and commerce ... Profitable investment there seems to be none. The rate of interest continues to decline. The Funds maintain an unnatural buoyancy. Deposits in Savings Banks are rapidly accumulating.

By October the investors had found their solution:

> Capital clamours for profitable investment; confidence has become eager, and may shortly become blind; railroads present the first tangible form of investment ...[72]

Railways appeared to offer the desirable combination of minimum risk and maximum anticipated return, and became by far the main attraction for idle funds.

By 1844 railways had apparently proved themselves. In the previous decade they had adjusted to their enhanced status as important business enterprises in their own right, no longer subordinate to other concerns in mining or manufacturing. Promoters had discovered, largely by trial and error, firstly how to estimate for and construct a railway, and then how best to manage it. In neither area was the record one of conspicuous success, and the comparatively satisfactory results achieved by many railways were a tribute more to the appearance, particularly after 1842, of an unexpectantly high level of demand for their services than to the efficiency of the companies. Paradoxically, the shortcomings of management may even have encouraged investors, for if profits could be made after such miscalculations, how much more might be achieved once the lessons of past experience had been learned?

As it was, some major English lines like the London & Birmingham and Grand Junction were already paying 10% dividends,[73] and Gladstone's 1844 act clearly envisaged such figures continuing indefinitely into the future. While it was possible to take various views about the direction and the details of future economic expansion, railways seemed certain to benefit under any circumstances. Transport facilities would obviously be required; the railways had already seen off competing road carriers, coastal shipping was only a rival on a limited number of routes, and the canals had been reduced to fighting a rearguard action. The demand for passenger travel was sure to increase, and a larger network would be more attractive for freight. Even if investors considered the pessimistic possibility that their chosen scheme might fail, railway companies were among the few investments which offered the security of limited liability once they had passed Parliament. For the capitalist with money to spare but with no desire to be closely involved in the management of his investment, this offered a considerable attraction.

Figure 4. *Movement of Share Prices, April 1841 – July 1845*

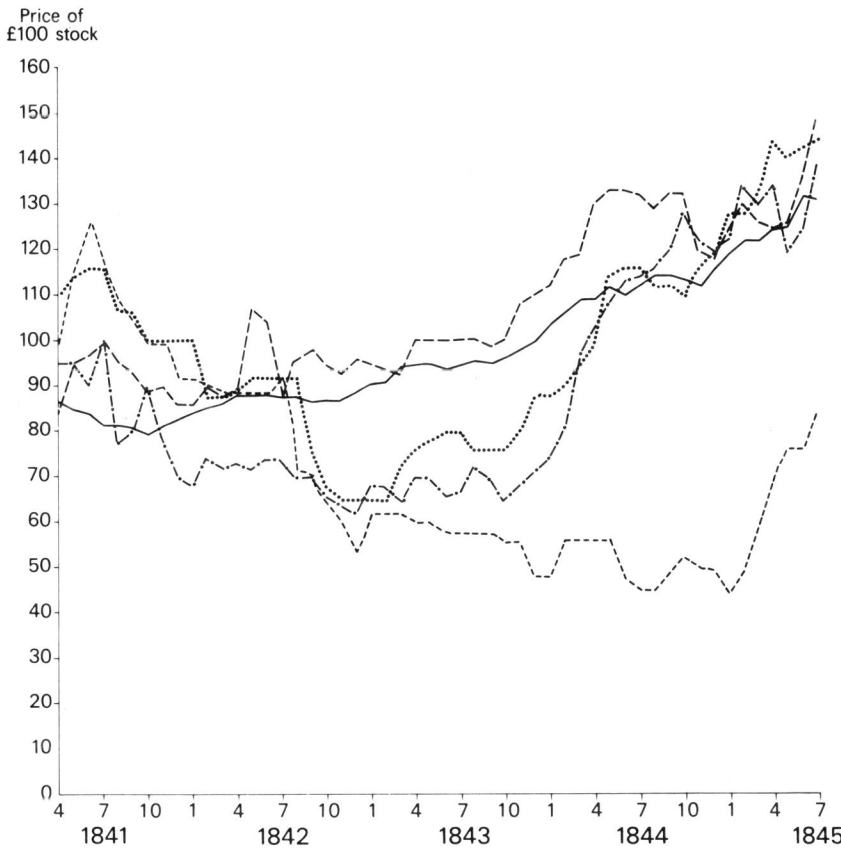

Sources: prices on Scottish stock exchanges at the beginning of each month as given in the *Scotsman, Edinburgh Evening Courant* and *Glasgow Herald;* A. D. Gayer, W. W. Rostow & A. J. Schwartz, *The Growth and Fluctuation of the British Economy 1790-1850*, 368.

In Scotland investors were influenced by these general considerations, but also by the record of existing Scottish companies. The coal railways were not greatly influential. Their financial record had been mixed, and the most profitable routes, between Glasgow and the Monklands coalfield, had been constructed. No new coal railways were projected in 1844-45, although in the following year speculative fever and the hope that Parliament might sanction competing lines led to a spate of unsuccessful Lanarkshire proposals. The plans of 1844-45 were almost all for more inter-urban lines, extending the main network to Aberdeen, Dundee, Carlisle and Hawick; if precedents were to be considered, they would be those of the existing inter-urban lines, and particularly of the two largest. The Edinburgh & Glasgow and the Ayrshire, which received much more press attention and had much more actively traded shares than other Scottish companies, both reached 1844 in tolerable condition. Dividends of four or five per cent. might not be exciting, but the trend was upward, the prospects bright, and the return as good as could be obtained elsewhere.

Figure 4 indicates the trend of share prices in the pre-media period for the only Scottish companies whose shares were traded with any regularity. The all-share index compiled by Gayer, Rostow and Schwartz touched bottom in October 1841, although their sub-index of railway shares (based for this period on a sample of a mere ten companies of which only the Edinburgh & Glasgow was Scottish) continued to fall until October 1842.[74] The Edinburgh & Glasgow, after a brief post-opening jump, started to rise slowly from July 1842, while the other Scottish shares continued to fall until the end of the year. In all cases a rapid rise developed from the autumn of 1843, with the exception of the Greenock company, whose difficulties with high costs and intense competition were reflected in share values which fell steadily until even they were caught up in the unthinking enthusiasm of 1845. In the key year from October 1843 the shares of the two leading companies, as well as those of many lesser lines here represented by the Dundee & Arbroath, rose much more sharply than the all-share index. While it put on 17.9% in twelve months, the railway-share figures were 31.3% for the Edinburgh & Glasgow, 51.3% for the Dundee & Arbroath, and 96.9% for the Ayrshire, which moved up from a lowly starting position to virtual parity with the Edinburgh & Glasgow (possibly because of optimism about its link to Carlisle). Such growth both reflected confidence in existing companies and encouraged speculation in new schemes on the assumption that capital values would continue to rise in the future. The judgement of the stock market was that the railways had come successfully through the depression, and that they offered an investment which was on the whole more attractive than the available alternatives. Conditions were right for an outbreak of expansionist enthusiasm; within the year it had degenerated into mania.

Appendix

Analysis by Location and Occupation of Subscriptions to Scottish Railway Companies authorised in 1837-38

Table 80. Glasgow Paisley Kilmarnock & Ayr, 1837: Amount of Subscriptions (£)

	Non-Occupational	Land	Trade	Industry	Law	Finance	Professions	Contracting	Miscellaneous	Women	Unknown	Total
Glasgow	17,600		235,350	7,250	1,000	3,500	20,000	3,000	500	5,500	1,000	294,700
Edinburgh	13,000		8,500		14,850		3,500		250			40,100
Renfrewshire	1,000		18,850	2,400	4,100		2,500		100	1,000		29,950
Ayrshire	16,350	2,100	14,600	3,300	10,150	3,000	6,050			5,850		61,400
Local Counties	17,350	2,100	33,450	5,700	14,250	3,000	8,550		100	6,850		91,350
Local Total	34,950	2,100	268,800	12,950	15,250	6,500	28,550	3,000	600	12,350	1,000	386,050
Other Scotland	500		1,000	11,500						2,750		15,750
Total Scotland	48,450	2,100	278,300	24,450	30,100	6,500	32,050	3,000	850	15,100	1,000	441,900
London	2,500		4,500	2,500	3,500		500	500				14,000
Lancashire	2,000		42,500	2,500	2,000		2,000		1,000			52,000
Yorkshire					4,250							4,250
Other England			5,000	1,500	3,000							9,500
Total England	4,500		52,000	6,500	12,750		2,500	500	1,000			79,750
Wales							2,500					2,500
Ireland							3,000					3,000
Other	1,000											1,000
Unknown												
Total	53,950	2,100	330,300	30,950	42,850	6,500	40,050	3,500	1,850	15,100	1,000	528,150

Table 81. Glasgow Paisley Kilmarnock & Ayr, 1837: Amount of Subscriptions (% of Total)

	Non-Occupational	Land	Trade	Industry	Law	Finance	Professions	Contracting	Miscellaneous	Women	Unknown	Total
Glasgow	3.3		44.6	1.4	0.2	0.7	3.8	0.6	0.1	1.0	0.2	55.8
Edinburgh	2.5		1.6		2.8		0.7		0.0			7.6
Renfrewshire	0.2		3.6	0.5	0.8		0.5		0.0	0.2		5.7
Ayrshire	3.1	0.4	2.8	0.6	1.9	0.6	1.1			1.1		11.6
Local Counties	3.3	0.4	6.3	1.1	2.7	0.6	1.6		0.0	1.3		17.3
Local Total	6.6	0.4	50.9	2.9	2.9	1.2	5.4	0.6	0.1	2.3	0.2	73.1
Other Scotland	0.1		0.2	2.2						0.5		3.0
Total Scotland	9.2	0.4	52.7	4.6	5.7	1.2	6.1	0.6	0.2	2.9	0.2	83.7
London	0.5		0.9	0.5	0.7		0.1	0.1				2.7
Lancashire	0.4		8.0	0.5	0.4		0.4		0.2			9.8
Yorkshire					0.8							0.8
Other England			0.9	0.3	0.6							1.8
Total England	0.9		9.8	1.2	2.4		0.5		0.2			15.1
Wales							0.5					0.5
Ireland							0.6					0.6
Other	0.2											0.2
Unknown												
Total	10.2	0.4	62.5	5.9	8.1	1.2	7.6	0.7	0.4	2.9	0.2	100.0

Table 82. Glasgow Paisley Kilmarnock & Ayr, 1837: Number of Subscriptions

	Non-Occupational	Land	Trade	Industry	Law	Finance	Professions	Contracting	Miscellaneous	Women	Unknown	Total
Glasgow	7		99	8	2	2	9	1	2	2		133
Edinburgh	7		5		17		2		1			32
Renfrewshire	1		18	3	5		4		2	1		34
Ayrshire	20	3	28	8	14	5	14			15		107
Local Counties	21	3	46	11	19	5	18		2	16		141
Local Total	28	3	145	19	21	7	27	1	4	18		274
Other Scotland	1		1	3						2		7
Total Scotland	36	3	151	20	38	7	29	1	5	20		313
London	1		4	1	2		1	1				10
Lancashire	2		19	2	2		2	1	1			28
Yorkshire					2							2
Other England			1	1	3							5
Total England	3		24	4	9		3	1	1			45
Wales							1					1
Ireland	1											1
Other							1					1
Unknown											1	1
Total	40	3	175	24	47	7	34	2	6	20	1	361

Table 83. *Glasgow Paisley Kilmarnock & Ayr, 1837: Number of Subscriptions (% of Total)*

	Non-Occupational	Land	Trade	Industry	Law	Finance	Professions	Contracting	Miscellaneous	Women	Unknown	Total
Glasgow	1.9		27.4	2.2	0.6	0.6	2.5	0.3	0.6	0.6	0.3	36.8
Edinburgh	1.9		1.4		4.7		0.6		0.3			8.9
Renfrewshire	0.3		5.0	0.8	1.4		1.1		0.6	0.3		9.4
Ayrshire	5.5	0.8	7.8	2.2	3.9	1.4	3.9			4.2		29.6
Local Counties	5.8	0.8	12.7	3.0	5.3	1.4	5.0		0.6	4.4		39.1
Local Total	7.8	0.8	40.2	5.3	5.8	1.9	7.5	0.3	1.1	5.0		75.9
Other Scotland	0.3		0.3	0.8						0.6		1.9
Total Scotland	10.0	0.8	41.8	5.5	10.5	1.9	8.0	0.3	1.4	5.5		86.7
London	0.3		1.1	0.3	0.6		0.3	0.3				2.8
Lancashire	0.6		5.3	0.6	0.6		0.6		0.3			7.8
Yorkshire					0.6							0.6
Other England			0.3	0.3	0.8							1.4
Total England	0.8		6.6	1.1	2.5		0.8	0.3	0.3			12.5
Wales							0.3					0.3
Ireland	0.3											0.3
Other							0.3					0.3
Unknown												
Total	11.1	0.8	48.5	6.6	13.0	1.9	9.4	0.6	1.7	5.5	0.3	100.0

Table 84. *Glasgow Paisley & Greenock, 1837: Amount of Subscriptions (£)*

	Non-Occupational	Land	Trade	Industry	Law	Finance	Professions	Contracting	Miscellaneous	Women	Unknown	Total
Glasgow	100		6,400	1,000	1,125		2,375	125				11,025
Edinburgh			500		5,500		200			500		6,800
Renfrewshire	11,750	4,600	93,500	22,700	10,275	10,400	12,225	200	4,450	3,075		173,175
Local Counties	11,750	4,600	93,500	22,700	10,275	10,400	12,225	200	4,450	3,075		173,175
Local Total	11,750	4,600	99,900	23,700	11,400	10,400	14,600	325	4,450	3,075		184,200
Other Scotland	6,550		1,450	100			2,125	125		100		10,450
Total Scotland	18,400	4,600	101,850	23,800	16,900	10,400	16,925	450	4,450	3,675		201,450
London	2,625		5,250		5,500				250			13,625
Lancashire	11,750	500	72,550	4,625	1,250	2,250	4,625		250			97,800
Yorkshire												
Other England	625		4,250	2,875		3,000	2,250					13,000
Total England	15,000	500	82,050	7,500	1,250	5,250	6,875		500			124,425
Wales												
Ireland	625		1,175		2,000		975		175			4,950
Other												
Unknown												
Total	34,025	5,100	185,075	31,300	26,650	15,650	24,775	450	5,125	3,675		330,825

Table 85. *Glasgow Paisley & Greenock, 1837: Amount of Subscriptions (% of Total)*

	Non-Occupational	Land	Trade	Industry	Law	Finance	Professions	Contracting	Miscellaneous	Women	Unknown	Total
Glasgow	0.0		1.9	0.3	0.3		0.7	0.0				3.3
Edinburgh			0.2		1.7		0.1				0.2	2.1
Renfrewshire	3.6	1.4	28.3	6.9	3.1	3.1	3.7	0.1	1.3		0.9	52.3
Local Counties	3.6	1.4	28.3	6.9	3.1	3.1	3.7	0.1	1.3		0.9	52.3
Local Total	3.6	1.4	30.2	7.2	3.4	3.1	4.4	0.1	1.3		0.9	55.7
Other Scotland	2.0		0.4	0.0			0.6	0.0			0.0	3.2
Total Scotland	5.6	1.4	30.8	7.2	5.1	3.1	5.1	0.1	1.3		1.1	60.9
London	0.8		1.6		1.7				0.1			4.1
Lancashire	3.6	0.2	21.9	1.4	0.4	0.7	1.4		0.1			29.6
Yorkshire												
Other England	0.2		1.3	0.9		0.5	0.7					3.9
Total England	4.5	0.2	24.8	2.3	2.0	1.6	2.1		0.2			37.6
Wales												
Ireland	0.2		0.4		0.6		0.3					1.5
Other												
Unknown												
Total	10.3	1.5	55.9	9.5	7.8	4.7	7.5	0.1	1.5		1.1	100.0

Table 86. Glasgow Paisley & Greenock, 1837: Number of Subscriptions

	Non-Occupational	Land	Trade	Industry	Law	Finance	Professions	Contracting	Miscellaneous	Women	Unknown	Total
Glasgow			11	1	4		2	1			1	19
Edinburgh	1		1		6		1					10
Renfrewshire	5	12	198	56	18	11	31	2	22	18		373
Local Counties	5	12	198	56	18	11	31	2	22	18		373
Local Total	5	12	209	57	22	11	33	3	22	18		392
Other Scotland	3		6	1			5	1		1		17
Total Scotland	9	12	216	58	28	11	39	4	22	20		419
London	4		3		2				1			10
Lancashire	14	1	95	7	3	3	8		1			132
Yorkshire												
Other England	1		8	4		3	4			1		21
Total England	19	1	106	11	5	6	12		2	1		163
Wales												
Ireland	1		4		4		3		2			14
Other												
Unknown												
Total	29	13	326	69	37	17	54	4	26	21		596

Table 87. Glasgow Paisley & Greenock, 1837: Number of Subscriptions (% of Total)

	Non-Oc-cupational	Land	Trade	Industry	Law	Finance	Pro-fessions	Con-tracting	Miscel-laneous	Women	Unknown	Total
Glasgow	0.2		1.8	0.2	0.7		0.3	0.2				3.2
Edinburgh			0.2		1.0		0.2			0.2		1.7
Renfrewshire	0.8	2.0	33.2	9.4	3.0	1.8	5.2	0.3	3.7	3.0		62.6
Local Counties	0.8	2.0	33.2	9.4	3.0	1.8	5.2	0.3	3.7	3.0		62.6
Local Total	0.8	2.0	35.1	9.6	3.7	1.8	5.5	0.5	3.7	3.0		65.8
Other Scotland	0.5		1.0	0.2			0.8	0.2		0.2		2.9
Total Scotland	1.5	2.0	36.2	9.7	4.7	1.8	6.5	0.7	3.7	3.4		70.3
London	0.7		0.5		0.3				0.2			1.7
Lancashire	2.3	0.2	15.9	1.2	0.5	0.5	1.3		0.2			22.1
Yorkshire												
Other England	0.2		1.3	0.7		0.5	0.7			0.2		3.5
Total England	3.2	0.2	17.8	1.8	0.8	1.0	2.0		0.3	0.2		27.3
Wales												
Ireland	0.2		0.7		0.7		0.5		0.3			2.3
Other												
Unknown												
Total	4.9	2.2	54.7	11.6	6.2	2.9	9.1	0.7	4.4	3.5		100.0

Table 88. Edinburgh & Glasgow, 1838: Amount of Subscriptions (£)

	Non-Occupational	Land	Trade	Industry	Law	Finance	Professions	Contracting	Miscellaneous	Women	Unknown	Total
Glasgow	4,800		122,600	7,800	3,000	2,400	1,400	2,100		1,800		143,500
Edinburgh	11,000		49,400	900	39,250		10,550	300	2,100	600		116,500
Lanarkshire					350	600			350			1,300
Stirlingshire					1,200	1,200						2,400
West Lothian				100								100
Midlothian				100	1,550	1,800			350			3,800
Local Counties	15,800		172,000	8,800	43,800	4,200	11,950	2,400	2,450	2,400		263,800
Local Total		600	5,750	1,800	1,200		9,250					18,600
Other Scotland	15,800	600	177,750	10,600	45,000	4,200	21,200	2,400	2,450	2,400		282,400
Total Scotland												
London	57,100		31,800		11,600	9,500	8,200	1,200	6,600	2,400	3,000	100,100
Lancashire	30,300		228,750	50,400	600		20,850				660	362,200
Yorkshire	3,000		12,500									16,100
Other England	16,200		3,600		16,200	6,600	16,500	1,500	1,500	2,400		60,600
Total England	106,600		276,650	50,400	28,400	16,100	45,550	1,200	8,100		3,600	539,000
Wales	1,800			300	6,000		1,200					9,300
Ireland	3,000		1,200		1,200		17,700					23,100
Other			2,400									2,400
Unknown	1,800						1,500					3,300
Total	129,000	600	458,000	61,300	80,600	20,300	87,150	3,600	10,550	4,800	3,600	859,500

Appendix

Table 89. Edinburgh & Glasgow, 1838: Amount of Subscriptions (% of Total)

	Non-Occupational	Land	Trade	Industry	Law	Finance	Professions	Contracting	Miscellaneous	Women	Unknown	Total
Glasgow	0.6		14.3	0.9	0.3		0.2	0.2		0.2		16.7
Edinburgh	1.3		5.7	0.1	4.6	0.3	1.2	0.0	0.2	0.1		13.6
Lanarkshire					0.0	0.1			0.0			0.2
Stirlingshire					0.1	0.1						0.3
West Lothian												0.0
Midlothian				0.0	0.2							0.4
Local Counties				0.0	0.2	0.2	1.4		0.0			
Local Total	1.8		20.0	1.0	5.1	0.5	1.1	0.3	0.3	0.3		30.7
Other Scotland		0.1	0.7	0.2	0.1							2.2
Total Scotland	1.8	0.1	20.7	1.2	5.2	0.5	0.5	0.3	0.3	0.3		32.9
London	6.6		3.7				1.0					11.6
Lancashire	3.5		26.6	5.9	1.4	1.1	1.1	0.1	0.8	0.3	0.3	42.1
Yorkshire	0.3		1.5		0.1						0.1	1.9
Other England	1.9		0.4		1.9	0.8	1.9		0.2			7.1
Total England	12.4		32.2	5.9	3.3	1.9	5.3	0.1	0.9	0.3	0.4	62.7
Wales	0.2				0.7		0.1					1.1
Ireland	0.3		0.1	0.0	0.2		2.1					2.7
Other			0.3									0.3
Unknown	0.2						0.2					0.4
Total	15.0	0.1	53.3	7.1	9.4	2.4	10.1	0.4	1.2	0.6	0.4	100.0

Table 90. Edinburgh & Glasgow, 1838: Number of Subscriptions

	Non-Occupational	Land	Trade	Industry	Law	Finance	Professions	Contracting	Miscellaneous	Women	Unknown	Total
Glasgow	1		69	3	3	3	3	1	1	1		81
Edinburgh	10		31	2	24		6	1	2	1		79
Lanarkshire					1	1						4
Stirlingshire					1	1						2
West Lothian												
Midlothian				1								1
Local Counties				1	2							7
Local Total	11		100	6	29	5	9	2	3	2		167
Other Scotland		1	7	2	3		6					19
Total Scotland	11	1	107	8	32	5	15	2	3	2		186
London	15		9				3					28
Lancashire	13		143	40	8	4	16	1	6	6	1	238
Yorkshire	1		3		1							5
Other England	7		4		4	2	2		3			22
Total England	36		159	40	13	6	21	1	9	6	2	293
Wales	1			1	1		2					5
Ireland	1		1		1		3					6
Other			3									3
Unknown	1						1					2
Total	50	1	270	49	47	11	42	3	12	8	2	495

Table 91. *Edinburgh & Glasgow, 1838: Number of Subscriptions (% of Total)*

	Non-Occupational	Land	Trade	Industry	Law	Finance	Professions	Contracting	Miscellaneous	Women	Unknown	Total
Glasgow	0.2		13.9	0.6	0.6	0.6	0.6	0.2	0.2	0.2		16.4
Edinburgh	2.0		6.3	0.4	4.8	0.2	1.2	0.2	0.4	0.2		16.0
Lanarkshire					0.2	0.2						0.8
Stirlingshire					0.2							0.8
West Lothian												
Midlothian				0.2								0.2
Local Counties				0.2	0.4	0.4						1.4
Local Total	2.2		20.2	1.2	5.9	1.0	1.8	0.4	0.6	0.4		33.7
Other Scotland		0.2	1.4	0.4	0.6		1.2					3.8
Total Scotland	2.2	0.2	21.6	1.6	6.5	1.0	3.0	0.4	0.6	0.4		37.6
London	3.0		1.8				0.6					5.7
Lancashire	2.6		28.9	8.1	1.6	0.8	3.2	0.2	1.2	1.2	0.2	48.1
Yorkshire	0.2		0.6		0.2							1.0
Other England	1.4		0.8		0.8	0.4	0.4		0.6		0.2	4.4
Total England	7.3		32.1	8.1	2.6	1.2	4.2	0.2	1.8	1.2	0.4	59.2
Wales	0.2			0.2	0.2		0.4					1.0
Ireland	0.2		0.2		0.2		0.6					1.2
Other			0.6									0.6
Unknown	0.2						0.2					0.4
Total	10.1	0.2	54.5	9.9	9.5	2.2	8.5	0.6	2.4	1.6	0.4	100.0

Sources: HLRO Deposited Plans, HC 1837, Glasgow Paisley Kilmarnock & Ayr Railway; HC 1837, Glasgow Paisley & Greenock Railway; HC 1838, Edinburgh & Glasgow Railway.

Notes

In each chapter, full details of each source are given at its first appearance. In subsequent references to the same source, use is made of *op.cit., loc.cit.*, an abbreviated title or, for Parliamentary Papers and SRO and NLS collections, a reference number as appropriate.

Chapter 1. The Century of the Waggonways

1. G. Dott, *Early Scottish Colliery Waggonways*, passim; C. F. Dendy Marshall, *A History of British Railways down to the Year 1830*, 112-35; M. J. T. Lewis, *Early Wooden Railways*, 133-5, 189-90, 254-6, 300-1; B. Baxter, *Stone Blocks and Iron Rails*, 15-64, 226-36, 259; B. F. Duckham, *A History of the Scottish Coal Industry*, I, 203-39.
2. For information on Scottish industrialisation, see, among others, R. H. Campbell, *Scotland since 1707*, ch.1-10; S. G. E. Lythe & J. Butt, *An Economic History of Scotland 1100-1939*, ch.7-10; B. P. Lenman, *An Economic History of Modern Scotland*, ch.4-5; T. C. Smout, *A History of the Scottish People 1560-1830*, ch.10; A. Slaven, *The Development of the West of Scotland 1750-1960*, ch.2-6.
3. E. Pawson, *Transport and Economy: The Turnpike Roads of Eighteenth Century Britain*, 22.
4. H. Hamilton, *Economic History of Scotland in the Eighteenth Century*, 213-4; J. Lindsay, *The Canals of Scotland*, 210; R. H. Campbell, *op.cit.*, 49-50; A. Slaven, *op.cit.*, 29-30.
5. J. Lindsay, *op.cit.*, 212-3. Small canals, none more than three miles long, were opened in the eighteenth century at Campbeltown and Stevenston, and near Alloa and Castle Douglas (*ibid.*, 210-13).
6. E.g. E. Pawson, *op.cit.*; W. Albert, *The Turnpike Road System in England 1663-1840*; G. L. Turnbull, *Traffic and Transport: an Economic History of Pickfords*; G. L. Turnbull, 'Provincial road carrying in England in the eighteenth century', *Jnl Trans. Hist.* NS IV (1977), 17-39. The more traditional view, classically expressed in S. & B. Webb, *English Local Government: the Story of the King's Highway*, is also found in, for instance, H. J. Dyos & D. H. Aldcroft, *British Transport*, and C. I. Savage, *An Economic History of Transport*.
7. R. H. Campbell, *op.cit.*, 51; T.C. Smout, *op.cit.*, 120.
8. A. Slaven, *op.cit.*, 36-7.
9. *OSA* I, 3 (Holywood, Dumfriesshire).
10. *OSA* III, 352 (Hoddom, Dumfriesshire); A. Slaven, *op.cit.*, 39.
11. *OSA* XIX, 552 (Longforgan, Perthshire).
12. *OSA* II, 352-3 (Whittingham, East Lothian).
13. E. Pawson, *op.cit.*, 153, 160. Pawson's list, however, omits some important Scottish turnpike creations, such as those of the major roads in Perthshire (29 Geo.III c.17, 1789) and Forfarshire (29 Geo.III c.20, 1789). The Highlands had been much improved by the famous, if economically insignificant, military roads built by General Wade and his successors (see W. Taylor, *The Military Roads of Scotland*). The distinguished engineer Robert Stevenson was impressed with their quality:

'The government having, with the most enlightened policy, advanced one-half of the necessary funds for opening roads in the north of Scotland while the landed proprietors contributed the other, — by this happy union of objects and interests, together with the open and unenclosed state of the country, the hands of the engineer have been perhaps less hampered in this than in any other district of the kingdom. Hence it is that the highways of the North are extremely well laid out, . . . no country can boast of better roads, either in regard to their line of draught or general fabric.' (*Trans. Highland Society* 6 (1824), 2-3.)

14. *OSA* I, 84 (Ayton, Berwickshire).

15. *OSA* II, 211 (Hamilton).

16. *OSA* XV, 607 (Foulis Wester, Perthshire).

17. A. Slaven, *op.cit.*, 40. John Louden McAdam, who first came to prominence as an Ayrshire road trustee, earlier commented on the difficulties of the turnpikes: 'it was universally allowed that the roads in Scotland were in a deplorable state and in these circumstances bankrupt'. (H. Hamilton, *op.cit.*, 227.)

18. H. J. Dyos & D. H. Aldcroft, *op.cit.*, 34-5.

19. B. F. Duckham, *op.cit.*, I, 206; *OSA* VII, 84 (Barr, Ayrshire).

20. *OSA* III, 353 (Hoddom, Dumfriesshire).

21. R. H. Campbell, *op.cit.*, 48.

22. J. U. Nef, *The Rise of the British Coal Industry*, I, 19-20, 23; B. F. Duckham, *op.cit.*, I, 23-32.

23. For Scottish canal development, see J. Lindsay, *op.cit.*; for canals in Britain, see C. Hadfield, *The Canal Age*; H.J. Dyos & D. H. Aldcroft, *op.cit.*, ch.3.

24. M. J. T. Lewis, *op.cit.*, 103.

25. C. F. Dendy Marshall, *op.cit.*, 113.

26. *OSA* VI, 409 (Old Cumnock, Ayrshire).

27. M. J. T. Lewis, *op.cit.*, 103.

28. B. F. Duckham, *op.cit.*, I, 210.

29. D. Murray, *The York Buildings Company: a Chapter in Scottish History*, 3, 19-23, 24, 34, 65.

30. Alexander Scott of Ormiston, essay in R. Stevenson (ed.), 'Essays on rail-roads', *Trans. Highland Soc.* 6 (1824), 23-4.

31. D. Murray, *op.cit.*, 65-6.

32. B. F. Duckham, *op.cit.*, I, 209-10.

33. *Ibid.*, I, 210; G. Dott, *op cit.*, 15.

34. *OSA* X, 83 (Tranent).

35. C. Maclaren, 'Railways compared with canals and common roads, and their uses and advantages explained', *Pamphleteer*, vol. XXVI, no. 51 (1826), 61.

36. R. Stevenson, *Report relative to Various Lines of Railway from the Coal-Fields of Mid-Lothian to the City of Edinburgh and Port of Leith*, 10.

37. M. J. T. Lewis, *op.cit.*, 255; W. Nimmo (*History of Stirlingshire*, II, 297) gives 1766 as the date of the first Carron line.

38. B. Baxter, *op.cit.*, 17; M. J. T. Lewis, *op.cit.*, 255.

39. *OSA* VIII, 617-8 (Alloa).

40. M. J. T. Lewis, *op.cit.*, 133. Dixon sank new coal-pits at Knightswood and Gartnavel, north-west of Glasgow, and ran the first waggonway in the west of Scotland from them to the Clyde at Yoker. Barges then floated the coal fifteen miles to Dumbarton with the tide (J. A. Fleming, *Scottish and Jacobite Glass*, 154-5).

41. B. F. Duckham, *op.cit.*, I, 181; *NSA* VI (Lanarkshire), 697.

42. W. W. Tomlinson, *The North Eastern Railway*, 82.

43. M. J. T. Lewis, *op.cit.*, 189-90; *OSA* VIII, 617 (Alloa); A. Scott in R. Stevenson (ed.), *loc.cit.*, 24; R. Buchanan, *Report relative to the Proposed Rail-way from Dumfries to Sanquhar*, 14; C. Maclaren, *loc.cit.*, 60.

44. J. C. & F. Inglis, *The Fordell Railway*, 14; G. Dott, *op.cit.*, 7; M. J. T. Lewis, *op.cit.*, 190; R. Stevenson (ed.), *loc. cit.*, 141.

45. B. Baxter, *op.cit.*, 232.

46. *OSA* XIII, 467 (Dunfermline); B. F. Duckham, *op.cit.*, I, 145.

47. *OSA* XIII, 467 (Dunfermline).
48. *NSA* IX (Fife), 867.
49. For instance, B. Baxter, *op.cit.*, 232; C. F. Dendy Marshall, *op.cit.*, 117; B. F. Duckham, *op.cit.*, I, 212; G.Dott, *op.cit.*, 2. Dott considers Dendy Marshall's analysis 'anything but satisfactory', but his own text and map contradict each other. No author's version is totally convincing.
50. *OSA* XIII, 471 (Dunfermline).
51. M. J. T. Lewis, *op.cit.*, 134-5. He does, however, place Pitfirrane to the north of Dunfermline, rather than correctly to the west.
52. B. F. Duckham, *op.cit.*, I, 212; B. Baxter, *op.cit.*, 232.
53. G. Dott, *op.cit.*, 12; P. Chalmers, *History and Statistical Account of Dunfermline*, I, 403.
54. P. Chalmers, *op.cit.*, I, 35.
55. *Ibid.*, I, 403.
56. *Ibid.*, I.
57. G. Dott, *op.cit.*, 13.
58. Established respectively by 41 Geo.III c.33, 1801, and 48 Geo.III c.46, 1808.
59. A. Scott in R. Stevenson (ed.), *loc.cit.*, 9-10.
60. G. Robertson in *ibid.*, 70; J. Baird in *ibid.*, 129; G. Dott, *op.cit.*, 31.
61. H. J. Dyos & D. H. Aldcroft, *op.cit.*, 112.
62. B. Baxter, *op.cit.*, 33. On page 228 Baxter confirms that the company used the edge rail.
63. *Ibid.*, 226.
64. *OSA* II, 270 (Newton-upon-Ayr).
65. B. F. Duckham, *op.cit.*, I, 205, 209; *OSA* II, 89-90 (Kilmarnock).
66. NLS, Ms 9814 f.24, Marquis of Titchfield to Earl of Eglinton, 22 Apr 1806.
67. W. Aiton, *General View of the Agriculture of the County of Ayr*, 559.
68. NLS, Ms 9814 f.24.
69. HLRO Deposited Plan, HC 1808 Kilmarnock & Troon Railway, subscription contract. Highet suggests, without giving his source, that Titchfield held 67 shares, his daughter ten, and the other three landowners one each (C. Highet, *The Glasgow and South-Western Railway*, 4-5).
70. *NSA* V (Ayrshire), 684.
71. W. Aiton, *op.cit.*, 558.
72. R. Stevenson (ed.) *loc.cit.*, 132; Sir D. Brewster et al., *The Edinburgh Encyclopaedia*, XVII, 304; B. Baxter *op.cit.*, 29-30.
73. W. Aiton, *op.cit.*, 558.
74. NLS, Ms 9814 f.24.
75. *NSA* V (Ayrshire), 683.
76. B. F. Duckham, *op cit.*, I, 216; *NSA* V (Ayrshire), 554. In 1818 Alexander Scott of Ormiston appeared to raise a bizarre possibility which, among other things, suggested that the Kilmarnock & Troon could not have had a great deal of traffic: 'The distance between the two railroads or courses, being the same as the width of each, a horse may travel in the middle space, with a wheel on the inner range of each of the roads'. Fortunately for the railway, however, this was a direct and verbatim plagiarism from William Aiton, and was first written before the line was open (A. Scott in R. Stevenson (ed.), *loc.cit.*, 10; W. Aiton, *op.cit.*, 556).
77. C. Highet, *op.cit.*, 7; HLRO Deposited Plan, HC 1808 Kilmarnock & Troon Railway; W. McAdam, *The Birth, Growth and Eclipse of the Glasgow and South-Western Railway*, 16.
78. C. F. Dendy Marshall, *op.cit.*, 134.
79. R. Stevenson, *Report relative to . . . Mid-Lothian*, 38; C. Highet, *op.cit.*, 6.
80. PP 1839 (222,517) X, *Select Committee on Railways*, appendix 20; W. W. Tomlinson, *op.cit.*, 114. According to a company official, the company's act did not allow it to charge individual passenger fares (SRO, BR/LIB(S)/6/224 p.7, George Douglas to Select Committee on Railway Communication (n.d.)).
81. C. Dupin, *Relation succinte d'un second voyage fait en 1817 et 1818 dans le ports d'Angleterre, d'Ecosse et d'Irlande*, 17.
82. J. Duncan, *Itinerary of Scotland*, quoted in A. C. O'Dell & P. S. Richards, *Railways and*

Geography, 181.

83. John Bailey to John Watson, 4 Oct 1813; Watson to Bailey, 11 Oct 1813: quoted in R. M. Gard & J. R. Hartley, *Railways in the Making*.

84. J. K. Hunter, *The Retrospect of an Artist's Life*, 98-100.

85. *NSA* V (Ayrshire), 554.

86. J. K. Hunter, *op.cit.*, 100.

87. G. Buchanan, *An Account of the Glasgow and Garnkirk Railway and Other Railways in Lanarkshire*, 4; C. Highet, *op.cit.*, 7.

88. C. F. Dendy Marshall, *op.cit.*, 134.

89. A story given in, for example, W. J. Gordon, *Our Home Railways*, 190; C. Highet, *op.cit.*, 44; J. Thomas, *A Regional History of the Railways of Great Britain: VI, Scotland*, 20; F. Fernyhough, *The History of Railways in Britain*, 127; *Scotland's S.M.T. Magazine*, Jan 1948. The earliest of these sources, Gordon, is frequently unreliable.

90. W. Aiton, *op.cit.*, 558.

91. R. Buchanan, *op.cit.*, 16.

92. B. F. Duckham, *op.cit.*, I, 210. Baxter identifies some ninety miles of waggonway and horse railway by about 1830 (including such lines as the Monkland & Kirkintilloch, Edinburgh & Dalkeith, Dundee & Newtyle, and Wishaw & Coltness), as against 1075 miles in England and 400 miles in Wales (B. Baxter, *op.cit.*, 34).

93. A. Scott in R. Stevenson (ed.), *loc.cit.*, 14.

94. J. Lindsay, *op.cit.*, 181-8.

95. B. F. Duckham, *op.cit.*, I, 228.

96. *NSA* IX (Fife), 186, 834.

97. B. F. Duckham, *op.cit.*, I, 224-5.

98. *Ibid.*, I, 235; R. Bald, *A General View of the Coal Trade of Scotland*, 3.

99. R. Stevenson, *Report relative to ... Mid-Lothian*, 27; H. Baird, *Report on the Proposed Edinburgh and Glasgow Union Canal*, 13, 17.

100. R. Bald, *op.cit.*, 29-30, 43-4.

101. *Ibid.*, 31-4.

102. *Ibid.*, 34; B. F. Duckham, *op.cit.*, I, 204-5.

103. *OSA* VIII, 637-8 (Alloa): X, 507 (Inverkeithing): XII, 516-7 (Dysart): XIII, 477 (Dunfermline): XVIII, 14 (Kirkcaldy); B. F. Duckham, *op.cit.*, I, 228-9.

104. R. Bald, *op.cit.*, 39; H. Baird, *op.cit.*, 17.

105. R. Stevenson, *Report relative to ... Mid-Lothian*, 32.

106. John Fernie (1815), quoted in R. L. Galloway, *Annals of Coal Mining and the Coal Trade*, I, 367.

107. *OSA* VIII, 617 (Alloa); *NSA* VIII (Clackmannanshire), 31; *NSA* IX (Fife), 186.

108. B. F. Duckham, *op.cit.*, I, 151, 215.

109. J. A. Hassan, 'The supply of coal to Edinburgh, 1790-1850', *Transport Hist.* 5 (1972), 136.

110. R. Stevenson, *Report relative to ... Mid-Lothian*, 29; H. Baird, *Report on the Proposed Railway from the Union Canal at Ryal, to Whitburn, Polkemmet, and Benhar: or the West Lothian Railway*, 11.

111. NLS, Ms 9814 f.24. Thomas mistakenly gives this figure as the actual cost of construction (J. Thomas, *op.cit.*, 19).

112. B. F. Duckham, *op.cit.*, I, 215.

113. R. Stevenson, *Report relative to ... Mid-Lothian*, 8; C. MacLaren, *loc.cit.*, 61.

114. SRO, GD26/15/94, Leven and Melville Muniments, Gibson & Oliphant W.S., *Memorial Humbly Submitted to the Right Honourable the Lord Provost and Magistrates of Edinburgh, respecting the Projected Railways*, 5. According to Nimmo, the original Carron rails, which were wood with an iron strip on top, were replaced after a year with iron rails from Coalbrookdale (W. Nimmo, *op.cit.*, II, 297).

115. B. F. Duckham, *op.cit.*, I, 211; J. C. & F. Inglis, *op.cit.*, 14.

116. J. C. & F. Inglis, *op.cit.*, 13-14; *NSA* IX (Fife), 245.

117. B. F. Duckham, *op.cit.*, I, 211; G. Dott, *op.cit.*, 7.

118. According to Dott, this was in fact written by Robert Bald, who was mining engineer first to the earl of Mar, and later to the Alloa Coal Company (Dott, *op.cit.*, 28).
119. R. Stevenson, *Report relating to . . . Mid-Lothian*, 8.
120. T. Telford, *Report relative to the Proposed Rail-Way from Glasgow to Berwick-upon-Tweed*, 4-17 and appendix; W. J. Gordon, *op.cit.*, 201. South of the Highland line, the central Borders was the principal area of Scotland which had neither coal supplies nor water transport: if carting coal cost 1/- per ton-mile, Borders towns were paying up to £2 per ton for the transport of coal alone.
121. T. Telford, *op.cit.*, 2.
122. T. Telford, section on canals in J. Plymley, *General View of the Agriculture of Shropshire*, 313-4. According to Joseph Mitchell, later engineer of the Highland Railway, Telford declined to engineer the Liverpool & Manchester as 'he felt it unfair to his friends and to the canal interests to give the benefit of his great experience to this novel mode of locomotion' (J. Mitchell, *Reminiscences of my Life in the Highlands*, I, 101).
123. NLS, Acc.5111, box 23 f.2, Rennie Papers, Circular, *Railway from Berwick to Kelso*.
124. J. Rennie, *Report concerning the Proposed Rail-Way from Kelso to Berwick*; W. W. Tomlinson, *op.cit.*, 32; Act 51 Geo.III c.133, 1811. Rennie's report, which reduced the anticipated coal traffic to 12,000 tons, hoped that the local authorities would pay half of the £20,000 cost of a joint rail and road bridge across the Tweed, thus reducing the capital to be raised by the company to £80,000. Since the act retained the figure of £90,000, agreement was presumably not reached (J. Rennie, *op.cit.*, 11, 14-16). The parliamentary subscription contract covered only £54,000 of the required capital, with one man (George Leyburn) subscribing £5000 and five others, including the Duke of Buccleuch, subscribing £2000 each (HLRO Deposited Plan, HC 1811 Berwick & Kelso Railway, subscription contract). The company's act was later used as the model for that of the Stockton & Darlington (W. W. Tomlinson, *op.cit.*, 69).
125. SRO, GD104/346, Scott of Raeburn Papers, Circular, *To the Chairman and Committee of the Berwick and Kelso Rail-Way Company; Scotsman*, 22 Dec 1824. The Berwick group concluded that 'a well-planned Rail-road would be a most important benefit to the country — and they are fearful that if one prove unsuccessful, no other may, even at a future period, be attempted.' It was in fact to be twenty years before the two towns were linked by rail. One version of the 1824 plan also surveyed a possible extension via Melrose and Dalkeith to Edinburgh (H. G. Lewin, *Early British Railways*, 6).
126. W. W. Tomlinson, *op.cit.*, 293; H. G. Lewin, *op.cit.*, 6.
127. R. Buchanan, *op.cit.*, 40-2.
128. D. Stevenson, *Life of Robert Stevenson*, 74-90; A. Stevenson, *Biographical Sketch of the Late Robert Stevenson*, 3-16. The two authors were sons of Robert Stevenson, who was described by Sir Walter Scott as 'a most gentlemanlike and modest man, and well known by his scientific skill' (*DNB* XVIII, 1131).
129. Quoted in D. Stevenson, *op.cit.*, 68.
130. G. Robertson in R. Stevenson (ed.), *loc.cit.*, 96. A letter to the *Scotsman* from 'C.G.S.M.' — the initials are those of early railway enthusiast Charles Menteath of Closeburn — wanted Baird's 'iron rail-road, suitable to the wheels of carts' to be placed on all major road ascents (*Scotsman*, 22 Dec 1824).
131. G. Stephenson to R. Stevenson, 28 June 1821, quoted in D. Stevenson, *op.cit.*, 128-9.
132. C. F. Dendy Marshall, *op.cit.*, 152-3.
133. W. W. Tomlinson, *op.cit.*, 58-63; *Trans. Highland Soc.* 6 (1824), 1-146.
134. C. F. Dendy Marshall, *op.cit.*, 115.
135. G. Dott, *op.cit.*, 22-3; A. Scott in R. Stevenson (ed.), *loc.cit.*, 15; G. Robertson in R. Stevenson (ed.), *loc.cit.*, 69; C. F. Dendy Marshall, *op.cit.*, 115.
136. R. Stevenson, *Report relative to Lines of Railway surveyed from the Ports of Perth, Arbroath, and Montrose, into the Valley of Strathmore*, 11.
137. *Ibid.*, 1-4; R. Stevenson, *A Memorial regarding the Propriety of Opening the Great Valleys of Strathmore and Strathearn by means of a Railway or Canal*. The survey indicated a possible level line of 104 miles from Aberdeen to Crieff, 'a circumstance of which there is, perhaps, no other instance in the United Kingdom' (R. Stevenson, *Memorial regarding . . . Strathmore and Strath-*

earn, 2-4). In 1818 Stevenson reported on a smaller project in the same area, estimating that a nine-mile line from Brechin to Montrose harbour would cost about £20,000, and yield a return of about 7½%. Although a group headed by the Hon. William Maule of Panmure immediately subscribed £4075, the plan lapsed (SRO, BR/PROS(S)/1/1/13, Projected Railway between the Harbour of Montrose and the City of Brechin).

138. R. Stevenson, *Report on the Roxburgh and Selkirk Railway*, 2-3.
139. D. Stevenson, *op.cit.*, 125-6; R. Stevenson, *Memorial regarding . . . Strathmore and Strathearn*, 11.
140. W. W. Tomlinson, *op.cit.*, 97.
141. Act 6 Geo.IV c.169, 1825.
142. R. Stevenson, *Report relating to . . . Mid-Lothian*, 1.
143. *Ibid.*, 3.
144. H. Baird, *West Lothian Railway*, 3, 9.
145. R. Stevenson, *Report relative to . . . Strathmore*, 2; R. Stevenson, *Roxburgh and Selkirk Railway*, 1-2, 5-6.
146. M. Dalrymple, *Letter to the Gentlemen of the Counties of Lanark, Peebles, Selkirk, Roxburgh, and Berwick*, 1.
147. R. Stevenson, *Roxburgh and Selkirk Railway*, 5-7.
148. R. Stevenson, *Report relating to . . . Mid-Lothian*, 38. This nuisance was to affect Edinburgh for a long time. Edwin Chadwick paid it particular attention in his great Sanitary Report of 1842, in which he praised the use of town sewage suspended in water to irrigate the meadows near Holyrood Palace, in spite of the opposition of his colleague Dr Arnott, the numerous complaints about the smell, and the refusal of the government to allow the queen to stay at Holyrood for fear of infection (E. Chadwick, ed. M. W. Flinn, *Report on the Sanitary Condition of the Labouring Population of Great Britain*, 59-60, 120-4).
149. SRO, GD26/15/94, Gibson & Oliphant W.S., *op.cit.*, 1-3.
150. NLS, Ms 642 f.99, Melville Papers, Circular, *To the Interested Heritors in the Counties of Mid-Lothian, Roxburgh, Selkirk, and Berwick*, 1. The eight original projectors included three Scotts and three Pringles; their proposed line ran from Dalkeith to St Boswells, whence it forked to Selkirk and Hassendean, near Hawick. Sir Walter Scott also subscribed to the Berwick & Kelso (M. Robbins, 'Sir Walter Scott and two early railway schemes', in *Points and Signals*, 40-6). Outside his own locality, Sir Walter also subscribed for £100 on the contract of the Monkland & Kirkintilloch (HLRO Deposited Plan, HC 1824, Monkland & Kirkintilloch Railway, subscription contract).
151. NLS, Ms 642 f.99, 1-2.
152. R. Stevenson, *Roxburgh and Selkirk Railway*. The committee was now headed by the Duke of Roxburghe, the Marquis of Lothian and the Earl of Minto.
153. M. J. T. Lewis, *op.cit.*, 134; B. F. Duckham, *op.cit.*, I, 211. Outside users were charged from 1/9d to 2/- per trip; probably two waggons were allowed each time, since it was calculated that receipts of 9½d per waggon would cover costs.
154. R. Buchanan, *op.cit.*, 37.
155. NLS, Ms 8914 f.24; C. Highet, *op.cit.*, 7.
156. Anon., *Calculations of the Probable Benefit to the Neighbouring Country, and to the Proprietors of an Iron Railway from Berwick to Kelso*.
157. R. Buchanan, *op.cit.*, 28.
158. R. Stevenson, *Report relating to . . . Mid-Lothian*, 19.
159. H. Baird, *West Lothian Railway*, 6; SRO, GD30/2172, Sharp of Houston Papers, 'Report by Thomas Lawrie on the proposed West Lothian Railway through the Houston estate of Thomas Shairp Esq., 24 March 1825'.
160. M. Dalrymple, *op.cit.*, 1.
161. SRO, GD224/689, Buccleuch Muniments, Thomas Telford to George Robinson, 12 Nov 1809.
162. R. Stevenson, *Report relating to . . . Mid-Lothian*, 36-8; R. Stevenson, *Roxburgh and Selkirk Railway*, 5.
163. T. Telford, *Report relative to . . . Glasgow to Berwick*, 16; H. Baird, *West Lothian Railway*, 9-11.

164. R. Buchanan, *op.cit.*, 33.
165. R. Stevenson, *Roxburgh and Selkirk Railway*, 7; SRO, BR/LIB(S)/6/224 p.123, unidentified cutting; T. Telford, *Report relative to . . . Glasgow to Berwick*, 16-17. Rennie's proposed charges varied from 2d to 4½d per ton-mile depending on the commodity. Before the Kilmarnock & Troon opened, Titchfield suggested a rate of 3d per ton-mile: in 1839 the company was actually carrying at 3½d (J. Rennie, *op.cit.*, 10; NLS, MS 9814 f.24; *NSA* V (Ayrshire), 554).
166. John Francis believed in 1851 that the West Lothian had been constructed, and even gave a figure of £40,700 for its cost of construction (*History of the English Railway*, I, 63).
167. D. Stevenson, *op.cit.*, 130-54.

Chapter 2. The Coal Railways

1. *Prize Essays and Transactions of the Highland Society of Scotland*, 6 (1824), 1-146. Quotation on p.3.
2. *Ibid.*, 24.
3. *Ibid.*, 4, 59-60, 83, 103-4, 140.
4. *Ibid.*, 128-9, 133, 143-6. Such tracks came within at least some contemporary views of what constituted a railway: one 1830 definition was merely 'a road formed by laying distinct tracks of timber, iron or stone, for wheel carriages' (Sir David Brewster *et al.*, *Edinburgh Encyclopaedia*, XVII, 303). Stone road tracks on Stevenson's plan were later tried in Edinburgh (D. Stevenson, *Life of Robert Stevenson*, 71-3).
5. *Ibid.*, 57, 75, 104, 110-14, 119, 136.
6. See, for example, C. F. Dendy Marshall, *A History of British Railways down to the Year 1830*, 181-4, or, for a mid-nineteenth century view, John Francis, *A History of the English Railway 1820-1845*, I, 71-91. Gray had been anticipated in 1813 by Sir Richard Phillips, who complained of 'the inconceivable millions of money which have been spent about Malta, four or five of which might have been the means of extending double lines of iron railway from London to Edinburgh, Glasgow, Holyhead, Milford, Falmouth, Yarmouth, Dover, and Portsmouth . . . We might ere this have witnessed our mail coaches running at the rate of ten miles an hour, drawn by a single horse, or impelled fifteen miles an hour by Blenkinsop's steam engine'. (R. Pike, *Railway Adventures*, 20-1.)

For James, see also the propagandist volume by his daughter Mrs. Paine: E.M.S.P., *The Two James's and the Two Stephensons*. James has some claim to be regarded as one of the inventors of the locomotive; a quotation from his 1821 report to the subscribers to his projected Stratford & Moreton Railway gives the flavour of his enthusiasm:

> When the operation of the mind is not arrested by the novelty of the proposition, the prudence and practicability of steam engine power will be universally admitted . . . I beg to state that it will draw on a level railroad 80 tons . . . That when fully employed it will work at less than one quarter the expense of horses; that when it has no work, it incurs no expense . . . Upon the whole it may be stated, as a general principle not to be controverted, that the moving articles upon railroads by steam engines can be done cheaper, more certain, and twice as expeditiously as by boats on navigable canals, and neither the repair of locks and banks, the want of water in the summer's heat or the winter's frost, will retard its operation . . .

The shareholders, unconvinced, opted for horse-power and a level line, and James left the company in disgust (J. Simmons, 'For and against the locomotive', *Jnl Transport Hist.*, 2 (1955-6), 146.
7. *Scotsman*, 1 Dec 1824, 4 Dec 1824, 11 Dec 1824, 22 Dec 1824, 12 Jan 1825; Charles MacLaren, *Railways Compared with Canals and Common Roads, and their Uses and Advantages Explained* (published separately in 1825, and in *The Pamphleteer*, XXVI, 51, in 1826); J. Grant, *Cassell's Old and New Edinburgh*, I, 293-5. MacLaren (1782-1866) was co-founder of the *Scotsman* in 1817, and edited the first few issues, relinquishing the position to J. R. McCulloch since such a connection with a Whig paper was incompatible with his tenure of a post in the customs service. He resumed the editorship in 1820, and later edited the sixth edition of the *Encyclopaedia Britannica* (*DNB*, XII, 638; J. Grant, *op.cit.*, I, 283-4).

8. Nicholas Wood (1795-1865), a distinguished Northumbrian coal viewer who eventually became co-proprietor of several collieries, was arguably the most respected expert on early nineteenth-century waggonways and railways. A vigorous proponent of the locomotive, he published his influential *Practical Treatise on Railroads* in 1825. John Blenkinsop, also a coal viewer, was manager of Middleton Colliery, near Leeds, where in 1758 a waggonway had been established with, for the first time, the supporting authority of an act of Parliament. In 1811 Blenkinsop laid out a form of rack railway for use with locomotives, designed to overcome the supposed problem of insufficient adhesion between locomotive wheels and iron rails. Sir John Leslie (1766-1832), professor of mathematics at Edinburgh from 1805 to 1819 and then professor of natural philosophy, published volume I of *Elements of Natural Philosophy*, including much work on mechanics, in 1823. He was then working on 'a series of experiments projected on the constitution and power of steam' for a second volume which never in fact appeared. MacLaren, as a fellow member of intellectual and Liberal circles in Edinburgh, may well have been given information on the progress of this work (C. F. Dendy Marshall, *A History of British Railways down to the Year 1830*, 7, 38; Frederic Boase, *Modern English Biography*, III, 1474; J. Francis, *op.cit.*, I, 51; R. Chambers (ed.), *Biographical Dictionary of Eminent Scotsmen* (1872 ed.), II, 500; *DNB* XXXIII, 105-7).

9. C. MacLaren, *op.cit.*, 59.
10. *Ibid.*, 75-9.
11. *Ibid.*, 85-7.
12. [John Barrow], 'Canals and rail-roads', *Quarterly Review*, XXXI (March 1825), 361.
13. *Glasgow Herald*, 5 Nov 1824, quoted in Frank Ferneyhough, *The History of Railways in Britain*, 128.
14. John Rennie, *Report Respecting the Proposed Rail-Way from Kelso to Berwick*, 10.
15. C. Maclaren, *op.cit.*, 83.
16. R. Stevenson (ed.), 'Essays on Rail-roads', *Trans. Highland Soc.* 6 (1824), 2.
17. C. MacLaren, *op.cit.*, 64-5; Robert Stevenson, *Report Relative to Lines of Railway Surveyed from the Ports of Perth, Arbroath, and Montrose, into the Valley of Strathmore . . .*, 3. Stevenson claimed that even a horse railway 'possesses nearly all the advantages of the ditch or small canal, at a much less expence, and offers much greater facilities to a country whose trade is of a desultory nature.' (*Ibid.*, 3). The report marks the effective end of Stevenson's railway activities, perhaps because he was not sufficiently in sympathy with the locomotive. He returned to work on harbours and river improvements (D. Stevenson, *op.cit.*, 130-54).
18. Rev. J. Adamson, *Sketches of our Information as to Rail-Roads*, 25-6.
19. C. MacLaren, *op.cit.*, 71, 88; Thomas Telford (ed. J. Rickman), *Life of Thomas Telford, Civil Engineer, Written by Himself*, xxi.
20. ???, *The Fingerpost, or, Direct Road from John-o'-Groat's to the Land's End*, 15, 41.
21. C. MacLaren, *op.cit.*, 83; [J. Barrow], *loc.cit.*, 365.
22. [J. Barrow], *loc.cit.*, 366-8; C. MacLaren, *Scotsman*, 11 Dec 1824.
23. Rev. J. Adamson, *An Account of the Stockton and Darlington Rail-Way . . .*, 26.
24. C. MacLaren, *op.cit.*, 80.
25. Jean Lindsay, *The Canals of Scotland*, 41-2, 45; Charles Hadfield, *British Canals: An Illustrated History*, 166; Hugh Baird, *Report on the Proposed Edinburgh and Glasgow Union Canal*, 16.
26. [J. Barrow], *loc.cit.*, 369.
27. C. MacLaren, *op.cit.*, 80.
28. Quoted in J. Francis, *op.cit.*, I, 104.
29. [J. Barrow], *loc.cit.*, 362.
30. *Ibid.*, 369; C. Maclaren, *op.cit.*, 79.
31. Rev. J. Adamson, *Sketches*, 26. One opponent accused Adamson of merely repeating inaccurate information fed to him by the Stockton & Darlington management, and claimed that water transport was in fact much cheaper (Thomas Grahame, *A Treatise on Internal Intercourse and Communication in Civilised States, and particularly in Great Britain*, 50-1).
32. [J. Barrow], *loc.cit.*, 361.
33. A. Slaven, *The Development of the West of Scotland*, 1750-1960, ch.5.

34. George Buchanan, *An Account of the Glasgow and Garnkirk Railway and Other Railways in Lanarkshire*, 4.
35. A. Slaven, *op.cit.*, 115-18. By 1840, 65 of the 88 existing Scottish iron furnaces were in or immediately adjacent to Old Monkland parish. In 1848, at a time of recession, there were 54 iron furnaces on the line of the Monkland & Kirkintilloch, producing about 300,000 tons of pig iron and requiring over two million tons of coal, ironstone and limestone per year (*NSA*, VI (Lanarkshire), 658; GUEH, 'Monklands Amalgamation Bill. Notes of Evidence of George Knight.').
36. G. Buchanan, *op.cit.*, 4-5. The main subscribers to the railway included three members of the Dixon family, and the major local landowners J. H. Colt of Gartsherrie and G. More Nisbett of Cairnhill (HLRO Deposited Plan, HC 1824, Monkland & Kirkintilloch Railway).
37. SRO, BR/FCN/1/68, Forth & Clyde Canal Company Minutes, 6 Nov 1828.
38. HLRO Deposited Plan, HC 1824, Monkland & Kirkintilloch Railway. The company's engineer, Thomas Grainger, described his original commission as a survey of routes to link the two canals between Gartsherrie and Kirkintilloch (More Nisbett Papers, Bundle 4, Grainger to George Brown, 24 Mar 1823).
39. SRO, BR/FCN/1/68, 6 Nov 1828.
40. *Ibid.*, p.342, 'Amount of the annual revenue and expenditure ... of the Forth and Clyde Navigation, from 1800 to 1828 inclusive ...'; J. Lindsay, *op.cit.*, 41.
41. G. Buchanan, *op.cit.*, 5.
42. *Ibid.*, 5-6.
43. J. Lindsay, *op.cit.*, 41. Grainger's report to the subscribers reassured them that he at least still saw railways as adjuncts to canals: 'In by far the greater number, it is evident that, instead of being hurtful to Canal undertakings they will not only be the means of extending their usefulness, but also add prodigiously to the trade, as they will serve to connect many mineral districts with Canals, which must otherwise have remained entirely excluded.' (Thomas Grainger, *Report to the Subscribers for a Survey of a Railway from the Monkland and Kirkintilloch Railway to that part of the Monkland Coalfield situated north & east of Airdrie*).
44. GUEH, 'Monkland Amalgamation Bill. Brief for the Promoters. 1848.' (draft), 9.
45. Thomas Grainger & John Miller, *Report Relative to the Proposed Railway to connect the Clydesdale, or Upper Coal Field of Lanarkshire, with the City of Glasgow, and the East and West Country Markets*, 11; GUEH, Robert Dodds, 'General Remarks on the Wishaw and Coltness Railway from its Commencement to the Present Date.'
46. Acts 4 & 5 Will IV c.41, 7 Will IV & 1 Vict c.100.
47. *NSA*, VI (Lanarkshire), 610.
48. Acts 2 & 3 Vict c.58, 4 & 5 Vict c.11; GUEH, 'Memorial for the Wishaw and Coltness Railway Company, 22 February 1845' (in law case against Sir Henry Steuart of Allanton). 'The expense,' noted the minister of Bothwell, 'has been found greatly to exceed the original calculation.' (*NSA*, VI (Lanarkshire), 798).
49. T. Grainger & J. Miller, *Report Relative to the Proposed Railway . . .*, 4-7; G. Buchanan, *op.cit.*, 10. Belhaven (£5000), Steuart Denham (£5000) and Sir Henry Steuart of Allanton (£3000), the three largest subscribers to the project, all held estates at the south end of the proposed line, furthest away from alternative rail or canal transport (HLRO Deposited Plan, HC 1829, Garturk and Garion Railway).
50. The main subscribers include Tennant, his son, and their firm in its corporate capacity; three members of the coal-and-iron owning Sprot family of Garnkirk; three Dixons; and a number of other prominent mineral proprietors including Merry, Henry Houldsworth, James Jeffray and George More Nisbett (HLRO Deposited Plan, HC 1826, Garnkirk & Glasgow).
51. G. Buchanan, *op.cit.*, 7, 9-10.
52. *Ibid.*, 10.
53. Thomas Grainger & John Miller, *Report to the Proprietors of, and Traders on the Canals and Railways terminating on the North Quarter of Glasgow, and Other Gentlemen interested therein and in the Harbour and Streets of Glasgow*, 1-4; 'A Friend to the Tunnel', *Letter to the Author of Remarks on the Report by Messrs. Grainger & Miller on the Subject of the Proposed Railway & Tunnel from the North Quarter of Glasgow to the Broomielaw*, 4.

54. A. Slaven, *op.cit.*, 29-30; T. Grainger & J. Miller, *Report to the proprietors* . . ., 2.
55. See below, chapter 3, pp. 99 ff.
56. SRO, BR/EGR/1/1, Edinburgh & Glasgow Minutes, 12 Jan 1831.
57. T. Grainger & J. Miller, *Report to the Proprietors* . . ., 1-4; 'A Friend to the Tunnel', *op.cit.*, 5-11. Grainger and Miller estimated that the railway could be built for under £42,000, would carry 200,000 tons of coal per year, with the downhill run being powered by gravity, and would pay at least 10% on its capital. The idea for the tunnel had originally come from engineer John Gibb of Aberdeen, and owed much to the Liverpool & Manchester Railway's entry to Liverpool (*Report to the Proprietors* . . ., 3-4).
58. T. Grainger & J. Miller, *Report to the proprietors* . . ., 1; *Scotsman*, 10 Nov 1830. In the 1830 election for the Glasgow Burghs, Campbell defeated Finlay by the casting vote of the returning officer (H. S. Smith, *The Parliaments of England*, 649).
59. G. Buchanan, *op.cit.*, 7.
60. GUEH, 'Monkland Amalgamation Bill. Brief for the Promoters. 1848.' (draft), 9. No individual appeared among the original subscribers to all five of the Monklands area coal railways (Monkland & Kirkintilloch, Ballochney, Garnkirk & Glasgow, Wishaw & Coltness, Slamannan). Land- and coalowner George More Nisbett, Glasgow merchant Kirkman Finlay and Forth & Clyde Canal director Thomas Grahame subscribed to four each, while among the several who subscribed to three was William Dixon (HLRO Deposited Plans, HC 1824 Monkland & Kirkintilloch, HC 1826 Ballochney HC 1826 Garnkirk & Glasgow, HC 1829 Garturk & Garion, HC 1835 Slamannan).
61. GUEH, 'Monkland Amalgamation Bill. Brief for the Promoters. 1848.' (draft), 9; GUEH, 'Interim Report by the Committee of Inquiry to the Directors of the Wishaw and Coltness Railway, 10 July 1844'; GUEH, 'Draft statement by Mr. Glasgow 1844' (reply to previous item); SRO, BR/WIC/1/4, Wishaw & Coltness Minutes, 5 Feb 1844.
62. SRO, BR/WIC/1/3, Wishaw & Coltness Minutes, 24 Feb 1842.
63. *NSA*, VI (Lanarkshire), 665. The Garnkirk & Glasgow claimed in 1833 that it was taking two-thirds of the traffic originating at the southern end of the Monkland & Kirkintilloch (SRO, BR/CAL/4/25, House of Commons debate on the Monkland & Kirkintilloch Bill, 18 June 1833, reprinted from the *Mirror of Parliament*, ccxxiii).
64. More Nisbett Papers, bundle 6, *Prospectus of the Slamannan Railway*, Oct 1835, p.2. Company engineer Sir John Macneill confirmed that 'along great portions of it, there are no cart roads, and in some places the farmers are obliged to carry the produce of their farms on horses' backs to the nearest parish road, where carts can be had . . .' (SRO, GD1/579/25, *Reports &c regarding the proposed Slamannan Railway*, 3).
65. J. Lindsay, *op.cit.*, 81-2.
66. More Nisbett Papers, bundle 6, *Prospectus of the Slamannan Railway*, Oct 1835,2.
67. SRO, GD1/579/25, *Reports &c regarding the proposed Slamannan Railway*, 2 (Union Canal Committee of Management Report, 3 March 1835); 'A Canal Proprietor', *Letter to the Proprietors of the Edinburgh and Glasgow Union Canal, 29th January 1836*.
68. More Nisbett Papers, bundle 6, *Reports of the Committee of Management of the Slamannan Railway*, 8 Feb 1838, 7 Feb 1839.
69. GUEH, 'Monklands Amalgamation Bill. Brief for the Promoters. 1848.' (draft); S. C. Brees (ed.), *Railway Practice*, 78.
70. Hugh Baird, *Report on the Proposed Edinburgh and Glasgow Union Canal*, 14; Hugh Baird, *Report on the Proposed Railway from the Union Canal at Ryal, to Whitburn, Polkemmet, and Benhar: or the West Lothian Railway*, 11.
71. John A. Hassan, 'The supply of coal to Edinburgh, 1790–1850', *Transport Hist.* 5 (1972), 126.
72. John Grieve & James McLaren, *Report on the Utility of a Bar-Iron Railway from the City of Edinburgh to Dalkeith and the Harbour of Fisherrow* . . ., 3-12, 16-18. Between July and September 1824, Grieve moved from the belief that 'it is a point not yet determined, whether a Canal or a Railway is the cheapest conveyance', to a conviction of the advantages of a railway (*ibid.*, 4, 12). He estimated that Dalkeith 'great coal' could be sold in Edinburgh for 8/- per ton. An anonymous pamphlet of reprints from the *Caledonian Mercury*, either written by James Adamson or largely plagiarised from him, confirms Grieve's figure of 4/- for cartage from Dalkeith to Edinburgh, adding

that cartage cost 3/6 from Leith and 1/3 from the Union Canal basin at Port Hopetoun. Coal sold in Edinburgh for 11/- to 17/- depending on quality; it was claimed that an Edinburgh to Glasgow railway (or presumably Baird's West Lothian Railway) would allow coal from Benhar to be sold for 7/- (*Remarks on the Comparative Merits of Cast Iron and Malleable Iron Rail-ways; and an Account of the Stockton and Darlington Rail-Way, &c. &c., 17-18, 20-1*).

73. H. G. Lewin, *Early British Railways*, 16. In January 1825 a committee chaired by the Marquess of Tweedale and including Belhaven and the Cadell brothers of Tranent and Cockenzie proposed an East Lothian Railroad from the Edinburgh & Dalkeith to Haddington and Dunbar. They proposed as engineer 'Mr. Robert Stevenson or any other person that they may think fit'; Stevenson had of course already surveyed the routes of both the Dalkeith line and this possible extension at the end of the previous decade (*Scotsman*, 8 Jan 1825).

74. SRO, GD 224/554, Buccleuch Muniments, Minute of Meeting of Edinburgh & Dalkeith Railway Company, 5 Aug 1840; J. A. Hassan, *loc.cit.*, 137, 139.

75. HLRO Deposited Plan, HC 1826, Edinburgh and Dalkeith Railway; SRO, GD 135, Box 82, Stair Muniments, Roll of Proprietors of the Edinburgh & Dalkeith Railway, January 1834. Stenhouse, along with his landlord John Wauchope and his associate Laing, originally opposed the railway, ostensibly for running through Wauchope's land on its way to helping rival coalowners. Without the railway, Stenhouse gained a considerable advantage from the Edmonston waggonway, and both he and Laing claimed that the railway company should compensate them for their loss. All three men, however, appear as assenters to the railway in the parliamentary records (G. Dott, *Early Scottish Colliery Waggonways*, 23; HLRO Deposited Plan, HC 1826, Edinburgh and Dalkeith Railway). Buccleuch also built a private branch line into the town of Dalkeith, opened in 1838 (*NSA*, I (Edinburghshire), 512).

76. Joseph Priestley, *Historical Account of the Navigable Rivers, Canals, and Railways, of Great Britain*, 226, 279. It is possible that a large number of initial subscribers might imply some difficulty in raising capital, while the Garnkirk had simply been able to raise its requirements from twelve supporters, with no need to look further. There is however no evidence that the Edinburgh & Dalkeith had problems in finding subscribers.

77. SRO, GD 135, Box 82, Roll of Proprietors . . . The baronets rather surprisingly included Admiral Sir P. C. H. Durham, who as proprietor of the Fordell waggonway might have been expected to oppose a rival for the Edinburgh market.

78. NLS, Ms 3421 f.42, *Prospectus of the Proposed Branch Railway to Leith from the Main Line of the Edinburgh and Dalkeith Railway*.

79. NLS, Ms 3421 f.40, List of Proprietors of the Leith Branch of the Edinburgh and Dalkeith Railway. Of 931 shares whose owners' status can be determined, 24.9% were held by aristocracy, 18.4% by landed gentry, 18.4% by merchants, 10.4% by bankers, 9.0% by lawyers, 10.8% by manufacturers, 2.1% by Leith Town Council and 6.0% by others.

80. Quoted in David Bremner, *The Industries of Scotland*, 83-4. The supposed freedom from fatal accident is accepted by, for example, G. Dow, *The First Railway across the Border*, 8, and H. H. Meik, 'The Edinburgh & Dalkeith Railway', *Railway Mag*. 52 (1923), 179. Parliamentary returns, however, note the death of a driver in 1840 and two children in 1843 and 1844 (PP 1841 (116) XXV, *Reports, Returns, &c relative to Railways*, 126; PP 1843[440] XLVII, *Third Report of Railway Department*, p.iv; PP 1846 [698] XXXIX, *Report of Railway Department for 1844-45*, p.xvi.

81. PP 1839 (222,517) X, *S. C. on Railways*, Appendix 2, 101; evidence, Q.4672.

82. *NSA*, I (Edinburghshire), 512.

83. S. G. E. Lythe, 'The Dundee & Newtyle Railway', *Railway Mag*. 97 (1951), 547; W. M. Duncan, *Newtyle: a Planned Manufacturing Village*, 72.

84. NLS, Ms 9944 f.63, Menzies papers, Hugh Watson of Keillor to Sir Neil Menzies, 20 Feb 1838. Of the 187 original subscribers, almost all were local; Airlie, Hon. J. A. S. Wortley (later Wharncliffe) and Dundee Town Council led the list with £1500 each (HLRO Deposited Plan, HC 1826, Dundee & Newtyle).

85. *Dundee Advertiser*, 27 Aug 1829.

86. PP 1843 [440] XLVII, *Third Report of the Railway Department*, appx. IV, 175.

87. In 1832 Lord Wharncliffe laid out a fifteen-acre village round the Newtyle terminus; ten years

later it had a population of 505 (*NSA*, XI (Forfarshire), 561).
88. PP 1843 [440] XLVII, *Third Report of the Railway Department*, appendix IV, 176.
89. C. E. Lee, 'The Dundee & Newtyle Railway', *Railway Mag.* 97 (1951), 848; PP 1844 [551] XLI, *Reports of the Officers of the Railway Department ... for 1843*, appendix III, 182. Parliamentary authority for the harbour branch was obtained retrospectively (S. G. E. Lythe, *loc.cit.*, 548). Of the three locomotives supplied by the Dundee firm of J. & C. Carmichael in 1833, Hamilton Ellis notes that 'though monstrous in design, they deserve record as the first bogie engines to run a British public railway' (C. Hamilton Ellis, *British Railway History*, I, 139).
90. SRO, GD16/38/77, bundle 3, Airlie Muniments, Dundee & Newtyle Railway Committee of Mangement, *Vidimus of the State of the Affairs and Prospects of the Dundee and Newtyle Railway Company*, 3.
91. *Ibid.*, 4.
92. S. G. E. Lythe, *loc.cit.*, 547-8. In the parliamentary estimate Landale allowed a further £2000 for land (HLRO Deposited Plan, HC 1826, Dundee & Newtyle). The surveyor Matthias Dunn, commenting on Landale's scheme for the company, suggested £27,000 excluding land (SRO, GD16/38/77, bundle 3, Airlie Muniments, Matthias Dunn, *Observations upon the Line of Railroad projected by Mr. Landale, from Dundee to the Valley of Strathmore*, 4).
93. SRO, GD16/38/77, bundle 1, Airlie Muniments, David Dobson to John Kerr, 25 Feb 1829; Alex Meldrum to John McNicoll, 24 Apr 1829; James Paterson to McNicoll, 24 June 1829.
94. Ibid., John Kerr to McNicoll, 21 Jan 1841; Christopher Kerr to McNicoll, 24 Oct 1832; Nicholas Wood, *Report to the Committee of Management relative to the Dundee and Newtyle Railway, and the Extension of that Railway in Strathmore*, 3-8.
95. PP 1843 [440] XLVII, *Third Report of the Railway Department*, appendix IV, 176; William Blackadder, *Report relative to the Strathmore Railway, being the Extension of the Dundee and Newtyle Railway from Newtyle along Strathmore between Coupar and Forfar*. Lord Airlie contributed £4000 of the £12,150 parliamentary subscription of the Glammis company (HLRO Deposited Plan, HC 1835, Newtyle & Glammis).
96. S. G. E. Lythe, *loc.cit.*, 548-9.
97. PP 840 [474] XIII, *Select Committee on Railway Communication, Fifth Report*, evidence, QQ.4893, 4904 (George Kinloch); S. G. E. Lythe, *loc.cit.*, 549.
98. Shiell & Small, Dundee & Newtyle Sederunt Book, 18 Mar 1842, 21 Feb 1843. Shareholders' apathy was accurately reflected in a letter dutifully minuted by the directors in 1843: 'I have never derived a farthing of benefit from the Coys Stock, nor is it likely I ever shall — such a valueless concern to the Shareholders is not I should think worth *their* while contending about' (ibid., 3 Jan 1843).
99. Ibid., 17 June 1841, 8 Aug 1841, 9 Nov 1841, 31 May 1844.
100. Ibid., 14 May 1841, 12 Oct 1841, 9 Nov 1841, 1 Apr 1842, 17 Jan 1843, 7 Mar 1843, 21 Mar 1843, 5 May 1843, 20 June 1843, 24 Dec 1844.
101. *NSA*, XI (Forfarshire), 566.
102. *NSA*, V (Ayrshire), 202-3; J. Lindsay, *op.cit.*, 86-8.
103. J. Lindsay, *op.cit.*, 88-90; J. Priestley, *op.cit.*, 286. When construction work stopped in 1815, Telford and Rennie agreed that some £300,000 more would be required (W. J. Gordon, *Our Home Railways*, I, 190).
104. More Nisbett Papers, bundle 1, 'Ardrossan Railway Company: Draft Memorandum by James Moffat, secretary, October 1840', 1; J. Priestley, *op.cit.*, 286-7; J. Lindsay, *op.cit.*, 95. Of the £28,950 subscribed, £20,000 came from the Earl of Eglinton's trustees (HLRO Deposited Plan, HC 1827, Ardrossan Railway)
105. Thomas Grainger & John Miller, *Report relative to a Proposed Railway Communication between the City of Glasgow and the Towns of Paisley and Johnstone*, 13-15. The construction of the railway from Kilwinning to Johnstone 'must, it is thought, depend in a great measure upon the spirit of railway speculations among the gentlemen of Renfrew and Lanarkshire', who in the event showed little enthusiasm (*Scotsman*, 27 Oct 1830).
106. More Nisbett Papers, bundle 1, 'Ardrossan Railway Company: Draft Memorandum by James Moffat, secretary, October 1840'; ibid., *Report of the Committee of Management of the Ardrossan*

Railway to the Annual General Meeting, 27 Jan 1841; J. Lindsay, *op.cit.*, 95. The canal was eventually converted into a railway under a Glasgow & South Western act of 1881 (J. Lindsay, *op.cit.*, 98).

107. J. Priestley, *op.cit.*, 525-6. Five men signed the original £8500 subscription contract, with £6500 coming from Dixon, who was in effect the sole proprietor (HLRO Deposited Plan, HC 1830, Polloc & Govan; Anon., *Memoirs and Portraits of One Hundred Glasgow Men* . . ., 105).

108. SRO, GD46/13/60, Seaforth Muniments, Petition of the Royal Incorporation of Hutcheson's Hospital against the Polloc and Govan Railway Bill. Dispute between the railway company and Hutcheson's Hospital continued over the price of land to be taken for the railway, leading to a law case which lasted from 1838 to 1845, and another one in 1853. Dixon was notoriously litigious: 'it is said that he had more lawsuits in his day than any peer or commoner in Scotland ever had' (SRO, BR/HRP(S)/1/55, Oliphant, Maxwell & Steven to Alexander Ferguson, treasurer of the Caledonian, 7 Nov 1853; *Memoirs and Portraits of One Hundred Glasgow men* . . ., 104).

109. More Nisbett Papers, bundle 7, *Reports of Committee of Management of the Wishaw & Coltness*, 6 Feb 1837, 4 Feb 1839.

110. Ibid., bundle 5, *Report of the Directors of the Paisley & Renfrew Railway Company*, 10 June 1841.

111. Ibid., bundle 5, Alexander Gibson to Hugh Morton, 8 Dec 1846; Francis Whishaw, *The Railways of Great Britain and Ireland*, 391; G. H. Robin, 'Railways of Paisley, Renfrew and Barrhead', *Railway Mag.* 104 (1958), 158.

112. SRO, BR/EGR/1/1, Edinburgh & Glasgow Minutes, 12 Jan 1831.

113. T. Grainger & J. Miller, *Report relative to a Proposed Railway Communication between . . . Glasgow . . . Paisley and Johnstone*, 3, 5.

114. More Nisbett Papers, bundle 1, *Report of the Committee of Management of the Ardrossan Railway*, 27 Jan 1841.

115. Harold Pollins, *Britain's Railways: An Industrial History*, 21. The original Stockton & Darlington act allowed haulage 'with men or horses, or otherwise', but, since a specific clause to allow 'locomotive or moveable engines' was included in an amending act in 1823, it may have been unclear whether they were covered by the original generalisation (J. S. Jeans, *Jubilee Memorial of the Railway System*, 43).

116. G. Buchanan, *op.cit.*, 5.

117. E. H. Ahrons, *The British Steam Locomotive 1825-1925*, 11.

118. F. Whishaw, *op.cit.*, 14. The Ballochney main lines inclines were worked by gravity; there was also one worked by a stationary steam engine on the Rochsoles branch. Though the Garnkirk & Glasgow ran trains over the Ballochney to Airdrie, they had to use horse haulage on this section (PP 1840 [474] XIII, *Select Committee on Railway Communication*, evidence, QQ.4082-6 (Charles A. King)).

119. Charles Landale, *Report Submitted to the Subscribers for Defraying the Expence of a Survey to Ascertain whether it is Practicable and Expedient to Construct a Railway between the Valley of Strathmore and Dundee*, 5; N. Wood, *op.cit.*, 3-4. Landale's doubts about locomotives had led him to construct his railway with as much level line as possible (S. G. E. Lythe, *loc.cit.*, 547).

120. H. G. Lewin, *op.cit.*, 17.

121. C. E. Lee, *loc.cit.*, 693-4; SRO, BR/LIB(S)/6/224, cutting from *Weekly News*, 5 Nov 1898, giving the recollections of an aged inhabitant of Meigle. A similar experiment is claimed for the early days of the Monkland & Kirkintilloch: 'The first trial was successful; the wind was strong in their favour, and the waggon on which they were seated rattled along at rapid pace for some distance. They wished to return by the same method, but the wind being contrary to their desires, and tacking scarcely practicable on a railway, they were forced to dismount, and apply their shoulders to the machine, to get it back to the place whence they started. This one trial was sufficient to show the impracticability of such a propeller.' (Andrew Miller, *The Rise and Progress of Coatbridge and Surrounding Neighbourhood*, 11). This problem was solved in the Coupar Angus experiments by carrying a horse in a small open truck or 'dandy-cart' attached to the train, for use when the wind was unfavourable. A dandy-cart was also used on the Ballochney, to carry the horse on sections where the power of gravity was used (W. W. Tomlinson, *The North Eastern Railway*, 158).

122. S. G. E. Lythe, *loc.cit.*, 549. The poor quality of the locomotives and the lack of money for repairs did not help. In 1841 on the top level 'it was necessary to have the Steam up on both the Engines as Neither of them are to be trusted to for a single day', while the engine on the other level, which belonged to the Coupar Angus company, damaged the track. According to the manager, 'for some time back we have been very irregular with the trains owing to the inefficient state of the Engines'. In 1844 the trouble was a shortage of waggons, over 30% of which were unfit for use. This had meant that Hill of Newtyle, 'our principal customer', who imported coal through Dundee, had had to sell it in the city rather than transporting it by the railway to the Strathmore markets (Shiell & Small, Dundee & Newtyle Sederunt Book, 10 Sept 1841, 9 Nov 1841, 4 Oct 1844).

123. SRO, BR/WIC/1/2, Wishaw & Coltness Minutes, 6 Feb 1837; BR/WIC/4/3, 'Garturk and Garion Railway. Objections by Major James Drysdale to the Bill, 1829.'

124. SRO, BR/WIC/4/4, Copy Contract & Agreement between the Wishaw & Coltness Railway Company and Thomas Houldsworth Esquire of Coltness, 1837.

125. J. Grieve & J. McLaren, *op.cit.*, 14-15.

126. Stephenson Locomotive Society, *The Glasgow and South-Western Railway*, 6; G. H. Robin, *loc.cit.*, 158.

127. G. Buchanan, *op.cit.*, 8.

128. Thomas Grainger & John Miller, in Edinburgh and Glasgow Railway, *Reports by Mr. George Stephenson of Liverpool, and Messrs. Grainger and Miller of Edinburgh, Civil Engineers*, 18.

129. SRO, BR/CAL/4/25, Report of House of Commons Debate on the Monkland & Kirkintilloch Bill, 18 June 1833 (reprinted from *Mirror of Parliament*, CCXXIII). Quotation from speech by Capt. John Dunlop, MP for Kilmarnock Burghs.

130. More Nisbett Papers, bundle 7, *Report of Committee of Management of Wishaw & Coltness*, 5 Feb 1838. At this meeting the Committee agreed to allow Garnkirk locomotives on their line without extra toll, and to buy locomotives of their own. Two years later they had ordered three locomotives for £798 each ibid., *Report of Committee of Management of Wishaw & Coltness*, 3 Feb 1840).

131. GUEH, 'Ballochney Railway. Draft Report of Committee of Management to General Meeting, 1 Feb 1831.'

132. GUEH, Robert Dodds, 'General Remarks on the Wishaw and Coltness Railway from its Commencement to the present state' 1844; GUEH, Anonymous notes of draft reply to the Interim Report by the Committee of Inquiry to the Directors of the Wishaw and Coltness Railway Company, 10 July 1844.

133. PP HL 1849(21)XXIX, *HL S.C. on Audit of Railway Accounts*, p.525, evidence of J. J. Hope Johnstone (quoting evidence given by Joseph Locke to HC Select Committee on the Caledonian Railway Purchase or Lease of the Wishaw & Coltness Railway Bill, 24 April 1849).

134. PP 1839 (222, 517)X, *S.C. on Railways*, appendix 2, 109.

135. *Ibid.*, 101.

136. GUEH, *Rules, Bye Laws and Regulations of the Ballochney Railway Co . . . 1836*.

137. PP 1841 [116] XXV, *Reports, Returns &c. relative to Railways, under the Provisions of the Act 3 & 4 Vict c.97*, 90-3.

138. More Nisbett Papers, bundle 7, *Report of the Committee of Management of the Wishaw & Coltness*, 9 Sept 1839.

139. Two companies where outside carrying took place, the Kilmarnock & Troon and the Dunfermline & Charlestown, made no reply to the inquiry.

140. PP 1846 [698] XXXIX, *Report of the Railway Department for 1844-45*, appendix 2.

141. GUEH, Accounts of the Monkland & Kirkintilloch Railway Companies, 1836-9; ibid., Accounts of the Ballochney Railway Company, 1828-35.

142. The debate about the role of independent carrying companies, plying for hire rather than carrying their own goods, was effectively decided before Scotland had any lines important enough to interest such companies. The 1839 Select Committee on Railways spent much time examining the feud between Pickford's and the Grand Junction Company, and commented that although 'it does not appear to have been the intention of Parliament to give a Railway Company the complete monopoly of the means of communication on their line of road', such a monopoly was in practice inevitable and for safety reasons desirable (PP 1839(517)X, *S.C. on Railways*, Second Report,

pp.vi-vii).
143. GUEH, John Moffat to C. D. Gairdner, 9 June 1852.
144. PP 1842 [360] XLI, *Second Report of the Railway Department,* appendix 10; PP 1846 [698] XXXIX, *Report of the Railway Department for 1844-45,* appendix 2.
145. Act 7 & 8 Vict c.98, 1844.
146. GUEH, 'Ballochney Railway Bill 1843: case for the Promoters.'
147. GUEH, Lord Belhaven to James Mitchell, secretary of the Wishaw & Coltness, 8 March 1842; SRO, BR/WIC/1/3, Wishaw & Coltness Minutes, 10 Mar 1842, 25 Mar 1842.
148. GUEH, List of subscribers to Ballochney Railway; ibid., List of Shareholders in Ballochney Railway, 17 Jan 1831. In 1831 three individuals, Charles Tennant, Thomas Grahame and Robert Grahame, all of whom were directors, held £1000 each. Thomas Grainger held £250, an early example of an engineer investing in one of his own projects — though by 1835 he had disposed of his shares.
149. GUEH, List of subscribers to Ballochney Railway; ibid., 'Wishaw & Coltness Railway; proprietors as at 24th July 1844.'
150. GUEH, Wishaw & Coltness Railway, Minute of Meeting of Shareholders, 31 Jan 1844; ibid., note of dividends paid by Wishaw & Coltness to 1844.
151. GUEH, Slamannan Railway, Bathgate & Jawcraig Branches, Notes of Evidence of A. J. Adie, 1846; ibid., rough note, 'As to raising the Capital for the Bo'ness Extension', 14 Nov 1842; ibid., Report by E. Dickson to Slamannan directors, 3 Nov 1842. Power to build was granted in an act of 1846, but two years later it was still a branch 'which they contemplate executing when times improve' (GUEH, 'Monkland Amalgamation Bill, 1848. Notes of Evidence of George Knight').
152. *NSA,* VI (Lanarkshire), 411.
153. J. Butt & J. T. Ward, 'The promotion of the Caledonian Railway', *Transport Hist.* 3 (1970), 225.
154. GUEH, 'Monklands Amalgamation Bill. Brief for the Promoters. 1848.', 26, 68.
155. GUEH, 'Monklands Amalgamation Bill. Notes of Evidence of George Knight.'
156. PP 1846(159)XLI, *Return of Moneys Raised under Railway Acts, 1826-43;* J. Priestley, *op.cit.,* 228.
157. GUEH, 'Monklands Amalgamation Bill. Brief for the Promoters. 1848,' 16-17.
158. SRO, GD16/38/78, Airlie Muniments, *Report of Committee of Management of Monkland & Kirkintilloch,* 3 Feb 1830.
159. GUEH, Robert Dodds, 'General remarks on the Wishaw and Coltness Railway from its Commencement to the present state', 1844.
160. Ibid.
161. GUEH, 'Answers for Sir Henry Stewart [sic] of Allanton Bart, Defender, to the Minute for the Wishaw & Coltness Railway Co, Pursuers', 1836.
162. See above, p. 75.
163. Act 4 & 5 Vict c.43, 1841.
164. GUEH, List of arrears by Ballochney subscribers, 8 Aug 1827.
165. GUEH, A. Graham, Moncrieff & Weems (Slamannan agents) to Capt. Harness (secretary to Railway Commissioners), 3 Feb 1848.
166. GUEH, Accounts of Monkland & Kirkintilloch and Ballochney Railways, 1836 and 1843. As early as 1826, the Royal Bank lent the Monkland & Kirkintilloch £10,000, but on that occasion insisted on the directors pledging their personal security (Anon., 'Early Scottish Railways', *Three Banks Rev.* 74 (1967), 29.
167. SRO, GD16/38/77/3, Airlie Muniments, Hugh Watson & John McNicoll, *Report to the Committee of the Dundee & Newtyle Railway on the State of the Company's Affairs, at 4th November, 1828.*
168. More Nisbett Papers, bundle 7, *Reports of Committee of Management of Wishaw & Coltness,* 30 June 1835 and 6 Feb 1837. In spite of their problems, the Committee declared a dividend of 2% in 1835, claiming to have reached dividend-earning status more quickly than any similar railway. This was justified by assuming that all the arrears of subscription calls, amounting to £12,805, would be paid up, while the debt to the bank was £4288 (*Report . . . 30 June 1835*). A loan of £20,000 was

eventually obtained from the bank in 1838, in return for an assignation on the company and a personal guarantee from Thomas Houldsworth that the interest would be paid (SRO, BR/WIC/1/2, Wishaw & Coltness Minutes, 5 Feb 1838).

169. GUEH, 'Answers for Sir Henry Stewart Bart, Defender, to the Minute for the Wishaw & Coltness Railway Co, Pursuers.'

170. SRO, BR/WIC/4/3, Letter (author unknown) to Sir James Steuart Denham, Sept 1828.

171. SRO, BR/WIC/4/3, 'Proposals for raising Funds for making a Railway from Allanton Coltness Cleland &ca join the Glasgow Railways and Canals.'

172. GUEH, 'Mem^m. as to case W. & C. Ry Co v Sir H Stewart & others.'

173. GUEH, 'Answers for Sir Henry Stewart . . .'; ibid., 'Remarks for the Wishaw and Coltness Railway Coy on the Answers for Sir Henry Stewart of Allanton to the Minute for them, 14 March 1836.'

174. More Nisbett Papers, bundle 7, *Reports of Committee of Management of Wishaw & Coltness*, 6 Feb 1837, 5 Feb 1838, 4 Feb 1839; SRO, BR/WIC/1/2 Wishaw & Coltness Minutes, 22 Mar 1837, 29 Jan 1838.

175. More Nisbett Papers, bundle 7, *Report of Committee of Management of Wishaw & Coltness*, 3 Feb 1840.

176. GUEH, 'Interim Report by the Committee of Inquiry to the Directors of the Wishaw and Coltness, 10 July 1844'; ibid., draft note of reply to above.

177. GUEH, 'Monklands Amalgamation Bill, 1848. Notes of evidence of James Wright.'

178. GUEH, ibid., notes of evidence of George Knight. The neighbouring Wishaw & Coltness also had all the problems of a coal line with multiple branches and short-distance traffic. According to Joseph Locke,

> Of all the railways I have had to do with, it is the most difficult to manage. I have never seen a line with so many juts on each side, and carrying such short distances; the average would not be above four or five miles; and there is no ordinary six-mile clause . . . so that, in fact, the mere charge of 2d. or 3d. or $1\frac{1}{2}$d a ton is a very small sum of money, and there is the inconvenience of the junctions and sidings, and things of that kind, upon the line, for this large traffic, producing not a very large amount of profit. (PP HL 1849(21)XXIX, *HL S.C. on Railway Accounts*, evidence, Q.3660, J. J. Hope Johnstone, quoting Locke's evidence to the HC S.C. on the Caledonian Railway Purchase or Lease of the Wishaw & Coltness Railway Bill, 24 April 1849).

179. GUEH, 'Monklands Amalgamation Bill, 1848. Notes of evidence of George Knight.'

180. GUEH, 'Slamannan Railway, Bathgate & Jawcraig Branches: notes of evidence of Edward Woods C.E.'

181. This may be implied in the comment of the Monklands lines manager/engineer A. J. Adie: 'There is no similarity whatever between the trade on this Line [the Monklands] & on the Garnkirk line and the rates which would be profitable on the one would be ruinous on the other' (GUEH, 'Monklands Amalgamation Bill, 1848. Notes of Evidence of A. J. Adie').

182. GUEH, 'Slamannan Railway, Bathgate & Jawcraig Branches, notes of evidence of James Thomson.'

183. GUEH, 'Monklands Amalgamation Bill, Brief for the Promoters, 1848', 104.

184. Robertson Buchanan, *Report relative to the Proposed Railway from Dumfries to Sanquhar*, 33.

185. S. G. E. Lythe, *loc.cit.*, 548. In 1834 the Garnkirk started excursion trips — possibly the first railway to do so (T. Gourvish, *Mark Huish and the London & North Western Railway; a Study of Management*, 69). Lee is certainly wrong to claim the first excursions for the Whitby & Pickering in 1839 (C. E. Lee, *Passenger Class Distinctions*, 9)

186. T. Grainger & J. Miller, *Report relative to the Proposed Railway to connect . . . Lanarkshire with . . . Glasgow . . .*, 11.

187. *Chambers' Edinburgh Journal*, 5 Oct 1833, 15 Dec 1832. The minister of Dalkeith confirmed that passenger services had not originally been contemplated (*NSA*, I (Edinburghshire), 512).

188. *Ibid.*, 5 Oct 1833. This article also praised the 'extremely moderate' fare of 6d for $8\frac{1}{2}$ miles: 'It proves . . . how effectually the cheapness of a thing tends to promote its prosperity, and how indispensable it now is that all ordinary things should be cheap, in order that the public may avail

themselves of their use.'
189. G. Buchanan, *op.cit.*, 9.
190. *NSA*, VI (Lanarkshire), 205; Anon., 'Early Scottish railways', *loc.cit.*, 34.
191. GUEH, *Rules, Bye Laws and Regulations of the Ballochney Railway Co . . . 1836*, bye-law 19; *NSA*, VIII (Dunbartonshire), 203.
192. *NSA*, V (Ayrshire), 203.
193. PP 1843 [440] XLVII, *Third Report of the Railway Department*, appendix 2.
194. The opening of the Slamannan in 1840 had itself caused a 57% fall in the passenger traffic of the Forth & Clyde Canal (J. Lindsay, *op.cit.*, 43).
195. SRO, BR/RAC(S)/1/113, *Report of the Directors of the Ardrossan Railway*, 25 Jan 1843.
196. J. Grieve & J. McLaren, *op.cit.*, 15; J. Priestley, *op.cit.*, 228.
197. More Nisbett Papers, bundle 7, *Reports of Committee of Management of Wishaw & Coltness*, 5 Feb 1828, 3 Feb 1840.
198. GUEH, 'Proposal for Compensation on Account of Wayleave claimed by Messrs Bairds from the M & K Ry Coy in the matter of the Chapelhall Branch'; ibid., Grahame & Weems to G. H. Lang, 19 May 1846. Bairds had been among the main beneficiaries of railway expansion. Between 1826 and 1828 they leased the minerals of Gartsherrie, Gargill and Cairnhill on the Monkland & Kirkintilloch line, and in the latter year commenced construction of the great Gartsherrie ironworks, which the opening of the Garnkirk railway made 'the best situated in the county as regarded to access' (A. M. [Andrew MacGeorge], *The Bairds of Gartsherrie*, 52-5; Anon., *Memoirs and Portraits of One Hundred Glasgow Men*, 14).
199. GUEH, Ballochney Railway, draft report of Committee of Management, 1 Feb 1831.
200. GUEH, Ballochney Railway, report of Directors, 2 Feb 1836.
201. SRO, BR/WIC/1/2, Wishaw & Coltness Minutes, 21 Jan 1837, 3 May 1837, 5 Feb 1838; BR/WIC/4/4/1, 'Copy Contract & Agreement between the Wishaw & Coltness Railway Company and Thomas Houldsworth Esquire of Coltness'; More Nisbett Papers, bundle 7, *Report of Committee of Management of Wishaw & Coltness*, 6 Feb 1837; GUEH, Robert Dodds, 'General Remarks on the Wishaw and Coltness Railway from its Commencement to the Present Date.'
202. GUEH, 'Monklands Amalgamation Bill. Brief for the Promoters, 1848.'
203. GUEH, Robert Dodds to James Donaldson, 8 Sept 1841.
204. GUEH, Robert Dodds to Wishaw & Coltness Committee, 16 Jan 1844.
205. SRO, BR/WIC/1/4, Wishaw & Coltness Minutes, 8 Feb 1845, 5 Mar 1845; GUEH, Memorial for the Wishaw & Coltness Railway Company, 22 Feb 1845; ibid., opinion on above by Andrew Rutherford and D. MacKenzie. In the following year Wilson was in negotiation with the Monklands railways again, as he demanded compensation for alterations to his waggons made necessary by the proposed change of gauge on the Slamannan from $4'6''$ to $4'8\frac{1}{2}''$ (GUEH, 'Petition of John Wilson Esq against Slamannan Ry Co, Bathgate & Jawcraig branches bill').

Chapter 3. From Town to Town: Planning and Authorisation

1. H. J. Dyos & D. H. Aldcroft, *British Transport*, 122-3.
2. R. C. O. Matthews, *A Study in Trade Cycle History. Economic Fluctuations in Great Britain, 1833-42*, 111. Of the 300 new companies established 1834-36, railways formed the largest proportion with 88, followed by mines with 71 (B. C. Hunt, *The Development of the Business Corporation in England, 1800-67*, 76).
3. J. R. McCulloch, *Dictionary of Commerce*, 898.
4. H. Pollins, *Britain's Railways*, 24-5. Some opponents continued to see railway development as speculative folly: 'To those who remember the bubble year of 1825, it would be a matter of astonishment that they should have had to witness, within the short period of a dozen years, the recurrence of a similar commercial fever, were they not aware that people never learn wisdom until it is too late' (*Fraser's Magazine* 17 (April 1838), 422).
5. *Circular to Bankers*, 6 Nov 1835, quoted in B. C. Hunt, *op.cit.*, 74.
6. PP 1839(517)X, *S.C. on Railways, Second Report*, p.iii.

7. F. Whishaw, *The Railways of Great Britain and Ireland*, 102.
8. A. Slaven, *The Development of the West of Scotland 1750-1960*, 118, 127; R. H. Campbell, *Scotland since 1707*, 122-3, 159.
9. *Scotsman*, 20 Oct 1830.
10. B. R. Mitchell & P. Deane, *Abstract of British Historical Statistics*, 21, 24.
11. Edinburgh & Glasgow Railway, *Reports by Mr. George Stephenson of Liverpool, and Messrs. Grainger and Miller of Edinburgh, Civil Engineers*, 5-6.
12. *Ibid.*, 15.
13. S. G. Checkland, *The Rise of Industrial Society in England, 1815-85*, 13; Anon., *Remarks on the Reports of the Committee to the Subscribers for . . . a Railway from Edinburgh and Leith to Glasgow, and on the Report of James Jardine, Esq., Civil Engineer*.
14. Anon. (probably Rev. James Adamson), *Remarks on the Comparative Merits of Cast Iron and Malleable Iron Rail-Ways; and an Account of the Stockton and Darlington Rail-Way, &c., &c.*, 21-4. An estimate in 1830 put Edinburgh's coal consumption at 400,000 tons and Glasgow's at almost 800,000 (*Remarks on the Reports . . .*, 16). Neither the newly built Monkland & Kirkintilloch nor the newly authorised Ballochney were seen as threats; it was not clear whether the Garnkirk was to be taken into the system or to be ignored.
15. *Remarks on the Reports . . .*, 1-2. Hamilton himself did envisage a line from Leith by Edinburgh, Glasgow and Paisley to Ardrossan and Troon (J. Thomas, *Regional History of the Railways of Great Britain, VI: Scotland*, 61).
16. J. Jardine, *Report to the Subscribers for a Survey and Plan of a Railway from Edinburgh and Leith to Glasgow . . .*, 3, 11-14.
17. *Remarks on the Reports . . .*, 2.
18. *Ibid.*, 17.
19. SRO, BR/EGR/1/1, Edinburgh & Glasgow Minute Book, 14 Oct 1830, 1 Nov 1830.
20. SRO, GD46/13/59, Seaforth Muniments, Edinburgh Glasgow & Leith Prospectus, 28 Dec 1831. Grainger and Miller made three surveys — a preliminary rough observation early in 1830, a full survey later in the year, and a third one which resulted in a revised line in 1831.
21. SRO, BR/EGR/1/1, 25 Nov 1830, Letter from Edinburgh Glasgow & Leith Provisional Committee to J. Gibson Craig, chairman of subscribers in 1825 to the Jardine line.
22. Edinburgh Glasgow & Leith Railway, *Reports by Messrs. Grainger & Miller of Edinburgh, and Mr. George Stephenson of Liverpool, Civil Engineers*, II, 1 (minute of provisional committee, 12 Jan 1831).
23. *Scotsman*, 27 Oct 1830.
24. J. Jardine, *op.cit.*, 1.
25. Edinburgh Glasgow & Leith Railway, *op.cit.*, II, 2 (minute of provisional committee, 12 Jan 1831); Edinburgh & Glasgow Railway, *op.cit.*, 1; SRO, GD/46/13/59, Edinburgh Glasgow & Leith Prospectus.
26. *Remarks on the Reports . . .*, 2-3; T. Grainger & J. Miller, *Observations on the Formation of a Railway Communication between the Cities of Edinburgh & Glasgow . . .*, 7.
27. SRO, BR/EGR/1/1, 12 Jan 1831.
28. Edinburgh Glasgow & Leith Railway, *op.cit.*, III, 2 (report by Stephenson); J. Jardine, *op.cit.*, 22 (letter of 2 Nov 1830); Edinburgh & Glasgow Railway, *op.cit.*, 12. Grainger and Miller also emphasised their advantage in having no inclined planes on their line (SRO, BR/EGR/1/1, 12 Jan 1831).
29. SRO, GD46/13/59, Edinburgh Glasgow & Leith Prospectus; SRO, BR/EGR/1/1, 12 Jan 1831; Edinburgh Glasgow & Leith Railway, *op.cit.*, IV, 4.
30. Pro Jardine advertisement in *Scotsman*, 10 Nov 1830.
31. J. Jardine, *op.cit.*, 6, 11. Jardine believed that Parliament would not allow a railway to cross the Monkland Canal, thus making it necessary to approach the harbour on the south side of the Clyde. The existence of the Polloc & Govan, even though its condition might be unsatisfactory, gave his route an important advantage in obtaining harbour access (*ibid.*, 22; *Scotsman*, 10 Nov 1830).
32. *Ibid.*, 11.
33. *Scotsman*, 10 Nov 1830.

34. *Remarks on the Reports . . .*, 10, 8.
35. Edinburgh & Glasgow Railway, *op.cit.*, 8.
36. SRO, BR/EGR/1/1, 23 Oct 1830, 29 Nov 1830, 3 Nov 1830.
37. Edinburgh & Glasgow Railway, *op.cit.*, 13-14; SRO, BR/EGR/1/1, 12 Jan 1831; and see Table 25. Stephenson rejected the valley route used by the canals as the entrances to the two cities 'could be effected only with an immense sacrifice of directness, to the extent probably of ten miles, and by a ruinous expenditure of money' (Edinburgh & Glasgow Railway, *op.cit.*, 12).
38. SRO, BR/EGR/1/1, 29 Sept 1831; *Scotsman*, 27 Nov 1830.
39. Edinburgh Glasgow & Leith Railway, *op.cit.*, II, 1.
40. Edinburgh & Glasgow Union Canal Company, *Report relative to the Railways between Edinburgh and Glasgow*, 24 (claimed that Grainger intended to charge fares of 9/-, 6/- and 4/-); Edinburgh & Glasgow Railway, *op.cit.*, 7; Edinburgh Glasgow & Leith Railway, *op.cit.*, IV, 11-12.
41. *Scotsman*, 30 Oct 1830, 27 Nov 1830.
42. SRO, BR/EGR/1/1, 24 Feb 1832.
43. Edinburgh Glasgow & Leith Railway, *op.cit.*, IV, 13.
44. Edinburgh & Glasgow Union Canal Company, *op.cit., passim* (quote p.43).
45. *Scotsman*, 10 Nov 1830. The canal company may well also have been behind a letter calling on all subscribers to the railway to withdraw while they still legally could (*ibid.*, 20 Nov 1830).
46. *Remarks on the Reports . . .*, 16.
47. Edinburgh & Glasgow Union Canal Company, *op.cit.*, 8-14.
48. *Remarks on the Reports . . .*, 14.
49. Edinburgh & Glasgow Railway, *op.cit.*, 22-4.
50. 'Detector', *Edinburgh, Glasgow and Leith Railway*, 1.
51. GUEH, 'Monkland Amalgamation Bill. Brief for the Promoters. 1848', 26.
52. SRO, BR/EGR/1/1, 14 Oct 1830.
53. *Scotsman*, 27 Nov 1830.
54. Edinburgh & Glasgow Union Canal Company, *op.cit.*, 31, 43.
55. SRO, GD46/13/59, 'Case of the Parliamentary Trustees and Creditors of the Bathgate Road against the Edinburgh & Glasgow Railway Bill, 1831', 2.
56. SRO, BR/EGR/1/1, 1 May 1832.
57. GUEH, 'Protest by William Davidson W.S., Glasgow, at the General Meeting of the Ballochney Railway Company, 1 February 1831'; ibid., draft reply to above by Thomas Grahame, engineer.
58. GUEH, Draft report by Ballochney Committee of Management to General Meeting, 1 Feb 1831. There is no problem in identifying Tennant, who was the only man to serve on the committees of the Monkland & Kirkintilloch, the Ballochney, and the new project. The only other member of the committees of the two established railways, Robert Grahame, signed this report, and would in any case, as a member of the council of the Forth & Clyde Navigation, have been hostile to any inter-city railway (SRO, BR/FCN/1/20, Minute Book of the Forth & Clyde Navigation Company).
59. *Remarks on the Reports . . .*, 6.
60. J. Mitchell, *Reminiscences of my Life in the Highlands*, II, 149.
61. J. Copeland, *Roads and their Traffic 1750-1850*, 165-70.
62. Edinburgh & Glasgow Railway, *op.cit.*, 18.
63. *Scotsman*, 7 Jan 1832; J. Copeland, *op.cit.*, 177; PP 1839(295)IX, *S.C. on Turnpike Trusts*, evidence, QQ. 502-3 (McAdam).
64. G. S. Emmerson, *John Scott Russell*, 8-12; J. Copeland, *op.cit.*, 172-3; D. Bremner, *The Industries of Scotland*, 90; C. H. Ellis, *British Railway History*, I, 46; *NSA*, VI (Lanarkshire), 202.
65. PP 1839(295)IX, evidence, Q.319 (Col. Maceroni); J. Copeland, *op.cit.*, 183.
66. SRO, BR/EGR/1/1, 29 Nov 1831, 23 Jan 1832.
67. Anon., *Statement and Opinion for the Subscribers to the Edinburgh and Glasgow Railway, who are Desirous of Withdrawing from that Concern*.
68. SRO, BR/EGR/1/1, 7 Feb 1832, 30 Mar 1832.
69. *Statement and Opinion for the Subscribers . . .*
70. SRO, BR/EGR/1/1, 24 Feb 1832, 17 May 1832, 24 May 1832, 4 June 1832.

71. Ibid., 29 Oct 1832 (report), 9 Nov 1832.
72. Ibid., 25 Aug 1832, 29 Oct 1832, 10 Apr 1833, 28 May 1833, 13 Nov 1833.
73. *Chambers' Edinburgh Journal*, 6 Dec 1834.
74. B. W. Crombie & W. S. Douglas, *Modern Athenians: a Series of Original Portraits of Memorable Citizens of Edinburgh, 1837 to 1847*, 152-3. Other notable subscribers, apart from those who actually joined the Edinburgh & Glasgow board, included James McCall, later chairman of the Ayrshire, Charles F. Davidson, later secretary and key promoter of the North British, and Erskine Sandford, later chairman of the Edinburgh Leith & Granton (SRO, BR/EGR/2/53, list of subscribers to Edinburgh & Glasgow Railway, 1835).
75. More Nisbett Papers, bundle 13, *Edinburgh & Glasgow Prospectus*, Oct 1835.
76. SRO, BR/EGR/1/2, 5 July 1836, 18 Apr 1837.
77. More Nisbett Papers, *Edinburgh & Glasgow Prospectus*.
78. J. Thomas, *A Regional History of the Railways of Great Britain, VI: Scotland*, 62-3.
79. More Nisbett Papers, *Edinburgh & Glasgow Prospectus*.
80. PP HL 1837-8(185)XX, *HL S.C. on the Edinburgh and Glasgow Railway Bill*, evidence, Q.29.
81. Ibid., QQ.63-5.
82. Edinburgh & Glasgow Junction Railway, *Report of a Sub-Committee of the Union Canal Company, 4th December, 1835*.
83. 'A Canal Proprietor', *Letter to the Proprietors of the Edinburgh and Glasgow Union Canal*, 29 Jan 1836.
84. SRO, BR/EGR/1/2, 17 Feb 1837, 22 Feb 1837.
85. SRO, BR/HRP(S)/1/15, John Macneill, Draft Report on the Proposed Plan of Conveying Passengers between Edinburgh and Glasgow, 14 Dec 1838.
86. SRO, BR/EGR/1/2, 18 Apr 1837, 26 Apr 1837, 18 May 1837, 26 May 1837; *RT*, 25 Aug 1838.
87. Ibid., 5 July 1836.
88. *RT*, 20 Jan 1838.
89. Act 1 & 2 Vict. c.58, 4 July 1838.
90. NLS, Ms 9944 f.63, Menzies Papers, Hugh Watson to Sir Neil Menzies, 20 Feb 1838.
91. Ibid.
92. NLS, Acc 3737, Chalmers of Auldbar Papers, Adam Burnes to Patrick Chalmers MP, 16 Oct 1837.
93. S. G. E. Lythe, 'Early Days of the Arbroath & Forfar Railway', *Railway Mag.* 99 (1953), 55; SRO, BR/AFR/1/1, Arbroath & Forfar Minutes, 19 Aug 1835; J. M. McBain, *Eminent Arbroathians*, 213-4. Earlier surveys for a canal along the route had been made by Whitworth in 1788 and Stevenson in 1817, and for a waggonway by Stevenson in 1820 (J. M. McBain, *op.cit.*, 213).
94. J. M. McBain, *op.cit.*, 218.
95. Ibid., 205; PP 1839(222,517)X, *S.C. on Railways*, evidence, QQ.3454-5 (Grainger); Act 6 & 7 Will.IV c.34, 19 May 1836.
96. SRO, BR/AFR/1/1, 26 Aug 1835; S. G. E. Lythe, *loc.cit.*, 56.
97. J. M. McBain, *op.cit.*, 209, 215; *RT*, 12 Oct 1839.
98. PP 1839(222,517)X, evidence, QQ.3406-7, 3421, 3436-9 (Grainger); SRO, BR/AFR/1/1, 19 Aug 1835; HLRO Deposited Plans, HC 1836, Arbroath & Forfar Railway; J. M. McBain, *Arbroath: Past and Present*, 205-6.
99. SRO, BR/AFR/1/1, 19 Aug 1835; PP 1839(222,517)X, evidence, Q.3439; Acts 6 & 7 Will.IV c.34, 29 May 1836; 3 & 4 Vict. c.14, 3 Apr 1840.
100. NLS, Acc 3737, Minutes of Annual General Meeting of Arbroath & Forfar, 13 June 1842. The guaranteed dividend was necessary if there was to be any hope of disposing of the shares; lack of confidence in the company meant that a nominal £25 of ordinary stock was selling in the market for £9 (J. M. McBain, *Arbroath: Past and Present*, 206).
101. NLS, Acc 3737, John Macdonald to Patrick Chalmers MP, 4 July 1840. Chalmers took the hint, adding £10,000 of the new stock to his existing £3000 holding (ibid., William Johnston, Arbroath Bank, to Chalmers, 4 Aug 1841).
102. J. M. McBain, *Eminent Arbroathians*, 219; C. E. Lee, 'The Dundee & Arbroath Railway',

Railway Mag. 104 (1958), 24; HLRO Deposited Plans, HC 1836, Dundee & Arbroath Railway.
103. PP 1839(222,517)X, evidence, QQ.3434-5 (Grainger); HLRO Deposited Plans, HC 1836, Dundee & Arbroath Railway; J. M. McBain, *Arbroath: Past and Present*, 198.
104. HLRO Deposited Plan, HC 1836, Dundee & Arbroath Railway.
105. *RT*, 24 Dec 1842.
106. C. E. Lee, *loc.cit.*, 25, 27. Quote from director William Thom.
107. Shiell & Small, Dundee & Perth Minute Book, 7 Nov 1835, 18 Nov 1835.
108. Ibid., 4 Feb 1836.
109. Ibid., meeting of Carse of Gowrie Turnpike Road Trustees and subcommittee of Dundee & Perth Railway, 18 Jan 1836; report of subcommittee, 4 Feb 1836.
110. Ibid., 17 May 1836, 13 June 1836, 5 Apr 1838.
111. SRO, GD112/53 Box 1, Breadalbane Muniments, John Richardson of Pitfour to Marquis of Breadalbane, 4 May 1836.
112. Shiell & Small, Dundee & Perth Minute Book, Hunter & Coning, Perth, to Dundee & Perth committee, 23 May 1836.
113. J. Milne, *Aberdeen*, 338.
114. PP 1845(120)XXXIX, *Report of Railway Department on Schemes for Extending Railway Communication in Scotland*.
115. In 1841 its population was 25,984 (*Oliver & Boyd's New Edinburgh Almanac 1845*, 489).
116. PP 1847-8 [938] XXVI, *Report of Railway Commissioners for 1847*, appx.10, Report of Capt. Coddington before the opening of the Edinburgh Leith & Granton.
117. B. W. Crombie & W. S. Douglas, *op.cit.*, 116; P. Neill, *Considerations regarding the Edinburgh, Leith, and Newhaven Railway*, 5.
118. P. Neill, *op.cit.*, 4-5, 12-18. Other hostile landowners included John Learmonth (*ibid.*, 20).
119. H. G. Lewin, *The Railway Mania and its Aftermath*, 506, 508.
120. P. Neill, *Remarks on the Progress and Prospects of the Edinburgh, Leith, & Newhaven Railway, in January 1839*, 14-16; Act 2 & 3 Vict. c.51, 1839.
121. *RT*, 17 Aug 1844; Act 7 & 8 Vict. c.81, 1844.
122. Letter from 'Verax' in *RT*, 20 Feb 1841.
123. *Railway Chronicle*, 20 May 1848.
124. 'An Original Proprietor', *An Authentic Statement of the Affairs of the Edinburgh, Leith, and Newhaven Railway Company, from the Period of its Projection in 1835, to the Close of the Year 1840*, 6; *RT*, 8 Aug 1840.
125. *RT*, 19 June 1841; SRO, GD224/554, Henry Wait Hall to Duke of Buccleuch, 10 June 1841. The decision to extend the line from Trinity to Granton had also been taken principally in the hope of attracting the duke's support (SRO, GD224/554, Richard Dawson to (?)Lord Sandon, c.22 Aug 1840).
126. *RT*, 4 Sept 1841.
127. *RT*, 23 Apr 1842.
128. SRO, GD224/554, John Gibson to Duke of Buccleuch, 4 Sept 1840.
129. Ibid., Henry Wait Hall to Duke of Buccleuch, 2 Sept 1840; extract from minutes of a meeting of Bristol shareholders in the Edinburgh Leith & Newhaven, 1 Sept 1840. Hall was one of the English directors of the company who resigned in 1841, after which he again wrote to the duke: 'After the exposure which has taken place of the proceedings of the company, I should hope that no person will again be imposed on by any representations emanating from that source ... The company has proved itself unworthy the support of any honest man.' (Ibid., Hall to Buccleuch, 10 June 1841).
130. Ibid., John Gibson to Duke of Buccleuch, 5 Feb 1844.
131. H. Cockburn, *A Letter to the Lord Provost on the Best Ways of Spoiling the Beauty of Edinburgh*, 1. Edinburgh Leith & Newhaven director Joseph Macgregor agreed: 'No city in the world offers the same facility for bringing a railway into its very centre.' (*RT*, 30 Apr 1842)
132. SRO, GD224/554, John Gibson to Duke of Buccleuch, 24 Aug 1840.
133. D. Robertson, *The Princes Street Proprietors*, 4-5, 11-18; A. J. Youngson, *The Making of Classical Edinburgh*, 275-6.

Notes 373

134. SRO, BR/EGR/1/1, 12 Jan 1831 (report by Grainger and Miller, Dec 1830).
135. D. Robertson, *op.cit.*, 37-9.
136. *Ibid.*, 40-1.
137. H. Cockburn, *Journal*, I, 130, 9 Jan 1837.
138. *Chambers' Edinburgh Journal*, 30 June 1838.
139. PP 1837(259)LI, *Correspondence between the Trustees for Manufactures in Scotland, the Commissioners of the Treasury, and the Promoters of the Edinburgh and Glasgow Railway, on the Subject of the Tunnel projected through the Mound in Edinburgh . . .*, 2-3.
140. *Ibid.*, 4-6.
141. SRO, GD224/554, John Gibson to Duke of Buccleuch, 24 Aug 1840.
142. *RT*, 4 Feb 1843.
143. SRO, GD224/554, Circular by Alex Douglas; D. Robertson, *op.cit.*, 5.
144. *RT*, 4 Mar 1843.
145. SRO, GD224/554, Duke of Buccleuch to Sir John Hope, 10 Sept 1840; John Gibson to Buccleuch, 6 Jan 1844.
146. D. Robertson, *op.cit.*, 44.
147. SRO, BR/EGR/3/1, Edinburgh & Glasgow Railway: Copies of Contracts, Agreements &c., Agreement with Princes Street Proprietors; Act 7 & 8 Vict. c.58, 1844.
148. H. Cockburn, *Letter to the Lord Provost*, 1.
149. *RT*, 8 Aug 1840.
150. H. Cockburn, *Journal*, II, 213-14, 6 Apr 1848.
151. H. G. Lewin, *op.cit.*, 508; J. Thomas, *op cit*, 272.
152. *Railway Chronicle*, 20 May 1848. Other estimates of total expenditure included £340,000 (SRO, BR/GSW/4/4, cutting from *Daily Review*, 17 Nov 1864) and £380,000 (*HRJ*, 8 Jan 1848).
153. Act 8 & 9 Vict. c.82, 1845.
154. See above, Table 2.
155. T. Grainger, *Observations addressed to the Committee of the Chamber of Commerce of the City of Edinburgh, relative to the Proposed Railway . . . to be called the Edinburgh, Dundee, and Northern Railway . . .*, 4-5.
156. *Ibid.*, 5.
157. *Ibid.*, 6.
158. Sir J. B. Paul (ed), *The Scots Peerage*, II, 245; SRO, GD224/554, John Gibson to Duke of Buccleuch, 24 June 1844.
159. S. G. Checkland, *The Gladstones: a Family Biography 1764-1851*, p.xii.
160. *Ibid.*, 327, 339-40.
161. *Ibid.*, 327-8, 340; SRO, GD224/554, Report of Committee appointed by the Commissioners of Supply of the County of Fife . . . re Burntisland Pier and Ferry, 1846; Act 5 & 6 Vict. c.91, 1842.
162. T. Grainger, *Observations addressed to the Committee . . .*, 29; J. Milne, *Reply to the Observations addressed by Mr. Thomas Grainger to the Committee of the Chamber of Commerce of the City of Edinburgh on the Proposed Railways through Fife*, 4-6; *RT*, 2 Jan 1841.
163. T. Grainger, *Observations addressed to the Committee . . .*, 8-9, 22-6; J. Milne, *Reply to the Observations . . .*, 5-6, 11; Anon., *Remarks on the Forth and Tay Railway*, 1.
164. SRO, GD224/554, John Gibson to Duke of Buccleuch, 26 Apr 1844.
165. Gladstone Papers, Letter Book 14, John Gladstone to Lord William Douglas, 15 Jan 1844; John Gladstone to Sir Charles Gordon, 15 Jan 1844.
166. SRO, BR/NBR/1/1, North British Minute Book, 19 Jan 1842.
167. Act 10 & 11 Vict. c.239, 1847.
168. Glasgow Paisley Kilmarnock & Ayr Railway, *Proceedings at the First General Meeting of the Shareholders . . . November 9, 1836*, 3.
169. T. Grainger & J. Miller, *Report relative to a Proposed Railway Communication between the City of Glasgow and the Towns of Paisley & Johnstone*, 2, 4-5, 11-16.
170. *Ibid.*, 14; *Scotsman*, 27 Oct 1830.
171. SRO, BR/PROS(S)/1/1, *Prospectus of the Glasgow, Paisley, Kilmarnock, & Ayr Railway*.
172. Scott, Stephen & Gale, *An Examination of Mr. G. Stephenson's Report on the Two Lines of*

Railway projected between Glasgow and Ayrshire, 16-18, 23.
173. *Ibid.*, 37-9 (Stephenson's report, annexed to Scott, Stephen & Gale's reply).
174. *Ibid.*, 5-7.
175. SRO, BR/PROS(S)/1/1, 1.
176. PP HL 1837(146)XVIII, *HL Committee on the Glasgow Paisley Kilmarnock & Ayr Bill*, evidence, pp.4, 9-10 (P. Blair).
177. *Ibid.*, evidence, pp.176 (J. Miller), 192 (G. Stephenson), 278 (Alex. McKellar), 296-7 (William Gale); J. Francis, *op.cit.*, I, 106-12.
178. PP HL 1837(146)XVIII, evidence, pp.57-67 (Alex. Guthrie), 17-52 (Robert Farquharson), 70-6 (Miller).
179. *Ibid.*, evidence, pp.19-20, 24 (Farquharson), 76 (James Drummond).
180. PP 1837-8(257)XVI, *S.C. on Railroad Communication*, evidence, QQ.1080, 1100, 1070 (W. Chaplin).
181. More Nisbett Papers, bundle 3, Glasgow Paisley Kilmarnock & Ayr, *Report of Meeting of Shareholders*, 17 Feb 1841, 2; ibid., *Report of Meeting of Shareholders*, 20 Dec 1837, 4.
182. Ibid., *Report of Meeting of Shareholders*, 20 Dec 1837, 5-7, 9.
183. PP 1844(318)XI, *S.C. on Railways, Fifth Report*, evidence, Q.5431 (Wyndham Harding).
184. PP HL 1837(146)XVIII, evidence, pp. 259 (Robert Ewing), 278 (Alex. McKellar).
185. Anon., *Memoirs and Portraits of One Hundred Glasgow Men . . .*, 105.
186. Glasgow Town Council Minutes, 6 Apr 1802 (I owe this reference to Mr Callum Brown). The plan had earlier been mentioned by James Gordon, managing director of the Muirkirk Ironworks, in a letter to the Royal Bank of Scotland, 13 Dec 1801 (Anon., 'Early Scottish Railways', *Three Banks Rev.* 74 (1967), 30).
187. SRO, BR/HRP(S)/51, p.110, cutting from *Greenock Telegraph*, 28 Mar 1898.
188. See below, Table 85.
189. J. Thomas, *op.cit.*, 24.
190. HLRO Minutes of Evidence, HC 1837 vol.12, Glasgow Paisley & Greenock Railway, QQ.45,65-6,69-72 (John Poynter). Poynter later became a director of the railway company (SRO, BR/GPG/1/1, Glasgow Paisley & Greenock Minutes, 17 Oct 1837).
191. Ibid., QQ.90-1 (Poynter).
192. *NSA* VII (Renfrewshire), 453.
193. HLRO Minutes of Evidence, HC 1837 vol.12, Glasgow Paisley & Greenock Railway, QQ.1-11 (Poynter).
194. *RT*, 23 June 1838.
195. *RT*, 12 Dec 1839.
196. SRO, BR/HRP(S)/51 p.110, cutting from *Greenock Telegraph*, 23 Nov 1898.
197. R. C. O. Matthews, *A Study in Trade Cycle History. Economic Fluctuations in Great Britain, 1833-42*, 113.
198. J. Reid, *Manual of the Scottish Stocks and British Funds . . .*, 10.
199. Sir W. Acworth, *The Railways of England*, 1.
200. J. Milne, *Aberdeen*, 338 (survey from Aberdeen to Perth in 1837); *RT*, 31 Aug 1839 (plans to link Aberdeen to Edinburgh and Inverness).
201. Stephenson Locomotive Society, *The Highland Railway*, 2.
202. *RT*, 27 Mar 1841, 3 Apr 1841.
203. *RT*, 10 Apr 1841.
204. H. A. Vallance, *The Highland Railway*, 12. The line later became the Morayshire Railway.
205. Acts 6 & 7 Vict. c.50, 1843 (Ballochney), 6 & 7 Vict. c.63, 1843 (Drumpellar), 4 & 5 Vict. c.43, 1841 (Wilsontown Morningside & Coltness).
206. HLRO Deposited Plan, HC 1841, Glasgow & Dundyvan Junction Railway.
207. More Nisbett Papers, bundle 13, *Prospectus of the Glasgow & Hamilton Railway*; PP 1843(571)XLIV, *Return of Railway Bills and Acts, 1839-43*; R. H. Campbell, *op.cit.*, 93.
208. Railway stock dealings as the mania approached stimulated the creation of provincial stock exchanges; both the Glasgow and Edinburgh exchanges were formed in the second half of 1844. Previously it was noted that Scottish stocks were difficult to price, with quotations depending either

on the last transaction (perhaps several months earlier) or on 'vague hearsay' (M. C. Reed, 'Railways and the growth of the capital market', in *Railways in the Victorian Economy*, 179).
209. *Ibid.*, 164-5.
210. M. C. Reed, *Investment in Railways in Britain, 1820-1844*, 84.
211. HLRO Deposited Plan, HC 1838, Edinburgh & Glasgow Railway, subscription contract.
212. SRO, BR/GPG/1/1, 1 Mar 1836, 19 Nov 1836, 4 Jan 1837; H. Pollins, 'The marketing of railway shares in the first half of the nineteenth century', *Econ. Hist. Rev.* II 7 (1955), 230, 239.
213. M. C. Reed, *Investment in Railways . . .*, 85-6.
214. T. R. Gourvish & M. C. Reed, 'The financing of Scottish railways before 1860 — a comment', *Scot. Jnl of Pol. Econ.* 18 (1971), 211.
215. SRO, BR/EGR/1/1, 1 Mar 1833.
216. M. C. Reed, 'A note on subscriptions to the Glasgow, Paisley & Greenock Railway', *Transport Hist.* 6 (1973), 268-9.
217. For an analysis of the Glasgow Paisley & Greenock where confusion is caused by dependence on the version of the subscription contract printed in a parliamentary return, see W. Vamplew, 'Sources of Scottish railway share capital before 1860', *Scot. Jnl of Pol. Econ.* 17 (1970), and the correction made by M. C. Reed in 'A note on subscriptions . . .', *loc.cit.*
218. M. C. Reed, *Investment in Railways . . .*, 135, 154, 164. The subscription contract requirement was eventually abandoned by the Board of Trade in 1858 (S. A. Broadbridge, 'The sources of railway share capital', in M. C. Reed (ed), *Railways in the Victorian Economy*, 185).
219. See above, pp. 194-6.
220. SRO, BR/AFR/1/1, 16 Mar 1836.
221. P. Neill, *Remarks on the progress . . .*, 13, 18; 'An Original Proprietor', *op.cit.*, 1.
222. HLRO Deposited Plan, HC 1836, Edinburgh Leith & Newhaven Railway, subscription contract; P. Neill, *Remarks on the Progress . . .*, 17.
223. 'An Original Proprietor', *op.cit.*, 8.
224. T. R. Gourvish & M. C. Reed, *loc.cit.*, 210; and see above, pp. 76, 136.
225. Shiell & Small, Dundee & Perth Minute Book, 13 June 1836.
226. T. R. Gourvish & M. C. Reed, *op.cit.*, 214.
227. There was also a small counterflow of money from Scottish residents into English railways; by the later 1830s this accounted for about 1% of the shares in such lines as the London & Birmingham, Grand Junction and Liverpool & Manchester (*ibid.*, 213).
228. PP 1837(537)XVIII pt.2, *S.C. on Railway Subscription Contracts*, evidence, Q. 10596.
229. M. C. Reed, *Investment in Railways . . .*, 116.
230. HLRO Deposited Plan, HC 1844, North British Railway, subscription list.
231. T. R. Gourvish, *Mark Huish and the London & North Western Railway*, 204-5, 293-4; J. Simmons, *The Railway in England and Wales, 1830-1914*, I, 83-4.
232. A. Slaven, *The Development of the West of Scotland 1750-1960*, 106, 118; W. Vamplew, 'Sources of Scottish railway share capital . . .', *loc.cit.*, 435-7.
233. *RT*, 11 Apr 1839.
234. For the Edinburgh & Glasgow, see *HRJ*, 15 Aug 1846, 22 Aug 1846, 5 Sept 1846, 26 Sept 1846, and SRO, BR/EGR/1/10, Edinburgh & Glasgow Minutes, Aug-Sept 1846, *passim*. The debates within other companies occupy much space in their minute books and in the railway press for 1849-50.

Chapter 4. From Town to Town: Construction and Operation.

1. SRO, BR/RAC(S)/1/35, John E. Errington, *Report on the Proposed Railway from Perth by Stirling to Edinburgh and Glasgow*, 1841.
2. More Nisbett Papers, bundle 3, *Proceedings at the First General Meeting of the Shareholders of the Glasgow, Paisley, Kilmarnock & Ayr*, 20 Dec 1837: *Report of Meeting of Shareholders*, 20 Aug 1842.
3. *RT*, 9 July 1842.

4. *RT*, 14 Jan 1843.
5. More Nisbett Papers, bundle 13, *Edinburgh & Glasgow prospectus*, Oct 1835; ibid., bundle 3, *Glasgow Paisley Kilmarnock & Ayr prospectus*, 1836.
6. PP 1839(517)X, *S.C. on Railways, Second Report*, p. v.
7. *Ibid.*, p. vi.
8. Act 9 & 10 Vict. c.17, 1845.
9. T. R. Gourvish, *Mark Huish and the London & North Western Railway: a Study of Management*, 92.
10. PP 1849(421)X, *HL S.C. on the Audit of Railway Accounts, Third Report;* H. Pollins, 'Aspects of railway accounting before 1868' in M. C. Reed (ed.), *Railways in the Victorian Economy*, 146.
11. GUEH, 'Brief for the Promoters of the Edinburgh and Glasgow, Scottish Central, and other Amalgamation Bills . . .', 15.
12. H. Pollins, 'A note on railway construction costs, 1825-50', *Economica*, N.S. 19 (1952).
13. PP 1846(687)XIV, *S.C. on Railways Acts Enactments, 'Second Report'*, p. xv and appx. I.
14. PP 1857 s.1(226)XXIII, *Return of Parliamentary Expenditure by Railway Companies to 31st December 1857*.
15. *Ibid.*
16. H. Parris, *Government and the Railways in Nineteenth-Century Britain*, passim.
17. Acts 2 & 3 Will.IV c.120, 1832, and 1 & 2 Vict. c.98, 1838.
18. H. Parris, *op.cit.*, 18-27; H. J. Dyos & D. H. Aldcroft, *British Transport. An Economic Survey from the Seventeenth Century to the Twentieth*, 133.
19. Speech of 5 Feb 1844, quoted in F. Clifford, *A History of Private Bill Legislation*, I, 97n.
20. H. Parris, *op.cit.*, 56-7; F. Clifford, *op.cit.*, 96, 99.
21. O. C. Williams, *The Historical Development of Private Bill Procedure and Standing Orders in the House of Commons*, I, 67.
22. H. Parris, *op.cit.*, 19.
23. O. C. Williams, *op.cit.*, I, 43-4.
24. *Ibid.*, I, 62-4, 78, 83-7. The committees were instructed to obtain detailed information on the amount of capital required; the directors; the subscribers; the amount, adequacy, and rates of existing transport; anticipated traffic and income; whether the railway was a complete plan or part of a larger project; competing lines; requirements for assistant engines; engineering difficulties; tunnels; gradients and curves; mileage; level crossings; adequacy of estimates of costs and revenue; working expenditure; assent or dissent of landowners; names of engineers for and against the project; petitioners against the project; any other relevant circumstances.
25. F. Clifford, *op.cit.*, I, 83-4; H. Pollins, *Britain's Railways: an Industrial History*, 29, 36; O. C. Williams, *op.cit.*, I, 64, 81-3; HLRO, 1845 A3, *Standing Orders of the House of Lords*, 66-7, orders 223 and 226.
26. HLRO, Evidence 1844 vol.31, North British Bill.
27. H. Parris, *op.cit.*, 61.
28. M. C. Reed, *Investment in Railways in Britain, 1820-1844*, 57.
29. PP 1844(159)XLI, *Return of Moneys raised under Railway Acts, 1st January 1826 to 1st January 1844*.
30. *Ibid.*; J. Priestley, *Historical Account of the Navigable Rivers, Canals, and Railways, of Great Britain*, 285.
31. GUEH, 'Monkland & Kirkintilloch Railway. Return required by the Railway Commissioners . . . in reference to the Monkland Railways Amalgamation Bill. 1848.'
32. PP 1844(159)XLI.
33. PP HL 1837(146)XVIII, *HL S.C. on the Glasgow, Paisley, Kilmarnock and Ayr Railway Bill*, evidence, p.12.
34. See Table 37.
35. Shiell & Small, Dundee & Perth Minute Book, meeting of sub-committee, 5 April 1838.
36. Anon., *Memoirs and Portraits of One Hundred Glasgow Men*, 25-6.
37. H. Pollins, 'Railway construction costs', *loc.cit.*, 402.
38. PP 1845(420)X, *HL S.C. on the Practicability and Expediency of Establishing some Principle*

of Compensation to be made to the Owners of Real Property whose Lands, &c., may be compulsorily taken for the Construction of Public Railways, 4.

39. SRO, BR/RAC(S)/1/35.
40. J. Mitchell, *Reminiscences of my Life in the Highlands*, I, 152.
41. C. E. R. Sherrington, *Economics of Rail Transport in Great Britain*, I, 16.
42. W. J. Gordon, *Our Home Railways*, 163.
43. SRO, GD224/689, Buccleuch Muniments, Thomas Telford to G. Robinson, 12 Nov 1809: the landowner was Logan of Fishwick. See also above, p.39.
44. *RT*, 6 Oct 1838; J. M. McBain, *Arbroath: Past and Present*, 198; and see above, p.124. Panmure feued fifty to sixty acres at £40 per year, most of which was to cover compensation to tenants. About two-thirds of the line was on his land (Dundee & Arbroath Railway, *Report by the Committee of Management to a Special General Meeting of the Company*, 13).
45. GUEH, 'Case for the Edinburgh and Glasgow Railway Company relative to Lord Belhaven's Claim for Land and Minerals on the Wishaw & Coltness Railway'.
46. *RT*, 1 Sept 1838.
47. More Nisbett Papers, bundle 6, *Report of the Directors of the Slamannan Railway*, 8 Feb 1838.
48. Ibid., *Report of the Directors of the Slamannan Railway*, 7 Feb 1839.
49. Ibid. The hint was meant particularly for landowners on the proposed Bathgate branch.
50. F. Whishaw, *The Railways of Great Britain and Ireland*, 400; GUEH, E. Dickson, Report to the Directors of the Slamannan, 3 Nov 1842: James Mitchell to Thomas Burns, 22 Nov 1842.
51. J. Francis, *A History of the English Railway*, I, 186-8.
52. SRO, GD206/1/63A/5, Hall of Dunglass Muniments, 'Case of Sir John Hall Bart. against the North British Railway Bill' 1844: Agreement between Sir John Hall and the North British Railway Company, 29 Mar 1844.
53. J. Locke, *Report on the Proposed Line of Railway from Carlisle to Glasgow and Edinburgh by Annandale*, 6; SRO, GD224/554, Buccleuch Muniments, J. E. Errington to William Brooks, 2 Oct 1839.
54. SRO, GD224/554, Provost Robert Kemp to Duke of Buccleuch, 23 Sept 1837: Buccleuch to Kemp, 30 Sept 1837. A Lancaster & Carlisle Proprietor who, having read a *Carlisle Journal* report that Buccleuch intended to prosecute for trespass any surveyor found on his land, informed the duke that 'from what I know of your character for *Patriotism* & *Public Spirit* I feel convinced that you are incapable of acting as . . . the Editor asserts', received an even more abrupt answer (ibid., Thomas Brockbank to Buccleuch, 1 Nov 1837: Buccleuch to Brockbank, 5 Nov 1837).
55. Ibid., Provost Henry Dunlop of Glasgow to Duke of Buccleuch, 5 Oct 1839: Buccleuch to Dunlop, 14 Oct 1839.
56. Ibid., John Gibson to Duke of Buccleuch, 22 Feb 1845.
57. SRO, GD224/689, Thomas Telford to G. Robinson, 28 Sept 1809.
58. SRO, GD206/1/63A/5, David Smith to Sir John Hall, 19 Mar 1839.
59. SRO, BR/HRP(S)/51, cutting from *Glasgow Herald*, 5 Jan 1889.
60. *Proceedings of the Third General Meeting of Shareholders of the Glasgow, Paisley, and Greenock Railway . . . 27th January, 1837* (reprint from *Greenock Advertiser*), 5; T. R. Gourvish, op.cit., 94.
61. See above, pp.120, 125-6.
62. D. Spring, 'The English landed estate in the age of coal and iron: 1830-1880', *Jnl Econ. Hist.*, 11 (1951), 8. See also J. T. Ward, 'West Riding landowners and the railways', *Jnl Transport Hist.*, 4 (1960) for some English examples.
63. SRO, GD206/1/63A/5, Tod & Romanes W. S. to Sir John Hall, 24 Nov 1838.
64. J. Reid, *Manual of the Scottish Stocks and British Funds*, 93.
65. *Chambers' Edinburgh Journal*, 28 Dec 1839.
66. Acts 9 & 10 Vict. c.19 (Land Clauses Consolidation (Scotland) Act) and 9 & 10 Vict. c.33 (Railway Clauses Consolidation (Scotland) Act). For a discussion of the effects of these acts, see J. R. Kellett, 'Urban and transport history from legal records: an example from Glasgow solicitors' papers', *Jnl Transport Hist.* 6 (1964), 223-5.
67. H. Cockburn, *Journal*, 130 (28 Nov 1845).

68. G. Alderman, *The Railway Interest*, 25. Alderman's figure after the 1841 election is only 13, but many members became associated with railways during the course of the 1841-7 Parliament. His figures represent a minimum, since they exclude Irish and radical MPs, and apparently also those associated only with unsuccessful projects. *Bradshaw's* estimated that there were 86 MPs in the railway interest in 1847.
69. SRO, GD30/2172, Shairp of Houston Muniments, 'Report by Thomas Lawrie on the proposed West Lothian Railway through the Houston estate of Thos. Shairp esq', 24 Mar 1825.
70. PP HL 1837 (146) XVIII, evidence p.158 (James Adams); GUEH, 'Slamannan and Bo'ness Railway: valuation by Robert Brown'.
71. Shiell & Small, Dundee & Newtyle Sederunt Book, 18 Oct 1844; *RT*, 25 Aug 1838.
72. J. R. Kellett, 'Glasgow's railways, 1830-80: a study in "natural growth"', *Econ. Hist. Rev.*, II 17 (1964), 356.
73. SRO, GD224/554, Gibson & Home to secretary of Edinburgh Leith & Granton, 1 May 1845.
74. J. R. Kellett, *The Impact of Railways on Victorian Cities*, 231.
75. See above, pp.91-2.
76. SRO, GD16/38/78, Airlie Muniments, *Report of the Committee of Management of the Monkland & Kirkintilloch Railway*, 3 Feb 1830.
77. More Nisbett Papers, bundle 6, *Report of the Directors of the Slamannan Railway*, 6 Feb 1840.
78. Mark Huish to James Turner, 18 May 1838, quoted in T. R. Gourvish, *op.cit.*, 4.
79. Glasgow Paisley & Greenock Railway, *Answer by the Directors to the Report of the Committee of Investigation*, 23 Dec 1850, 8.
80. SRO, BR/GPG/1/1, Glasgow Paisley & Greenock Minutes, 9 Sept. 1837.
81. Ibid., 2 July 1838.
82. *RT*, 9 Jan 1841.
83. More Nisbett Papers, bundle 3, Glasgow Paisley Kilmarnock & Ayr, *Report of Meeting of Shareholders*, 22 Aug 1839.
84. *Chambers' Edinburgh Journal*, 20 Oct 1838.
85. T. R. Gourvish, *op.cit.*, 94.
86. SRO, BR/GSW/4/4, cutting from *Daily Review*, 19 Oct 1864.
87. Glasgow Paisley & Greenock Railway, *op.cit.*.
88. S. Pollard, *The Genesis of Modern Management*, 153-4.
89. *Bradshaw's Railway Manual*, 1847 and 1857.
90. L. T. C. Rolt, *Victorian Engineering*, 239.
91. See above, pp.100, 108, 119, 138-9.
92. J. Francis, *op.cit.*, II, 150; prospectuses of various projects in *RT*, *HRJ*, *Scotsman*, *Scottish Railway Gazette* and *Glasgow Herald*, 1844-47 *passim*.
93. Checkland notes that an engineer 'had to perform as a public relations officer, spending much of his time in attending upon Parliamentary Committees' (S. G. Checkland, *The Rise of Industrial Society in England, 1815-85*, 136).
94. PP 1844 [551] XLI, *Report of the Officers of the Railway Department . . . for 1843*, appx. II: PP 1846 [698] XXXIX, *Report of the Railway Department for 1844-45*, appx. II.
95. *PICE*, 74(1883), 287. Note however that Miller's signature does appear on one of the documents deposited in the HLRO for Grainger's Edinburgh Leith & Newhaven.
96. *PICE*, 12 (1852), 159; SRO, BR/EGR/1/1, Edinburgh & Glasgow Minutes, 14 Oct 1830.
97. *RT*, 14 July 1838. Referring to the Dundee & Arbroath, Whishaw observed in 1840 that 'we object to any alteration of gauge, where a line is likely to become a link of any long chain of railway-communication already established; but whether this railway is likely in future years to be so circumstanced, we have not at present the means of judging' (F. Whishaw, *op.cit.*, 79).
98. *RT*, 4 Aug 1838.
99. PP 1846 (489) XIII, *HL S.C. on Railways*, evidence, QQ. 1009-12 (Miller).
100. *RT*, 23 Jan 1841.
101. Gladstone Papers, John Gladstone to John Learmonth, 20 Jan 1844: John Gladstone to Charles Gordon, 1 Feb 1844: John Gladstone to Capt William Ramsay, 6 Feb 1844.

102. J. Francis, op.cit., I, 234.
103. See above, p.79 and Table 17.
104. Prospectuses in railway press 1845 and 1846. His other mania projects included the Avon Water Mineral, Banffshire, Edinburgh & Eskdale Junction, Edinburgh Portobello & Musselburgh Direct, Fife Central Junction, and Inverness & Ross-shire, few of which achieved more than the most preliminary planning.
105. *PICE*, 12 (1852), 161; W. W. Tomlinson, *The North Eastern Railway*, 532.
106. *PICE*, 74 (1883), 286-7; GUEH, Eglinton Estate Papers, notes of evidence of C. D. Gairdner re Ardrossan Railway; *HRJ*, 23 Mar 1844 (prospectus of Glasgow Dumfries & Carlisle).
107. Dundee & Newtyle Railway, *op.cit.*, 8, 17.
108. SRO, BR/RAC(S)/1/11, *Report of Glasgow Paisley Kilmarnock & Ayr*, 24 Feb 1840; *RT*, 7 July 1838.
109. Gladstone Papers, John Gladstone to Charles Gordon, 22 Mar 1844.
110. See below, p.202.
111. SRO, GD224/554, Robert Scott Moncrieff to Duke of Buccleuch, 20 Mar 1844.
112. Ibid., Duke of Buccleuch to Sir James Graham, 19 Sept 1846.
113. More Nisbett Papers, bundle 3, *Glasgow Paisley Kilmarnock & Ayr, Report of Meeting of Shareholders*, 17 Feb 1841; undated circular to shareholders [1837-8] containing parliamentary estimates.
114. SRO, BR/RAC(S)/1/11, *Report of Glasgow Paisley Kilmarnock & Ayr*, 20 Mar 1849; *HRJ*, 13 Apr 1850.
115. *HRJ*, 29 Sept 1849.
116. Miller's other mania projects which remained unbuilt included the Ayrshire & Bridge of Weir, Dundee & Forfar Direct, Edinburgh & Hamilton Direct, Edinburgh & Glasgow & Dunbartonshire Junction, Glasgow Airdrie & Monklands Junction, Glasgow & Coatbridge Direct Mineral, Granton Junction, Lochryan Harbour & Railway, Newcastle Hawick Edinburgh & Glasgow Junction, Perth & Crieff Direct, Scottish Central & Caledonian Junction, Scottish North Western, Scottish Southern, Scottish South Midland Junction, Tyne Valley and West of Scotland Junction.
117. *PICE*, 74 (1883), 288-9.
118. J. Bateman, *The Great Landowners of Great Britain and Ireland*, 311; J. Foster, *Members of Parliament, Scotland . . . 1357-1882*, 252.
119. J. Simmons, *The Railways of Britain*, 72.
120. See above, ch.2, sect.III and ch.3, sect.II; Jardine, who taught mathematics at Edinburgh University before becoming a civil engineer about 1806, was most noted for the creation of the Union Canal (*DNB*, X 687).
121. H. G. Lewin, *The Railway Mania and its Aftermath*, 234; HLRO Deposited Plans, HC 1835, Slamannan Railway, and HC 1841, Shotts & Wilsontown Railway; GUEH, Robert Dodds, 'General remarks on the Wishaw and Coltness Railway from its commencement to the present state' 1844; F. Whishaw, *op.cit.*, 107.
122. Anon., *Memoirs and Portraits of One Hundred Glasgow Men*, 277-8; *PICE*, 30 (1869), 456-7.
123. See above, pp.100, 108, 138-9.
124. NLS, Menzies of Castle Menzies Papers, Ms 9944 f.63, Hugh Watson to Sir Neil Menzies, 20 Feb 1838.
125. *Proceedings of the Third General Meeting . . . of the Glasgow, Paisley & Greenock Railway*, 5.
126. Ibid., 5; SRO, BR/HRP(S)/51 p.10, cutting from *Greenock Telegraph*, 28 Mar 1898.
127. J. Devey, *Life of Joseph Locke*, 98-106, 117-8; *Proceedings of the Third General Meeting . . . of the Glasgow, Paisley, and Greenock Railway*, 5; SRO, GD112/53, Breadalbane Muniments, P.M. Stewart MP to Marquess of Breadalbane, 12 Apr 1844. For Locke's career see also N. W. Webster, *Joseph Locke: Railway Revolutionary*.
128. *HRJ*, 8 Oct 1845; 5 Apr 1845 (prospectuses).
129. Glasgow Paisley & Greenock Railway, *op.cit.*, 15. Locke was also the engineer for the improvements at Greenock harbour — his principal non-railway work (N. Webster, *op.cit.*, 168-9).
130. *HRJ*, 7 Dec 1850.
131. SRO, BR/HRP(S)/72/7, Glasgow Paisley & Greenock circular, 10 Nov 1840.

132. SRO, GD112/53, William L. Colquhoun to Marquess of Breadalbane, 8 Mar 1844.
133. *HRJ*, 7 Dec 1850.
134. Glasgow Paisley & Greenock Railway, *op.cit.*, 1.
135. *Ibid.*, 4.
136. *HRJ*, 7 Dec 1850.
137. *Ibid.*; Glasgow Paisley & Greenock Railway, *op.cit.*, 7.
138. L. T. C. Rolt, *op.cit.*, 38.
139. Brassey was employed to construct the Glasgow to Paisley joint line (SRO, BR/HRP(S)/51 p.10, cutting from *Greenock Telegraph*, 28 Mar 1898).
140. SRO, GD112/53, P. M. Stewart MP to Marquess of Breadalbane, 23 Mar 1844.
141. T. R. Gourvish, *op.cit.*, 60.
142. PP 1841 (116) XXV, *Reports, Returns, &c, relative to Railways, under the Provisions of the Act 3 & 4 Vict. c.97*, 79.
143. Glasgow Paisley & Greenock Railway, *op.cit.*, 15; SRO, BR/GPG/1/2, Glasgow Paisley & Greenock minutes, 31 Aug 1843. Errington himself suggested a cut to £100 or even £50.
144. SRO, BR/RAC(S)/1/35, J. E. Errington, *Report on the Proposed Railway from Perth by Stirling to Edinburgh and Glasgow*, 1841.
145. PICE, 22 (1862), 628-9; J. Mitchell, *op.cit.*, I, 155.
146. *Blackwood's Mag.*, 41 (June 1837), 733. The problem of over-estimation has already been noted above in the cases of the Kilmarnock & Troon (p.38), the Wishaw & Coltness (p.79), the Arbroath & Forfar (p.124), and the Edinburgh Leith & Newhaven (p.128).
147. *HRJ*, 29 Sept 1849.
148. H. Pollins, *Britain's Railways*, 30.
149. *PICE*, 33 (1871-2), quoted in H. Pollins, 'Railway contractors and the finance of railway development in Britain', *Jnl Transport Hist.*, 3 (1957), 41.
150. J. Handley, *The Navvy in Scotland*, 101.
151. GUEH, 'Protest against Adam Hosie and Willox by Dan. Paul attorney for the Ballochney Railway Company', 11 Apr 1828.
152. More Nisbett Papers, bundle 6, *Report by the Directors of the Slamannan*, 7 Feb 1839.
153. SRO, BR/LIB(S)/6/224, p.107, *Report of Edinburgh & Dalkeith Railway Company*, 27 July 1832.
154. Dundee & Newtyle Railway, *op.cit.*, 4-6.
155. SRO, BR/RAC(S)/1/11, *Report of Glasgow Paisley Kilmarnock & Ayr*, 24 Feb 1840.
156. See above, p.198.
157. PP 1847-8 [938] XXVI, *Report of the Commissioners of Railways*, appx. X, Report of Captain J. Coddington on the Edinburgh, Leith and Granton Railway, 14 April 1847. An alternative view of the line claimed that not only were the estimates 'utterly fallacious' but 'the engineering difficulties of the Leith branch were insurmountable' (*Railway Chronicle*, 20 May 1848).
158. PP 1846 [752] XXXIX, *Report of the Railway Department for 1844-45*, appx. III, p.530.
159. PP 1847 (436) LXIII, *Report by Gen. Pasley to Lord Clarendon, President of the Board of Trade, on the North British Railway*; PP 1847 (248) LXIII, *Report by Capt. Coddington and James Walker to the Railway Commissioners on the State of the North British Railway*. Back in 1841, the Railway Department had refused the Edinburgh & Glasgow board permission to open as soon as they wanted, imposing a two-month delay while some necessary works were completed (PP 1842 [360] XLI, *Second Report of the Officers of the Railway Department to the President of the Board of Trade*, appx. V, p.163).
160. SRO, BR/SNR/1/1, St Andrews Railway Minutes, 28 Aug 1850, 25 July 1851: PBR/1/1, Peebles Railway Minutes, 4 Feb 1853.
161. SRO, BR/NBR/1/1, North British Minutes, 25 Jan 1842,: BR/GDU/1/1, Glasgow Dumfries & Carlisle Minutes, 21 July 1844.
162. *Athenaeum*, 22 Apr 1843, quoted in W. Acworth, *The Railways of Scotland*, 13.
163. PP 1842 [360] XLI, appx. V, p.163.
164. Dundee & Newtyle Railway, *op.cit.*, 7-10.
165. S. G. E. Lythe, 'Early days of the Arbroath & Forfar Railway', *Railway Mag.*, 99 (1953), 57;

PP 1839 (222, 517) X, *Select Committee on Railways*, evidence, QQ.3434-9 (Thomas Grainger); SRO, BR/AFR/1/1, Arbroath & Forfar Minutes, 1 Feb 1836. Grainger also blamed the rising cost of property in Arbroath for part of the increase in the estimate.

166. More Nisbett Papers, bundle 6, *Report by the Directors of the Slamannan*, 7 Feb 1839.
167. F. Whishaw, *op.cit.*, 8, 84, 97, 105, 110, 117, 404; J. M. McBain, *Arbroath: Past and Present*, 199.
168. See, for instance, pp.79-80 above.
169. F. Whishaw, *op.cit.*, 79.
170. HLRO Deposited Plan, HC 1836, Dundee & Arbroath Railway; *RT*, 3 July 1841.
171. PP 1842 [360] XLI, appx. VI, pp. 196, 206, 214; F. Whishaw, *op.cit.*, 118.
172. *RT*, 10 Dec 1842; W. Acworth, *op.cit.*, 10.
173. More Nisbett Papers, bundle 3, *Report of Glasgow Paisley Kilmarnock & Ayr Committee of Inquiry*, 22 Feb 1844.
174. PP 1842 [360] XLI, appx. VI, pp.206, 223.
175. S. G. E. Lythe, 'The Dundee & Newtyle Railway', *Railway Mag.*, 97 (1951), 548; C. H. Ellis, *British Railway History*, 139; E. L. Ahrons states the cost at £700 each (*The British Steam Locomotive, 1825-1925*, 27).
176. C. E. Lee, 'The Dundee & Arbroath Railway', *Railway Mag.*, 104 (1958), 27; E. L. Ahrons, *op.cit.*, 40, 115.
177. F. Whishaw, *op.cit.*, 85.
178. S. G. E. Lythe, 'Arbroath & Forfar' *loc.cit.*, 57.
179. F. Whishaw, *op.cit.*, 117.
180. *Ibid.*, 109-111. A Garnkirk bye-law insisted that 'Passengers shall take and keep seats allotted to them, and shall not stand up in the carriages or on the seats, nor lean upon the doors, or over the sides' (*RT*, 19 May 1841).
181. PP 1842 [360] XLI, appx. VII, p.240.
182. F. Whishaw, *op.cit.*, 86; S. G. E. Lythe, 'Dundee & Newtyle', *loc.cit.*, 548.
183. Stephenson Locomotive Society, *The Glasgow and South-Western Railway*, 13.
184. F. Whishaw, *op.cit.*, 98.
185. *Ibid.*, 118; *RT*, 10 July 1841.
186. PP 1842 [360] XLI, appx. 7, pp.230, 241. According to Highet, the Ayrshire's third-class carriages originally had seats which were removed in 1840 (C. Highet, *The Glasgow and South-Western Railway*, 63).
187. F. Whishaw, *op.cit., passim*.
188. Shiell & Small, Dundee & Newtyle Sederunt Book, 13 Dec 1844.
189. GUEH, 'Interim Report of the Committee of Inquiry to the Directors of the Wishaw and Coltness Railway', 10 July 1844.
190. F. Whishaw, *op.cit.*, 98.
191. W. Albert, *The Turnpike Road System in England, 1663-1840*, 190; W. Vamplew, 'Railways and the Scottish transport system in the 19th century', *Jnl Transport Hist.*, N.S. 1 (1972), 135-6.
192. PP 1837 (456) XX, *S.C. on Internal Communication Taxation, Report*, p.iv.
193. See above, pp.113-5.
194. SRO, GD135 Box 82, Stair Muniments, *Report of Edinburgh & Dalkeith*, 29 Jan 1834: Viscount Melville to Clerk of Edinburgh & Dalkeith, 27 Feb 1833: Melville to Earl of Rosebery, 15 Sept 1833.
195. SRO, GD224/554, Buccleuch Muniments, 'Statement in regard to compensation paid by Railway Companies to Trustees of Turnpike Roads in Scotland, 1845'.
196. Ibid., SRO, BR/EGR/3/1, 'Contracts, Agreements, &c, entered into by the Edinburgh and Glasgow Railway Company', p.41.
197. SRO, BR/NBR/1/2, North British Minutes, 31 July 1844.
198. SRO, BR/NBR/1/1, North British Minutes, Report of delegation appointed to attend the progress of the Bill through Parliament, 4 Apr 1844.
199. SRO, BR/NBR/1/2, 31 July 1844.
200. W. Vamplew, *loc.cit.*, 144n. The fears of one trustee were thus fulfilled: 'if made decidedly

objectionable to the Railway interest of which your Lordship knows the influence in the House of Commons it may by their disagreement be lost altogether' (SRO, GD224/554, John Richardson to Earl of Rosebery, 31 Mar 1845).

201. There is no obvious explanation for the fall in the cumulative amount recorded for these payments.
202. SRO, EGR/3/1, p.41.
203. GUEH, Minute of Meeting of Stirling and Carlisle Road Trustees, 28 Aug 1837.
204. GUEH, John Torrance to James Mitchell, clerk of Wishaw & Coltness, 18 Oct 1842, 19 Nov 1842, 16 Jan 1843, 5 Jan 1844; Mitchell to Torrance, 25 Oct 1842, 31 Nov 1842: Mitchell, Henderson & Mitchell, 'Case for the Wishaw and Coltness Railway Company', 1844. Another dispute between the company and road trustees, which added an unexpected £2,455 to the costs of construction, has been noted already, p.79.
205. S. G. E. Lythe, 'Arbroath & Forfar, *loc.cit.*, 56.
206. PP 1842 [360] XLI, appx. IV.
207. See above, pp.81-2.
208. *RT*, 29 June 1839.
209. T. R. Gourvish, *op.cit.*, 91.
210. GUEH, John Gladstone to James Mitchell, 31 Oct 1842.
211. SRO, GD224/554, John Gibson to Duke of Buccleuch, 5 Feb 1844; and see above, p.124.
212. SRO, BR/GPG/1/2, Glasgow Paisley & Greenock Minutes, 7 Mar 1844.
213. NLS, Acc 3737, Chalmers of Auldbar papers, Minutes of Annual General Meeting of Arbroath & Forfar, 13 June 1842.
214. PP 1847-8 (731) LXIII, *Return of Calls made by Railway Companies*.
215. GUEH, *Interim Report by the Committee of Inquiry to the Directors of the Wishaw and Coltness Railway*, 10 July 1844.
216. SRO, BR/RAC(S)/1/11, *Reports of the Glasgow Paisley Kilmarnock & Ayr*, 28 Aug 1843, 16 Aug 1844, 20 Feb 1845.
217. T. R. Gourvish, *op.cit.*, 91.
218. SRO, BR/RAC(S)/1/73, *Abstracts of the Accounts of Income and Expenditure ... of the Edinburgh and Glasgow Railway, 1835-48*.
219. Company reports in More Nisbett Papers, bundle 3, and in SRO, BR/RAC(S)/1/11.
220. W. Vamplew, 'Banks and railway finance: a note on the Scottish experience, *Transport Hist.*, 4 (1971), 179.
221. SRO, BR/GPG/1/2, 7 Feb 1840.
222. W. Vamplew, 'Banks and railway finance ...', *loc.cit.*, 179.
223. T. R. Gourvish, 'The Bank of Scotland, 1830-45', *Scot. Jnl Pol. Econ.*, 16 (1969), 299; T. R. Gourvish & M. C. Reed, 'The financing of Scottish railways before 1860 — a comment', *Scot. Jnl Pol. Econ.*, 18 (1971), 217.
224. C. A. Malcolm, *The History of the British Linen Bank*, 207-8; T. R. Gourvish & M. C. Reed, *loc.cit.*, 217.
225. N. Munro, *The History of the Royal Bank of Scotland 1727-1927*, 402; Anon., 'Early Scottish railways', *Three Banks Rev.*, 74 (1967), 34.
226. W. Vamplew, 'Banks and railway finance ...', *loc.cit.*, 179; S. G. Checkland, *Scottish Banking: a History, 1695-1973*, 420.
227. T. R. Gourvish & M. C. Reed, *loc.cit.*, 217; and see above, p.60.
228. See above, pp.81-2.
229. Anon., 'Early Scottish railways', *loc.cit.*, 34. These loans included £20,000 to the Scottish North-Western, £15,000 to the Aberdeen Banff & Elgin, £12,000 to the Ayr & Dumfries, and £10,000 each to the Edinburgh & Hamilton Direct and the Kinross Junction.
230. SRO, BR/EGR/1/7, 12 Oct 1841.
231. H. Pollins, 'Railway contractors and the finance of railway development in Britain', *Jnl Transport Hist.*, 3 (1957), 107.
232. SRO, BR/RAC(S)/1/73, 31 Dec 1841, 30 June 1842, 31 Jan 1844; GPG/1/2, 1 July 1842, 5 Sept 1842.

233. PP 1844 (318) XI, *S.C. on Railways, Fifth Report*, pp. xvii-xviii; H. Parris, *op.cit.*, 15.
234. F. S. Williams, *Our Iron Roads*, 238-9. The figures for the Glasgow Paisley & Greenock include only the wages of the policemen.
235. For Scottish companies, the change in the method of levying passenger duty in 1842 helped to reduce this burden (see below, ch.4,sect.X).
236. See Table 19.
237. F. Whishaw, *op.cit.*, 102, 394.
238. More Nisbett Papers, bundle 13, *Edinburgh & Glasgow prospectus*, Oct 1835; ibid., bundle 3, *Glasgow Paisley Kilmarnock & Ayr prospectus*, 1836.
239. PP 1837-38 (257) XVI, *S.C. on Railroad Communication*, evidence, QQ. 926-30 (Locke).
240. More Nisbett Papers, bundle 3, *Report of the Committee of Inquiry into the Expenditure of the Glasgow Paisley Kilmarnock & Ayr Railway Company*, 22 Feb 1844.
241. Glasgow Paisley & Greenock Railway, *op.cit.*, 20-2; G. F. Gold (ed.), *The Glasgow, Paisley, and Greenock Railway*, 44-5 (extract from *Glasgow Herald*, 13 Sept 1850).
242. P. W. Kingsford, *Victorian Railwaymen: the Emergence and Growth of Railway Labour, 1830-1870, passim*.
243. *RT*, 22 Oct 1842. One week later, another anonymous letter claimed that the Edinburgh & Glasgow carriages were 'as good and commodious in every respect as those of any English railway whatsoever . . . and the servants of all classes, are to the full as attentive and civil' (*RT*, 29 Oct 1842).
244. *RT*, 15 Oct 1842.
245. *RT*, 11 Nov 1840, 12 Dec 1840, 22 Oct 1842.
246. SRO, BR/HRP(S)/1/15, p 13, J. Macneill, *Report on the Proposed Plan of Conveying Passengers between Edinburgh and Glasgow*, 14 Dec 1838.
247. P. W. Kingsford, *op.cit.*, 148-9.
248. Shiell & Small, Dundee & Newtyle Sederunt Book, 12 June 1841, 30 June 1841, 18 Mar 1842, 10 June 1843, 24 June 1843.
249. *RT*, 4 Mar 1843.
250. Shiell & Small, Dundee & Newtyle Sederunt Book, 13 Mar 1843.
251. *RT*, 4 Mar 1843.
252. More Nisbett Papers, bundle 3, Glasgow Paisley Kilmarnock & Ayr, *Report of Meeting of Shareholders*, 22 Feb 1844; M. Robbins, 'The North Midland Railway and its enginemen, 1842-3', *Jnl Transport Hist.*, 4 (1960), 180-5. The driver of the train which crashed had been a fireman on the North Midland until August 1841, and returned as a driver in the following December; in the interim he had been sacked from employment as an engine cleaner by both the London & South Western and the Great Western (*RT*, 18 Mar 1843).
253. PP 1843 [440] XLVII, *Third Report of the Railway Department to the President of the Board of Trade*, 1843, appx. IV, 236-7, and report, p.x.
254. More Nisbett Papers, bundle 3, *Report of the Committee of Inquiry* . . .
255. *RT*, 3 Mar 1844.
256. More Nisbett Papers, bundle 3, *Report of the Committee of Inquiry* . . .; M. Robbins, *loc.cit.*, 182.
257. *RT*, 16 Mar 1844, 16 June 1842, 10 July 1841.
258. GUEH, 'Interim Report by the Committee of Inquiry to the Directors of the Wishaw & Coltness Railway', 10 July 1844: draft notes of reply to the Interim Report. The directors claimed in reply that the salaries bill had already been cut by about £230 from 1843 to 1844 (including a reduction of Dodds' commission from 3% to $1\frac{1}{2}$% and the withdrawal of his horse since the line was completed); any further reductions might make it impossible to get good staff (draft notes of reply).
259. *RT*, 1 Sept 1838; More Nisbett Papers, Glasgow Paisley Kilmarnock & Ayr, *Report to Meeting of Shareholders*, 29 Aug 1842.
260. *RT*, 29 Dec 1838, 16 July 1842.
261. *RT*, 27 Aug 1842.
262. See above, pp.85-6, and below, pp.240-1.
263. J. Francis, *op.cit.*, I, 202-3.
264. See below, p.260.

265. PP 1846 (687) XIV, evidence, Q.1116 (Miller).
266. See Table 21.
267. See Tables 33 and 34.
268. PP HL 1837 (146) XVII, evidence, pp. 57-67.
269. *RT*, 26 Feb 1842. In the previous month it had been announced that the fares would be 9/-, 6/6d and 4/- (*RT*, 15 Jan 1842).
270. The Dundee & Newtyle rejected a petition for a reduction in fares, blaming 'the great Expence caused by the Law Incline, the heavy Tax exacted by Government on passengers and the impossibility upon a single Railway of carrying more than a limited number of Passengers with Safety, particularly upon the Law Incline' (Shiell & Small, Dundee & Newtyle Sederunt Book, 15 Jan 1841).
271. *RT*, 8 Jan 1842.
272. See above, p.93.
273. More Nisbett Papers, bundle 13, *Edinburgh & Glasgow prospectus*, Oct 1835.
274. More Nisbett Papers, bundle 3, Glasgow Paisley Kilmarnock & Ayr, *Report to Meeting of Shareholders*, 31 Aug 1841.
275. *NSA*, V (Ayrshire), 683.
276. PP 1839 (517) X, evidence, Q.3454 (Grainger).
277. L. T. C. Rolt, *op.cit.*, 17-19.
278. PP 1842 [360] XLI, p.xii.
279. PP 1844 (318) XI, appx. II.
280. *Ibid.*
281. PP 1844 (166) XI, *S.C. on Railways, Third Report*, 3.
282. PP 1839 (517) X, evidence, Q.4762 (R. Stephenson).
283. *RT*, 22 Jan 1842.
284. PP 1840 (474) XIII, *S.C. on Railway Communication*, evidence, Q.4225 (Huish).
285. *Ibid.*, QQ.4810-2 (Lindsay Carnegie).
286. *Ibid.*, Q.4199 (Huish). See table 66, and below pp.262-3.
287. S. G. E. Lythe, 'Dundee & Newtyle', *loc.cit.*, 548.
288. PP 1845 (614) XXXIX, *Return of Charges made by Railway Companies*.
289. PP 1839 (517) X, evidence, Q.4650 (David Rankine).
290. PP 1842 [360] XLI, p.xii.
291. PP 1840 (437) XIII, *S.C. on Railway Communication, Fourth Report*, 3-4.
292. F. Whishaw, *op.cit.*, 2. From 1844 the Arbroath & Forfar also offered Saturday-to-Monday return tickets at the price of a single ticket (*RT*, 14 Sept 1844).
293. PP 1839 (517) X, evidence, Q.3454 (Grainger).
294. F. Whishaw, *op.cit.*, 2.
295. PP 1840 (474) XIII, evidence, Q.4192 (Huish).
296. *RT*, 31 Aug 1844. At the end of the year the company claimed that 'the Edinburgh & Glasgow Company can state, with the utmost confidence, that in extent of accommodation, safety, and cheap fares, the public cannot expect to be better served than they have all along been by them since the opening of their line' (SRO, GD224/555, 'Case for the Edinburgh and Glasgow Railway against the proposed Caledonian Railway', 22 Dec 1844).
297. *Chambers' Edinburgh Journal*, 21 Sept 1844.
298. *RT*, 15 Oct 1842, 22 Oct 1842, 20 July 1844.
299. *RT*, 30 Oct 1841.
300. PP 1844 (318) XI, evidence, QQ.5398-9 (Wyndham Harding).
301. *Chambers' Edinburgh Journal*, 29 June 1844, 21 Sept 1844. An earlier correspondent to the *Railway Times* had made the same point: 'The most wealthy and influential merchants of Glasgow are daily to be seen crowding into the *third-class* carriages; indeed, it is nothing uncommon to see them drive up to the station in their own carriages, and then step into a "stand-up", as they are called, at 6d. for Greenock' (*RT*, 22 Oct 1842).
302. One English witness emphasised how unpleasant third-class conditions could be. The poor removal officer of Liverpool considered that 'the risk and exposure of the poor people in the open

stand-up third-class carriages, particularly in the winter time, is so severe, that I would sooner pay the difference out of my own pocket than subject the poor in my charge to the danger of that conveyance' (PP 1844 (318) XI, appx. IV).

303. *RT*, 11 Dec 1841. The chairman subsequently claimed that his suggestion was meant as a joke (*RT*, 18 Dec 1841).
304. *Fraser's Mag.*, 29 (May 1844), 513-4.
305. PP 1842 [360] XLI, appx. VII, p.226. In 1840 Lindsay Carnegie had observed that 'cheapness, regularity, frequency, security . . . all have been attended to . . . one remaining want has not been attended to, comfort' (S. G. E. Lythe, 'Arbroath & Forfar Railway, *loc.cit.*, 130).
306. *RT*, 29 Aug 1840.
307. PP 1844 (318) XI, evidence, Q.5361 (Wyndham Harding).
308. *Ibid.*, QQ.5354-6 (Wyndham Harding). At the beginning of the year it was reported that the Greenock company 'have enclosed and glazed their third-class carriages' (*RT*, 20 Jan 1844).
309. *Ibid.*, Q.5360.
310. *RT*, 17 Feb 1844.
311. More Nisbett Papers, bundle 3, Glasgow Paisley Kilmarnock & Ayr, *Report to Meeting of Shareholders*, 21 Feb 1840.
312. *RT*, 14 Jan 1843.
313. PP HL 1837 (146) XVIII, evidence, p.27 (Robert Farquharson). The Kilmarnock & Troon had earlier been credited with increasing the tourist trade of Troon (see above, p.24).
314. *RT*, 18 July 1840.
315. PP 1840 (474) XIII, evidence, QQ.4813 (Lindsay Carnegie), 4864 (George Duncan). The abortive Brechin plan of 1839 hoped to attract tourists in the opposite direction, to 'the bosom of the lower Grampians' (*RT*, 10 Aug 1839).
316. *RT*, 31 July 1841.
317. *RT*, 7 July 1838.
318. *Arbroath Journal*, quoted in *RT*, 12 May 1838.
319. *Aberdeen Banner*, quoted in *RT*, 10 Apr 1841.
320. *Blackwood's Mag.*, 41 (June 1837), 'The world we live in', 735.
321. T. R. Gourvish, *op.cit.*, 69; S. G. E. Lythe, 'Dundee & Newtyle', *loc.cit.*, 548.
322. *Caledonian Mercury*, quoted in *RT*, 30 Apr 1842.
323. See below, p.251.
324. More Nisbett Papers, bundle 3, Glasgow Paisley Kilmarnock & Ayr, *Report to Meeting of Shareholders*, 20 Aug 1842.
325. *Chambers' Edinburgh Journal*, 21 Sept 1844.
326. *Times*, 29 Aug 1840.
327. *RT*, 10 Sept 1842, 8 Oct 1842.
328. *RT*, 22 July 1843.
329. PP 1846 [698] XXXIX, pp. xxvi-xxvii.
330. Act 2 & 3 Will.IV, c.120; S. Dowell, *A History of Taxation and Taxes in England from the Earliest Times to the Present Day*, III, 48.
331. S. Dowell, *op.cit.*, III, 51-4; PP 1837 (456) XX, report, p.iv. The committee's report disliked the taxes: 'your committee earnestly recommend the abolition of all taxes on public conveyances and on carriages generally, at the earliest period consistent with a due regard to the financial arrangements of the country' (ibid., p. v.).
332. PP 1839 (222,517) X, report, p.xiv.
333. *Ibid.*, appx. XXVII.
334. *Ibid.*, evidence, QQ.3454-5 (Grainger).
335. PP 1840 (474) XIII, evidence, QQ.4807, 4845 (Lindsay Carnegie).
336. *Ibid.*, evidence, QQ.4857, 4881-5 (Duncan).
337. *Ibid.*, evidence, QQ.4830-5 (Lindsay Carnegie), 4874-5 (Duncan).
338. *Ibid.*, evidence, QQ.4199, 4212 (Huish).
339. PP 1839 (222,517) X, evidence, QQ.3454-5, 3476-7 (Grainger), 3578 (Henry Wickham): Pender's comments are inserted in the record of the evidence after Q.3578; *RT*, 22 Feb 1840, 18 Apr

1840. In the following year the secretary of the Ardrossan was accused of over-reacting to the tax: 'You have raised your fares 2d. when the whole charge for taxation was three-farthings; is that the whole you have to complain of in respect to taxation? — Yes' (PP 1840 (474) XIII, evidence, Q. 4066 (James Moffat).
340. PP 1839 (222,517) X, evidence, Q.3558 (Henry Wickham).
341. *RT*, 21 Jan 1840, 18 Apr 1840.
342. PP 1840 (437) XIII, 3. The six witnesses were Huish, Lindsay Carnegie, George Duncan, Charles A. King (Garnkirk & Glasgow), James Moffat (Ardrossan), and George Kinloch (Dundee & Newtyle).
343. *Ibid.*, report, 4: appx. XIV. In 1838 the Scottish proportion was 1.5%.
344. *Ibid.*, report, 5-6; *RT*, 9 Apr 1842. In his evidence Charles King had stated that 'an increase to a small extent upon the fares of railways carrying the richer classes of passengers, would not diminish the quantity of travelling upon these railways'; he, Lindsay Carnegie and Wickham all expressed support for the committee's recommendation (ibid., evidence, QQ.4173, 4177 (King), 4838-40 (Lindsay Carnegie), 4717-9 (Wickham)).
345. G. H. Lang, *Letter to the Right Hon. Henry Goulburn, M.P. . . . on the unequal Pressure of the Railway Passenger Tax*, 4.
346. *Ibid.*, 5, 8.
347. *Ibid.*, 10-11.
348. Act 5 & 6 Vict. c.79; *RT*, 14 Oct 1843.
349. *RT*, 27 Jan 1844. It was not quite true that conveyances for the poor had been untaxed: those who could afford the railway could also have afforded the 'caravan' waggons travelling at less than three miles per hour which were taxed from 1822 (S. Dowell, *op.cit.*, III, 52-3).
350. Act 7 & 8 Vict. c.85; H. Parris, *op.cit.*, 57.
351. Glasgow Paisley & Greenock Railway, *op.cit.*, 24.
352. This section is based on C. J. A. Robertson, 'Early Scottish railways and the observance of the sabbath', *Scottish Hist. Rev.*, 57 (1978), which continues the discussion through the mania and into the 1850s.
353. Rev. D. Macfarlan, *Railway Travelling on the Lord's Day Indefensible*, 11.
354. Sir A. Agnew, *The Observance of the Lord's Day*, 9; PP 1847 (167) LXIII, *Regulations of Every Railway Company on the Subject of Travelling on Sunday*.
355. T. McCrie, *Memoirs of Sir Andrew Agnew of Lochnaw, Bart.*, 315-22, 345, 354, 362.
356. *Ibid.*, 207.
357. Act 1 & 2 Vict. c.98.
358. Sir A. Agnew, *op.cit.*, 1-2.
359. *RT*, 24 Oct 1840, 7 Nov 1840.
360. A. Marjoribanks, *Mistaken Views regarding the Observance of the Sabbath*, 6-7; *RT*, 11 Aug 1838.
361. PP 1841 (116) XXV, 90.
362. *Discussion on Sabbath Trains at the Annual General Meeting of the Newcastle and Carlisle Railway Company*, (1847), 8; *RT*, 11 July 1840, 9 Jan 1841.
363. *RT*, 19 June 1841.
364. Anon., *Edinburgh and Glasgow Railway: Sabbath Passenger Trains*, 5.
365. SRO, EGR/1/7, Edinburgh & Glasgow Directors' Minutes, 1841 *passim*; *RT*, 27 Aug 1842, 10 Sept 1842.
366. Anon., *Speeches delivered at the Town Hall of Manchester, January 23rd, 1847 . . .* , 12; Rev. D. Macfarlan, *op.cit.*, 29: *RT*, 19 Dec 1840. The Manchester meeting was called specifically to discuss Sunday trains on the Edinburgh & Glasgow.
367. *RT*, 10 Sept 1842.
368. *RT*, 2 Feb 1842, 10 Sept 1842; J. Thomas, *A Regional History of the Railways of Great Britain: VI, Scotland*, 68.
369. *RT*, 18 Mar 1843. Agnew persuaded the committee not to call for a complete boycott of the railway, realising that sympathisers would not obey it; he himself never used the line while the Sunday trains ran (T. McCrie, *op.cit.*, 271-2).

370. T. McCrie, *op.cit.*, 374, 378.
371. *RT*, 2 Dec 1843.
372. *RT*, 27 Aug 1842.
373. Anon., *Speeches delivered at . . . Manchester*, 39.
374. *RT*, 15 July 1843.
375. See C. J. A. Robertson, *loc.cit.*, 155-9.
376. [G. F. Lewis(?)], 'The financial state of the public roads', *Edinburgh Rev.*, 72 (Jan 1841), 480-1. An earlier contributor to the journal had equally little sympathy for the roads: 'A turnpike trust is, or ought to be, held only for the public good; and if it should be affected by the establishment of a railroad, this would only prove that the one was found more beneficial than the other' ([D. Lardner], 'Improvements in inland transport — railroads', *Edinburgh Rev.*, 60 (Oct 1834), 98).
377. PP 1837 (456) XX, report, p.iv; and see above, p.141.
378. SRO, GD135 Box 82,. Stair Muniments, John Hay, 'Report upon the present State of the Different Turnpike Trusts in the County of Mid-Lothian, submitted to the coach proprietors, farmers, carriers and others, interested in the Toll Duties of that County', 3, 10.
379. Ibid., William MacKenzie W.S. to Sir John H. Dalrymple, 24 Dec 1828.
380. PP 1840 (474) XIII, evidence, QQ.4876-7 (George Duncan), 4825-6 (Lindsay Carnegie). The Arbroath & Forfar Company, on the other hand, welcomed the proposal to turnpike the Forfar-Kirriemuir road, since the existing road was so bad that Kirriemuir traffic went to the railway at Glamis. The company and the Dundee & Arbroath also subscribed £250 each towards the turnpiking of the Friockheim-Brechin road (NLS, Acc 3737, Chalmers of Auldbar MSS, Minutes of Annual General Meeting of Arbroath & Forfar, 18 June 1841).
381. See above, pp.208-10. In 1839 a select committee noted that 'although the lines of Turnpike Road running parallel to steam communication are less frequented, it appears that nearly all Roads or Highways leading to stations, or termini, of steam communication, have increased their traffic, although the aggregate of this increase may not probably be found equal to the amount of decrease on the lines of Road first mentioned' (PP 1839 (295) IX, *S. C. on Turnpike Trusts*, Report, p.iii).
382. J. Lindsay, *The Canals of Scotland*, 182-4; *HRJ*, 8 Oct 1845 (Scottish Grand Junction prospectus).
383. More Nisbett Papers, bundle 13, J. Macneill and N. Robson, *Prospectus of a Canal intended to connect the Town of Stirling with the cities of Edinburgh & Glasgow, and the towns of Paisley, Greenock and Port Glasgow*, 1835, 2.
384. *Ibid.*, 2-4.
385. More Nisbett papers, bundle 13, *Edinburgh & Glasgow prospectus*, Oct 1835.
386. PP HL 1837-8 (185) XX, *H.L.S.C. on the Edinburgh & Glasgow Railway Bill*, evidence, p.27 (Leadbetter).
387. PP 1840 (299) XIII, *S.C. on Railway Communication, Third Report*, 9.
388. PP 1840 (474) XIII, evidence, QQ.4123-4 (King).
389. See above, pp.140-1.
390. PP HL 1837 (144) XVIII, p.315 (William Giffen); More Nisbett Papers, bundle 13, John Macneill and N. Robson, *op.cit.*, 4-5; J. Lindsay, *op.cit.*, 94.
391. *RT*, 11 Sept 1841.
392. More Nisbett Papers, bundle 3, Glasgow Paisley Kilmarnock & Ayr, *Report to Meeting of Shareholders*, 30 Aug 1841.
393. T. R. Gourvish, *op.cit.*, 68.
394. J. Lindsay, *op.cit.*, 96; More Nisbett Papers, bundle 3, Glasgow Paisley Kilmarnock & Ayr, *Report to Meeting of Shareholders*, 28 Aug 1843; *RT*, 2 Sept 1843. Formally, the canal did not abandon all claim to passengers, but agreed to limit its speed to four miles per hour and to pay the railway 2d for each passenger carried, which in effect meant an end to the competition. The understanding on goods traffic was later made more formal, until ended by law in 1850 (Stephenson Locomotive Society, *The Glasgow & South Western Railway*, 7).
395. See Table 65.

396. PP 1846 (275) XIII, *S.C. on Railway and Canal Amalgamations, Second Report*, evidence, QQ.776-7 (B. P. Gregson); J. Lindsay, *op.cit.*, 44, 82-3.
397. J. A. Hassan, 'The supply of coal to Edinburgh, 1790-1850', *Transport Hist.*, 5 (1972), 142.
398. *RT*, 13 Jan 1844; GUEH, 'Brief for the Promoters of the Edinburgh and Glasgow, Scottish Central, and other Amalgamation Bills', 1846.
399. *HRJ*, 15 Aug 1846, 22 Aug 1846, 5 Sept 1846, 26 Sept 1846; SRO, EGR/1/11, Edinburgh & Glasgow Minutes, 2 May 1848, 7 May 1848.
400. J. Lindsay, *op.cit.*, 47, 63, 83-4, 97.
401. *RT*, 16 Apr 1842.
402. PP 1844 (318) XI, evidence, QQ.5366-7 (Wyndham Harding).
403. *RT*, 16 Mar 1844.
404. *RT*, 10 July 1841, 14 Jan 1843.
405. SRO, BR/GPG/1/2, Glasgow Paisley & Greenock Minutes, 18 Dec 1838.
406. T. R. Gourvish, 'The railways and steamboat competition in early Victorian Britain', *Transport Hist.*, 4 (1971), 10-11; *RT*, 23 July 1842. Of 843 scheduled connections between the Railway Steampacket Company and the trains during three months in 1843, only three were missed (*RT*, 30 Dec 1843).
407. *RT*, 23 Sept 1843.
408. SRO, GD115/53, Breadalbane Muniments, Box 1, G. H. Lang to Marquess of Breadalbane, 1 Jan 1842. The Greenock railway was planning a large ferryboat to Dumbarton, to connect with the proposed Leven Valley railway to Loch Lomond (ibid.).
409. PP 1842 [360] XLI, pp.xx-xxi.
410. *RT*, 23 Sept 1843.
411. *RT*, 16 Mar 1844.
412. T. R. Gourvish, 'Railways and steamboat competition', *loc.cit.*, 11-12; *RT*, 14 Jan 1843.
413. *RT*, 23 Feb 1839.
414. More Nisbett Papers, bundle 3, Glasgow Paisley Kilmarnock & Ayr, *Report to the Meeting of Shareholders*, 19 Aug 1840; T. R. Gourvish, 'Railways and steamboat competition', *loc.cit.*, 12.
415. More Nisbett Papers, bundle 3, Glasgow Paisley Kilmarnock & Ayr, *Report to Meeting of Shareholders*, 17 Feb 1841; *RT*, 9 Mar 1844. A letter to the *Railway Times* claimed in December 1843 that Fleetwood harbour was still incomplete, that the *Fire King*, was an unsuitable and expensive converted pleasure yacht, and that J. & G. Burns were much more interested in promoting their service between Glasgow and Liverpool (*RT*, 2 Dec 1843).
416. More Nisbett Papers, bundle 3, Glasgow Paisley Kilmarnock & Ayr, *Report to Meeting of Shareholders*, 30 Aug 1841.
417. Ibid., Glasgow Paisley Kilmarnock & Ayr, *Report to Meeting of Shareholders*, 28 Aug 1843: Circular to Shareholders, 11 Aug 1843; *RT*, 4 Mar 1843. Another beneficiary was the Ardrossan railway: on its behalf Lord Eglinton observed that 'it will prove in the interest of all Clyde vessels, and chiefly those of the first class, to unship and land their cargoes at Ardrossan' (SRO, BR/RAC(S)/1/113, *Report of Ardrossan Railway*, 31 Jan 1844).
418. *RT*, 2 Dec 1843.
419. More Nisbett Papers, bundle 3, Glasgow Paisley Kilmarnock & Ayr, *Report to Meeting of Shareholders*, 22 Sept 1844; *RT*, 18 Nov 1843, 3 Mar 1844; and see below, ch.5.
420. SRO, BR/RAC(S)/1/113, *Report to Ardrossan Railway*, 29 Jan 1845.
421. PP 1844 (318) XI, p.xviii.
422. T. R. Gourvish, 'Railways and steamboat competition', *loc.cit.*, 16.
423. *Ibid.*, 16-17. Chartering vessels was no cheaper; in the early days a combination of the two methods was costing the company £123 per week (*HRJ*, 5 June 1841).
424. *Ibid.*, 13, 17; GUEH, 'Estimate of Weekly Expense for Running a Steamer, about 300 Tons Reg^r, exclusive of Engine Room, & about 200 Horse Power, between Ardrossan & Fleetwood', 1843. The main items of expenditure were wages (£33-17/- for 29 men), depreciation (£27), repairs (£27), coal (£24) and lights (£18).
425. More Nisbett Papers, bundle 3, Glasgow Paisley Kilmarnock & Ayr, *Report to Meeting of Shareholders*, 26 Feb 1846.

Chapter 5. The Battle for the Border

1. Anonymous, quoted in W. Acworth, *The Railways in Scotland*, 4.
2. More Nisbett Papers, bundle 3, Glasgow Paisley Kilmarnock & Ayr, *Reports to Shareholders' Meetings*, 28 Aug 1843, 22 Feb 1844, 20 Feb 1845, 26 Feb 1846.
3. C. H. Grinling, *The History of the Great Northern Railway*, 5; PP 1867 [3844-II] XXXVIII pt. 2, *Royal Commission on Railways*, appx. EK.
4. J. Thomas, *A Regional History of the Railways of Great Britain, VI: Scotland*, 103, 259. The company was authorised by 17 & 18 Vict. c.212, 1854.
5. Quoted in G. Graham, *The Caledonian Railway: Account of its Origins and Completion*, 10.
6. *RT*, 9 June 1838.
7. C. F. Dendy Marshall, *A History of British Railways down to the Year 1830*, 183.
8. C. MacLaren, *Railways compared with Canals and Common Roads, and their Uses and Advantages Explained*, 56-7.
9. D. Stevenson, *Life of Robert Stevenson*, 125-6.
10. *Tyne Mercury*, 25 Jan 1825, quoted in W. W. Tomlinson, *The North Eastern Railway*, 97.
11. W. W. Tomlinson, *op.cit.*, 97-8. The 1825 proposal by a number of Dumfriesshire gentlemen to build a line from Brampton, on the newly planned Newcastle & Carlisle, to Port Annan would also technically have qualified as a trans-border railway (*ibid.* 191).
12. The Border Union, promoted by the North British, authorised by 22 & 23 Vict. c.24, 1859, and opened on the same day as the Border Counties, 1 June 1862 (J. Thomas, *op.cit.*, 100; PP 1867 [3844-II] XXXVIII pt.2, appx. EK).
13. *RT*, 11 May 1839.
14. SRO, GD104/348, Scott of Raeburn Muniments, J. Blackmore, *Report to the Directors of the Newcastle-on-Tyne and Carlisle Railway*, Dec 1838.
15. *RT*, 11 May 1839; SRO, GD104/349, Scott of Raeburn Muniments, 'Resolutions adopted at a Meeting of Landed Proprietors in the County of Mid-Lothian and Others . . .', 9 July 1839.
16. PP 1841 (132) XXV, *Commission on Railway Communication between London, Dublin, Edinburgh, and Glasgow, Fourth Report*, 53; *RT*, 8 Sept 1838.
17. N. Wood & G. Johnson, *Report on a Central Line of Railway into Scotland for the Directors of the Newcastle and Carlisle Railway*.
18. *HRJ*, 16 Mar 1844.
19. Prospectus in *HRJ*, 18 Oct 1845.
20. W. W. Tomlinson, *op.cit.*, 517.
21. C. H. Grinling, *op.cit.*, 6.
22. J. Devey, *Life of Joseph Locke*, 97-124.
23. J. Locke, *Report on the Proposed Line of Railway from Carlisle to Glasgow and Edinburgh by Annandale*, 3; G. Graham, *op.cit.*, 10; W. Acworth, *op.cit.*, 30-1.
24. G. Graham, *op.cit.*, 11-12.
25. J. Butt & J. T. Ward, 'The promotion of the Caledonian Railway Company', *Transport Hist.*, 3 (1970), 164-5; Charles Stewart to J. J. Hope Johnstone, 23 Feb 1836, quoted in G. Graham, *op.cit.*, 12. Stewart was later described in an obituary as 'the most active of the original promoters of the Caledonian system of railways' (*Glasgow Herald*, 28 July 1875).
26. J. Butt & J. T. Ward, *loc.cit.*, 165. Hope Johnstone was appointed convener of the survey committee at a meeting in Dumfries in September 1836 (G. Graham, *op.cit.*, 17-18).
27. NLS, Ms 6354 f.5, John Moss to J. J. Hope Johnstone, 19 April 1837.
28. J. Locke, *op.cit.*, 3-4.
29. *Ibid.*, 6-7; J. Devey, *op.cit.*, 138-9.
30. Anon., *Western Railway between Scotland & England*, 1; J. Locke, *op.cit.*, 6, 9-10.
31. *Glasgow Herald*, 3 Nov 1839.
32. G. Graham, *op.cit.*, 26; J. Butt & J. T. Ward, *loc.cit.*, 166.
33. Aberdeen Univ., OD.MISC.25, O'Dell Papers, 'Notes on early border schemes'; W. W. Tomlinson, *op.cit.*, 292.
34. See above, pp.37-8, 144-5.

35. G. Graham, *op.cit.*, 16-17.
36. *RT*, 7 July 1838.
37. *Fraser's Mag.*, 28, July 1838, p.45.
38. PP 1836 (511) XXI, *S.C. on Standing Orders for Railroad Bills, Report*.
39. H. Parris, *Government and the Railways in Nineteenth Century Britain*, 22-4.
40. Act, 7 & 8 Vict. c.85, 1844. Among other clauses, the act gave the Board of Trade greater control over railway charges, introduced parliamentary trains, and allowed (under complex and unlikely circumstances) for state control of new companies after the lapse of 21 years.
41. *Blackwood's Mag.*, 41, (June 1837), 'The world we live in', p.734.
42. J. Locke, *op.cit.*, 7-8.
43. W. W. Tomlinson, *op.cit.*, 323.
44. *RT*, 9 June 1838.
45. *Scotsman*, 12 June 1838; *RT*, 16 June 1838.
46. PP 1841 (132) XXV, 3.
47. *Ibid.*, 62-4.
48. *Ibid.*, 53-4; NLS, Ms 6354 f.68, James R. Grant to Charles Stewart, 11 Mar 1844; W. W. Tomlinson, *op.cit.*, 447.
49. PP 1841 (132) XXV, 8.
50. PP 1841(8)XXV, *Commission on Railway Communication between London, Dublin, Edinburgh, and Glasgow, Third Report*, 3-6.
51. PP 1841 (132) XXV, 54.
52. SRO, BR/GDU/1/1, Glasgow Dumfries & Carlisle Minute Book, 29 Mar 1844.
53. S. G. Checkland, *The Rise of Industrial Society in England 1815-1885*, 18-20.
54. *HRJ*, 18 Oct 1845.
55. Gladstone Papers, Letter Book 14, 'Fifeshire Railroad', John Gladstone to Sir Charles Gordon, 15 Jan 1844. The reference should have been to Patrick Chalmers of Auldbar, Angus; could Gladstone's mind have slipped back to the time of Thomas Chalmers' ministry in Anstruther?
56. *RT*, 13 Apr 1844.
57. W. F. Spackman, *An Analysis of the Railway Interest of the United Kingdom*, 14.
58. SRO, BR/PROS(S)/1/1, *Prospectus of the Glasgow, Paisley, Kilmarnock, & Ayr Railway Company*.
59. *HRJ*, 23 Mar 1844; SRO, BR/GDU/1/1, 9 Mar 1844, 11 Mar 1844, 13 Mar 1844.
60. SRO, BR/GDU/1/1, 1 Apr 1844, 8 Apr 1844; *HRJ*, 13 Apr 1844, 20 Apr 1844, 27 Apr 1844.
61. *HRJ*, 23 Mar 1844.
62. *Edinburgh Evening Courant*, 14 Aug 1843.
63. SRO, GD206/1/63A/5, Hall of Dunglass Muniments, Smith & Kinnear W. S. to Sir John Hall Bt., 3 Sept 1838; SRO, BR/NBR/1/2, North British Minutes, 2 Aug 1844.
64. SRO, BR/NBR/1/1, North British Minutes, 19 May 1843; C. H. Ellis, *British Railway History*, 110; J. Thomas, *The North British Railway*, I, 18, 22.
65. SRO, BR/NBR/1/2, 2 Aug 1844, 7 Aug 1844.
66. J. Thomas, *North British Railway*, I, 33-7.
67. *RT*, 22 June, 1838.
68. *HRJ*, 16 Mar 1844, 23 Mar 1844. George C. Mounsey, the ex-mayor of Carlisle, was also a director of the Newcastle & Carlisle from its authorisation (J. Butt & J. T. Ward, *loc.cit.*, 167; J. S. MacLean, *The Newcastle & Carlisle Railway 1825-1862*, 111; W. W. Tomlinson, *op.cit.*, 201).
69. NLS, Ms 6354 f.68, James R. Grant to Charles Stewart, 11 Mar 1844.
70. SRO, BR/CAL/1/7, Caledonian Directors' Minutes, 19 Feb 1844 (copy minute of 15 Mar 1841).
71. H. Glynn, *Reference Book of the Incorporated Railway Companies of Scotland*, 7; O. S. Nock, *The Caledonian Railway*, 33; W. J. Gordon, *Our Home Railways*, 168.
72. SRO, BR/CAL/1/7, 19 Feb 1844.
73. *HRJ*, 16 Mar 1844.
74. SRO, GD224/554, Buccleuch Muniments, Duke of Montrose to Duke of Buccleuch, 4 Mar 1844; Duke of Buccleuch to Duke of Montrose, 5 Mar 1844.

75. *HRJ*, 16 Mar 1844; SRO, BR/CAL/1/1, Caledonian Minutes of General Meetings, 19 Nov 1844; J. Butt & J. T. Ward, *loc.cit.*, 167-8.
76. *RT*, 30 Mar 1844, 6 Apr 1844.
77. SRO, BR/CAL/1/1, 20 Nov 1844.
78. SRO, BR/GDU/1/1, 24 Apr 1844.
79. 'N'Oubliez', *The Caledonian Railway and its Originators*, 14 (quote from minute of meeting at Lockerbie).
80. More Nisbett Papers, bundle 3, Glasgow Paisley Kilmarnock & Ayr, *Reports to Shareholders' Meetings*, 27 Feb 1843, 28 Aug 1843. The inquiry found nothing seriously wrong and declared that increased profit would have to come from increased revenue (*ibid.*, 22 Feb 1844).
81. Anon., *Remarks on the Proposed Railway Communication between England and Scotland*, 1-2. The proposed line was referred to in the draft of the pamphlet as the Clydesdale; this was then changed to the Western, presumably to concentrate the readers' attention on comparisons with the east coast line while ignoring the existence of the Nithsdale alternative.
82. J. Butt & J. T. Ward, *loc.cit.*, 167.
83. *Chambers' Edinburgh Journal*, 14 Oct 1843, pp.311-2.
84. Anon., *Remarks on the Proposed Railway . . .*, 3.
85. NLS, Ms 6354 f.15, Henry Dunlop to J. J. Hope Johnstone, 30 Mar 1839.
86. G. Graham, *op.cit.*, 33.
87. NLS, Ms 6354 f.44, Mark Huish to Col. William Graham, 3 Mar 1843.
88. G. Graham, *op.cit.*, 53.
89. *RT*, 1 Jan 1842.
90. SRO, BR/CAL/1/7, 19 Feb 1844 (copy minute of 15 Mar 1841).
91. Ibid., 1 Mar 1844, 9 Mar 1844. In 1843 Graham had become a director of the Lancaster & Carlisle ('N'Oubliez', *op.cit.*, 7).
92. G. Graham, *op.cit.*, 49-50; J. Butt & J. T. Ward, *loc.cit.*, 167.
93. SRO, BR/NBR/1/1, 19 Apr 1843; *RT*, 11 Mar 1843, 2 Sept 1843.
94. SRO, BR/CAL/1/1, 30 Mar 1844; *RT*, 3 Aug 1844.
95. SRO, GD104/348, J. Blackmore, *Report* . . .
96. SRO, GD206/1/63A/5, Tod & Romanes, W. S. to Sir John Hall Bt., 24 Nov 1838.
97. The original survey had been made by Thomas Grainger. Both Grainger and Stephenson preferred the inland route by Haddington to the alternative by the coast, mainly because of its greater potential traffic (PP 1841 (132-II) XXV, appx. p.30).
98. PP 1841 (132-II) XXV, appx. pp.44-77; BR/NBR/1/1, 21 Jan 1842.
99. *Ibid.*, appx. p.75.
100. SRO, BR/NBR/1/1, 21 Jan 1842, 9 Feb 1842.
101. *Tait's Edinburgh Mag.*, Feb 1842, 138-9.
102. SRO, BR/NBR/1/1, 17 Jan 1842, 18 Jan 1842, 9 Feb 1842, 14 Feb 1842, 9 July 1845.
103. Ibid., 16 Feb 1842.
104. Ibid., 15 June 1842, 9 Oct 1842.
105. Ibid., 19 May 1843; W. W. Tomlinson, *op.cit.*, 453; C. H. Ellis, *North British Railway*, 4.
106. SRO, BR/NBR/1/1, 7 July 1843; G. Dow, *The First Railway across the Border*, 6.
107. SRO, BR/CAL/1/7, 6 Mar 1844.
108. NLS, Ms 6354 f.66, Mark Huish to Hope & Oliphant, 6 Mar 1844.
109. *RT*, 3 Aug 1844.
110. SRO, GD206/1/63A/5, 'Case of Sir John Hall Bt against the North British Railway Bill' 1844: Alexander Thomson W. S. to John Spottiswoode, 20 Mar 1844: 'Agreement between Sir John Hall Bt. and the North British Railway Company', 29 Mar 1844.
111. SRO, BR/NBR/1/1, 4 Apr 1844, 26 Nov 1844.
112. PP 1859 s.1 (226) XXIII, *Return of Parliamentary Expenditure by Railway Companies to the end of 1857*; SRO, BR/NBR/1/1, 22 June 1844.
113. SRO, BR/NBR/1/1, 22 June 1844.
114. SRO, GD45/7/59, Dalhousie Muniments, G. R. Porter, 'Memorandum on Scottish Railways', 13 Feb 1845.

115. Ibid.
116. *RT*, 26 Oct 1844.
117. SRO, GD224/555, Buccleuch Muniments, *Report by the Provisional Committee of the Glasgow, Dumfries and Carlisle Railway Company*, Nov 1844.
118. SRO, BR/GDU/1/1, 29 Mar 1844.
119. SRO, BR/NBR/1/2, 4 Sept 1844, 18 Sept 1844, 11 Oct 1844.
120. Ibid., 19 Oct 1944, 17 Dec 1844, 19 Feb 1845, 21 Feb 1845. If the joint opposition to the Caledonian succeeded, the Glasgow Dumfries & Carlisle was to pay three-quarters and the Edinburgh & Glasgow one-quarter of the costs. In the event of failure, these two companies would pay one-third each, and the North British and Glasgow Paisley Kilmarnock & Ayr one sixth each (ibid., 21 Feb 1845).
121. SRO, GD45/7/59, [Earl of Dalhousie], draft memorandum, 'Railways in the South of Scotland', 17 Feb 1845.
122. SRO, BR/NBR/1/2, 8 Jan 1845, 6 Mar 1845.
123. Ibid., 17 Dec 1844, 23 Apr 1845; PP 1845 (317) XL, and PP 1845 (625) XL, *Alphabetical Lists of ... all Persons subscribing to any Railway Subscription Contract deposited ... during the Present Session of Parliament*.
124. SRO, GD224/555, James Hope to Duke of Buccleuch, 4 Jan 1845.
125. Ibid., 'Case for the Edinburgh and Glasgow Railway Company against the proposed Caledonian Railway', 12 Dec 1844.
126. Ibid., 'Statement by the Promoters of the Glasgow, Dumfries and Carlisle Railway to the Board of Trade', 1 Jan 1845; *John Leadbetter to Earl of Dalhousie*, 27 Jan 1845 (printed open letter). The Glasgow Junction proposed to link the Ayrshire to the Edinburgh & Glasgow by bridging the Clyde in Glasgow.
127. *RT*, 25 Jan 1845; SRO, BR/CAL/1/7, 30 Oct 1844, 5 Nov 1844; NLS, Ms 6355 ff.27-9, Heads of Agreement between Caledonian Railway and Edinburgh Water Company, Oct 1844.
128. *RT*, 25 Jan 1845.
129. SRO, BR/CAL/1/7, 30 Oct 1844, 5 Nov 1844.
130. Directors of Glasgow Merchants' House to Directors of Glasgow Paisley Kilmarnock & Ayr, 31 Jan 1845, quoted in J. M. Reid, *A History of the Merchants' House in Glasgow*, 49. This decision was taken even though John Leadbetter, as Dean of Guild, was the chairman of the Merchants' House directors (G. MacGregor, *The History of Glasgow from the Earliest Period to the Present Time*, 522).
131. *RT*, 30 Mar 1844.
132. PP 1845 (120) XXXIX, *Report of the Railway Board on the Schemes for Extending Railway Communication in Scotland*, 3-10.
133. H. Parris, *op.cit.*, 61.
134. SRO, GD45/7/59, [Earl of Dalhousie], draft memorandum, 'Railways in the South of Scotland', 17 Feb 1845.
135. SRO, GD224/554, Earl of Morton to Duke of Buccleuch, 8 Jan 1845.
136. PP 1845 (120) XXXIX, 11; SRO, GD45/7/59, [Earl of Dalhousie], draft memorandum, 'Railways in the South of Scotland', 17 Feb 1845.
137. *RT*, 15 Mar 1845.
138. SRO, BR/CAL/1/7, 8 Mar 1844, 14 Mar 1844.
139. J. Butt & J. T. Ward, *loc.cit.*, 167.
140. SRO, BR/CAL/1/7, 24 Jan 1845.
141. GUEH, J. H. Humfrey (secretary of Caledonian) to James Mitchell (secretary of Monkland & Kirkintilloch), 21 Mar 1844; John Marr (Caledonian provisional committeeman) to James Mitchell (clerk of Wishaw & Coltness), 20 Mar 1844; *HRJ*, 23 Mar 1844. Quotation from Marr's letter. Note Mitchell's role, as secretary or clerk of the Monkland & Kirkintilloch, the Ballochney, the Wishaw & Coltness and the Slamannan, in coordinating the views of the Lanarkshire coal railways.
142. SRO, BR/CAL/1/7, 5 Aug 1844.
143. Ibid., 20 Aug 1844; GUEH, report by George Lish to Monkland & Kirkintilloch Committee of Management 1844: Archibald Grahame to James Mitchell, 31 Aug 1844; James Mitchell to

Archibald Grahame, 3 Sept 1844: Archibald Grahame to Alexander O. Mitchell, 9 Oct 1844: Memorandum of Heads of Agreement between Caledonian and Monkland & Kirkintilloch, 1844: Memorandum of Heads of Agreement between Caledonian and Glasgow Garnkirk & Coatbridge, 16 Aug 1844; More Nisbett Papers, bundle 4, David Rankine (Caledonian secretary) to James Mitchell, 5 Oct 1844. Rankine succeeded Humfrey as Caledonian secretary during 1844.

144. SRO, BR/CAL/1/7, 25 Jan 1845, 14 Apr 1845.
145. Ibid., 28 Feb 1845, 8 Mar 1845, 14 Apr 1845, and *passim* Feb to July 1845.
146. PP 1845 (120) XXXIX p.3. The authorising acts for the 'Caledonian system' railways were 8 & 9 Vict. c.153 (Aberdeen), c.160 (Clydesdale Junction), c.161 (Scottish Central), c.162 (Caledonian), c.170 (Scottish Midland Junction) — all in 1845 — and 9 & 10 Vict. c.81, 1846 (Caledonian & Dunbartonshire Junction).
147. SRO, BR/CAL/1/7, 17 Mar 1845, 27 Aug 1845. The plan may well have weakened Hudson's support for the Glasgow Dumfries & Carlisle, since it offered him a direct link, via Kelso, to Glasgow and west of Scotland trade.
148. Ibid., 27 Aug 1845; J. Butt & J. T. Ward, *loc.cit.*, 178.
149. SRO, BR/CAL/1/7, 23 Jan 1845, 7 May 1845, 3 July 1845.
150. Ibid., 5 Feb 1845.
151. SRO, GD224/554, John Gibson to Duke of Buccleuch, 5 Feb 1845.
152. Ibid., D. Bannatyne to Duke of Buccleuch, 9 Nov 1844; Duke of Buccleuch to Hon. J. C. Talbot, 21 Feb 1845. Buccleuch still opposed the Caledonian's Edinburgh branch, since one of his principal concerns was that any direct connection from Edinburgh to Carlisle should go by Hawick (ibid., Duke of Buccleuch to John Gibson, 15 Apr 1845).
153. Ibid., John Gibson to Duke of Buccleuch, 27 Feb 1845, 22 Apr 1845.
154. Ibid., John Leadbetter to Duke of Buccleuch, 26 Feb 1845; Buccleuch to Leadbetter, 27 Feb 1845.
155. ibid., John Gibson to Duke of Buccleuch, 2 Jan 1845; Buccleuch to Gibson, 15 Apr 1845. The two landowners kept in close touch: according to the duke, 'I am bound to support & act with Sir Jas. Graham in this matter' (ibid., Buccleuch to Gibson, 15 Apr 1845).
156. Respectively Lord Privy Seal and Home Secretary.
157. SRO, GD224/554, John Gibson to Duke of Buccleuch, 13 Mar 1845; SRO, BR/CAL/1/7, 15 May 1845.
158. Ibid., John Gibson to Duke of Buccleuch, 7 Apr 1845.
159. SRO, BR/CAL/1/7, 27 May 1845.
160. Ibid., 28 May 1845, 29 May 1845, 4 June 1845; O. S. Nock, *op.cit.*, 17.
161. SRO, BR/CAL/1/1, 27 Aug 1845.
162. Acts 8 & 9 Vict. c.163 (Newcastle & Berwick) and c.164 (Edinburgh & Hawick).
163. Act 22 & 23 Vict. c.24, 1859.
164. Act 9 & 10 Vict. c.372, 1846.

Chapter 6. Summary and Conclusions

Note: Frequently in this chapter reference is made to points more fully discussed in earlier chapters. In these instances the footnote references have not been repeated.

1. W. Vamplew, 'Railways and the transformation of the Scottish economy', unpublished Ph.D. thesis, University of Edinburgh (1969), 15.
2. See S. Smiles, *Lives of the Engineers*, vol.III.
3. A. Slaven, *The Development of the West of Scotland 1750–1960*, 43.
4. E.g. G. Dott, *Early Scottish Colliery Waggonways*, 3; B. Baxter, *Stone Blocks and Iron Rails*.
5. More Nisbett Papers, bundle 13, *Edinburgh & Glasgow prospectus*. Oct 1835.
6. The Wishaw & Coltness was leased to the Caledonian in 1846; the Wilsontown Morningside & Coltness, after prolonged negotiation, amalgamated with the Edinburgh & Glasgow in 1849.
7. Of the long-distance waggonway projects, the Midlothian, the Dumfries & Sanquhar and the West Lothian all expected that at least two-thirds of their revenue would come from coal. The three projects intended to serve the eastern Borders all anticipated that over 70% of their traffic would be either coal or lime (see Table 9).

8. See Table 21, page 90.
9. The Edinburgh to Glasgow schemes designed by Grainger and Jardine in the early 1830s had both given priority to coal traffic, and selected routes through the middle of the coalfield.
10. *RT*, 18 July 1840 (statement by Greenock chairman R. D. Ker at the opening of the joint line).
11. *RT*, 10 Apr 1841.
12. B. R. Mitchell, 'The coming of the railways and United Kingdom economic growth', *Jnl Econ. Hist.*, 24 (1969), 323, 316.
13. T. R. Gourvish, 'Railway enterprise', in R. Church (ed.), *The Dynamics of Victorian Business*, 129
14. H. Pollins, *Britain's Railways: an Industrial History*, 60. Even where through goods traffic between companies was physically possible, difficulties arose over the regulation of inter-company accounts and the movement of rolling stock. A start to solving these problems was made in England with the creation of the Railway Clearing House in 1842; no analogous institution existed in Scotland, and naturally no Scottish company joined the clearing house until the trans-border links were opened (P. S. Bagwell, *The Railway Clearing House in the British Economy, 1842-1922*, ch.2 and p.293).
15. *RT*, 24 Aug 1844.
16. P. Deane and W. A. Cole, *British Economic Growth, 1688-1959*, 230-1.
17. PP 1847 (278) LXIII, *Amounts expended in the Cost of Construction and of Working Stock of all Railways* . . . The Scottish proportions were 8.3% of capital authorised and 8.4% of expenditure.
18. W. Vamplew, *op.cit.*, 224-5; W. Vamplew, 'The railways and the iron industry: a study of their relationship in Scotland', in M. C. Reed (ed.), *Railways in the Victorian Economy*, 45.
19. W. Vamplew, 'Railways and the transformation of the Scottish economy', *Econ. Hist. Rev.* II 24 (1971), 40, 42.
20. See Tables 62–65. For the Ayrshire and Greenock companies, the figure of over 70% includes their shares of the joint line traffic.
21. *RT*, 31 Aug 1844.
22. J. Thomas, *A Regional History of the Railways of Great Britain: VI, Scotland*, 200–1.
23. R. H. Campbell, *Scotland since 1707*, 96.
24. See Appendix.
25. HLRO, HC Deposited Plans of the various railways. The allowances per mile for land included on the Newtyle & Glammis £142, Dundee & Newtyle £190, Monkland & Kirkintilloch £259, Slamannan £282, Garnkirk & Glasgow £562, Paisley & Renfrew £926 and Polloc & Govan £3571.
26. See Tables 44 and 45.
27. See Tables 36 to 39.
28. *Chambers' Edinburgh Journal*, 30 June 1838, 20 Oct 1838.
29. H. Parris, *Government and the Railways in Nineteenth-Century Britain*, 28–31, 50–57.
30. See Table 8.
31. HLRO, HC Deposited Plans of the various railways.
32. HLRO, Deposited Plans, HC 1835 Paisley & Renfrew, HC 1835 Slamannan.
33. See Tables 7 and 25.
34. See, for example, the case of the Wishaw & Coltness, detailed in Table 17.
35. HLRO Deposited Plans, HC 1826 Garnkirk & Glasgow, HC 1826 Dundee & Newtyle.
36. See Tables 34 and 35.
37. *Circular to Bankers*, 19 Aug 1836, quoted in S. A. Broadbridge, 'The sources of railway share capital', in M. C. Reed (ed.), *op.cit.*, 210.
38. HLRO Deposited Plans, HC 1836 Dundee & Arbroath Railway; HC 1836 Arbroath & Forfar Railway; PP 1845 (317) XL and 1845 (625) XL, *Alphabetical List of . . . all Persons subscribing . . . to any Railway Subscription Contract deposited . . . during the Present Session of Parliament.*
39. For a much more complete discussion of the points touched upon in this paragraph, see R. C. Michie, *Money, Mania and Markets*, chs. 3–6.
40. H. Pollins, *op. cit.*, 55.
41. Perhaps the last serious horse versus locomotive debate on a main line took place on the

Newcastle & Carlisle, where a decision in favour of steam had still not been taken in 1834, five years after construction started (H. J. Dyos & D. H. Aldcroft, *British Transport*, 116).

42. SRO, BR/GPK/1/1, Glasgow Paisley Kilmarnock & Ayr minutes; BR/EGR/1/7 and 1/8, Edinburgh & Glasgow minutes.

43. T. R. Gourvish, *Mark Huish and the London & North Western Railway*, 51.

44. T. R. Gourvish, 'Railway enterprise', *loc. cit.*, 138.

45. G. L. Turnbull, *Traffic and Transport: an Economic History of Pickfords*, chs.6-7.

46. H. Pollins, *op.cit.*, 67; P. S. Bagwell, *The Railwaymen*, 20.

47. SRO, BR/CAL/4/5, 'List of Directors and Officials of the Caledonian Railway Company from 1844'; H. Parris, *op.cit.*, 232; PP 1847 (248) LXIII, *Report by Captain Coddington R.E. and James Walker to the Railway Commissioners on the State of the North British Railway*.

48. H. G. Lewin, *Early British Railways*, 115-6.

49. T. R. Gourvish, 'Railway enterprise', *loc.cit.*, 138.

50. The North British lines included the Border Counties (authorised 1854, opened 1862), the Wansbeck (1859-65), the Northumberland Central (1863-70, never completed), the Port Carlisle (1853-54) and the Carlisle & Silloth Bay (1854-56) (J. Thomas, *The North British Railway*, I, 97, 103-4, 196-7). According to Simmons, 'the railways might almost be said to have erected a new version of Hadrian's Wall between England and Scotland' (J. Simmons, *The Railway in England and Wales, 1830-1914*, I, 16).

51. SRO, BR/SCC/1/1, Scottish Central minutes, *passim*.

52. *RT*, 29 Dec 1838.

53. *RT*, 8 Aug 1840, 10 July 1841, 22 Jan 1842.

54. P. Mathias, *The First Industrial Nation*, 278-9.

55. J. Adamson, *An Account of the Stockton and Darlington Railway* . . .; T. Grainger and J. Miller, *Report relative to the Proposed railway to connect the Clydesdale, or Upper Coal Field of Lanarkshire, with the City of Glasgow* . . .; R. Stevenson (ed.), 'Essays on Rail-roads', *Trans. Highland Soc. of Scotland*, 6 (1824).

56. H. Pollins, *op.cit.*, 57.

57. SRO, GD/46/13/59, Seaforth Muniments, Edinburgh Glasgow & Leith Prospectus.

58. Jessop was also called in by Telford to give a second opinion on the Glasgow to Berwick project (T. Telford, *Report relative to the Proposed Railway from Glasgow to Berwick-upon-Tweed*, appx.).

59. J. Simmons, *The Railways of Britain*, 222.

60. *RT*, 25 Aug 1838.

61. W. Vamplew, *op.cit.*, 82.

62. T. R. Gourvish and M. C. Reed, 'The financing of Scottish railways before 1860 — a comment', *Scot. Jnl of Pol. Econ.*, 18 (1971), 213.

63. See Appendix.

64. *RT*, 25 Aug 1838, 16 Mar 1844.

65. If the company subscriptions of £200,000 from the Ayrshire and £100,000 from the Edinburgh & Glasgow are discounted, 32.2% of the remaining money was English. (PP 1845 (317) XL and 1845 (625) XL; HLRO Deposited Plans, HC 1844, North British Railway).

66. For the Edinburgh & Glasgow, see *HRJ*, 15 Aug 1846, 22 Aug 1846, 5 Sept 1846, 26 Sept 1846, and H. G. Lewin, *The Railway Mania and its Aftermath*, 272-4. The troubles of the other companies are fully documented in the railway press and in their minute books in the SRO.

67. See above, p.155.

68. H. J. Dyos and D. H. Aldcroft, *op.cit.*, 126.

69. L. H. Jenks, *The Migration of British Capital to 1875*, 128; J. Francis, *History of the Bank of England*, quoted in D. M. Evans, *The Commercial Crisis, 1847-1848*, 2.

70. L. H. Jenks, 101-15; J. Francis, *A History of the English Railway, 1820-1845*, I, 134.

71. A. Slaven, *op.cit.*, 120.

72. *Morning Chronicle*, 22 Jan 1844, 12 Oct 1844, quoted in B. C. Hunt, *The Development of the Business Corporation in England, 1800-1867*, 103-4.

73. J. Francis, *English Railway*, 137.

74. A. D. Gayer, W. W. Rostow & A. J. Schwartz, *The Growth and Fluctuation of the British Economy, 1790-1850*, I, 368, 375.

Bibliography

A. Manuscript Sources
1. Scottish Record Office, Edinburgh.
   a. British Railways (Scotland) records.
   Surviving minute books, accounts, letter books, and other records of Scottish companies, particularly under references:

BR/AFR	Arbroath & Forfar Railway
BR/CAL	Caledonian Railway
BR/DAR	Dundee & Arbroath Railway
BR/EGR	Edinburgh & Glasgow Railway
BR/ENO	Edinburgh & Northern Railway
BR/FCN	Forth & Clyde Navigation
BR/GDU	Glasgow Dumfries & Carlisle Railway
BR/GPG	Glasgow Paisley & Greenock Railway
BR/GPK	Glasgow Paisley Kilmarnock & Ayr Railway
BR/GSW	Glasgow & South-Western Railway
BR/HRP(S)	Historical Records and Papers
BR/LIB(S)	Library
BR/NBR	North British Railway
BR/PYB(S)	Parliamentary Bills
BR/RAC(S)	Reports and Accounts
BR/WIC	Wishaw & Coltness Railway

   b. Family papers.

GD1/532, GD1/579	A. G. Dunbar papers
GD16	Airlie muniments
GD18	Clerk of Penicuik muniments
GD26	Leven & Melville muniments
GD30	Shairp of Houston muniments
GD45	Dalhousie muniments
GD46	Seaforth muniments
GD104	Scott of Raeburn muniments
GD112	Breadalbane muniments
GD135	Stair muniments
GD206	Hall of Dunglass muniments
GD224	Buccleuch muniments

2. National Library of Scotland, Edinburgh.

MS 642	Melville papers
MS 3421	Dalrymple letters

MS 6354-5   Caledonian Railway papers
MS 9814   Eglinton papers
MS 9944   Menzies of Castle Menzies papers
ACC 3737   Chalmers of Auldbar papers
ACC 5111   Rennie papers

3. Department of Economic History, Adam Smith Building, University of Glasgow.
   A substantial collection, now catalogued UGD/8 but uncatalogued at the time of use, of records and correspondence relating primarily to the Monkland & Kirkintilloch, Ballochney, Slamannan, Garnkirk & Glasgow and Wishaw & Coltness railways. There is also some material from the Ardrossan, Caledonian, Edinburgh & Glasgow and Wilsontown Morningside & Coltness railways, and the Forth & Clyde, Monkland and Union canals.
4. Messrs. Shiell & Small, Solicitors, Dundee.
   Dundee & Perth minute book, 1835-38
   Dundee & Newtyle sederunt book, 1841-46
5. University Library, University of Aberdeen.
   OD.MISC.25 Notes by Professor A. C. O'Dell on early trans-border projects.
6. House of Lords Record Office, London.
   HC Deposited Plans of the various railways, including subscription contracts.
   Evidence, 1837 vol. 12, Glasgow Paisley & Greenock Railway.
7. Hawarden Castle, Flintshire.
   Sir John Gladstone papers (material kindly shown to me by Professor and Mrs S. G. Checkland)
8. G. A. More Nisbett, Esq., Edinburgh.
   Deed Box marked 'G. M. Nisbett Esqr of Cairnhill', containing reports and papers on:
   Bundle 1   Ardrossan Railway
   Bundle 3   Glasgow Paisley Kilmarnock & Ayr Railway
   Bundle 4   Monkland & Kirkintilloch Railway
   Bundle 5   Paisley & Renfrew Railway
   Bundle 6   Slamannan Railway
   Bundle 7   Wishaw & Coltness Railway
   Bundle 13   Miscellaneous Prospectuses and Reports

B. Parliamentary Printed Records
   1. Public General Acts.
   2. Local and Personal Acts — individual acts for the various railways.
   3. Parliamentary Reports and Accounts:
       1836(511)XXI   *S.C. on Standing Orders for Railroad Bills, Report*
       1837(95)XLVIII   *Subscription Contracts Deposited 1837*
       1837(226)XVIII pt.1
       1837(428)XVIII pt.1
       1837(429)XVIII pt.1
       1837(519)XVIII pt.2   *S.C. on Railway Subscription Contracts, First to Sixth Reports*
       1837(520)XVIII pt.2
       1837(537)XVIII pt.2
       1837(259)LI   *Correspondence between the Trustees for Manufactures in Scotland, the Commissioners of the Treasury, and the Promoters of the Edinburgh and Glasgow Railway, on the Subject of the Tunnel projected through the Mound in Edinburgh. . . .*
       1837(456)XX   *S.C. on Internal Communication Taxation, Report*
       HL 1837(146)XVIII   *Evidence to the Committee on the Glasgow, Paisley, Kilmarnock and Ayr Railway Bill*

HL1837-38(185)XX	Evidence to the Committee on the Edinburgh and Glasgow Railway Bill
1837-38(257)XVI	S.C. on Railroad Communication, Report
1839(222)X 1839(517)X	S.C. on Railways, First and Second Reports
1839(295)IX	S.C. on Turnpike Trusts, Report
1840(92)XIII 1840(299)XIII 1840(437)XIII 1840(474)XIII	S.C. on Railway Communication, Second to Fifth Reports
1841(116)XXV	Reports, Returns, &c, relative to Railways, under the Provisions of the Act 3 & 4 Vict. c.97.
1841(132)XXV	Commission on Railway Communication between London, Dublin, Edinburgh and Glasgow, Fourth Report
1841[287]XXV	Report of the Officers of the Railway Department to the President of the Board of Trade
1841(354)VIII	S.C. on Railways, Report
1842[360]XLI	Second Report of the Officers of the Railway Department to the President of the Board of Trade
1843(151)XXX	Passenger Duty paid by Railway Companies
1843[440]XLVII	Third Report of the Railway Drtment to the President of the Board of Trade
1843(571)XLIV	Return of Railway Bills and Acts, 1839-1842/3
1844(37)XI 1844(79)XI 1844(166)XI 1844(283)XI 1844(318)XI	S.C. on Railways, First to Fifth Reports
1844(159)XLI	Return of Moneys raised under Railway Acts, 1st January 1826 to 1st January 1844
1844[551]XLI	Report of the Officers of the Railway Department ... for 1843
1845(120)XXXIX	Report of the Railway Department on Schemes for Extending Railway Communication in Scotland
1845(135)X	S.C. on Railway Bills, Second Report
1845(420)X	HL S.C. on the Practicability and Expediency of establishing some Principle of Compensation to be made to the Owners of Real Property whose Lands, &c, may be Compulsorily Taken for the Construction of Public Railways ...
1845(317)XL 1845(625)XL	Alphabetical List of ... all Persons subscribing ... to any Railway Subscription Contract deposited ... during the Present Session of Parliament
1845(614)XXXIX	Return of Railway Company Charges
1845(637)XXXIX	Return of Bills in the Present Session, with Estimates, Capital, &c.
1846(489)XIII	HL S.C. on Railways, Report
1846(530)XIII	S.C. on Railway Labourers, Report
1846(687)XIV	S.C. on Railway Acts Enactments, 'Second Report' (report drawn up by committee chairman James Morrison, and accidentally published without the agreement of the committee)

Bibliography 399

1846[698]XXXIX } 1846[752]XXXIX }	Report of the Railway Department for 1844-45 and appendices
1847(248)LXIII	Report by Captain Coddington R.E. and James Walker to the Railway Commissioners on the State of the North British Railway
1847(278)LXIII	Amounts expended in the Cost of Construction and of Working Stock of all Railways . . .
1847(436)LXIII	Report by General Pasley to the President of the Board of Trade on the North British Railway
1847-48[938]XXVI	Report of the Commissioners of Railways
1867 [3844] XXXVIII	Royal Commission on Railways, Report

C. Newspapers and Journals
   Blackwood's Magazine
   Bradshaw's Railway Almanack, Directory, Shareholders' Guide and Manual
   Caledonian Mercury
   Chambers' Edinburgh Journal
   Edinburgh Evening Courant
   Edinburgh Review
   Fraser's Magazine
   Glasgow Courier
   Glasgow Herald
   Herapath's Railway Journal
   Oliver & Boyd's New Edinburgh Almanack (1845)
   Proceedings of the Institution of Civil Engineers
   Quarterly Review
   Railway Times
   Scotsman
   Scottish Railway Gazette
   Scottish Railway Shareholders' Manual (1848)
   Times

D. Contemporary Books, Articles and Pamphlets (published before 1860)
Adamson, James, An Account of the Stockton and Darlington Rail-Way; and Extracts from the Report of the Committee to the Proprietors, at their Annual Meeting, held at Yarm, on Tuesday, the 10th day of July, 1827 (n.d., n.p.)
— Sketches of our Information as to Rail-Roads (n.d. [1825], n.p.)
Agnew, Sir Andrew, The Observance of the Lord's Day. A Respectful Appeal to the most reverend the Archbishop and Bishops of the United Free Church of England and Ireland, as instructed by a Public Meeting in Edinburgh held on the 2nd of March, 1841 (Carlisle, 1841)
Aiton, William, General View of the Agriculture of the County of Ayr (Glasgow, 1811)
Anon., Calculations of the Probable Benefit to the Neighbouring Country, and to the Proprietors, of an Iron Rail Way from Berwick to Kelso (Kelso, 1809)
Anon., Discussion on Sabbath Trains at the Annual General Meeting of the Newcastle and Carlisle Railway Company (Newcastle, n.d. [1847])
Anon., The Edinburgh and Glasgow Railway, and the Post-Office (n.d., n.p.)
Anon., Edinburgh and Glasgow Railway: Sabbath Passenger Trains (n.p., 1846)
Anon., Prospectus of a Railway between Edinburgh and the Port of Leith As projected by Mr John Strachan (Edinburgh, n.d.)
Anon., Remarks on the Comparative Merits of Cast Iron and Malleable Iron Rail-Ways; and an Account of the Stockton and Darlington Rail-Way, &c, &c. (Newcastle, 1827)
Anon., Remarks on the Forth and Tay Railway (Edinburgh, 1840)
Anon., Remarks on the Proposed Railway Communication between England and Scotland (Glasgow, 1841)

Anon., *Remarks on the Reports of the Committee to the Subscribers for a Survey and Plan of a Railway from Edinburgh and Leith to Glasgow, and on the report of James Jardine, Esq., Civil Engineer* (Edinburgh & Glasgow, 1830)

Anon., *Running of Railway Trains on Sabbath. Report of the Proceedings of a Great Public Meeting of the Inhabitants of Edinburgh* (Edinburgh, n.d. [1846])

Anon., *Speeches delivered at the Town Hall of Manchester, January 23rd, 1847, at a Meeting to promote the Better Observance of the Sabbath; and to adopt a Memorial to the Directors of the Edinburgh & Glasgow Railway in Approbation of their conduct in Discontinuing Passenger Trains on the Lord's Day* (Manchester, n.d.)

Anon., *Statement and Opinion for the Subscribers to the Edinburgh and Glasgow Railway, who are desirous of withdrawing from that concern* (n.p., 1832)

Anon., *Western Railway between Scotland & England* (Lockerbie, 1837)

Baird, Hugh, *Report on the Proposed Edinburgh and Glasgow Union Canal* (Glasgow, 1813)

— *Report on the Proposed Railway from the Union Canal at Ryal, to Whitburn, Polkemmet, and Benhar: or, the West Lothian Railway* (Edinburgh, 1824)

Bald, Robert, *A General View of the Coal Trade of Scotland* (Edinburgh, 1808)

[Barrow, John], 'Canals and Rail-Roads', *Quarterly Rev.*, 31 (1825)

Blackadder, William, *Report relative to the Strathmore Railway, being the Extension of the Dundee and Newtyle Railway from Newtyle along Strathmore between Coupar and Forfar* (Dundee, 1833)

Brees, S. C., *Railway Practice: a Collection of Working Plans and Practical Details of Construction in the Public Works of the Most Celebrated Engineers* (London, 1856)

Brewster, Sir David, et al., *The Edinburgh Encyclopaedia*, vol. 17 (Edinburgh, 1830)

Buchanan, George, *An Account of the Glasgow and Garnkirk Railway and Other Railways in Lanarkshire* (Edinburgh, 1832)

Buchanan, Robertson, *Report relative to the Proposed Rail-Way from Dumfries to Sanquhar* (Dumfries, 1811)

'Canal Proprietor, A', *Letter to the Proprietors of the Edinburgh Glasgow Union Canal* (n.p., 1836)

Chalmers, Peter, *History and Statistical Account of Dunfermline* (Edinburgh & London, 2 vols., 1844 & 1859)

Chambers, William, *Peebles and its Neighbourhood, with a run on the Peebles Railway* (Edinburgh, 1856)

Chattaway, E. D., *Railways: their Capital and Dividends, with Statistics of their Working in Great Britain, &c., &c.* (London, 1855-56)

Cockburn, Henry, *A Letter to the Lord Provost on the Best Ways of Spoiling the Beauty of Edinburgh* (Edinburgh, 1849)

Dalrymple, Morton, *Letter to the Gentlemen of the Counties of Lanark, Peebles, Selkirk, Roxburgh, and Berwick* (n.p., 1809)

'Detector', *Edinburgh Glasgow and Leith Railway* (n.p., 1832)

Dod, Charles R., *Electoral Facts from 1832 to 1853 impartially stated* (Brighton, 1972 reprint of 1853 edn.)

Dundee & Arbroath Railway Company, *Report by the Committee of Management to a Special General Meeting of the Company* (Dundee, 1838)

Dupin, Charles, *Relation succincte d'un second voyage fait en 1817 et 1818 dans les ports d'Angleterre, d'Ecosse et d'Irlande* (Paris, 1818)

Edinburgh & Glasgow Railway Company, *Reports by Mr. George Stephenson of Liverpool and Messrs. Grainger and Miller of Edinburgh, Civil Engineers* (Edinburgh, 1831)

Edinburgh Glasgow & Leith Railway Company, *Reports by Messrs. Grainger and Miller of Edinburgh, and Mr George Stephenson of Liverpool, Civil Engineers* (n.p., 1831)

Edinburgh & Glasgow Union Canal Company, *Report of a Sub-Committee on the Edinburgh & Glasgow Junction Railway* (n.p., 1835)

— *Report relative to the Railways between Edinburgh and Glasgow* (Edinburgh & Glasgow, 1830)

'Edinensis', *Sunday Railway Travelling. The History and Nature of the Sabbath, as contained in the*

Scriptures, and in the writings of Calvin, Luther, Milton, Jeremy Taylor, Paley, and others (Edinburgh & Glasgow, 1847)
Errington, John E., *Report on the Proposed Railway from Perth by Stirling to Edinburgh and Glasgow* (n.p., 1841)
Evans, D. Morier, *The Commercial Crisis 1847-1848* (Newton Abbot, 1969 reprint of 1849 edn.)
Findlay, James, *Directory to Gentlemen's Seats, Villages, &c. in Scotland* (Edinburgh & Glasgow, n.d. [1845])
Francis, John, *A History of the English Railway: its Social Relations and Revelations, 1820-45* (Newton Abbot, 1967 reprint of 1851 edn.)
'Friend to the Tunnel, A', *Letter to the Author of Remarks on the Report by Messrs. Grainger and Miller on the subject of the Proposed Railway & Tunnel from the North Quarter of Glasgow to the Broomielaw* (Glasgow, 1829)
Gibson & Oliphant, W. S., *Memorial humbly submitted to the Right Honourable the Lord Provost and Magistrates of Edinburgh respecting the Proposed Railways* (n.p., n.d., [c.1818])
Glasgow Paisley & Greenock Railway Company, *Answers by the Directors to the Report of the Committee of Investigation* (n.p., 1850)
Glasgow Paisley Kilmarnock & Ayr Railway Company, *Proceedings at the First General Meeting of the Shareholders, 9th November, 1836* (Glasgow, 1836)
Glynn, Henry, *Reference Book to the Incorporated Railway Companies of Scotland* (London & Newcastle, 1847)
Grahame, Thomas, *A Letter on the Present System of Legislation which regulates Internal Intercourse in Great Britain, addressed to W. D. Gillon, Esq., M.P.* (London, 1836)
— *A Treatise on Internal Intercourse and Communication in Civilised States, and particularly in Great Britain* (London, 1834)
Grainger, Thomas, *Observations addressed to the Committee of Commerce of the City of Edinburgh, relative to the Proposed Railway ... to be called the Edinburgh, Dundee, and Northern Railway. With Remarks on the Proposed Line ... called the Forth and Tay Western Line of Railway* (Edinburgh, 1841)
— *Report to the Subscribers for a Survey of a Railway from the Monkland and Kirkintilloch Railway to that part of the Monkland Coalfield situated north & east of Airdrie* (n.p., 1825)
Grainger, Thomas and Miller, John, *Observations on the Formation of a Railway Communication between the Cities of Edinburgh and Glasgow, with branches to the Firth of Forth at Leith and the River Clyde at Glasgow* (Edinburgh, 1830)
— *Report relative to a Proposed Railway Communication between the City of Glasgow and the Towns of Paisley and Johnstone* (Edinburgh, 1831)
— *Report relative to the Proposed Railway to connect the Clydesdale, or Upper Coal Field of Lanarkshire, with the City of Glasgow, and the East and West Country Markets* (Edinburgh, 1828)
— *Report to the Proprietors of, and Traders on the Canals and Railways terminating on the North Quarter of Glasgow, and Other Gentlemen interested therein, and in the Harbour and Streets of Glasgow* (n.p., 1829)
Grieve, John and McLaren, James, *Report on the Utility of a Bar-iron Railway from the City of Edinburgh to Dalkeith and to the Harbour of Fisherrow* (Edinburgh, 1824)
[Jardine, James], *Report to the Subscribers for a Survey and Plan of a Railway from Edinburgh and Leith to Glasgow* (n.p., 1830)
Landale, Charles, *Report on the Proposed Railway between the Valley of Strathmore and Dundee* (Dundee, 1825)
Lang, Gabriel H., *Letter to the Right Hon. Henry Goulburn, M.P. on the Unequal Pressure of the Railway Passenger Tax* (Glasgow, 1842)
Lardner, Dionysius, *Railway Economy* (London, 1850)
[Lardner, Dionysius], 'Improvements in inland transport — railroads', *Edinburgh Rev.*, 60 (1834)
— 'Inland transport', *Edinburgh Rev.*, 56 (1832)
— 'Railways at home and abroad', *Edinburgh Rev.*, 89 (1846)
Leadbetter, John, Open Letter *To the Earl of Dalhousie* (n.p., 1845)

[Lewis, G. F.(?)], 'The financial state of the public roads', *Edinburgh Rev.*, 72 (1841)
Locke, Joseph, *Report on the Proposed Line of Railway from Carlisle to Glasgow and Edinburgh by Annandale* (Liverpool, 1837)
McCrie, Thomas, *Memoirs of Sir Andrew Agnew of Lochnaw, Bart.* (London, 1850)
Macfarlan, Duncan, *Railway Travelling on the Lord's Day Indefensible* (Glasgow, 1841)
MacLaren, Charles, *Railways Compared with Canals and Common Roads, and their Uses and Advantages Explained* (London, 1825; also in *The Pamphleteer*, 26 (1826))
Marjoribanks, Alexander, *Mistaken Views regarding the Observance of the Sabbath* (London, 1840)
Milne, John, *Reply to the Observations addressed by Mr. Thomas Grainger to the Committtee of the Chamber of Commerce of the City of Edinburgh on the Proposed Railways through Fife* (Edinburgh, 1841)
Neill, Patrick, *Considerations regarding the Edinburgh, Leith, and Newhaven Railway* (Edinburgh, 1837)
— *Remarks on the Progress and Prospects of the Edinburgh, Leith, & Newhaven Railway, in January 1839 (Edinburgh & London, 1839)*
New Statistical Account of Scotland (Edinburgh, 1845)
'Original Proprietor, An', *An Authentic Statement of the Affairs of the Edinburgh, Leith, and Newhaven Railway Company, from the Period of its Projection in 1835, to the Close of the Year 1840* (Bristol, 1840)
Poole, Braithwaite, *Statistics of British Commerce* (London, 1852)
Porter, G. R., *The Progress of the Nation, in its Various Social and Economic Relations, from the Beginning of the Nineteenth Century to the Present Time* (London, 1847 edn.)
Priestley, Joseph, *Historical Account of the Navigable Rivers, Canals, and Railways, of Great Britain* (London, 1831)
Rankine, David, *A Popular Exposition of the Effects of Forces applied to Draught* (Glasgow, 1828)
Reid, John, *Manual of the Scottish Stocks and British Funds, with a List of the Joint Stock Companies in Scotland, arranged in a Tabular Form* (Edinburgh, 3rd edn. 1841)
— *Remarks on the Scottish Stocks and British Funds* (Edinburgh, 1841)
Rennie, John, *Report respecting the Proposed Rail-Way from Kelso to Berwick* (Kelso, 1810)
Salt, Samuel, *Facts and Figures principally relating to Railways and Commerce* (London, 1848)
— *Railways and Commercial Information* (London, 1850)
— *Statistics and Calculations essentially necessary to Persons connected with Railways or Canals* (London, 2nd edn. 1846)
Scott, Stephen & Gale, Messrs., *An Examination of Mr. G. Stephenson's report on the Two Lines of Railway projected between Glasgow and Ayrshire* (Glasgow, 1837)
Scottish Society for Promoting the Due Observance of the Lord's Day, *Address No. 2: On the Desecration of the Sabbath by the Running of the Royal Mail* (n.p., n.d.)
Scrivenor, Henry, *The Railways of the United Kingdom Statistically Considered* (London, 1849)
'Scrutator', *Exposure of the Statements relative to the Edinburgh & Glasgow Railway* (Edinburgh & Glasgow, 1832)
Slaughter, Mihill, *Railway Intelligence,* (London, 1859)
Smith, Henry S., *The Parliaments of England* (Chichester, 1973 reprint of 3 vol. edn. of 1844-50)
Spackman, W. F., *An Analysis of the Railway Interest of the United Kingdom* (London, 1845)
[Spencer, Herbert], 'Railway morals and railway policy', *Edinburgh Rev.*, 100 (1854)
Statistical Account of Scotland (Edinburgh, 1791)
Stevenson, Alan, *Biographical Sketch of the late Robert Stevenson* (Edinburgh, 1851)
Stevenson, Robert, *A Memorial regarding the Propriety of Opening the Great Valleys of Strathmore and Strathearn by means of a Railway or Canal* (Edinburgh, 1820)
— *Report relative to the Lines of Railway surveyed from the Ports of Perth, Arbroath, and Montrose, into the Valley of Strathmore* (Edinburgh, 1827)
— *Report relative to various Lines of Railway from the Coal-Field of Mid-Lothian to the City of Edinburgh and Port of Leith* (Edinburgh, 1819)
— *Report on the Roxburgh and Selkirk Railway* (n.p., 1821)
Stevenson, Robert (ed.), 'Essays on Rail-Roads', *Prize Essays and Transactions of the Highland*

Society of Scotland, 6 (1824)

Telford, Thomas, 'Canals', in Joseph Plymley, *General View of the Agriculture of Shropshire* (London, 1803)

— *Report relative to the Proposed Railway from Glasgow to Berwick-upon-Tweed* (Edinburgh, 1810)

Telford, Thomas, ed. John Rickman, *Life of Thomas Telford, Civil Engineer, written by himself* (London, 1838)

Tooke, Thomas, *A History of Prices, and of the State of the Circulation, from 1839 to 1847 inclusive* (London, 1848)

Tuck, Henry, *Every Traveller's Guide to the Railway of England, Scotland, Ireland, Belgium, France and Germany* (London, 1844)

Whishaw, Francis, *The Railways of Great Britain and Ireland* (London, 1840)

Williams, F. S., *Our Iron Roads: their History, Construction, and Social Influences* (London, 1852, 6th edn. 1888)

Wood, Nicholas and Johnson, George, *Report on a Central Line of Railway into Scotland for the Directors of the Newcastle & Carlisle Railway* (n.p., 1843)

'Y', *The Rival Raids, or Reply to the "Remarks on the Proposed Railway between England and Scotland"* (Carlisle, 1841)

'???', *The Fingerpost, or, Direct Road from John-o'-Groat's to the Land's End* (n.p., n.d. [1825])

E. Books and articles published after 1860, and unpublished theses.

Acworth, Sir William, *The Railways of England* (London, 1889)

— *The Railways of Scotland* (London, 1890)

Ahrons, E. L., *The British Steam Locomotive 1825-1925* (London, 1927)

Albert, William, *The Turnpike Road System in England 1663-1840* (Cambridge, 1972)

Alderman, Geoffrey, *The Railway Interest*, (Leicester, 1973)

— 'The politics of the railway passenger duty', *Transport Hist.*, 3 (1970)

Anon., 'Early Scottish Railways', *Three Banks Rev.*, 74 (1967)

Anon., *Memoirs and Portraits of One Hundred Glasgow Men who have died during the Last Thirty Years, and in their Lives did much to make the City what it now is* (Glasgow, 1886)

Appleton, J. H., *The Geography of Communications in Great Britain* (Oxford, 1962)

Bagwell, Philip S., *The Railwaymen* (London, 1963)

— 'The railway interest: its organisation and influence, 1839-1914'. *Jnl Transport Hist.*, 7 (1965)

Barclay-Harvey, Sir Malcolm, *History of the Great North of Scotland Railway* (London, 2nd edn. 1949)

Bateman, John, *The Great Landowners of Great Britain and Ireland* (Leicester, 1971 reprint of 1883 edn.)

Baxter, Bertram, *Stone Blocks and Iron Rails* (Newton Abbot, 1966)

Bell, S. P., *A Biographical Index of British Engineers in the Nineteenth Century* (New York & London, 1975)

Best, Geoffrey, *Mid-Victorian Britain 1851-75* (London, 1971)

Black, David., *The History of Brechin to 1864* (Edinburgh, 1867)

Boase, Frederic, *Modern English Biography* (London, 1965 reprint of 1901 edn.)

Brand, Andrew, *Caledonian Railway. Index of Lines, Connections, Amalgamations, etc., Chronologically Arranged* (Glasgow, 1902)

Bremner, David, *The Industries of Scotland. Their Rise, Progress and Present Condition* (Edinburgh, 1869)

Broadbridge, Seymour A., 'The sources of railway share capital', in M. C. Reed (ed.), *Railways in the Victorian Economy* (Newton Abbot, 1969)

Brooke, D., 'The opposition to Sunday rail services in north-east England, 1834-1914', *Jnl Transport Hist.*, 6 (1963)

Brown, Kenneth, 'The first railway in Scotland', *Railway Mag.*, 82 (1938)

Bruce, William S., *The Railways of Fife* (Perth, 1980)

Butt, J. and Ward, J. T., 'The promotion of the Caledonian Railway Company', *Transport Hist.*, 3 (1970)
Caird, J. B., 'The making of the Scottish rural landscape', *Scot. Geog. Mag.*, 80 (1964)
Campbell, R. H., *Carron Company* (Edinburgh & London, 1961)
— *Scotland since 1707. The Rise of an Industrial Society* (Oxford, 1965)
— *The Rise and Fall of Scottish Industry* (Edinburgh, 1980)
Campbell, R. H. and Dow, J. B. A., *Source Book of Scottish Economic and Social History* (Oxford, 1968)
Carter, Ernest F., *An Historical Geography of the Railways of the British Isles* (London, 1959)
Chambers, William, *About Railways* (London & Edinburgh, 1865)
Checkland, S. G., *The Gladstones: a Family Biography, 1764-1851* (Cambridge, 1971)
— *The Rise of Industrial Society in England 1815-1885* (London, 1964)
— *Scottish Banking: a History, 1695-1973* (Glasgow & London, 1975)
Clapham, Sir J. H., *An Economic History of Modern Britain*, vol 1 (Cambridge, 1926)
Cleveland-Stevens, E., *English Railways. Their Development and their Relation to the State* (London, 1915)
Clifford, Frederick, *A History of Private Bill Legislation* (London, 1968 reprint of 1885-87 edn.)
Cockburn, Henry, *Journal 1831-1854* (Edinburgh, 1874)
Copeland, John, *Roads and their Traffic, 1750-1850* (Newton Abbot, 1968)
Coleman, Terry, *The Railway Navvies* (London, 1965)
Crombie, Benjamin W., *Modern Athenians: a Series of Original Portraits of Memorable Citizens of Edinburgh, 1837 to 1847* (Edinburgh, 1882)
Deane, Phyllis and Cole, W. A., *British Economic Growth 1688-1959* (Cambridge, 1962)
Dendy Marshall, C. F., *A History of British Railways down to the Year 1830* (Cambridge, 1938)
Devey, Joseph, *Life of Joseph Locke* (London, 1862)
Dictionary of National Biography (London, 53 vols., 1885-1900)
Dott, George, *Early Scottish Colliery Waggonways* (London, 1947)
Dow, George, *The First Railway across the Border* (London, 1946)
Dowell, Stephen, *A History of Taxation and Taxes in England from the Earliest Times to the Present Day* (London, 1884)
Drummond, Andrew L., and Bulloch, James, *The Scottish Church 1688-1843* (Edinburgh, 1973)
Duckham, Baron F., *A History of the Scottish Coal Industry*, vol. I (Newton Abbot, 1970)
Duncan, W. M., *Newtyle: a Planned Manufacturing Village* (Forfar, 1979)
Dyos, H. J. and Aldcroft, D. H., *British Transport. An Economic Survey from the Seventeenth Century to the Twentieth* (Leicester, 1969)
Ellis, C. Hamilton, *British Railway History*, vol. I (London, 1954)
— *The North British Railway* (London, 1955)
Emmerson, George S., *John Scott Russell* (London, 1977)
Ferneyhough, Frank, *The History of Railways in Britain* (Reading, 1975)
Fleming, J. Arnold, *Scottish and Jacobite Glass* (Glasgow, 1938)
Foster, Joseph, *Members of Parliament, Scotland . . . 1357-1882* (London, 1882)
Frew, Iain, D. O., 'The Brora colliery tramway', *Railway Mag.*, 102 (1960)
Galloway, Robert L., *Annals of Coal Mining and the Coal Trade* (London, 2 vols., 1898 & 1904)
Gard, R. M. and Hartley, J. R., *Railways in the Making* (Newcastle, 1969)
Gardner, John W. F., *Railway Enterprise* (n.p., 1934)
Gayer, Arthur D., Rostow, W. W., and Schwartz, Anna J., *The Growth and Fluctuation of the British Economy, 1790-1850* (Oxford, 2 vols., 1953)
Gilfillan, John B. S. and Moisley, H. A., 'Industrial and commercial developments to 1914', in Ronald Miller and Joy Tivy (eds.), *The Glasgow Region* (Glasgow, 1958)
Gold, G. F. (ed.), *The Glasgow, Paisley & Greenock Railway* (Glasgow, 1978)
Gordon, T. Crouther, *A Short History of Alloa* (Alloa, 1937)
Gordon, W. J., *Our Home Railways. How they began and how they are worked* (London & New York, 1910)
Gourvish, T. R., 'The Bank of Scotland, 1830-45', *Scot. Jnl of Pol. Econ.*, 16 (1969)

— *Mark Huish and the London & North Western Railway: a Study of Management* (Leicester, 1972)
— 'Railway enterprise', in Roy Church (ed.), *The Dynamics of Victorian Business* (London, 1980)
— *Railways and the British Economy, 1830-1914* (London & Basingstoke, 1980)
— 'The railways and steamboat competition in early Victorian Britain', *Transport Hist.*, 4 (1971)
Gourvish, T. R. and Reed, M. C., 'The financing of Scottish railways before 1860 — a comment', *Scot. Jnl of Pol. Econ.*, 18 (1971)
Graham, George, *The Caledonian Railway: Account of its Origins and Completion. Largely composed of extracts from information got from the original promoters* (Glasgow, 1888)
Grant, J., *Cassell's Old and New Edinburgh* (London & New York, 1884)
Hadfield, Charles, *British Canals: an Illustrated History* (London, 1959)
— *The Canal Age* (Newton Abbot, 1968)
Hamilton, Henry, *The Industrial Revolution in Scotland* (Oxford, 1932)
Handley, James, *The Navvy in Scotland* (Cork, 1970)
Harrison, J. F. C., *The Early Victorians 1832-1851* (London, 1971)
Hart, Harold W., 'Some notes on coach travel, 1750-1848', *Jnl Transport Hist.*, 4 (1960)
Hassan, John A., 'The supply of coal to Edinburgh, 1790-1850', *Transport Hist.*, 5 (1972)
Hawke, G. R., *Railways and Economic Growth in England and Wales, 1840-1870* (Oxford, 1970)
Hawke, G. R. and Reed, M. C., 'Railway capital in the United Kingdom in the nineteenth century', *Econ. Hist. Rev.*, II 22 (1969)
Highet, Campbell, *The Glasgow and South-Western Railway* (Lingfield, 1965)
Horne, John (ed.), *Kirkintilloch* (Kirkintilloch, 1910)
Hunt, B. C., *The Development of the Business Corporation in England, 1800-1867* (Boston, Mass., 1936)
Hunter, John Kelso, *The Retrospect of an Artist's Life* (Greenock, 1868)
Inglis, J. C. & F., *The Fordell Railway* (n.p., 1946)
Irvine, H. S., 'Some aspects of passenger traffic between Britain and Ireland, 1820-50', *Jnl Transport Hist.*, 4 (1960)
Irving, Joseph, *The Book of Scotsmen* (Paisley, 1881)
Jackman, W. T., *The Development of Transportation in Modern England* (Cambridge, 1916)
Jeans, J. S., *Jubilee Memorial of the Railway System. A History of the Stockton and Darlington Railway and a Record of its Results* (London, 1875)
Jenks, Leland H., *The Migration of British Capital to 1875* (London, paperback edn., 1971)
Kellett, John R., *The Impact of Railways on Victorian Cities* (London, 1969)
— 'Glasgow's railways 1830-80: a study in natural growth', *Econ. Hist. Rev.*, II 17 (1964)
Kenwood, A. G., 'Railway investment in Britain, 1825-75', *Economica*, 32 (1965)
Kingsford, P. W., *Victorian Railwaymen: the Emergence and Growth of Railway Labour, 1830-1870* (London, 1970)
Kininburgh, Ian A. G., 'Greenock: growth and change in the harbours of the town', *Scot. Geog. Mag.*, 76 (1960)
Lambert, R. S., *The Railway King 1800-71: a Study of George Hudson and the Business Morals of his Time* (London, 1934)
Lee, Charles, E., *The Evolution of Railways* (London, 1943)
— *Passenger Class Distinctions* (London, 1946)
— 'The Dundee & Arbroath Railway', *Railway Mag.*, 104 (1958)
Lenman, Bruce, *An Economic History of Modern Scotland* (London, 1977)
Lenman, Bruce and Gauldie, E. E., 'The industrial history of the Dundee region from the 18th to the early 20th century', in S. J. Jones (ed.), *Dundee and District* (Dundee, 1968)
Levitt, Ian and Smout, Christopher, *The State of the Scottish Working-Class in 1843* (Edinburgh, 1979)
Lewin, H. G., *Early British Railways* (London & New York, n.d. [1925])
— *The Railway Mania and its Aftermath, 1845-52* (London, 1936)
Lewis, M. J. T., *Early Wooden Railways* (London, 1970)
Lindsay, Jean, *The Canals of Scotland* (Newton Abbot, 1968)

Lythe, S. G. E., 'Early days of the Arbroath & Forfar Railway', *Railway Mag.*, 99 (1953)
Lythe, S. G. E. and Butt, J., *An Economic History of Scotland 1100-1939* (Glasgow, 1975)
Lythe, S. G. E. and Lee, C. E., 'The Dundee and Newtyle Railway', *Railway Mag.*, 97 (1951)
McAdam, W., *The Birth, Growth and Eclipse of the Glasgow and South-Western Railway* (Glasgow, 1924)
McBain, J. M., *Arbroath: Past and Present* (Arbroath, 1887)
— *Eminent Arbroathians* (Arbroath, 1897)
[McGeorge, Andrew], *The Bairds of Gartsherrie* (Glasgow, 1875)
MacGregor, George, *The History of Glasgow from the Earliest Period to the Present Time* (Glasgow & London, 1881)
Maclean, John S., *The Newcastle & Carlisle Railway* (Newcastle, 1948)
Malcolm, Charles A., *The Bank of Scotland 1695-1945* (Edinburgh, n.d., [1948])
— *The History of the British Linen Bank* (Edinburgh, 1950)
Martin, Don, *The Monkland & Kirkintilloch Railway* (n.p., 1976)
Marwick, W. H., *Scotland in Modern Times* (London, 1964)
Mathias, Peter, *The First Industrial Nation* (London, 1969)
Matthews, R. C. O., *A Study in Trade Cycle History. Economic Fluctuations in Great Britain, 1833-1842* (Cambridge, 1954)
Meik, H. H., 'The Edinburgh & Dalkeith Railway', *Railway Mag.*, 52 (1923)
Michie, R. C., *Money, Mania and Markets* (Edinburgh, 1981)
Miller, Andrew, *The Rise and Progress of Coatbridge and Surrounding Neighbourhood* (Glasgow, 1864)
Milne, John, *Aberdeen: Topographical, Antiquarian, and Historical papers on the City of Aberdeen* (Aberdeen, 1911)
Mitchell, B. R., 'The coming of the railway and United Kingdom economic growth', *Jnl Econ. Hist.*, 24 (1964)
Mitchell, B. R. and Deane, Phyllis, *Abstract of British Historical Statistics* (Cambridge, 1962)
Mitchell, John O., *Old Glasgow Essays* (Glasgow, 1905)
Mitchell, Joseph, *Reminiscences of my Life in the Highlands* (London, 2 vols., 1883)
Morgan, E. V., 'Railway investment, Bank of England policy and interest rates, 1844-48', *Econ. Hist.*, 4 (1940)
Munn, C. W., *The Scottish Provincial Banking Companies 1747-1864* (Edinburgh, 1981)
Munro, Neil, *The History of the Royal Bank of Scotland 1727-1927* (Edinburgh, 1928)
Murray, David, *The York Buildings Company: a Chapter in Scotch History* (Glasgow, 1883)
Nef, J. U., *The Rise of the British Coal Industry* (London, 1932)
Nimmo, William, *The History of Stirlingshire* (Glasgow & London, 3rd edn., 1880)
Nock, O. S., *The Caledonian Railway* (London, n.d. [1961])
— *Scottish Railways* (London & Edinburgh, 1950)
Norrie, W., *Dundee Celebrities of the Nineteenth Century* (Dundee, 1873)
'N'Oubliez', *The Caledonian Railway and its Originators* (Glasgow, 1890)
O'Dell, Andrew C., 'A geographical examination of the development of Scottish railways', *Scot. Geog. Mag.*, 55 (1939)
O'Dell, Andrew C. and Richards, P. S., *Railways and Geography* (London, revised edn., 1971)
Ottley, G., *A Bibliography of British Railway History* (London, 1965)
[Pagan, James (ed.)], *Glasgow Past and Present* (Glasgow, 1884)
Page, P. J., 'The influence of public transport on the growth of Edinburgh', unpublished M.Sc. thesis, Univ. of Edinburgh, 1972
'E. M. S. P.' [Mrs Paine], *The Two James's and the Two Stephensons* (Dawlish, 1961 reprint of 1861 edn.)
Parris, Henry, *Government and the Railways in Nineteenth-Century Britain* (London, 1965)
— 'Railway policy in Peel's administration, 1841-1846', *Bull. Inst. Hist. Research.*, 33 (1960)
Paul, Sir J. B. (ed.), *The Scots Peerage* (Edinburgh, 9 vols., 1904-14)
Pawson, Eric, *Transport and Economy: the Turnpike Roads of Eighteenth Century Britain* (London, New York & San Francisco, 1977)

Pike, Richard (ed.), *Railway Adventures and Anecdotes* (London, 1884)
Pollard, Sidney, *The Genesis of Modern Management* (London, 1965)
Pollins, Harold, *Britain's Railways: an Industrial History* (Newton Abbot, 1971)
— 'Aspects of railway accounting before 1868', in M. C. Reed (ed.), *Railways in the Victorian Economy* (Newton Abbot, 1969)
— 'The finances of the Liverpool and Manchester Railway', *Econ. Hist. Rev.*, II 5 (1952)
— 'The marketing of railway shares in the first half of the nineteenth century', *Econ. Hist. Rev.*, II 7 (1955)
— 'A note on railway construction costs 1825-50', *Economica*, N.S. 19 (1952)
— 'Railway contractors and the finance of railway development in Britain', *Jnl Transport Hist.*, 3, (1957)
Reed, M. C., 'A note on subscriptions to the Glasgow, Paisley & Greenock Railway', *Transport Hist.*, 6 (1973)
— *Investment in Railways in Britain, 1820-1844* (Oxford, 1975)
— 'Railways and the growth of the capital market', in M. C. Reed (ed.), *Railways in the Victorian Economy* (Newton Abbot, 1969)
Reid, J. M., *A History of the Merchant's House in Glasgow* (n.p., n.d.)
Richards, Eric, *The Leviathan of Wealth* (London & Toronto, 1973)
Robbins, Michael, 'The North Midland Railway and its enginemen, 1842-3' *Jnl Transport Hist.*, 4 (1960)
— *Points and Signals: a Railway Historian at Work* (London, 1967)
— *The Railway Age* (London, 1962)
Robertson, C. J. A., 'The cheap railway movement in Scotland: the St Andrews Railway Company', *Transport Hist.*, 7 (1974)
— 'Early Scottish railways and the observance of the sabbath', *Scot. Hist. Rev.*, 57 (1978)
Robertson, David, *The Princess Street Proprietors* (Edinburgh & London, 1935)
Rolt, L. T. C., *Isambard Kingdom Brunel* (London, 1957)
— *George and Robert Stephenson* (London, 1960)
— *Victorian Engineering* (London, 1970)
Sherrington, C. E. R., *The Economics of Rail Transport in Great Britain*, vol. I (London, 1928)
Simmons, Jack, 'For and against the locomotive', *Jnl Transport Hist.*, 2 (1955)
— 'The power of the railway', in H. J. Dyos and M. Wolff (eds.), *The Victorian City*, vol. I (London, 1973)
— *The Railway in England and Wales, 1830-1914*, vol. I (Leicester, 1978)
— *The Railways of Britain* (London, 1961)
Simpson, Michael, 'Urban transport and the development of Glasgow's West End, 1830-1914', *Jnl Transport Hist.*, N.S. 1 (1972)
Slaven, Anthony, *The Development of the West of Scotland, 1750-1960* (London, 1975)
Smiles, Samuel, *Lives of the Engineers*, vol. III (London, 1962)
Smout, T. C., *A History of the Scottish People, 1560-1830* (London, 1969)
Spring, David, 'The English landed estate in the age of coal and iron: 1830-1880', *Jnl Econ. Hist.*, 11 (1951)
Stenton, M., *Who's Who of British Members of Parliament*, vol. I (Brighton, 1976)
Stephen, William, *History of Inverkeithing and Rosyth* (Aberdeen, 1921)
Stephenson Locomotive Society, *The Glasgow and South-Western Railway, 1850-1923* (London, 1950)
Stevenson, David, *Life of Robert Stevenson* (Edinburgh, 1878)
Taylor, William, *Military Roads of Scotland* (Newton Abbot, 1976)
Thomas, John, *The North British Railway*, vol. I (Newton Abbot, 1969)
— *A Regional History of the Railways of Great Britain: vol. VI. Scotland: the Lowlands and Borders* (Newton Abbot, 1971)
Tomlinson, W. W., *The North Eastern Railway* (Newton Abbot, 1967 reprint of 1915 edn.)
Turnbull, Gerard L., 'A note on the supply of staff for the early railways', *Transport Hist.*, 1 (1968)

— 'Provincial road carrying in England in the eighteenth century', *Jnl Transport Hist.*, N.S. 4 (1977)
— *Traffic and Transport: an Economic History of Pickfords* (London, 1979)
Vallance, H. A., 'From Glasgow to the Clyde coast', *Railway Mag.*, 112 (1966)
Vamplew, Wray, 'Banks and railway finance: a note on the Scottish experience', *Transport Hist.*, 4 (1971)
— 'The financing of Scottish railways before 1860 — a reply', *Scot. Jnl of Pol. Econ.*, 18 (1971)
— 'Nihilistic impressions of British railway history', in D. N. McCloskey (ed.), *Essays on a Mature Economy: Britain after 1840* (London, 1971)
— 'Railways and the Transformation of the Scottish economy', unpublished Ph.D. thesis, Univ. of Edinburgh, 1969
— 'Railways and the Scottish transport system in the nineteenth century', *Jnl Transport Hist.*, N.S. 1 (1972)
— 'Sources of Scottish railway share capital before 1860', *Scot. Jnl of Pol. Econ.*, 17 (1970)
— 'Railways and the transformation of the Scottish Economy', *Econ. Hist. Rev.*, II 24 (1971)
— 'The railways and the iron industry: a study of their relationship in Scotland', in M. C. Reed (ed.), *Railways in the Victorian Economy* (Newton Abbot, 1969)
Ward, J. T., 'West Riding landowners and the railways', *Jnl Transport Hist.*, 4 (1960)
Ward-Perkins, C. N., 'The commercial crisis of 1847', *Oxford Econ. Papers*, II 1 (1950)
Warren, Kenneth, 'Locational problems of the Scottish iron and steel industry since 1760', *Scot. Geog. Mag.*, 81 (1965)
Webster, Norman W., *Joseph Locke: Railway Revolutionary* (London, 1970)
Williams, O. Cyprian, *The Historical Development of Private Bill Procedure and Standing Orders in the House of Commons* (London, 2 vols., 1948-49)
Wilson, James A., *A History of Cambuslang* (Glasgow, 1929)
Youngson, A. J., *The Making of Classical Edinburgh* (Edinburgh, 1966)

Index

Aberdeen 36, 99, 127, 144-5, 302, 308, 311, 338
Aberdour 4
Accounting practices 168-9
Acworth, Sir William (author) 144
Adamson, Rev James (railway propagandist) 48-9, 51
Adie, Alexander J. (secretary and engineer) 190
Agnew, Sir Andrew of Lochnaw, Bt, M.P. (sabbatarian) 252-6
Agreements between companies 155, 298
Agriculture 37, 59-60, 65, 99, 123
Airlie, 6th Earl of 65
Allanton 82-3
Alloa 26-8
Amalgamations 23, 62-3, 69-70, 77, 81, 89, 133, 137, 263, 293, 393; proposed 62, 81, 117
Amenity 130-2, 181-4
Annandale (see also Railway companies – Caledonian) 3, 267-76, 280-3, 289, 316, *Tables 74 5, Maps 15 16*
Arbroath 36, 123-4, *Map 11*
Ardrishaig 143
Ardrossan 68, 138, 142, 262, 264, 288
Atholl, 5th Duke of 126
Atmospheric railway 196
Auditing 168
Ayr 140-2, 262, 267, 288, 299

Baird, Hugh (engineer) 26-8, 36, 39, 63, 113, 321, *Tables 2, 7*
Baird, John, of Shotts Ironworks, 35, 46, *Table 1*
Baird, William & Co. of Gartsherrie (coal and ironmasters) 81, 92-6, 311, 368, *Table 23*
Bald, Robert (mining engineer) 18, 26-9
Ballantyne, Henry (cloth manufacturer) 305
Banks 82-3, 160, 211, 218-9; Bank of England 325; Bank of Scotland 218; British Linen 218, Commercial 218, Dundee Union 67, 218; National 219; Royal Bank of Scotland 81-2, 218; Union 218-19
Bannatyne, Andrew & Dugald (lawyers) 118, 180, 277
Barlow, Prof. Peter (engineer) (see also Smith-Barlow Commission) 278

Barstow, Charles (accountant) 298
Bathgate 86, 104, 114
Baxter, Bertram (author) 1, 18, 32
Beattock 268, 275-6, 316
Belfast 264
Belgium, railways in 306
Belhaven, 8th Lord (land and coalowner, director) 59-60, 62, 75, 81, 83, 182, *Table 23*
Bell, William (engineer) 36, 268-9
Benhar 100, 104, 107, 112, 114
Berwick 5, 32-4, 267, 293, 330
Birkinshaw, John (inventor) 35
Bishopton 187, 190, 201
Black, Adam (lord provost of Edinburgh) 129, 296
Blackadder, William (engineer) 123, 196-7, 333, *Tables 12, 43, Map 14*
Blackmore, John (engineer) 269, 280-2, 291, *Tables 74-5, Map 16*
Blair, Col., of Blair (landowner) 182
Blair, Patrick (law agent) 180
Blantyre, 12th Lord (landowner) 186-7
Blenkinsop, John (engineer) 24, 46, 359
Board of Trade (see also Railway Board, Railway Department): statistics 89, 168; supervision of railways 176, 178, 237, 251, 255, 300, 320
Border railway projects 267-304, 315-17, 320, 335, *Tables 74-6, Maps 15-17*
Borrowing 81-2, 179-80, 211-19, 366-7, 382, *Tables 46-9*; debentures 211-18; loan notes 219; overdrafts 81-2, 213, 218-19
Bouch, Thomas (engineer) 202, 333
Branches 68, 75, 86, 92, 119, 146, 302, 367
Breadalbane, 2nd Marquis of 258
Brechin 121-3
Bridgwater, 1st Duke of 5
Bristol, shareholders in 129, 152
Brokers 148-9, 162, 327
Broomielaw (see Glasgow harbour)
Brown, William (engineer) *Table 2*
Brucehaven 19
Brunel, Isambard K. (engineer) 116, 191
Buccleuch, 4th Duke of 28, 36
Buccleuch, 5th Duke of: and border projects

183-4, 268-9, 286-7, 292, 300-2, 377, 393; and Edinburgh & Dalkeith 63-4, *Table 23*; and Granton 127-9, 132, 135, 152, 186, 372; miscellaneous 84, 195, 209, 245, 362
Buchanan, George (engineer) 24, 59, 61, 274
Buchanan, Robertson (engineer) 18, 25, 34, 39, 87, 317, *Tables 2, 7*
Buckley, Edmund (promoter) 160
Burns, J. & G. (shipowners) 264
Burntisland 133-6
Bury, Edward (locomotive builder) 206
Butt, John & Ward, J.T. (authors) 290
Bye-laws 65, 73, 88

Cadell, John and Hugh (coal and ironmasters) 17, *Table 1*
'Caledonian system' of railways 194, 299, 302-3, 393, *Map 17*
Campbell, Archibald, of Blythswood (landowner) 60-1, 287
Campbell, Roy H. (author) 4
Campbell, W.F., of Islay 92
Canal companies
  Aberdeenshire 5
  Caledonian 38, 47, 145
  Forth & Cart 258
  Forth and Clyde 2, 5-6, 50, 56-60, 64, 99-104, 115, 117, 119-20, 131, 136, 159, 197, 258, 260, 307, *Tables 1, 20, Maps 6-7*
  Glasgow Paisley & Ardrossan 2, 68-9, 137, 140, 159, 259-60, 313, *Table 73, Maps 6, 13*
  Monkland 2, 5, 26, 30, 47, 56-7, 60, 81, 93, 112, 115, 146, 197, 306-8, 311, 324, *Tables 2, 20, Maps 6-7*
  Paisley & Johnstone (see Glasgow Paisley & Ardrossan)
  Union 26, 36, 39, 56, 62-3, 113, 115, 120, 260-1, 307-8, 311, *Table 2, Maps 6-7*
  others 25, 352
Canal projects, unsuccessful: Arbroath to Forfar 371; Breadalbane's 258; Dundee to Strathmore 65; Stirling 258-9; Strathmore 36
Canals: competition with railways 5, 25, 30-32, 36-7, 43-4, 47-50, 56-9, 69, 81, 88, 104, 112-3, 115, 117, 119-20, 137-8, 140-1, 163-4, 168, 236-7, 241, 249-50, 258-62, 268, 307, 310, 361, *Table 20*; cooperation with railways 37, 56-9, 62-3, 120, 234, 260-1; construction costs 113; passengers on 50, 88, 259-61, *Table 20*; rates and fares 112, 141, 259-61; steamboats on 49, 259; traffic 50, 112, 119, 252, 259-62, *Tables 20, 73*
Capital (see also shares, subscriptions): amounts authorised 59, 66, 68, 77-9, 124, 179, 323, *Tables 12-16, 25, 46-9, 78*; calls 76, 81, 212, *Table 79*; difficulties of raising 67-8, 81-3,

117, 124, 126, 211-19, 288-90, 316; paid up 12, *Tables 29, 46-9*; returns on, actual 76-7, *Tables 46-9*; returns on, estimated 32, 38, 100, 128-9, *Table 27*; sources of 23, 37-8, 75-6, 98, 105, 140, 142-3, 147-62, 211-19, 256, 295, 297, 323-7, *Tables 29-31, 80-91*
Carlisle 267, 275, 286, 330, 338
Carmichael, J. & C. (locomotive builders) 206-7, 362
Carr Glyn, George (director) 290-3
Carron Company 1, 6, 17, 29, *Table 1*
Carrying practices, by companies or private traders 72-5, *Table 15*
Cart, River 69, 143
Castle Steamship Co. 263
Chalmers, Patrick, of Auldbar, M.P. (landowner) 211, 283
Chalmers, Patrick, of Dunfermline (author) 19, 21
Chambers, Robert (author) 64-5
Chaplin, William (coach proprietor) 141
Charlestown 19, 21, 26-8
Clark, John, of Eldin 28
Clyde, River 2, 18, 60, 69, 99, 137, 162-3, 262-4
Clyde Shipping Company 143
Coaching departments, costs of 225, *Tables 34-7*
Coal 1, 3-28, 30-43, 51, 56-64, 68-9, 74, 77, 81, 86-7, 146-7, *Tables 1, 9-10*; bunkering trade 28; coastal trade 6, 22, 25-8, 63, 108, 127, 234; demand for 25-6, 59, 63, 77, 86-7, 306-8, 369; estimates of traffic *Table 27*; exports 5, 22, 25-8; internal trade 3-6, 22-3, 25-8, 47, 56-9, 62-4, 100, 107, 112, 119, 135, 137, 233-4, 306-9, 311; owners 61-4, 69, 76, 123, 266, 307, 323-5, 331; prices 22, 26-8, 32, 62-3, 100, 307, 356, 361-2; viewers 190
Coalmines: Cairnhill 57; Canonbie 268, 300, 303; Dalkeith 300; Elphinston 6; Hetton 48; Killingworth 48; Plashetts 268
Cochran, Joseph (manager and engineer) 190
Cockburn, Lord (conservationist) 130, 132, 185
Coddington, Capt. Joshua (secretary) 329
Colt, J.H., of Gartsherrie (landowner) 113
Committee for Opposing Sabbath Traffic on the Edinburgh & Glasgow Railway 255-6
Committees of inquiry, shareholders' 61, 85, 168, 190, 196, 198, 226, 229-30, 289, 329
Committees of management 66, 284, 298, 322-3
Committees, provisional 105, 283-7
Commuters 241-2, 244
Construction of railways (see also Works): costs of 17, 23, 28-9, 79, 100, 105, 124, 133, 166, 169-74, 202-3, 320-1, *Tables 4, 12, 18, 25, 33, 36-40*; estimates for 23, 28, 32, 38, 68, 79, 114, 119, 124-5, 132-3, 194, 198, *Tables 7-8, 17, 25-6*
Contractors 198-204
Cotton 1, 4, 159, 236
Cowlairs 119

Craufurd, W.S.S. (lawyer) 186
Crieff 3, 36
Cubitt, William (engineer) 131
Cumnock 297, 306
Cyclical fluctuations and railway development 51, 75-6, 97-8, 100, 118, 126, 136, 144, 159, 163-4, 236, 275, 282-3, 291-2, 306, 325-7, 332-8, 368

Dalhousie, 10th Earl of (Vice-President, later President of Board of Trade) 292, 295-7
Dalkeith 32, 36, 63, 65, 245, 253
Dalrymple, Martin of Fordell (landowner) 32
Dance, Sir Charles (pioneer of steam road vehicles) 116
Davidson, Charles F. (lawyer, secretary) 148, 180, 291-2, *Tables 74, 77*
Davidson (resident engineer) 226
Dawson, Richard (merchant, director) 129, 182
Deane, Phyllis & Cole, W.A. (authors) 313
Dendy Marshall, C.F. (author) 1
Denham, Sir James Stewart, of Coltness (land and coalowner) 32, 59, 82, *Table 2*
Dennistoun, John M.P. (merchant) 287, 300
Directors 132, 198, 204, 230, 236, 283-6, 328-9, 332, 334, 366; English 128, 160, 283-5, 372
Dividends and rates of return: canals 57, 113; railways, actual 67, 69, 76-7, 165-8, 308, 336, 338, *Table 32*; railways, estimated 21, 43, 99, 163
Dixon, Jacob (glass manufacturer) 105
Dixon, John (glass manufacturer) 18, *Table 1*
Dixon, William junior (coal and ironmaster, director) 18, 59-60, 69, 76, 83, 92, 104, 137, 142, 218, 287
Dixon, William senior (coal and ironmaster) 18, 26, 68, 353
Dobson, David (treasurer) 66-7
Dobson, John (engineer) *Table 74*
Dodds, David (manager) 79, 93
Dodds, Robert (resident engineer) 230
Dott, George (author) 1
Douglas, Alexander (of Princes Street Proprietors) 132
Douglas, George (author) 45
Douglas, Lord (landowner) 186
Drumlanrig Castle 183-4, 300
Drummond, James (provost of Paisley) 141
Duckham, Baron F. (author) 1, 4, 6, 9, 25
Dumbarton Glassworks 18, 105, *Table 1*
Dumfries 34, 277
Dumfries, 5th Earl of 5
Dunbar 277, 285
Duncan, George (director) 247
Dundas, J. (engineer) 276, *Table 75*
Dundas, Robert, of Arniston (coalowner) 64, *Table 23*
Dundee 65-8, 99, 125, 136, 186, 244, 302, 308, 338, *Map 11*
Dunfermline 18, 21, 28

Dunkeld 121, 126-7, 197
Dunlop, Henry (lord provost of Glasgow) 289
Dunlop, James (ironmaster) 146
Dunn, Mathias (engineer, surveyor) *Table 74*
Dupin, Charles (French traveller) 23-4
Dyos, H.J. & Aldcroft, D.H. (authors) 22

East coast projects (see also Railway companies – North British) 261-74, 276-7, 280-2, 285, 289, 291, 316, *Tables 74-6, Maps 15-16*
Edinburgh 2, 34, 99-100, 108, 255, 292, 311, *Map 9*; as source of capital 159-60, *Tables 30, 80-91*; coal consumption and prices 25-8, 36-7, 39, 62-4, 86, 100, 113, 127, 261, 296, 308, 369; Princes Street Gardens 128-33, 286, 305; Royal Institution 131; support for Caledonian 286-7, 296, 300; town council 37, 128, 130, 132-3, 195, 285, 296, 300; water supply 296
Eglinton, 12th Earl of 68
Eglinton, 13th Earl of 68, *Table 1*
Elgin, 5th Earl of 18-9, *Table 1*
Elgin, 7th Earl of 18-9
Employees (see also Wages and salaries) 221-8, 255-6
Engineers (see also under individual names) 21, 23, 64-7, 166, 178-80, 188-205, 321-3, 328, *Tables 1, 12, 25, 43, Map 14*; consulting 191, 196; costs and fees 118, 169, 196, 202, 321, *Tables 36-9, 42*; importance of English 108, 119, 140, 143-4, 191, 194, 197, 274, 332-3; resident 190-1, 230
England: committeemen and directors from 152, 284-6, 317; comparisons with Scotland 173-4, 306, 355; as source of capital 76, 99, 143, 152-62, 256, 283, 292-3, 324, 327, 333-5, *Tables 30, 80-91*; dividends 310; employment of engineers from 108, 119, 143, 274, 332-3; fares 240; influences from 4-5, 17-18, 44-5, 161, 316-7, 329-35; railway promotion in 97-9; trade with 99-100, 127, 263-8, 295, 315, 335
Errington, John E. (engineer) 143, 163, 182, 194, 197-200, 286, 299, 328
Erskine, John Francis, of Mar (land and coalowner) 17-8, *Table 1*
Estimates 77, 163, 204-5, 292, 295, 317-18; construction 23, 28, 32, 38-9, 57, 63, 66-8, 114, 119, 124-5, 128-9, 133, 137-8, 179-80, 194-6, 200, 259, 274, 320-3, *Tables 7-8, 26*; returns on capital 32, 38-9, 129, 137-8, 142, 147, 259, 292, 295, *Tables 9, 11, 34-5*; traffic 34, 37, 39, 86-7, 112-3, 119, 137-8, 147, 231, *Tables 9-10, 27, 34-5*; working expenditure 29, 83-6, 137-8, 147, 259, *Tables 11, 34-5*
Ewart, William, M.P. 284
Expenditure: capital 164-8, *Table 33*; current 219-30, *Tables 50-3*

Falkirk 113, 119-20
Fares (see rates)
Farey, John (surveyor) 28
Farquharson, Robert (manufacturer) 141, 243-4
Ferry, Granton to Burntisland 133-6, 303, *Map 12*
Fife 5-6, 26-8, 127, 133-7, 194
Finlay, Kirkman (merchant) 61, 146
'Fire King' (steamship) 264, 388
Fleetwood 89-90, 264-5, 288
Forfar 65, 121-3, *Map 11*
Forth, River 2, 262
Francis, John (auhor) 183, 194, 231
Fullarton, Col. (landowner) 22-3

Gairdner, C.D. (lawyer) 195
Galashiels 36, 274, 294, 299
Garnock valley 138, 142, 160, 310-11
Gartsherrie 56, 114
Gasworks 59, 141
Gauges 17-8, 46, 124-5, 194, 298, 315, 378, *Tables 12, 25*
Geddes, John (engineer) 135
Gibb, John (engineer) 108, 361
Gibson, John (lawyer) 129, 131-2, 135-6, 300-2
Gibson & Oliphant, W.S. 37
Gibson-Craig, Sir James (landowner) 120
Gibson-Craig, William, M.P. 132-3
Gladstone, Sir John, of Fasque (merchant) 135-6, 152, 194-5, 212, 283, 333
Gladstone, William M.P. (President of Board of Trade) 176, 251
Glasgow 99-100, 107, 255, 288, 298, *Maps 7, 10*; as source of capital 140, 155-61, *Tables 30, 80-91*; coal consumption and prices 5, 22, 25-6, 56-7, 59-62, 86, 100, 112, 307-8, 315; harbour 2, 60-1, 68, 100, 107, 114, 116, 142, 146, 155, 307, 315; merchants 105, 138, 140, 146, 155, 160-1, 267, 282-4, 290, 296, *Tables 74, 80-91*; railway links in 311, 315; roads 60; support for border projects 284-5, 287, 290, 300; town council 287, 290, 296, 300
Gofton, William (colliery manager) 29
Golborne, John (river improver) 2, 60
Goods departments, costs of 225, *Tables 54-7*
Goods traffic: canal 141, 259-62; estimates of 37, 112, 119, *Table 27*; railway 23, 25, 57, 66, 76, 79, 86-7, 233-6, 253, 260-2, 311, *Tables 21, 60-5, Figures 1-2*; road 2, 4, 258; rates 60, 141, 143, 234-6, 260-1, *Table 66, Figure 2*
Goulburn, Henry, M.P. (Chancellor of Exchequer) 250
Gourvish, T.R. (author) 152, 199, 329
Government (see also Board of Trade, Railway Board, Railway Commissioners, Railway Department): appealed to for money 37-8, 135, 144-5, 277; attitude to railways 176, 250, 277-8, 296-7, 306, 319-20; public works 38; stocks 97
Gowrie, Carse of 3, 125-7
Gradients (see also level line) 45-6, 274-6
Graham, George (engineer) 275
Graham, Sir James, of Netherby, M.P. (landowner) 302
Graham, Lt-Col. William of Mossknow (landowner, director) 290, 298
Grahame, Robert (director) *Table 13*
Grahame, Thomas (steamboat proprietor, director) 50, *Table 13*
Grain trade 34, 236
Grainger, Thomas (engineer): career and opinions 70, 184, 191-5, 237, *Table 43, Map 14*; in Angus 121-5, 202, 205, 318, 320-1; coal railways 57, 79, 82, 87, 179, 197, 332, *Tables 12, 17*; Edinburgh to Glasgow projects 60, 104-16, 118-19, 127, 130, 133, 195, 315, *Map 8*; estimates and costs 79, 82, 105, 112-14, 124, 133, 137, 179, 194-5, 318, 322, *Tables 17, 26-7, 34*; Granton and Fife 128, 133, 135-6, *Table 25, Maps 9, 12*; Greenock and Paisley 68, 137; passenger tax 240, 247; miscellaneous 29, 66, 315, 333, 391
Granton 125, 128-9, 132-5, *Maps 9, 12*
Gray, Thomas (railway propagandist) 45-6, 268, 358
Green, Benjamin & John (engineers) *Table 74*
Greenock 142-3, 159, 263, 289, *Map 13*
Grieve, John (factor and engineer) 63-4, *Table 1*
Grinling, Charles (author) 267, 274
Gurney, Goldsworthy (pioneer of steam road vehicles) 49, 116
Guthrie, Alexander (coal overseer) 141, 234

Haddington 292
Hall, Henry Wait, of Bristol (director) 129, 152, 372
Hall, Sir John, of Dunglass (landowner) 183, 185
Hamilton 256
Hamilton, 10th Duke of 100, 146, 183
Hamilton J.G. (merchant) 146
Hancock, Walter (pioneer of steam road vehicles) 49, 116
Harding, Wyndham (secretary) 328
Hawick 267-9, 294, 297, 300-3, 338
Hawthorn, Robert (engineer) 29, 206, *Table 74*
Hay, James, of Seggieden (landowner) 126, 184
Haymarket, Edinburgh 108, 129, 131, 255, *Map 9*
Helensburgh 263, 265
Henderson, Eagle (merchant, director) 285
Henderson, Sir John, of Fordell (land and coalowner) 29, 38
Henderson, Robert, of Fordell (land and coalowner) *Table 1*

Highland and Agricultural Society of Scotland 35, 45, 331
Highlands, railway projects in 144-6, 196-7, 258, 267
Hill, Laurence (lawyer) 83
Hinde, J. Hodgson, M.P. (director) 278
Hog, John M., of Newliston (landowner) 120
Hope, Sir Archibald, of Pinkie (land and coalowner) 28
Hope, Sir John, of Pinkie (land and coalowner) 28, 35, 64, *Tables 1, 23*
Hope Johnstone, John J., of Annandale, M.P. (landowner, director) 182, 275-6, 280, 290, 298, 300, *Table 74*
Hopetoun, 5th Earl of 120
Horses: pulling power of 30, 45-8, *Tables 5-6*; versus locomotives 48-9, 56, 70-4, 328, 394-5; railways designed for 63-4, 245; on branches 125, 328; on roads 26; buses 128, 143, 260, 310
Houldsworth, Thomas (ironmaster) 71, 75, 81, 93, 287
Houldsworth, William (manufacturer) 152
Houston 186
Houston, William (canal boat proprietor) 259
Hudson, George (director) 98, 229, 284-6, 288, 293, 295, 297, 299, 303, 306, 317, 335, *Table 74*
Huish, Mark (secretary) 186, 199, 239-41, 249, 260, 289, 328
Hunter, John Kelso (artist) 24
Hutcheson's Hospital, Glasgow 69, 364

Inchyra 126-7
Inclined planes 21, 65-6, 70-1, 107, 121
Income: capital account (see Borrowing, Capital, Shares, Subscriptions); current account 221, 230-3, *Tables 50-4, 60-5*
'Innocent Railway' (Edinburgh & Dalkeith) 64
Interest 67, 165, 169, 212-13, 226, 321-3, *Tables 37-9, 54-7*
Inverkeithing 26
Ireland, trade with 5, 22, 26, 89-90, 100, 262-5
Iron: industry 56, 79, 93, 99, 142, 159, 282, 308, 311-13, 336, 359; prices 203, 321, *Figure 3*; trade 4-5, 25, 45, 56, 60, 81
Ironstone 63, 79, 92
Ironworks 5, 25, 90, *Map 7*; Blair 163; Calder 18, 59, 69, *Table 1*; Carnbroe 94, *Table 23*; Carron 6, *Table 1*; Castlehill 93; Chapelhall 56, 79; Clyde 68, 146; Coltness 71, 93; Dalry 90; Dundyvan 93; Gartsherrie 56, 79; Govan 18, *Table 1*; Kilbirnie 90; Monkland 56; Omoa 59, *Table 1*; Shotts 4, 36, *Table 1*; Summerlee 94, *Table 23*; Wilsontown 4, 18, 146, *Table 1*
Irvine 22, 142, 234

James, William (railway propagandist) 45-6, 358

Jardine, James (engineer) 64, 68, 100-15, 131, 195-6, 274, 315, 333, 379, *Tables 12, 43, Maps 8, 14*
Jardine, Sir William (landowner) 299
Jerviston 71, 82, 93
Jessop, William (engineer) 23-4, 28, 32, 332, *Tables 1, 43, Map 14*
Johnson, George (engineer) 269, 286, 332, *Tables 1, 74*
Johnstone 68, 237

Keith Dick, Sir Robert, of Prestonfield (landowner) 71, 91, 108
Kelso 299
Kemp, Robert (provost of Dumfries) 183
Ker, Robert D. (director) 244, 263
Kilmarnock 22, 138-40, 234, 267, 288, *Map 13*
King, Charles A. (secretary) 259
Kingsford, P.W. (author) 226
Kinmond, Hutton & Steel (locomotive builders) 206
Kinnaird, 9th Lord (landowner, promoter) 125-6, 153, 184
Kirkcaldy 27, 135, 146
Knight, George (manager) 84-5

Laing, Alexander, of Shawfair 35, 362, *Table 1*
Lanark 267, 299
Lancashire: as source of capital 140, 148-60, 289, *Tables 80-91*; shareholders 131-2, 256-7; trade 267
Land acquisition and cost of 39, 93, 108, 114, 117, 124-6, 133, 169, 181-8, 280, 293, 299, 317-19, 394, *Tables 8, 36-9, 41*; estimates for 114, 124-5, 132, *Table 26*; feuing of 82, 124, 377
Landale, Charles (engineer) 21, 65-7, 71, 196, *Tables 1, 12, 43, Map 14*
Landowners 21, 37, 64-5, 76, 82-3, 91-2, 123-6, 136, 184-5; opposition to railways 82-3, 116, 120, 124-6, 183, 186-7, 199; support for railways 123, 182-3
Lang, Gabriel H. (lawyer) 180, 250-1, 263
Lang, John (landowner) 104
Law and lawyers 82-3, 117, 166, 180-1
Leadbetter, John (merchant, director): Edinburgh & Glasgow 118-20, 194, 261; Glasgow & Hamilton 146; Glasgow Paisley Kilmarnock & Ayr 11, 229, 244, 255, 259, 311; Nithsdale line 118, 276, 282, 284, 290, 300, 392
Learmonth, John (merchant, director); career and ambitions 130, 136, 303-4, 306; Edinburgh & Glasgow 118, 129-30, 133, 161, 241, 255, 257, 261, 285; Edinburgh & Northern 118, 136-7, 303; Glasgow Dumfries & Carlisle 284, 286, 294, 303-4; North British 118, 133, 137, 209, 282, 285, 290, 292, 294, 299, 303, *Table 74*; Royal Bank of Scotland 218

Leases 67-8, 71, 308, 393; of tolls 95-6
Leather, George (engineer) 333
Leslie, James (engineer) *Table 12*
Leslie, Professor Sir John 36, 359
Leith 2, 26-8, 39, 100-4, 108, 127, 129, 133, 315, *Map 9*
Leith Hay, Sir Andrew, M.P. (director) 152
Level line, desire for 36, 43, 46, 136, 190, 202-3, 333
Lime 18-9, 25, 34, 37-9, 65, 135, *Tables 1, 9-10*
Limekilns 19, 21
Limited liability 319, 337
Lindsay Carnegie, William F., of Spynie and Boysack (landowner, director) 123-6, 150, 182, 239, 247, 258
Linen 65, 124
Liverpool: as source of capital 148, 152, 160, 290; representatives in 117-18, 159; sea trade with Scotland 264, 266, 288
Loans (see Borrowing)
Locke, Joseph (engineer): career and opinions 133, 191, 197-200, 202, 206, 225, *Table 43, Map 14*; estimates 194-5, 198-9, 322, 332; Glasgow Paisley & Greenock 143, 159, 184, 187, 194, 197-9, 206, 334; Edinburgh & Glasgow 119, 121, 133; border projects 143-4, 183, 199, 267, 274-80, 289-91, 316-17, 333, *Tables 74, 77*; miscellaneous 146
Lockhart, William, M.P. (promoter) 290, 298
Locomotive departments, costs of 225, *Tables 54-7*
Locomotives 21, 24, 36, 44-50, 70-5, 81, 85, 89, 107, 115, 206-7, 259, 308, 313, 318, 331, 365, *Table 14*; cost of 206-7; design 70, 362; opposition to 50-1, 132; restrictions on use of 71; speed 48, 50-1, 70, *Table 22*
London, as source of capital 148, 150, 153, 160, *Tables 30, 80-91*
Lossiemouth 145
Lothian, 7th Marquess of 64, *Table 23*
Lumsden, James (lord provost of Glasgow) 284, 287

McAdam, John Louden (road engineer) 4, 47, 116, 353
Macaulay, Thomas B., M.P. 300
McCall, James, of Daldowie (merchant, director) 138, 146, 229-30, 243, 260, 264, 284, 371
McCallum, Duncan (engineer) 275, *Table 75*
McCulloch, J.R. (economist) 97, 358
Macfie, William (provost of Greenock) 142-3
Mackenzie, Brassey & Stephenson (contractors) 199, 298-9
MacLaren, Charles (railway propagandist) 46-51, 268, 358
Macneill, Sir John (engineer) 116, 120, 196-7, 227, 258-9, *Tables 12, 25, 43, 74 Map 14*

Mail 253-6, 264
Maintenance of way, costs of 225, *Tables 54-7*
Management, organisation and practice of 72, 311, 327-9, 336
Manchester 152, 154, 160
Mania, railway 118, 125, 140, 166, 168, 178, 188, 191, 202, 286, 298, 304, 310, 329, 334-8
Marr, John (town clerk of Lanark) 289
Matthews, R.C.O. (author) 144
Maule, William Ramsay (see Panmure)
Maxwell, Sir John, of Pollock (landowner) 186
Menteath, Charles F.G., of Closeburn (landowner) 356, *Tables 1-2*
Merchants (see also Glasgow) 76, 105, 118, 123, 140, 142, 153, 155, 325
Merry, James (coal merchant) 59-60, 197, *Tables 13, 23*
Mileages of railways *Tables 1, 12, 25, 43*
Miller, John (engineer): career and opinions 57, 70, 184, 191-6, 202-3, 206, 233, 333, *Table 43, Map 14*; estimates 108-12, 119, 137-8, 165, 190, 195-6, 318, 322, 332, *Table 35*; fees 118, 196, 202; coal railways 57, 79, 87, 194, *Table 12*; in Angus 121-5, 203, 205, 318, 320, *Table 44*; Edinburgh & Glasgow 60, 104, 108-19, 127, 133, 190, 202, 225, 333, *Table 75*; Ayrshire and Nithsdale 68, 137-41, 165, 195-6, 201, 225, 274-6, 279-80, 288, *Tables 25, 74-5*; North British 195-6, 200, 202, 274, 294, *Table 74*; miscellaneous 29, 66, 135, 184, 315, 333
Miller, Cllr Hugh, of Ayr 141
Milne, John (engineer) 136
Mitchell, James (secretary) 62, 392
Mitchell, Joseph (engineer) 108, 116, 182, 199, 310, 333
Moffat, James (manager) 328
Monklands 51, 56-63, 84-5, 146, 307-9, 322, 324, *Map 7*
Montrose 36, 121-3
Montrose, 4th Duke of 286-7
More Nisbett, George (coal and landowner) 138, *Table 13*
Morrison, James, M.P. 153, 176, 277
Morrison, John (mining engineer) 29
Moss, John (director) 275, 287, 290, 317
Murdoch & Aitken (locomotive builders) 70
Murray, David (author) 6

Nef, J.U. (author) 4
Neill, Dr Patrick (conservationist) 128, 130
Neilson, John B. (ironmaster) 45, 56
Newcastle-on-Tyne 2, 26-8, 266-9, 286, *Table 75*
Newhaven 127-30, *Map 9*
New Lanark 4
Nithsdale projects (see also Railway companies – Glasgow Dumfries & Carlisle) 183, 264-9,

275-6, 280-2, 286, 288, 316, *Tables 74-6, Maps 15-16*

Ogilvie, Sir John (landowner) 186, 244
Oswald, Richard, of Auchencruive (waggonway proprietor) *Table 1*
Outram, George (engineer) 17

Paisley 116, 137, 141, 164, 245, 289, *Map 13*
Panmure, 1st Lord (landowner, promoter) 124-5, 143, 150, 182
Parliament
    Acts of: private railway acts 22, 51, 97-8, 132, 174, 319, *Tables 12, 24-5*; Railways (1844) 176, 243, 251-2, 278, 319, 336; Bank Charter (1844) 335; Companies (1844) 287-8; Clauses Acts (1845) 168, 185, 319
    attitude to railways 98, 146, 153, 176, 180, 184-5, 209, 277-8, 297
    Commission on Railway Communication (see Smith-Barlow Commission)
    expense of 62, 293, 300-2, 319, 321, *Table 40*
    procedure 72, 97, 117-9, 121, 174-81, 278, 290, 293-5, 302, 319, 376
    railway interest in 185, 378, 381-2
    Select Committees: Standing Orders (1836) 177, 277; Internal Communication Taxation (1837) 257; Railroad Communication (1837) 141, 278; Railways (1839) 65, 246-7, 315; Turnpike Trusts (1839) 117, 387; Railways (1840) 240, 246-7, 250, 259; Railways (1844) 219, 237-9, 242, 264-5, 390; Compensation to Property Owners (HL, 1845) 181-2; Railway Acts Enactments (1846) 169-74; Audit of Railway Accounts (HL, 1849) 168
    Sunday trains 252-3
Parliamentary trains 176, 239-40, 243, 251
Parris, Henry (author) 176
Pasley, General (railway inspector) 65-6, 229
Passenger duty 67, 176, 225, 239, 245-52, 262, 314, 320, 385-6, *Tables 54-7, 69-72*
Passenger traffic: 21, 23-5, 49-50, 66, 87-90, 107, 207-8, 226, 231-46, 309, 314, 331, *Tables 20-1, 60-5, 67*; authorisation of services 23, 34; estimates 87, 112, 119, 143, *Table 27*; fares 112, 141, 143, 233-5, 237-43, *Tables 66-7*; on canals 50, *Table 20*; on roads 3-4, 137; on shipping 143; on waggonways 23, 25, 37, 49, *Tables 9-10*; provision of cheap accommodation 237-51, 260, 262-3, 314-15, 384-5
Pawson, Eric (author) 3
Pease, Edward (promoter) 267
Peel, Sir Robert, Bt, M.P. (Prime Minister) 176
Pender, Thomas (tax office controller) 249
Perth 2-3, 36, 126-7, 135-6, 199, *Map 11*

Pickfords (carriers) 329
Plateways 22-3, *Table 6*
Playfair, W.H. (architect) 131-2
Pleasure, travel for 24, 89, 143, 243-6, 256, 262, 265, 367
Pollins, Harold (author) 98, 200, 328
Port Dundas, Glasgow 119, *Map 10*
Port Eglinton, Glasgow 260
Port Glasgow 146
Porter, George R. (of Railway Board) 294
Portland, 4th Duke of 22-4, 43, 68, 74, 323, *Table 1*
Poynter, John (merchant) 142-3
Prestonpans 6
Pritchard, W.B. (engineer) *Table 74*
Private carrying on railways 71-5, 88, 365, *Table 15*
Private property, rights of 319
Private railways 85, 309
Prospectuses 283

Queensferry, 6th Marquis of 284, 287

Rails: cost of 188, 203, 321, *Tables 8, 36-9, 42*; estimates for 114, *Table 26*; specifications of 22-3, 29-30, 34-5, 43, 45-6, 124, 203, *Tables 1, 5*
Railway Board 179; Report on Schemes for extending Railway Communication in Scotland 296-300
Railway Clearing House 394
Railway companies (see also Waggonways)
    Aberdeen 123, 161, 299
    Arbroath & Forfar *Table 25, Map 11*; promotion 123-4, 174-6, 187, 318, *Tables 36, 40-1*; construction 124, 190, 194, 200, 202-3, *Tables 33, 36, 42, 45*; capital 123-4, 149-50, 180, 211-13, 218, 305, 324, *Tables 29, 46, 79*; estimates 124, 163, 205, *Tables 33, 45*; in operation 123, 207, 221, 225-6, 233-4, 237, 258, 313, 330, *Tables 50, 54, 60, 66-7*; rates and fares 258, *Tables 66-7*; dividends 163, 310, *Table 32*; guaranteed shares 124; passenger duty 247, *Tables 69-70, 72*; rolling stock 206-7, 243; road trustees 210-11, 387
    Ardrossan *Table 12, Maps 6-7*: promotion and construction 68-9, 203, 219, 320-1, 362; capital 68, 77, 179, *Table 16*; in operation 70, 89-90, 240-2, 257, 388, *Tables 14, 21-2, 66, 69-70, 72*; and Glasgow Paisley Kilmarnock & Ayr 69-70, 89-90, 142, 219
    Ardrossan & Johnstone (see Ardrossan)
    Ayrshire (see Glasgow Paisley Kilmarnock & Ayr)

Ballochney *Table 12, Maps 6-7*: promotion and construction 57, 70-1, 201, 307, 331; capital 76, 79-82, 152, 179, *Tables 16, 18*; in operation 62, 72, 74-5, 79, 81, 88-9, 96, 234, 240, *Tables 14-15, 21-2, 66*; branches 92, 146; private carrying on 74-5, *Table 15*; dividends 77, 308; directors 116, *Table 13*; passenger duty 246, *Tables 69-70, 72*; road trustees 210; rolling stock 70-1; and Edinburgh to Glasgow projects 108, 115-6, 120

Border Counties 267, 395

Caledonian: capital 160, 290, 298, 334; committee and directors 161, 278, 284, 286-7, 297-8, 329; engineering 143, 197, 199-200; estimates 274; parliamentary proceedings 176, 178, 300-2; promotion 283, 286-99, 302-3, 316-17, 320, *Table 74*; amalgamations 62, 263; and other companies 269, 286, 288-304, 330, 335, 392, *Map 17*

Caledonian & Dunbartonshire Junction 299

Carlisle & Silloth Bay 395

Clydesdale Junction 69, 288-9, 302, 315

Drumpellar 146, 197, 218-9, 309, 323, *Tables 25, 28, 43, Map 7*

Dundee & Arbroath *Table 12, Map 11*: promotion and construction 124-5, 164, 182, 195, 201, 203-6, 318, 320-1, *Tables 39, 45*; capital 124, 150, 324, 362, *Tables 29, 79*; estimates 124-5, 205, *Tables 33, 45*; in operation 125, 233, 311, *Tables 61, 66-7*; passenger duty 247, 249, *Tables 69-70, 72*; dividends *Table 32*; share prices 338, *Figure 4*; rolling stock 206

Dundee & Newtyle *Table 12, Maps 6, 11*: promotion and construction 21, 65-6, 186, 320; capital 66-7, 179-80, 324, *Table 16*; financial problems 66-8, 76, 82, 308, 363; estimates 66, 363; in operation 66-8, 71, 84, 87, 225, 227-8, 322, 365, *Tables 21-2, 66*; harbour branch 66, 71, 123, *Table 12*; passenger duty *Tables 69-72*; lease 67-8; rolling stock 71, 206-8, 321, *Table 14*

Dunfermline & Charlestown (see also Waggonways – Elgin) 21, 65, 89, 246, 321, *Tables 1, 21-2, 70, Map 6*

Dundee & Perth 67-8, 196

Eastern Counties 98, 153, 240

Edinburgh & Bathgate 195

Edinburgh & Dalkeith *Table 12, Maps 4, 6, 9*: promotion and construction 28, 63-4, 113, 201, 203; amalgamation 133, 293; capital 64, 76, 79, 105, 218, *Table 16*; estimates 113; in operation 22, 64-5, 70-1, 74, 87-91, 96, 209, 225, 240, 308, 328, *Tables 14-15, 21-23, 66*; Leith branch 64, 73-4, 77-9, 133, 262, *Tables 21, 66*; passenger duty 251, *Tables 69-70, 72*; byelaws 65, 73; wayleave 91, 108; road trustees 209; rolling stock 207-8; and waggonways 28, 35-6, 44; and Edinburgh & Glasgow 108

Edinburgh Dundee & Northern (see Edinburgh & Northern)

Edinburgh & Glasgow *Table 25, Maps 8-9*: promotion 98, 118-21, 174, 180, 185, 309, *Tables 40-1*; construction 174, 190-1, 202-3, 321, 333, *Tables 33, 42*; estimates 119, 163, 165, 234, 259, 311, *Tables 33-4*; capital 131-2, 148, 154, 159-60, 213, 218-9, 256, 266, 323-4, 327, 330, 334, *Tables 34, 47, 79, 88-91*; in operation 85, 165, 208, 225-7, 231, 234-7, 241-2, 245, 260-1, 314, 322, *Tables 34, 47, 51, 55, 60-2, 66-7*; rates and fares 234-5, 237, 240-1, *Tables 66-7*; Sunday trains 255-7, 329; Princes Street Gardens 129-32; passenger duty *Tables 70, 72*; wages 229-30; dividends 163, 168, 261, *Table 32*; share prices 338, *Figure 4*; committee and directors 118, 161; road trustees 209-10; rolling stock 206, 209; amalgamation plans 81, 182; and canals 119-21, 241, 260-1, 307; and coal railways 81, 89, 108, 119, 311; and border projects 285-6, 295, 302-3

Edinburgh & Hawick 196, 294-5, 300, 303, *Table 74*

Edinburgh Leith & Granton *Table 25, Map 9*: promotion and construction 127-9, 133, 164, 169, 186, 194, 197, 201, 309-10, 380, *Table 33*; capital 150-2, 180, 327, 334, *Tables 29, 79*; estimates 128-9, 133, *Table 33*; in operation 133, 234, 239, 258, *Table 66*; passenger duty *Tables 70, 72*

Edinburgh Leith & Newhaven (see Edinburgh Leith & Granton)

Edinburgh & Northern 118, 135-7, 164, 194-5, 283-4, 303, 334, *Map 12*

Garion & Garturk (see Wishaw & Coltness)

Garnkirk & Glasgow *Table 12, Maps 6-7*: promotion and construction 59, 105, 107, 114, 203, 320; capital 64, 179, *Table 16*; in operation 61-2, 70-4, 77, 84-5, 87-8, 112, 115, 225, 231, 236, 240, 245, 307-9, *Tables 14-15, 19-22, 66, 68*; directors 59, *Table 13*; passenger duty 249, 253-5, *Tables 69-70, 72*; rolling stock 207, 321; and canals 44, 60-2, 81, 87, 113, 120, 259, 308; and other railways 59, 61-2, 72, 84-5, 88, 104, 117, 287, 298, 311

Glasgow & Ayrshire (see Glasgow Paisley Kilmarnock & Ayr)

Glasgow Barrhead & Neilston Direct 197, 324, 334, *Table 1*

Glasgow Dumfries & Carlisle (see also Nithsdale projects): promotion 118, 183-4, 196, 283-5, 288-9, 294-5, 300-4, 316-17, *Tables 74-6*; capital 334; committee and directors 195, 284, 294; and Caledonian 282, 288-91, 294-7, 299-304

Glasgow Garnkirk & Coatbridge (see Garnkirk & Glasgow)

Glasgow Paisley & Ardrossan (see Ardrossan)

Glasgow Paisley & Greenock *Table 25, Map 13*: promotion 142-4, *Table 40*; land 184, 186-8, 318, *Table 41*; construction 143-4, 174, 190, 197-9, 203, 274, 321, *Tables 33, 38, 42*; capital 142-3, 148-9, 155-61, 168, 211-13, 218-19, 323-4, 327, 334, 375, *Tables 30-1, 48, 79, 84-7*; estimates 143, *Table 33*; in operation 163-4, 206, 225-7, 231, 236-7, 241-4, 260, 262-3, 309, *Tables 52, 56, 63, 67*; management 328; fares 239-41, 243, *Tables 66-7*; wages 229-30; dividends *Table 32*; share prices 338, *Figure 4*; passenger duty 249, 252, *Tables 70-2*; Sunday trains 253-5; road trustees 209, 211, 243; rolling stock 206; and other railways 69, 71, 89, 311; and shipping 125, 137, 142-3, 234, 239-40, 262-3, 265

Glasgow & Paisley joint line 160, 184, 221, 226, 233-4, 244-5, 250, 260, 265, 286, 311, *Tables 65-6, 70, Map 13*

Glasgow Paisley Kilmarnock & Ayr *Table 25, Map 13*: promotion 138-42, 176, 180, *Table 40*; land 182, 184-5, 187, 318, *Table 41*; construction 138-40, 174, 188-91, 195-6, 201, 203, 321, *Tables 33, 39, 42*; capital 140, 155-62, 180, 212-13, 218, 266, 323-4, 334, *Tables 30-1, 49, 79-83*; committee and directors 118, 138-40; estimates 138, 141-2, 163, 165, 196, 234, 311, *Table 35*; in operation 163, 225-7, 234, 237, 245, 309, 314, 322, *Tables 35, 53, 57, 64, 66-7*; wages 225, 228-30, *Table 58*; dividends 142, *Table 32*; share prices 338, *Figure 4*; passenger duty *Tables 70, 72*; Sunday trains 253-5; road trustees 209-10; rolling stock 206-8; amalgamations 23-4, 69-70, *Table 1*; and Nithsdale projects 276-7, 280, 284-5, 288, 316, 335, *Tables 74-5*; and canals 140-1, 260; and shipping 89, 137, 141, 226, 262-5, 267, 288, 304

Glasgow & South-Western 121, 155, 335

Grand Junction 97, 136, 155, 197, 237, 240, 250, 267, 274-6, 280, 287, 289-90, 293, 299, 316-17, 328, 333, 335-6, *Table 74*

Great Northern 98

Great Western 98, 150, 174, 237, 250

Highland 105

Kilmarnock & Troon 18, 21-5, 28, 32, 38, 43, 73-4, 234, 309, 323, 331, 354, *Tables 1, 4, 6, 69-70, 72, Map 6*

Lancaster & Carlisle 199, 289, 297, 330, 335

Leeds Northern 195

Liverpool & Manchester 44, 49, 51, 81, 87, 97-8, 105-7, 113, 136, 140, 274, 309, 331-2; used as precedent by Scots 72, 114-15, 119, 137-8, 165, 225, 259, 276, 332

London & Birmingham 70, 97, 150, 153, 174, 206, 237, 240, 249-50, 274, 290, 297, 329, 336

London & Blackwall 125

London & Brighton 174, 240, 246, 329-30

London & Greenwich 125, 310

London & North-Western 320

London & Southampton 97, 197, 250

London & South-Western (see London & Southampton)

London & Woolwich 50

Manchester & Leeds 98, 174, 237-8, 240

Manchester & Sheffield 274

Midland 98

Monkland & Kirkintilloch *Table 12, Maps 6-7*: promotion and construction 22, 56-7, 63, 70, 72, 186, 307, 331; capital 77, 79-82, 179, *Tables 16, 18*; directors 61, *Table 13*; in operation 62, 70, 75, 79, 88, 92, 233-6, 240, 307, *Tables 14-15, 21-2, 66*; passenger duty 72, 246, *Tables 69-70*; dividends 77, 308; and canals 307, 309; and other railways 57-63, 72, 115-16, 120, 287, 298, 322

Monklands 84-6

Newcastle & Berwick 285, 299, 303, 335

Newcastle & Carlisle 104, 267, 269, 277, 280-2, 286, 290, *Table 74*

Newcastle & North Shields 278

Newtyle & Coupar Angus 67, 71, 77, 89, 121, 231, *Tables 12, 14, 16, 21-2, 66, 69-70, 72, Map 6*

Newtyle & Glammis 67, 71, 77, 89, 121, *Tables 12, 16, 70, 72, Map 6*

North British (see also East coast projects) *Map 17*; promotion 118, 129, 176, 178, 283, 285-6, 289-94, 303, *Table 74*; land 183, 293; construction 195-6, 200, 202, 329; capital 155, 160, 334; committee and directors 118, 285-6, 291-2; passenger duty *Table 70*; road trustees 209-10; amalgamations 62, 70, 133, 293, *Table 1*; and Caledonian 178, 269, 282, 289-95, 297, 299-300; and other railways 118, 129, 137, 267, 315, 330, 335

North Eastern 267

North Midland 150, 229, 329

Northern & Eastern 242

Northumberland Central 395
Paisley & Renfrew 69, 71, 77, 89, 143, 225, 231, 245, 308, 320, 322, *Tables 12, 14, 16, 21, 69-70, 72, Map 13*
Polloc & Govan 18, 69, 76, 81, 89, 96, 107, 123, 146-7, 218, 298, 311, 318, 323, 369, *Tables 1, 12, 16*
Port Carlisle 395
Preston & Wyre 264, 335
Scottish Central 161, 299, 303, 330, 334
Scottish Midland Junction 68, 299, 334
Shotts & Wilsontown (see Wilsontown Morningside & Coltness)
Slamannan *Table 12, Maps 6-7*; promotion and construction 62-3, 119, 201, 203, 320, 361; land 114, 183, 185-6; capital 77, 81, 179, 212, 218, 323, *Table 16*; in operation 63, 77, 79, 85-6, 88, 227, 308, *Tables 15, 21-2, 66*; and other railways 62, 113, 120, 311
Slamannan Junction 311, 315
Solway Junction 330
South-Eastern 98
Stirling & Dunfermline 196
Stockton & Darlington 21, 35, 43, 49, 81, 87, 100, 153, 236, 331
Wansbeck 395
Wilsontown Morningside & Coltness 81, 146, 201, 305, 308-9, 321, *Tables 25, 28, 66, Map 7*
Wishaw & Coltness *Table 12, Maps 6-7*: promotion 57-9, 107; construction 79, 82-3, 93, 197; capital 59, 76, 82-3, 93, 105, 212, 323, *Table 16*; estimates 57, 79, 87, *Table 17*; directors 104, *Table 13*; committee of inquiry 61, 77, 230; land 71, 92, 182; in operation 72-3, 81, 84, 89, 93-6, 104, 230, 234, *Tables 15, 17, 19, 21-3, 66, Figures 1-2*; dividends 77; road trustees 210; rolling stock 208; and other railways 61-2, 69, 72, 75, 104, 146, 308
York & North Midland 174, 237, 293
Railway Commissioners 243
Railway Department 178, 229, 245, 263, 320
Railway projects not constructed (see also Waggonway projects not constructed)
Aberdeen Banff & Elgin 195
Airdrie & Bathgate Junction 195
Bathgate & Slamannan *Table 28*
Blythswood tunnel, Glasgow (1829-30) 60, 107, 114, 116, 315, 361, *Map 10*
Brechin 385, *Table 28*
British & Irish Union 196
Caledonian Extension 268, 299-302, 304, 335, *Map 2*
Central Union 269-74, 280, 283, 286, *Table 74, Map 16*
Coupar Angus to Dunkeld 121
Direct Newcastle Edinburgh & Glasgow *Table 74*
Direct Northern 274
Dundee to Perth 125-7, 333, *Map 11*
Edinburgh to Glasgow (1824) 51, 60
Edinburgh to Glasgow (1829-33) 60, 71, 81, 104-18, 149, 325, 331-2, *Map 8*
Edinburgh Glasgow & Leith (see Edinburgh to Glasgow (1829-33))
Edinburgh & Glasgow Union 113
Edinburgh Haddington & Dunbar 276-7, 285, 333, *Table 74*
Edinburgh & Leith Atmospheric 196
Edinburgh to London (1823-5) 36, 268-9
Edinburgh & Perth 195
Elgin to Lossiemouth 146
Forth & Tay 136, *Map 12*
Galashiels 294
Gartsherrie & Stanrig 146, *Table 28*
Glasgow & Belfast Union 196
Glasgow & Dundyvan Junction 146, *Table 28*
Glasgow & Falkirk Junction 120
Glasgow & Hamilton 146-7, *Table 28*
Glasgow & Hamilton Junction 147, *Table 28*
Glasgow Junction 295, 302, 392
Glasgow to Johnstone (1830-1) 137-8
Glasgow to Kilmarnock 295
Glasgow to Newcastle *Table 74*
Grand Eastern Union 277, *Table 74*
Great Anglo-Caledonian *Table 74*
Great North British 291-2, *Table 74*
Halbeath & Lochgelly 146, *Table 28*
Hawick to Carlisle 195, 294-6, 300-3
Hexham to Edinburgh *Table 74*
Inverkeithing Halbeath & Lochgelly 146, *Table 28*
Leven Valley 388, *Table 28*
Lochgelly & Kirkcaldy 146, *Table 28*
London & Edinburgh Direct 274, 283, *Table 74*
Newcastle Hawick Edinburgh & Glasgow Junction *Table 74*
Paisley Barrhead & Hurlet 69
Perth to Aberdeen 127, 144-5
Perth to Crieff 127
Perth to Dunkeld 126-7
Perth & Inverness 144-5, 197
Rutherglen 146, 325, *Table 12*
Scottish Grand Junction 197, 258
Scottish Western 196
Stirling to Newburgh 127
Stirling to Perth (1841) 199
Tyne Edinburgh & Glasgow *Table 74*
United Central (see Central Union)
West Lothian (see Waggonway projects not constructed – West Lothian)
Railway Steampacket Company 265
Rankine, David (railway manager) 65

Rastrick, John U. (engineer) 119, 333
Rates and fares: goods 39, 43, 60, 63, 72, 85-6, 93-5, 112, 143, 234-6, 260-1, 313-14, 357-8, *Tables 9, 66, Figure 1*; passenger 112, 141, 234-43, 249-50, 260, 262-3, 314, 367, 384, *Tables 66-8*; on canals 141, 259; on roads 26, 60, 93, 100, 112, 141, 234; on shipping 141-3, 262; effects of cuts 112; short distance tolls 61-2
Rathbone, Theodore (Liverpool merchant) 117-8, 159-60
Reed, Malcolm (author) 150, 152
Reed, Stephen (engineer) *Table 74*
Reid, John (stockbroker) 144, 185
Remington, George (engineer) 269, 279, 282, *Tables 74-5, Map 16*
Rennie, John (engineer) 34, 38-9, 47, 60, 269, 332, *Tables 2, 7, 74*
Reoch, James (provost of Leith) 296
Richardson, Joshua (engineer) *Tables 74, 77*
Rivers, competition with railways 69, 239, 310
Road trustees 34, 79, 115, 117, 120, 124-6, 169, 208-11, 257-8, 321
Roads: carting on 26, 60, 100, 112, 126, 141, 234, *Table 6*; coaches 256; competition with railways 47, 115, 141, 208-9, 236-7, 246, 257-8, 262; plateways on 34-5, 45-6; rates 26, 60, 93, 100, 112, 141, 259; steam vehicles on 49, 116-7, 258; state of 2-4, 62, 387; turnpikes (see also road trustees) 3-4, 32, 257-8, 352-3, 387; passenger traffic 112, 236-7, 246, 256, 258, 260
Robertson, George (author) 45-6
Robson, Neil (engineer) 108, 197, 226, 258-9, 333 *Tables 25, 43*
'Rocket' (locomotive) 43
Rolling stock (see also Locomotives, Waggons) 49, 206-8, 241, 313, 321, 381, *Tables 36-9*
Rothesay 263, 265
Running powers 298, 303-4
Russell, John Scott (pioneer of steam road vehicles) 116-7

Sabbatarianism 252-7
Safety 50-1
St. Leonard's, Edinburgh 63-4, 108, 127, *Map 9*
St. Rollox, Glasgow 59, 107, *Map 10*
Salt 5-6, 25
Saltmarshe, Christopher (promoter) 152
Scotland Street station, Edinburgh 186, 234, *Map 9*
Scott, Alexander, of Ormiston (author) 25, 45
Scott, Sir Walter, Bt 37, 357
Scott, Stephen & Gale (engineers) 138-40
Scottish Railway Gazette 330
Scottish Society for Promoting the Due Observance of the Lord's Day 253, 256
Scrip 147, 160
Servants, railway (see employees)

Shairp, Thomas, of Houston (landowner) 185-6
Shap 276, 279, *Table 75*
Shares (see also Capital): prices 338, *Figure 4*; calls on 76, 81-3, 211; issued at discount 67, 81, 211-12; new issues 211-2; preference and guaranteed 124, 212-3, 295, 371
Shareholders 23, 64, 76, 116, 136, 211, 242, 256-7, 324, 329, 334-5
Sharp & Roberts (locomotive builders) 206
Shiell & Small (lawyers) 180
Sherrington, C.E.R. (author) 182
Shipbuilding 99, 282
Shipping 2, 6, 141-3, 233, 240-1, 243-4, 262-6, 292, 294, 330, 335; competition with railways 21, 71, 125, 137, 141-3, 234, 237-41, 262-5, 315-16; cooperation with railways 263-5, 288; operated by railway companies 226, 263-5, 304, 387-8; rates and fares 141-3, 262
Shotts 107, 112, 114
Shropshire, waggonways in 5
Simmons, Jack (author) 196
Sinclair, Sir John 36, 268
Skene, James (lawyer) 130-1
Sligo, John (director) 285
Smiles, Samuel (author) 306
Smith, Adam 2, 306
Smith, Sir Frederick (engineer – see also Smith-Barlow Commission) 199, 278
Smith, James, of Deanston (agricultural improver) 99
Smith-Barlow Commission 184, 278-82, 285, 288-92, 294, 296, 320, *Table 75, Map 16*
Spiers, Archibald, of Elderslie (landowner) 60, 104
Sprot, Mark, of Garnkirk (coal and ironmaster) 104, 287
'Stanhope' carriages 208
Steam engines, stationary 36, 66
Stenhouse, Alexander, of Whitehill (coalowner) 35, 64, 362, *Tables 1, 23*
Stephenson, George (engineer): career and opinions 17, 35, 202-3, 276, 305, 333; locomotives 24, 45-6, 70, 331; Stockton & Darlington 17, 21-2, 46; Liverpool & Manchester 140, 274, 276; Edinburgh & Glasgow 100, 107-8, 115, 119, 333; Glasgow Paisley Kilmarnock & Ayr 138-41, 191, 197, 333; border projects 274, 279, 291, 293, *Tables 74-5, 77, Map 16*, miscellaneous 60
Stephenson, Robert junior (engineer) 190-1, 197, 206-7, 239
Stephenson, Robert senior (engineer) 24
Steuart, Sir Henry, of Allanton (land and coalowner) 82-3
Steuart, Henry Seton (railway projector) 28, 35, *Table 2*
Steuart, Robert, of Carfin (land and coalowner) 83, *Table 23*

Stevenson Coal Company *Table 23*
Stevenson, Robert (engineer): career and opinions 18, 34-7, 45-6, 332, 356; and level line 36, 43, 46, 136; Edmonston waggonway 35, *Table 1, Map 4*; other waggonways 23, 29; Midlothian project 28, 36-7, 51, 63, 317, 320, *Tables 2, 7, Map 4*; long distance projects 36-7, 39, 43, 48, 65, 100, 123, 133-5, 144, 266, 268, 317, 320, 359, 362, 371, *Table 2, Map 5*; canals 60, 65, 371; miscellaneous 130
Stewart, Charles (promoter) 275, 289-90, 389
Stewart, Sir James (landowner) 83
Stewart, Sir Michael Shaw M.P. 159
Stewart, Patrick Maxwell M.P. 159, 197
Stirling 2, 135, 199, 258, 302
Stirling, James (locomotive builder) 207
Stock market 98, 147, 327, 335-8, 374-5, *Figure 4*
Stone 65, 124, 269
Strathaven 2
Strathmore 36-7, 65-7, 121-3, 320
Subscribers and subscriptions (see also Capital, Shares, Shareholders) 64, 68, 76, 117, 140, 147-62, 333-5, 362-4; amounts raised 118, 126, 140, 142, 161, *Tables 29-31, 80-91*; town councils 64, 150; location of subscribers 148-50, *Tables 30, 80-91*; occupations of subscribers 23, 64, 150, 154-5, 160, *Tables 30, 80-91*; by railway companies 169, 285, 287, 294, 297-8, 395, *Tables 37-9*; local support for projects 123, 149-50, 155-61, *Tables 30, 80-91*
Subscription contracts 117, 147-62, 177-8, 375
Summit levels 107-8, 136, 138
Sunday trains 252-7, 329
Sylvester, Charles (engineer) 49-50

Taylor, George & Company (waggonway proprietors) *Table 1*
Telford, Thomas (engineer) 4, 39, 43, 47-9, 113, 184, 275, 332, 356, *Tables 2, 7*
Tennant, Charles, of St Rollox (industrialist) 59-62, 83, 104-5, 116, 118, 137-8, 287, 324, 370, *Table 13*
Thomson, James (engineer) 86, *Table 74*
Time allowed for construction, extension of 59
Titchfield, Marquis of (see Portland)
Tolls, short distance 61
Town councils 64, 123, 283-5; Arbroath 123; Brechin 123; Dundee 65, 67; Edinburgh 37, 64, 128, 130, 132-3, 195, 285, 296; Glasgow 287, 296; Greenock 142; Leith 296
Traders, company dealings with 85-6, 91-6, 236, *Table 23*
Traffic (see Coal, Goods, Passengers)
Trains, speed of *Table 22*
Tranent 6, 17
Trinity 128-9

Troon 22-4, 68, 138, 264
Tulloch, Thomas (inventor) 5-6
Tunnels 127-9, 133, 200, 280, *Table 75*
Tweeddale, 8th Marquis of 292, *Table 2*
Tyneside (see also Newcastle) 5, 17-8, 25, 30, 330

Victoria, Queen 237, 245
Vignolles, Charles (engineer) 119, 333

Wages and salaries 82, 169, 199, 221-30, 321, 383, *Tables 37-9, 58-9*
Waggons 17-8, 72-4, 85, 208
Waggonways 1, 5-30, 76, 307, 320, 323, 327, 332, *Tables 1-11, Map 1*; construction costs 17, 23, 28-9, *Table 4*; running costs 29
  Alloa 17-8, 22, *Tables 1, 3-6, Map 2*
  Ardrossan 22, 26, 76, *Table 1*
  Auchencruive *Table 1*
  Ayr 18, 22, 26, *Tables 1, 3, 6*
  Brora *Table 1*
  Burntisland *Table 1*
  Calder *Table 1*
  Carron 17, 22, 29, 355, *Table 1*
  Edmonston 28, 35, 64, 362, *Tables 1, 4, 6, Map 4*
  Elgin (see also Railways – Dunfermline & Charlestown) 18-21, 26, 29, 65, 332, *Tables 1, 3, 6, Map 3*
  Fordell 18, 26, 28-9, 38, *Tables 1, 3, 6*
  Govan 18, 69, *Table 1*
  Halbeath 21, 28-9, 146, *Tables 1, 3, 6, Map 3*
  Hurlet *Table 1*
  Irvine *Table 1*
  Kenton 24
  Kilmarnock & Troon (see Railway companies – Kilmarnock & Troon)
  Kilwinning *Table 1*
  Knightswood 18, 353, *Table 1*
  Legbrannoch *Table 1*
  Leven *Table 1*
  Middleton Colliery 18, 45, 359
  Muirkirk *Table 1*
  New Cumnock *Table 1*
  Newton (see Edmonston)
  Omoa *Table 1*
  Pinkie 18, 28, 35, *Tables 1, 3-4, 6, Map 4*
  Port Dundas Road 35, *Tables 1, 6*
  Sauchie *Table 1, Map 2*
  Shotts 22, *Table 1*
  Sirhowy 25
  Surrey Iron Railway 18, 22
  Tanfield 18
  Tindall Fell 35
  Tranent & Cockenzie 5-6, 17-8, 305, *Tables 1, 3-4, Map 4*
  Venturefair *Table 1, Map 3*
  Wemyss 28, *Table 1*
  Wilsontown *Table 1*

Waggonway projects not constructed 30-43, 266, *Map 5*
  Balbirnie to Newburgh 135
  Berwick & Kelso 23, 34, 38-9, 182, 356, *Tables 2, 7-11*
  Bo'ness *Table 2*
  Brechin to Montrose 121-3, 357, *Table 2*
  Dumfries to Sanquhar 25, 34, 317, *Tables 2, 6-11*
  East Lothian 43, *Tables 2, 7*
  Edinburgh & Dundee (see Fife)
  Edinburgh to Glasgow 100, *Table 2*
  Fife 133-4, *Table 2*
  Glasgow to Berwick 32-4, 37, 184, *Tables 2, 7-11*
  Glasgow to Greenock 32, 142-3, 374, *Table 2*
  Midlothian 28, 36-9, 44, 51, 317, 320, *Tables 2, 7-11*
  Monkland Canal waggonways *Table 2*
  Perth to Stirling and Forth & Clyde Canal *Table 2*
  Port Annan to Brampton 389, *Table 2*
  Roxburgh & Selkirk 36-8, 43, 123, 317, 320, *Tables 2, 7-11*
  Strathmore 36, 43, 123, 356, *Tables 2, 7*
  Union Canal waggonways *Table 2*
  West Lothian 28, 36-7, 39, 63, 185, 307, 321, 325, *Tables 2, 7, 9-11*
Walker, James (engineer) 108, 333
Wallace, Robert, of Kellie M.P. 243
Walrond, Theodore (merchant) 146
Watson, Hugh (director) 121
Watt, James (engineer) 36, *Table 2*
Wauchope, Col., of Niddrie (landowner) 258
Wauchope, John, of Edmonston (landowner) 35-6, *Table 1*
Wayleaves 91-2, 108, 186
Wemyss, 7th Earl of 64, *Table 1*
West Wemyss 25
Wharncliffe, Lord 65, 67
Whishaw, Francis (author) 99, 204-5, 208, 240
Wickham, Henry (of Board of Stamps and Taxes) 249
Wilson, John, of Dundyvan (coal and ironmaster) 93, 95-6, 155, 183, 311, 368, *Table 1*
Wind power on railways 71, 364
Winton, 5th Earl of 6
Wood, Nicholas (engineer) 46, 48, 50, 67, 71, 108, 269, 286, 332, 359, *Table 74*
Woods, Edward (engineer) 85
Working expenditure: actual 71, 84-5, 225-30, 322-3, *Tables 19, 50-7*; estimated 39, 83-4, 119, 322-3, *Tables 11, 27*
Works (see also Construction) 114, 119, 127-8, 133, 169, 188-90, 198-203, 280, 320, *Tables 36-9, 42*
Wright, James (mining engineer) 84
Wright, William (coach operator) 23

York Buildings Company 6, 17, 320, *Table 1*
Yorkshire, as source of capital 284, *Tables 80-91*